# Fluid Mechanics and Its Applications

Volume 122

The purpose of this series is to focus on subjects in which fluid mechanics plays a fundamental role. As well as the more traditional applications of aeronautics, hydraulics, heat and mass transfer etc., books will be published dealing with topics, which are currently in a state of rapid development, such as turbulence, suspensions and multiphase fluids, super and hypersonic flows and numerical modelling techniques. It is a widely held view that it is the interdisciplinary subjects that will receive intense scientific attention, bringing them to the forefront of technological advancement. Fluids have the ability to transport matter and its properties as well as transmit force, therefore fluid mechanics is a subject that is particulary open to cross fertilisation with other sciences and disciplines of engineering. The subject of fluid mechanics will be highly relevant in such domains as chemical, metallurgical, biological and ecological engineering. This series is particularly open to such new multidisciplinary domains. The median level of presentation is the first year graduate student. Some texts are monographs defining the current state of a field; others are accessible to final year undergraduates; but essentially the emphasis is on readability and clarity.

**Springer and Professor Thess welcome book ideas from authors. Potential authors who wish to submit a book proposal should contact Dr. Mayra Castro, Senior Editor, Springer Heidelberg, e-mail: mayra.castro@springer.com**.

Indexed by SCOPUS, EBSCO Discovery Service, OCLC, ProQuest Summon, Google Scholar and SpringerLink.

More information about this series at http://www.springer.com/series/5980

Medhat A. Nemitallah · Ahmed A. Abdelhafez · Mohamed A. Habib

# Approaches for Clean Combustion in Gas Turbines

Springer

Medhat A. Nemitallah
TIC in CCS and Mechanical
Engineering Department
King Fahd University of Petroleum
and Minerals (KFUPM)
Dhahran, Saudi Arabia

Ahmed A. Abdelhafez
TIC in CCS and Mechanical
Engineering Department
King Fahd University of Petroleum
and Minerals (KFUPM)
Dhahran, Saudi Arabia

Mohamed A. Habib
TIC in CCS and Mechanical
Engineering Department
King Fahd University of Petroleum
and Minerals (KFUPM)
Dhahran, Saudi Arabia

ISSN 0926-5112 ISSN 2215-0056 (electronic)
Fluid Mechanics and Its Applications
ISBN 978-3-030-44076-3 ISBN 978-3-030-44077-0 (eBook)
https://doi.org/10.1007/978-3-030-44077-0



This Springer imprint is published by the registered company Springer Nature Switzerland AG
The registered company address is: Gewerbestrasse 11, 6330 Cham, Switzerland

# Preface

The motivation for writing this book originates from the current international thrust for reducing greenhouse gas emissions to the atmosphere to alleviate global warming. Many industries worldwide started modifying their existing processes/equipment to comply with the Paris Agreement (Paris Climate Conference, December 2015), which is adopted by 195 countries. Accordingly, it becomes essential for engineers and scientists to develop green combustion systems that are friendly to the environment. Currently, gas turbines used for power generation and aviation are one of the largest sources of greenhouse gas emissions. This book addresses the available clean combustion technologies in gas turbines and provides innovative combustion techniques and burner designs for implementation in future gas turbines for different applications. The feasibilities of different combustion technologies are also presented and discussed. Special emphasis is given to the basic formulation of various combustion processes and computational modelling of conventional combustors together with applications to gas turbines supported by numerical results and detailed discussions for a number of case studies.

This book is entitled *Approaches for Clean Combustion in Gas Turbines*. The material covered focuses on the development of combustion approaches and novel burner designs for clean power generation in gas turbines. The idea is mainly to help gas turbine manufacturers control the release of pollutants to the environment in an effort to reduce global warming. Novel ideas for clean combustion in gas turbine applications are introduced, which may help manufacturers develop environment-friendly gas turbines adopting such novel ideas. This book also aims to fill a gap in the existing literature on clean combustion in gas turbines. This gap was created due to the absence of a comprehensive textbook that covers the fundamentals, current challenges, combustion technologies, and novel design considerations for gas turbine applications. This book can be used as a textbook for a graduate-level course in the area of clean combustion and as a reference book for short courses to be offered to mechanical and aerospace engineers and young researchers worldwide. It is intended to be a basic reference for graduate students, practising engineers, and young researchers in the area of clean combustion. The last chapter of this book presents some basic concepts and fundamentals of combustion, supported by solved examples and problems, which are very important, especially for beginners and undergraduate students in the field of combustion. Therefore, this book can also be utilized in teaching the fundamentals of combustion to undergraduate students worldwide.

This book consists of six chapters, starting with a general introduction to the problem of global warming and the different proposed technologies to control it. The second chapter discusses in detail the application of premixed-combustion technology in gas turbines for the control of NOx emissions and its associated problems. This chapter also proposes different combustion technologies to manage such problems for the safe operation of gas turbine combustors. The third chapter presents different burner designs for clean power production in gas turbines. Supported with different case studies, the fourth chapter addresses the performance of gas turbine combustors considering those different burners designs. The fifth chapter discusses in detail the applicability and operability of the fuel- and oxidizer-flexibility approaches in gas turbine combustors, supported by experimental data and numerical simulation results. The last chapter introduces to a new generation of environment-friendly reactors for gas turbine applications, including porous plate and membrane reactors.

The authors wish to acknowledge the support received from King Fahd University of Petroleum and Minerals under grant number BW191002 for the preparation of this book. Dr. Medhat A. Nemitallah would like to dedicate this book to the spirit of his parents, may God forgive them and have mercy on them. Dr. Nemitallah also wishes to declare the continuous support and love he receives from his wonderful wife, Heba Abdelrahman Morshedy, as well as from his kids, namely Amira, Malak, and Mohammed.

Dhahran, Saudi Arabia

Medhat A. Nemitallah
Ahmed A. Abdelhafez
Mohamed A. Habib

# Contents

# Chapter 1
# Global Warming and Emission Regulations

## 1.1 Introduction

This chapter provides an overview about the global warming issue and the efforts exerted trying to reduce its effects through the control of greenhouse gas emissions and the use of clean renewable sources of energy. Continuous growth of using renewable energies in the different energy sectors is recorded. In 2017, the total share of renewable energy to the total energy consumption approached 18.1%. This chapter presents analysis of the growth of different sources of renewable energy and the distribution of renewable energy share among the different energy sectors. While the application of renewable energy is growing, the use of fossil fuels will continue providing a great portion of the required total energy demand, resulting in more $CO_2$ emissions, which is the main contributor greenhouse gas for the global warming. This necessitates the development of technologies for carbon capture and sequestration. This chapter discusses the different carbon capture technologies and their impact on controlling the issue of global warming. Also, the regulations for controlling greenhouse gas emissions became stricter. An emission regulatory overview is presented in this chapter focusing mainly on controlling emissions for gas turbine combustion applications. The last section of this chapter introduces the clean power production technologies for gas turbine applications, which is the main subject of this book.

While there has been much progress on renewables, energy efficiency, and access to electricity and clean cooking facilities over the past decade, the world is not on track to meet international goals. Most notably, it was decided to limit the average rise in global temperatures to 1.5 degrees Celsius (°C) as stipulated under the Paris Agreement. Released during 2018, the Intergovernmental Panel on Climate Change (IPCC) special report on the impacts of global warming of 1.5 °C above pre-industrial levels found that just over a decade it was difficult to keep global warming below this threshold. Any increase beyond that point would greatly intensify the risk of extreme climate events, such as drought, floods and very high temperatures for much of the



M. A. Nemitallah et al., *Approaches for Clean Combustion in Gas Turbines*,
Fluid Mechanics and Its Applications 122,
https://doi.org/10.1007/978-3-030-44077-0_1

world's population. In addition, the annual review of UN Sustainable Development Goal 7 (SDG7) found that the objectives for renewables, energy efficiency and energy access set out for 2030 under SDG7 would not be achieved unless efforts are significantly scaled up. The High-Level Political Forum on Sustainable Development held in 2018 emphasized this point, as well, and underscored that greater ambition was necessary. In 2018, global energy demand increased at an estimated value of 2.3%, the greatest rise in a decade. This was attributed to strong global economic growth (3.7%) and to higher heating and cooling demand in some regions. China, United States and India together accounted for almost 70% of the total increase in demand. Due to a rise in fossil fuel consumption, global energy related carbon dioxide ($CO_2$) emissions grew at an estimated percentage of 1.7% during the year. Such numbers assure that the use of fossil fuels will continue and effective treatment technologies for the emissions should be applied alongside with the development of renewables. Such treatment technologies include the carbon capture and utilization/storage (CCUS). Without CCUS, the cost to mitigate climate change would increase, and the likelihood of achieving $<2$ °C is significantly reduced.

## 1.2 Global Warming Issue

Greenhouse gas emissions and carbon dioxide ($CO_2$) in particular have become an increasing concern in the power generation industry. Energy production from fossil fuel combustion results in the emission of greenhouse gases, the dominant contributor being $CO_2$ [1–3]. Public awareness and the potential threat of climate change have led to proposing policies for reducing greenhouse gas emissions, with the regulations partially driven by (international) initiatives such as the Kyoto protocol and more recently the Copenhagen summit and Paris Meeting [4–6]. Greenhouse gas emissions from energy production can be reduced by the use of alternative energy sources such as nuclear power and renewable energy as well as through efficiency improvements. Until alternative energy sources can reliably and cost effectively produce significant amounts of energy, immediate energy demands are likely to continue to be met by combustion of conventional fossil fuels.

To reduce greenhouse gas emissions from fossil fuel-fired power generation, carbon capture technologies are being investigated. $CO_2$ capture technologies that are being developed for combustion and gasification include, among others, oxy-fuel technology. The development of $CO_2$ capture technologies for hydrocarbon-fired power plants can be divided into three broad categories. The first is separation of $CO_2$ from waste gas, known as post-combustion decarbonization. The second is combustion in oxygen ($O_2$) environment instead of air, known as oxy-fuel combustion, and the third is production of a carbon-free fuel as proposed in pre-combustion decarbonization. As described later in this chapter, each of these approaches has some advantages and disadvantages. These include the reduction of power generation efficiency and increase the production cost. However, before we proceed with the discussion of the different CCS technologies, let us discuss the status of the different renewable energies and their contributions to the global energy demand.

## 1.3 Status of Renewables

As of 2017 (see Fig. 1.1), renewable energy accounted for an estimated 18.1% of total final energy consumption (TFEC) [7]. Modern renewables supplied 10.6% of TFEC, with an estimated 4.4% growth in demand compared to 2016. Traditional use of biomass for cooking and heating in developing countries accounted for the remaining share. The greatest portion of the modern renewable share was renewable thermal energy (an estimated 4.2%) being 23.2% of TFEC, followed by hydropower (3.6%), other renewable power sources including wind power and solar PV (2%), biofuels for transport (about 1%) [8].

The overall share of renewable energy (both modern renewables and traditional biomass) in TFEC has increased only gradually, averaging 0.8% annually between 2006 and 2016 [9], as shown in Fig. 1.2. This modest rise is due to a negligible change in the traditional use of biomass coupled with overall growth in global energy demand since 2006 (annual average increase of 1.5%). Despite strong demand growth in modern renewables, especially renewable electricity, these two factors have slowed gains in the combined share of renewable energy in TFEC.

In the power sector, renewables are increasingly preferred for new electricity generation. Around 181 GW of renewable power capacity was added in 2018 setting a new record just above that of the previous year [10]. Overall, renewable energy now accounts for around one-third of total installed power generation capacity worldwide [11]. Nearly two-thirds (64%) of net installations in 2018 were from renewable sources of energy (see Fig. 1.3), marking the fourth consecutive year that net additions of renewable power were above 50% [12].

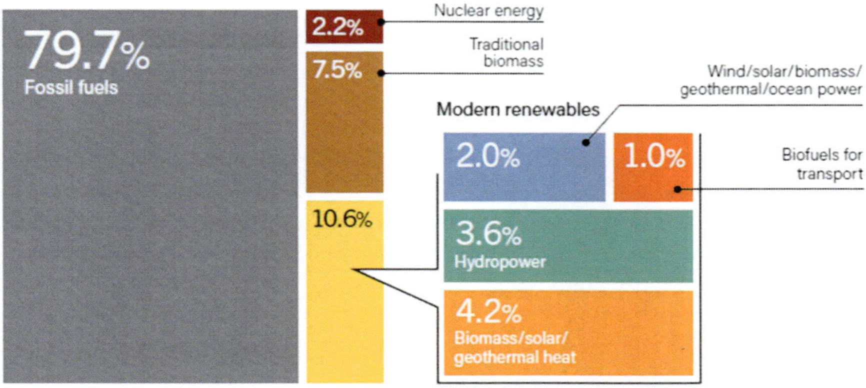


**Fig. 1.1** Renewable energy share of total energy consumption for the year 2017

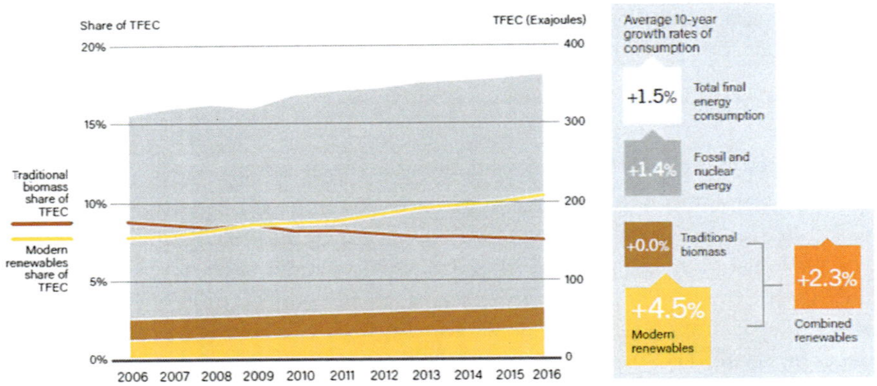


**Fig. 1.2** Growth of renewable energy as compared to total energy consumption for the period from 2006 to 2016

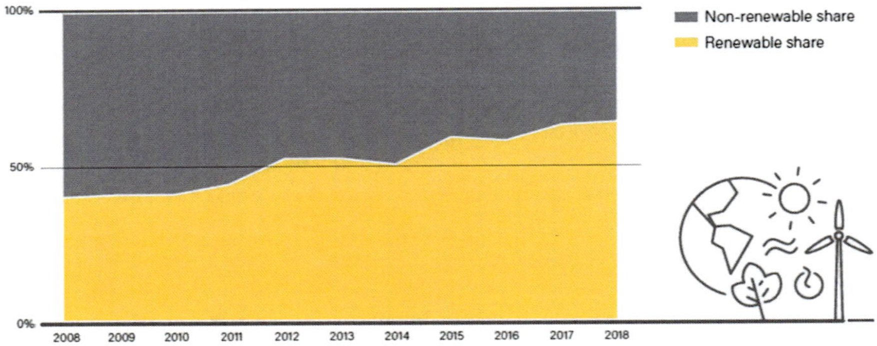


**Fig. 1.3** Renewable energy share in the net annually added power generation capacity for the period from 2008 to 2018

As of the end of 2016, heating and cooling accounted for around 51% of final energy use, transport for 32%, and final electricity demand (excluding the purposes of heating, cooling or transport) for around 17% [13], see Fig. 1.4. Modern renewable heat supplied around 10% of heating and cooling demand and has not grown significantly. While renewable electricity demand increased 25% between 2013 and 2017, modern renewable heat demand grew just under 5% during this period (around the same rate as global energy demand) [14]. In transport, consumption of biofuels (principally ethanol and biodiesel) increased around 18% between 2013 and 2017, although starting from a small base.

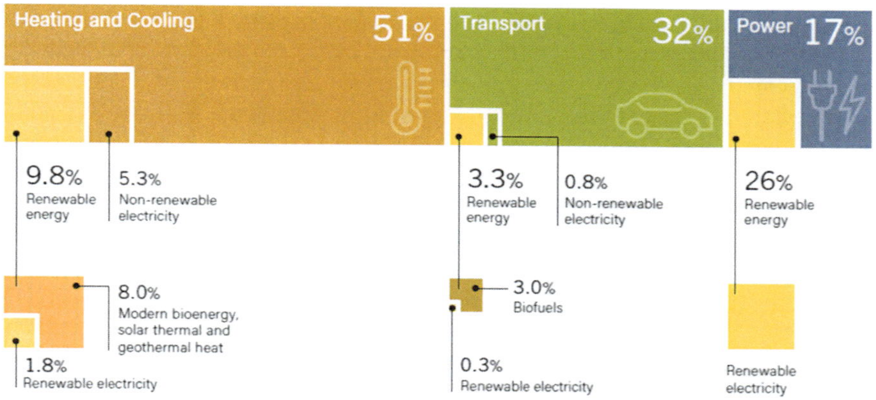


**Fig. 1.4** Share of renewables in total energy consumption by sector for the year 2016

## 1.4 Carbon Capture Technologies

Three main methods can be envisaged for the capture of $CO_2$ with these three systems as shown in Fig. 1.5 [15]: (1) pre-combustion capture: to capture $CO_2$ in a synthesis gas after conversion of CO into $CO_2$; (2) post-combustion capture: to capture $CO_2$ in the exhaust gases once the fuel has been fully burned with air; and (3) capture in oxy-combustion: consisting of combustion in oxygen with recycling of exhaust gases (therefore, composed mainly of $CO_2$ and water) and purification of the $CO_2$ flow, to eliminate incondensable gases. Carbon capture technologies (CCTs) including pre-combustion, oxy-combustion, and post-combustion capture are good tools toward the control of $CO_2$ emissions. Each of these technologies has its own advantages and disadvantages for controlling $CO_2$ emissions. Post-combustion decarbonization separates $CO_2$ from the flue gases and thus requires minimal modifications to the

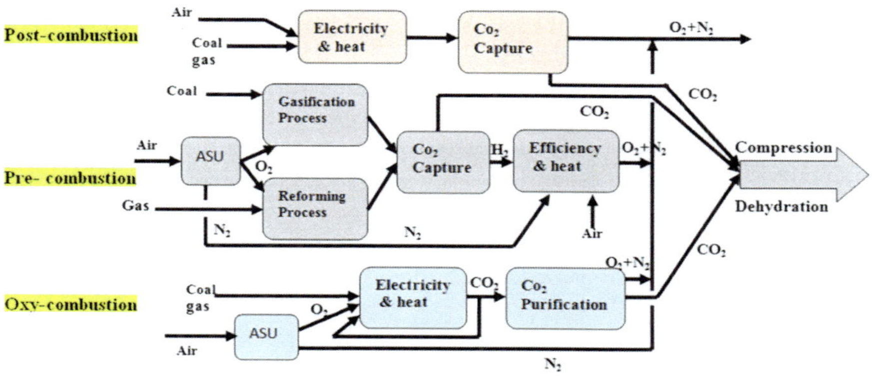


**Fig. 1.5** The three carbon capture technologies [15]

power cycle, but large gas quantities must be treated because $CO_2$ is diluted by the nitrogen that entered with the combustion air. The $O_2$ for oxy-fuel combustion is typically produced by a conventional (cryogenic) air separation unit. Fuel combustion in pure $O_2$ results in very high combustion temperatures. In order to reduce these temperatures, part of the flue gas is recycled back to the combustion chamber. Oxy-fuel cycles are typically based on near-stoichiometric combustion, where the fuel is burned in $O_2$ produced in an air separation unit diluted by recycled flue gas. Thus, combustion takes place in the absence of nitrogen. Oxy-fuel combustion produces only $CO_2$ and $H_2O$. The $CO_2$ separation is accomplished by condensing water ($H_2O$) from the flue gas, providing a byproduct of high-purity $H_2O$, while other impurities are removed from the $CO_2$ stream before compression and storage. However, a relatively large amount of energy is needed for the $O_2$ production. Based on a quantitative comparison of the three carbon capture strategies, those based on oxy-fuel combustion generally have higher thermodynamic efficiencies than those based pre- or post-combustion decarbonization [16].

Among these CCTs, oxy-combustion technology is considered as one of the most promising CCTs [17]. This technology can be integrated either with currently existing power plants with slight modifications or with the new power plants [18]. In this process, fuel is burned using oxidizer mixture consisting of high purity oxygen and recycled exhaust gases (consists mainly of $CO_2$). Portion of the exhaust stream s recycled to the combustor to control the combustion temperature. Theoretically if a hydrocarbon fuel is to be burned with pure oxygen under perfect complete combustion, the combustion products should be mainly $CO_2$ plus $H_2O$; however, some other species usually exist in the exhaust stream. This can be attributed to the dissociation of different species within the flame core, impurities in fuel or oxidizer mixture, air leak, combustion instabilities, and improper mixing and incomplete combustion. However, oxy-combustion process results in highly $CO_2$-concentrated exhaust stream which facilitates the capture process of $CO_2$ after $H_2O$ condensation [19].

There are two existing approaches for the application of oxy-combustion technology. The first approach is applied through the use of air separation unit (ASU) to separate oxygen from air and, then, the separated oxygen is used in oxidizing the fuel in conventional combustion systems. However, the ASU requires additional energy for the operation either in cryogenic separation or in terms of gas heating (to activate the oxygen separation membrane) in case of membrane separation [20]. The second approach is applied through the use of what is known as ion transport membrane reactor (ITMR). In ITMR, the two processes of oxygen separation and oxy-combustion are performed simultaneously within the same unit. Both approaches are investigated in detail in this book, considering novel environmental-friendly burners for clean power production in gas turbines.

## 1.5 Adaptation of Gas Turbines to Regulations of Pollutant Emissions

In the 1970s, when emission controls were first introduced, the pollutant of primary concern to regulators shifted to be concerned with $NO_x$. For the relatively low levels of $NO_x$ emission reduction initially targeted, the injection of steam into the flame zone produced the targeted reduction in $NO_x$ emissions with an insignificant effect on the performance. Also, the emissions of other pollutants including CO and volatile organic compounds (VOCs) did not increase in significant amounts. During the 1980s, stricter policies were imposed requiring a significant reduction in $NO_x$ emissions. Based on that, further attempts were made to utilize steam injection to ensure achieving the targeted emission values. As a result, cycle performance and part lives are affected badly, and the emission rates for other pollutants including CO and VOC increased significantly. At this stage, the research was forced to develop other techniques for controlling emissions, which led to the development of the lean premixed (LPM) combustion technique [21].

## 1.6 Emission Regulatory Overview

### *1.6.1 Clean Air Act (CAA)*

US Congress enacted the Clean Air Act (CAA) in 1970 to consider growing concerns about the quality of atmospheric air [22]. Based on this act, National Ambient Air Quality Standards (NAAQS) were established to control the emissions for criteria pollutants. Accordingly, areas of the country that exceed the NAAQS are considered non-attainment for that pollutant. In non-attainment areas, the US Regulatory Structure imposes more stringent air pollution control programs. The CAA is a federal law covering the entire country. However, states and local governments are permitted to create and implement more stringent air pollution strategies than those required by the CAA.

### *1.6.2 New Source Performance Standards (NSPS)*

In the early 1970s, the emissions control requirements for nitrogen oxides ($NO_x$) were first applied to control the emissions of gas turbines by the Los Angeles County Air Pollution Control District (LAAPCD) and the San Diego Air Pollution Control District (SDAPCD). In order to match these regulations, it was decided to inject water into the combustor in the flame zone in order to cool down the combustion temperature. $NO_x$ emissions were reduced by about 40% when water at a half rate of the fuel was injected into the flame zone. Under such operating conditions, emission level of approximately 75 ppmvd (parts per million by volume, dry) on oil was achieved. The US EPA used these data and other data to develop new source performance standards. These standards went into effect in September 1979. Turbines with heat input over 10 million Btu/h (generating less than 30 MW electrical output) and supplying less than one-third of their electrical output to an electric utility, are required to meet a $NO_x$ emission standard of 150 ppm, corrected for efficiency. The emergency turbines are excused from this standard, as are special types of turbines. Electric utility turbines having heat input above 100 million Btu/h should meet a $NO_x$ emissions standard of 75 ppm, corrected for efficiency. In fact, most of the currently working turbines can achieve $NO_x$ emissions of 42–25 ppm or less without any means for post-combustion control. Based on that, the existing NSPS is not typically a controlling regulation for gas turbines emissions [21].

In July 2004, EPA modified the NSPS for gas turbines in a direct final ruling. The noteworthy modification is that the new LPM turbines that originate construction after the date of July 8, 2004, are required to use a $NO_x$ Continuous Emissions Monitoring System (CEMS). Also, the owners can monitor continuously the engine parameters that give indication when the turbine is working out of the lean premixed mode of combustion. Lately, on February 18, 2005, EPA suggested standards of performance for new stationary gas turbines in 40CFR60 Subpart KKKK. These new standards reflect the changes in the turbine design and, accordingly, $NO_x$ emissions control technologies and are intended to make the emission limits up to date with the performance of present gas turbines.

### *1.6.3 New Source Review (NSR)*

In addition to the AAQS, CAA developed an air permitting program called New Source Review (NSR). The NSR is having two core programs: (1) Prevention of Significant Deterioration (PSD) and (2) Non-Attainment NSR. Each of these two programs applies to "major sources" and "major modifications." However, in Non-Attainment NSR at far lower thresholds, a source can be considered "major."

### *1.6.4 Best Available Control Technology (BACT)*

When a source triggers PSD review, then, the owner should define the appropriate level of emissions controls for the pollutants that exceed the specific limits. In attainment areas, the standard for evaluation is the BACT. The BACT determination sets limitations for the achievable emissions considering environmental, energy, and economic effects of applying the emission control technology required to meet that limitation. Currently, the EPA asks for a "top-down" BACT analysis to be followed for all PSD Permit applications [23]. Based on that, BACT is a dynamic standard, which is stricter over time unless new, low-cost, and more efficient emission control technologies are developed. For a certain application, BACT can only be defined in the form of current demonstrated technology. Generic BACT requirements do not exist, as BACT determinations are site-specific. However, for gas turbines with power greater than 25 MW, there have been recent BACT determinations for $NO_x$ as low as 5–2 ppm.

### *1.6.5 Lowest Achievable Emission Rate (LAER)*

Lowest achievable emission rate determination is used in major sources/modifications in non-attainment areas. For the cases of class or source category, LAER is defined as the severest emission control system accomplished in practice. The main difference between BACT and LAER is that LAER does not have much attention of economic effects when evaluating emission control technologies of pollutants.

## 1.7 Clean Power Production for Gas Turbine Applications

The increase in the strict environmental regulations encouraged researchers to develop combustion systems that could meet such restrictions. Combustors of gas turbine engines for power generation have been utilizing non-premixed flames thanks to their wider stability characteristics. Unfortunately, these kinds of combustors are no longer preferred for use in gas turbine applications due to associated unacceptable high concentrations of $NO_x$ pollutant emissions. One approach for controlling emissions of $NO_x$ is the flue gas recirculation (FGR) in which reactants are diluted with inert exhaust gases. However, mixing the reactants with inert exhaust gases results in lower flame burning velocity and, consequently, lower combustion efficiency, and the combustor is prone to loss of stability. Modern technologies and concepts such as lean premixed combustion (LPM) and catalytic combustion have been introduced in the last two decades to the gas turbine power generation industry. The catalytic combustion was found to be expensive, non-safe, and less durable. While in LPM, fuel and air are homogenously mixed upstream of the combustor. Lean burning is a usual practice in gas turbines to control the combustion temperature and, as a result, thermal $NO_x$ emissions (being the most significant $NO_x$ source over fuel and prompt $NO_x$) are significantly decreased. Nowadays, most manufacturers adopt the LPM technology in novel gas turbine combustors to replace non-premixed technology for the control of combustion temperature and $NO_x$ emissions. However, changing the flame from the non-premixed combustion mode to the premixed one causes unwanted combustion characteristics including increased levels of flame instabilities, auto-ignition and flashback. In addition, gas turbine operation under extra lean conditions may adversely affect flame stability, e.g., by enhanced tendency to blowout. These features bring their own challenges to burner designers and manufacturers.

These issues necessitate the development of novel combustor designs that can operate stably under extra lean conditions while keeping emissions at low levels. In this book, novel burner designs adopting premixed combustion are introduced for the sake of improving the combustor stability and performance for gas turbine combustion applications under different operating conditions.

Although approaching very low levels of $NO_x$ and CO emissions considering the above-mentioned techniques, the complete control of emissions is not achieved yet considering the lean premixed air combustion technique. Additional technologies should also be applied alongside with premixed combustion technology for the capture of $CO_2$. There are three main technologies for $CO_2$ capture; including pre-combustion, oxy-combustion, and post-combustion capture technologies. Among these technologies, oxy-fuel combustion is a very promising technology for the capture of $CO_2$ in gas turbine applications. In this technology, fuel is burned with pure oxygen to result in combustion products consisting mainly of $CO_2$ and $H_2O$, making the process of $CO_2$ separation easy and economic after $H_2O$ condensation. Combining both premixed and oxy-combustion technologies may be also a trending choice toward complete control of emissions while capturing $CO_2$. However, these technologies bring their own challenges in terms of combustion stability and combustor efficiency. Different fuels have different combustion and emission characteristics, which means different combustor stability behaviors at different loading conditions can be obtained using different fuels. This necessitates the development of fuel-flexible burners that can handle different fuels based on the loading conditions to sustain stable flame under premixed oxy-combustion conditions. Different fuel/oxidizer flexibility approaches are addressed in detail in this book. The utilization of captured $CO_2$ has become popular because of the raised concerns about its sequestration techniques in the underground and in the oceans. Two approaches have been adopted for $CO_2$ utilization, namely the direct utilization and the conversion of $CO_2$ into chemical and energy products. The two approaches are addressed in this book, and the current advances in the carbon capture/utilization and sequestration technologies (CCS/CCUS) technologies are presented.

The material covered in this book focuses on the development of combustion approaches and novel burner designs for clean power generation in gas turbines. The book consists of six chapters starting with general introduction to the problem of global warming and the different proposed technologies to control it. The second chapter discusses in detail the application of premixed combustion technology into gas turbines for the control of $NO_x$ emissions and its associated problems. This chapter also proposes the different combustion technologies to manage such problems for safe operation of gas turbine combustors. The third chapter presents different burner designs for clean power production in gas turbines. Supported with different case studies, the fourth chapter addresses the performance of gas turbine combustors considering the different burners designs. The fifth chapter discusses in detail the applicability and operability of the fuel/oxidizer flexibility approaches in gas turbine combustors, supported with experimental data and numerical simulation results. The final chapter introduces new generation of environmental-friendly reactors for gas turbine applications including porous-plate and membrane reactors. The book aims to fill a gap in the existing publications because of the absence of a good reference book in the area of clean combustion in gas turbines that covers the fundamentals, novel design considerations, combustion technologies, and current challenges. The book introduces novel ideas for lean combustion in gas turbine applications, which may help manufacturers to develop gas turbines adopting such novel ideas. The book can be used as a textbook for a graduate-level course in the area of clean combustion and as a reference book for short courses to be offered to mechanical and aerospace engineers and young researchers worldwide.

**Acknowledgements** The authors wish to acknowledge the support received from King Fahd University of Petroleum & Minerals under Grant number BW191002 for the preparation of this book chapter.

## References

1. OECD/IEA (2018) World energy outlook 2018. Paris, p 27. https://www.iea.org/weo2018/
2. Allen M et al (2018) Special report: global warming of 1.5 °C. Intergovernmental Panel on Climate Change (IPCC), Geneva. https://www.ipcc.ch/sr15/
3. United Nations (UN) Sustainable Development Goals Knowledge Platform (2019) Sustainable development goal 7. https://sustainabledevelopment.un.org/sdg7. Viewed 27 Apr 2019; IEA et al (2019) Tracking SDG7: the energy progress report 2019. World Bank, Washington, DC. https://www.worldbank.org/en/topic/energy/publication/tracking-sdg7-the-energy-progress-eport-2019
4. Embed the Essence of 2030 Agenda into Everything (2018) UN chief urges at close of Global Goals forum. UN News, 18 July 2018. https://news.un.org/en/story/2018/07/1015092
5. OECD/IEA (2019) Global energy and $CO_2$ status report. https://www.iea.org/geco/. Viewed 25 May 2019
6. Copenhagen Summit 2009 (2009) United Nations Climate Change Conference (COPIS). Copenhagen, 6–18 Dec 2009
7. Based on OECD/IEA (2018) World energy balances and statistics, 2018 edn. Paris. https://www.iea.org/statistics/balances/
8. Estimated shares based on the following sources: total final energy consumption in 2017 (estimated at 370.0 EJ) is based on 363.7 EJ for 2016, from OECD/IEA, op. cit. note 51, and escalated by the 1.74% increase in estimated global total final consumption (including non-energy use) from 2016 to 2017. Derived from OECD/IEA, op. cit. note 42
9. All values derived from OECD/IEA, op. cit. note 51
10. Total capacity based on data provided throughout this report. See Market and Industry chapter, Reference Table R1 and related endnotes for details and sources
11. Total capacity from Ibid. (2019) Data on non-renewable capacity provided by International Renewable Energy Agency (IRENA), personal communication with REN21, Apr 2019
12. Based on Ibid. In 2015, 2016, 2017 and 2018, renewable energy demonstrated shares of net installed power capacity of 59%, 58%, 63% and 64%, respectively
13. Based on OECD/IEA, op. cit. note 51; renewable electricity for heat from OECD/IEA, personal communication with REN21, 28 April 2019. In 2016, all electricity for heating and cooling met 7.1% of final energy demand in the sector (1.8% renewable and 5.3% non-renewable electricity). All electricity for transport met 1.1% of final energy demand in the sector (0.3% renewable and 0.8% non-renewable electricity). See Methodological Notes for details

14. Calculated from OECD/IEA, op. cit. note 51
15. Nemitallah MA, Habib MA, Badr HM, Said SA, Jamal A, Ben-Mansour R, Mokheimer EMA, Mezghani K (2017) Oxy-fuel combustion technology: current status, applications, and trends. Int J Energy Res 41:1670–1708
16. Kvamsdal H, Jordal K, Bolland O (2007) A quantitative comparison of gas turbine cycles with $CO_2$ capture. Energy 32(1):10–24
17. Gonzalez-Salazar MA (2015) Recent developments in carbon dioxide capture technologies for gas turbine power generation. Int J Greenhouse Gas Control 34:106–116
18. Habib MA, Nemitalah MA, Ben-Mansour R (2013) Recent development in oxy-combustion technology and its applications to gas turbine combustors and ITM reactors. Energy Fuels 27:2–19
19. Guo J, Liu Z, Wang P, Huang X, Li J, Xu P, Zheng C (2015) Numerical investigation on oxy-combustion characteristics of a 200 MWe tangentially fired boiler. Fuel 140:660–668
20. Nemitallah MA, Habib MA (2013) Experimental and numerical investigations of an atmospheric diffusion oxy-combustion flame in a gas turbine model combustor. Appl Energy 111:401–415
21. Bender WR (2007) Lean premixed combustion. Technology & Management Services Inc., Gaithersburg, MD, p 20879
22. Hydari NH, Yousuf AA, Ellis HM (2002) Comparison of the most recent BACT/LAER determinations for combustion turbines by state air pollution control agencies
23. Calcagni J (1989) U.S. Environmental protection agency, memorandum dated June 13, 1989. Transmittal of background statement on "top down" BACT

# Chapter 2
# Premixed Combustion for Gas-Turbine Applications

## 2.1 Introduction

Power-generation industries rely mainly on combined-cycle power plants (steam plus gas turbines) for high-efficiency energy production up to 60% [1], thanks to new heat-recovery technologies that utilize a portion of the heat rejected from gas turbine and convert it to power in the steam turbine. Hence, the development of durable and low-emission gas-turbine combustion systems is of significant importance. Many environment-friendly combustion technologies have been introduced. For instance, integrated gasification combined cycle (IGCC) is a recent technology that utilizes synthetic gas (syngas) fuel gasified from solid fuels, such as coal, waste products, or biomass [2–4]. If $CO_2$ is removed after the gasification process, the syngas becomes hydrogen-rich, and the power plant generates less $CO_2$ emissions.

Gas turbines for power generation have traditionally used combustors that host non-premixed flames due to their acceptable performance and superior stability characteristics. Unfortunately, this type of combustors is no longer preferred because they generate excessive $NO_x$ emissions and, thus, do not comply with recent environmental regulations. This encouraged researchers to develop combustors that can meet such restrictions [5, 6]. Hence, new technologies and concepts have been introduced to stationary gas turbines for power generation, such as lean premixed (LPM) and catalytic combustion [7, 8]. During the last four decades, LPM combustion became the standard technology in many power-generation applications. This can be ascribed to the fact that it generates significantly lower $NO_x$ emissions while maintaining compliant CO levels. Dynamic flame instabilities and limited operability window (static instability) are, however, two major issues in premixed combustion. Enriching the fuel with hydrogen is one of the promising approaches that mitigate combustion instabilities, improve combustor turndown (i.e., allow for leaner operation), and reduce $NO_x$ emissions. However, hydrogen enrichment brings its own challenges in terms of earlier flame flashback because of the greater flame speed of hydrogen [9–11].



M. A. Nemitallah et al., *Approaches for Clean Combustion in Gas Turbines*,
Fluid Mechanics and Its Applications 122,
https://doi.org/10.1007/978-3-030-44077-0_2

Accordingly, the current chapter focuses on the operability and stability of premixed combustors.

Dynamic combustion instabilities result from the constructive coupling of local heat-release fluctuations and unsteady flow oscillations, which amplifies certain resonant frequencies that depend on the combustor geometry, heat-release rate, mechanism of flame stabilization, and pressure drop across fuel-injection orifices (aka softness of fuel nozzles) [12]. The pressure oscillations associating the resonant frequencies can be high enough to induce severe hardware damage, as shown in Fig. 2.1 [13]. Dynamic combustion instabilities are not limited to gas turbines only; they have also been observed in many applications, such as liquid rocket engines [14, 15], solid rocket motors [16], ramjet and scramjet engines [17, 18], and boilers and furnaces [19] as well. Combustion instabilities can be attributed to two main reasons [20, 21]; the first one is the confined nature of combustor, where any confinement geometry has its specific acoustic frequencies, and the second reason is that the energy required to drive the detrimental pressure oscillations is a very small fraction of the heat released by combustion.

While instabilities occur in both LPM and non-premixed flames [21], the former are more prone to instabilities because of the following reasons. Firstly, LPM combustion systems are typically operated very close to their lean blow-off limits. At these conditions, a small change in equivalence ratio can cause a substantial change in the heat release rate and, consequently, the coupling with combustor acoustics [22, 23]. Secondly, LPM combustors split the compressor-discharge air significantly differently compared to non-premixed combustors, see Fig. 2.2. The latter have a rich primary flame zone that requires a small portion of the compressor air; the larger portion is thus used as secondary air for liner cooling, combustion completion, and downstream dilution. Non-premixed flames also have high-temperature stoichiometric zones, so the large flow of secondary air is critically needed for liner cooling. The large temperature (and density) gradients within the flame and near the wall suppress the propagation of resonant pressure waves [24, 25], which explains why

**Fig. 2.1** Dynamic combustion instabilities can induce severe hardware damage. Left: pristine combustor; right: damaged combustor [13]

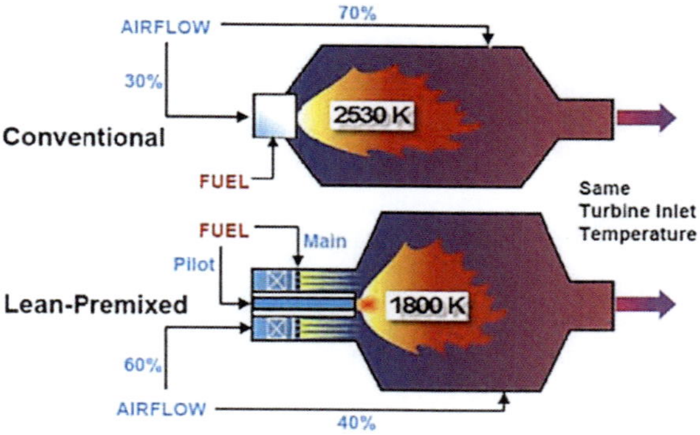


**Fig. 2.2** Split of compressor-discharge air in conventional (non-premixed) versus LPM combustors [27]

non-premixed combustors are less prone to instabilities. By comparison, LPM combustors feed most of the compressor air into the flame to make it lean, leaving a small portion for liner cooling, see Fig. 2.2. Besides, LPM flames are free of any stoichiometric zones, and the flame temperature is significantly lower than the average of a non-premixed flame, so the liner of a LPM combustor can be efficiently cooled with a small amount of air. The temperature (and density) gradients within the premixed flame and near the wall are too small, so the resonant pressure waves can propagate freely within the quasi-uniform temperature field inside the combustor. Finally, LPM flames are more compact (due to the abundance of air), i.e., the combustion heat is released in a smaller volume, which boosts the potential of resonant coupling of local heat-release fluctuations and unsteady flow oscillations [26]. All these issues will be discussed later in the upcoming sections of the present chapter.

Even though LPM combustion offers emission levels that comply with the ever-stringent environmental regulations, it comes with challenges that must be addressed, such as dynamic instabilities (hardware damaging), static instabilities (flashback and blow-off), as well as limitations on operability window. LPM combustors typically operate close to the blow-off limit to reduce $NO_x$, but CO increases as this limit is approached, see Fig. 2.3, which allows for a narrow flame-temperature range (typi-

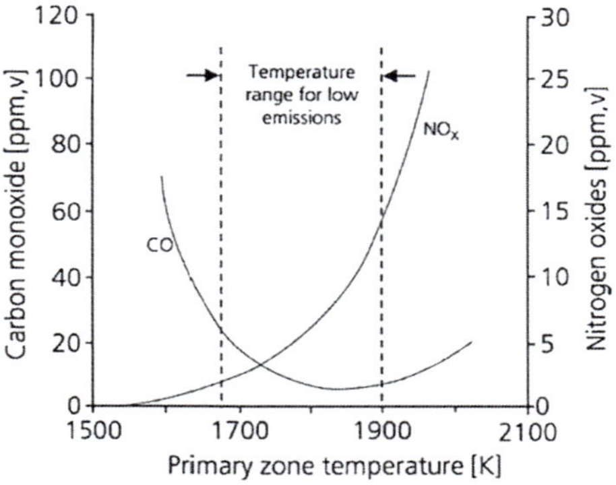


**Fig. 2.3** Typical flame-temperature range for low emissions in LPM gas turbines [27]

cally ~200 K). Extensive research has thus been motivated to understand the stability characteristics of LPM flames. Hence, this chapter focuses on discussing the instabilities associated with premixed combustion and aims at deepening the understanding of readers/researchers of the factors that significantly affect combustor operability. The chapter also discusses approaches/developments made to overcome instabilities while controlling emissions. The findings of recent experimental investigations and numerical modeling and simulations are presented. The discussion is not limited to LPM air-fuel applications but extends to premixed oxy-fuel combustion as well. The effects of oxidizer and fuel flexibility as well as hydrogen enrichment on oxy-fuel combustor operability and flame stability are analyzed. The influences of operating conditions and combustor design on flame characteristics are also discussed; the effect of reactant premixing on emissions is examined, considering a variety of combustor designs and operating conditions in different applications. Successful operational designs in current applications are presented, so as new promising designs that are still under development. The challenges regarding the application of premixed combustion in the industrial and power-generation fields are discussed. The role of numerical CFD modeling in predicting the reacting flow field and heat release in premixed combustion mode is also addressed by presenting a numerical case study of premixed combustion in a swirl-stabilized gas-turbine combustor.

The present chapter is organized in the following manner. Section 2.2 addresses the instabilities associated with premixed combustion, including static and dynamic instabilities. Section 2.3 discusses in detail the different approaches taken to overcome combustion instabilities, improve the combustion process, and reduce the exhaust emissions, with focus on premixed combustion. Different approaches are analyzed in this section, including fuel and oxidizer flexibility (such as hydrogen enrichment and oxy-fuel combustion), variable operating conditions, and variable flame type (non-premixed, partially premixed, and premixed flames). Special emphasis is made on the application of premixed combustion in swirl-stabilized combustors, mainly for gas-turbine applications, which is the topic of Sect. 2.4. The effect of swirl on flame stability and emissions is analyzed as one of the well-known approaches for overcoming the instabilities and operability issues associated with premixed combustion in gas-turbine combustors. Section 2.5 discusses advances in the numerical modeling of premixed combustion, towards better understanding of its characteristics in gas-turbine combustors. Furthermore, Sect. 2.6 presents a numerical case study of turbulent premixed combustion in a gas-turbine model combustor. Temperature, species concentrations, and flow field are investigated over a range of operating conditions. Section 2.7 finally lists the concluding remarks out of this chapter.

## 2.2 Instabilities of Premixed Combustion

Obtaining and/or maintaining a stable premixed flame to meet target combustor power and exit temperature is never a straightforward task. Therefore, this section focuses on the instabilities of premixed combustion. Flame stability is affected by many critical factors, namely burner and combustor design, fuel composition, oxidizer composition, turbulence level, the manner of introducing air and fuel into the combustor, and the interaction of neighboring flames. Combustion instabilities are classified into two types, namely static and dynamic instabilities [28, 29]. Static instabilities are related to flame blow-off and flashback, whereas the dynamic instabilities are primarily a thermo-acoustic phenomenon; both types will be discussed in the following subsections.

## 2.2.1 *Static Instabilities*

The static instability limits of a premixed flame are its blow-off and flashback. In either case, the flame leaves the location it was designed to anchor at, which shuts the combustor down and may induce hardware damage. Both limits are examined comprehensively in this subsection.

### 2.2.1.1 Blow-Off

Blow-off occurs when the flame detaches from the burner and propagates in the direction of the flow until it is physically blown off. This happens when the velocity of combustible mixture is higher than the flame burning velocity. In other words, blow-off occurs when the time required for chemical reaction is greater than the flow residence time. This is an issue because the rates of chemical kinetics/reactions and the flame speed are not constant but vary with flame temperature, fuel composition, and fuel transport properties. Recently, different theories for developing the blow-off correlations have been studied extensively [30–33]. The condition for blow-off is defined as function of the chemical and residence timescales; the former is defined as [34, 35]:

$$\tau_{chem} = \frac{\alpha}{S_L^2} \tag{2.1}$$

where $\alpha$ and $S_L$ are the thermal diffusivity and laminar flame speed, respectively [36, 37]. The residence timescale, on the other hand, is defined as:

$$\tau_{res} = \frac{d}{U_{ref}} \tag{2.2}$$

where d and $U_{ref}$ refer to the characteristic length and velocity scales of the flow, respectively. The ratio of residence to chemical timescales is known as the Damkohler number [38–41], i.e.

$$Da = \frac{\tau_{res}}{\tau_{chem}} = \frac{S_L^2 d}{\alpha U_{ref}} \tag{2.3}$$

The Damkohler number has been widely used to define the condition of blow-off. From one perspective, the Damkohler number characterizes the position of the flame by determining the balance between the flow and reaction timescales. This infers that any unbalance between these timescales will lead to change in the flame position. From another perspective, the Damkohler number also relates to the ratio of flame speed to combustible-mixture velocity. This is based on the logic that the flame will blow off if the burning velocity is lower than the combustible-mixture flow velocity. In other words, if the flow rate of the combustible mixture is too high, compared to the chemical reaction rate, the flow residence time becomes too short for chemical reactions to take place, and the flame is pushed downstream in the direction of the flow.

Lieuwen et al. [42] examined different blends of CO, $H_2$, and $CH_4$ to study the effect of fuel composition on combustor blow-off, flashback, and dynamic stability, see Fig. 2.4. For clear presentation, each fuel blend is depicted in a different color. The dark blue color represents a higher percentage of $CH_4$, the red color represents a higher percentage of hydrogen, and the yellow color represents a higher percentage of CO in the fuel blends. It can be observed that the Damkohler number defines the blow-off characteristics for the different fuel compositions; blow-off occurs at almost constant Da = 0.6, as shown in Fig. 2.4. However, the results also reveal that a flame with higher hydrogen concentration blows off at leaner equivalences ratios, as shown in Fig. 2.5. For the sake of comparison, at 0% $H_2$ blow-off occurs at $\phi$ = 0.3–0.5, while at 100% $H_2$ blow-off occurs at $\phi = 0.17$.

Zhang et al. [43] studied the lean blow-off limits of different $CO/H_2/CH_4$ fuel mixtures as an extension of the experiments conducted by Lieuwen et al. [42]. The blow-off residence time for well-stirred reactor was compared to the chemical timescale calculated from Eq. 2.1 for different $CO/H_2/CH_4$ mixtures. A correlation exists between the two timescales, as indicated by the fitting line in Fig. 2.6. In addition, Fig. 2.7 compares the anticipated and actual blow-off equivalence ratios for all

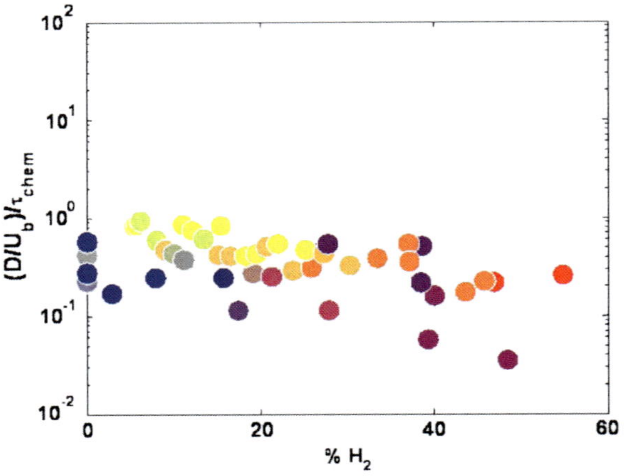


**Fig. 2.4** Damkohler number at blow-off versus %$H_2$ for different $CO/H_2/CH_4$ fuel blends. Combustor pressure = 1.7 atm, inlet temperature = 300 K [42]

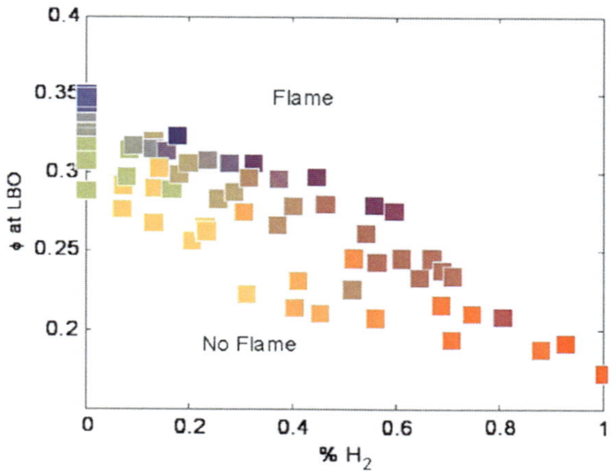


**Fig. 2.5** Equivalence ratio at lean blow-off versus %$H_2$ for different $CO/H_2/CH_4$ fuel blends. Combustor pressure = 4.4 atm, inlet temperature = 460 K [42]

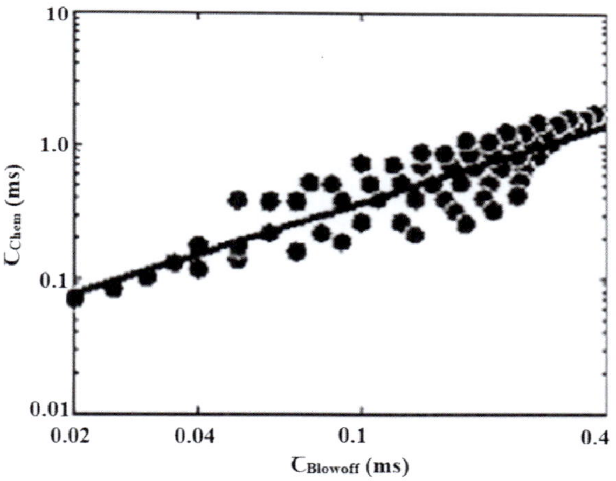


**Fig. 2.6** Relation between chemical time (calculated using Eq. 2.1) and blow-off residence time for different $CO/H_2/CH_4$ fuel blends [43]

the 300-K data. It can be seen that the error in the predicted equivalence ratio is less than 5%. It was reported that the highest error is associated with the highest %CO fuel mixtures.

Hudak and Alberto [44, 45] performed experiments and numerical modeling of the blow-off of oxy-methane ($CH_4/O_2/CO_2$) flames, oxygen-enriched air-methane ($CH_4/O_2/N_2$) flames, and methane-air ($CH_4$/air) flames. They calculated the chemical timescale based on adiabatic flame temperature. It can be seen from Fig. 2.8 that the $CH_4$/air flames have the smallest chemical time (fastest reaction), the $CH_4/O_2/CO_2$ flames have the largest chemical time (slowest reaction), while the $CH_4/O_2/N_2$ flames are in between. These results clearly demonstrate the adverse effect of $CO_2$ dilution in the combustion process. For further explanation, Fig. 2.9 clearly illustrates that higher flame temperature is required for the $CH_4/O_2/CO_2$ mixture to yield the same

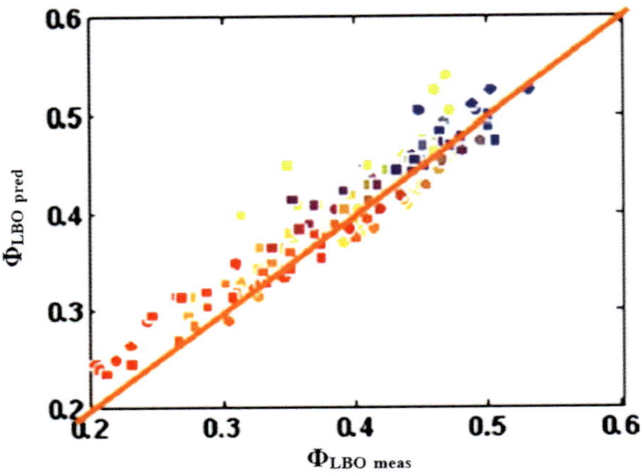


**Fig. 2.7** Comparison of the predicted and measured blow-off equivalence ratios for different $CO/H_2/CH_4$ fuel blends. Combustor pressure = 1.7 atm, inlet temperature = 300 K. Circle symbols ≡ 59 m/s flow velocity; square symbols ≡ 39 m/s [43]

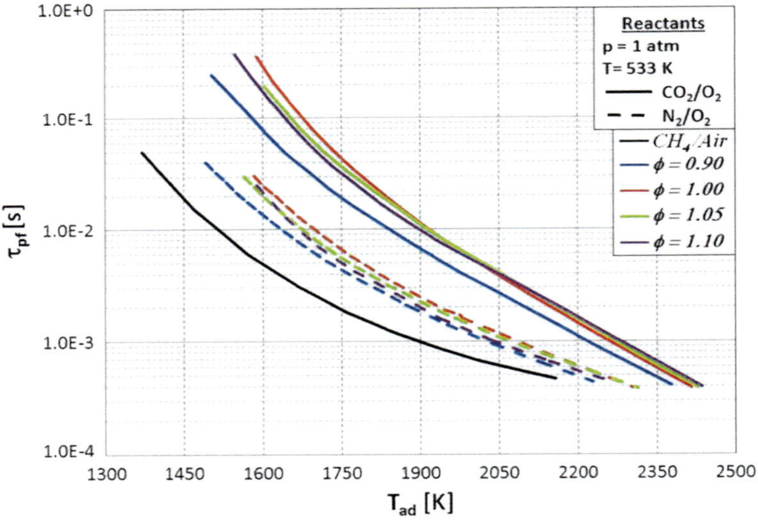


**Fig. 2.8** Chemical timescale versus adiabatic flame temperature for $CH_4/O_2/CO_2$, $CH_4/O_2/N_2$, and $CH_4$/air flames [44, 45]

chemical time of the $CH_4$/air mixture. A summary of relevant studies on the blow-off static instability is presented in Table 2.1.

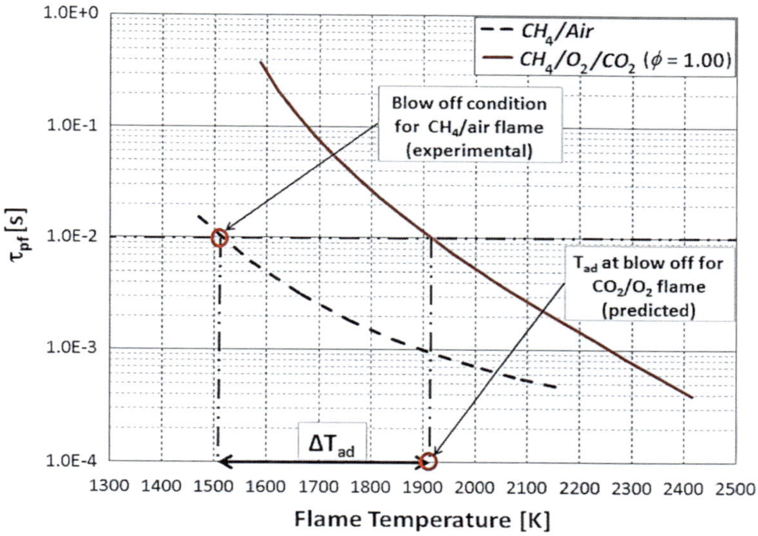


**Fig. 2.9** Chemical timescale versus adiabatic flame temperature for $CH_4$/air and $CH_4/O_2/CO_2$ flames [44, 45] Higher flame temperature is required for the $CH_4/O_2/CO_2$ mixture to yield the same chemical time of the $CH_4$/air mixture

#### 2.2.1.2 Flashback

Flashback is the opposite phenomenon of blow-off; flashback occurs when the flame propagates upstream of its anchor point against the direction of the flow, because the flow velocity is lower than the burning velocity. Similar to blow-off, flashback is an intricate problem, because the flame speed is affected significantly by the flame temperature and fuel and oxidizer compositions. However, unlike blow-off, flashback can induce serious hardware damages, as the flame propagates through flow passages that were not designed to withstand the flame temperature. Studying the potential for flashback is thus very crucial, because a safe margin has to be accounted for, in order to minimize the risk of hardware damage and increase the equipment lifespan.

Swirl is currently widely implemented as an effective stabilization technique to deter early blow-off in LPM flames, as will be seen in Sect. 2.4. Hydrogen enrichment offers additional resistance to blow-off and reduces $CO_2$ emissions (greenhouse gas) [47–50], as will be seen in Sect. 2.3. For instance, when hydrogen is blended with methane, it can produce several forms of alternative fuels that can be used to reduce pollutant emissions. However, hydrogen enrichment brings its own challenges, because the flashback risk increases with hydrogen concentration. This may be attributed to the fact that the flame speed of hydrogen is higher than most hydrocarbon fuels, which boosts the turbulent flame speed of hydrogen-fuel mixture. Flashback can be caused by one or more of the following scenarios:

**Table 2.1** Summary of relevant studies on the blow-off static instability

| Combustible mixture | Application Experimental/Numerical | References | Remarks/ main findings |
|---|---|---|---|
| Syngas fuel $CO/H_2/CH_4$ blends + air | Experimental lean premixed gas turbine model | Lieuwen et al. [42] | Blow-off occurs at almost constant Damkohler number of 0.6<br>Blow-off shifts to leaner equivalence ratios with $H_2$ enrichment |
| Propane/air | Experimental lean blow-off with bluff-body stabilizer | Shanbhogue et al. [46] | Comparison of computed chemical time scales |
| Syngas fuel $CO/H_2/CH_4$ blends + air | Experimental lean premixed gas turbine model | Zhang et al. [43] | A correlation exists between the blow-off residence time for well-stirred reactor and the chemical timescale |
| $CH_4/O_2/CO_2$, $CH_4/O_2/N_2$, and $CH_4$/air mixtures | Experimental and numerical investigation of the effect of $CO_2$ dilution on the chemical timescale as compared to air-fuel | Hudak et al. [44]<br>Alberto et al. [45] | $CO_2$ dilution has an adverse effect on the combustion process<br>Higher flame temperature is required for the $CH_4/O_2/CO_2$ mixture to yield the same chemical time of the $CH_4$/air mixture |

1. Low flow velocity of the combustible mixture allows upstream flame propagation within the boundary layer, which triggers flashback [51].
2. Turbulent flame propagation at the center of swirling flow; flashback can occur when the flame core/center velocity becomes higher than the local flow velocity. This allows for upstream flame propagation. Flashback is commonly associated with oxygen or hydrogen enrichment because of the greater chemical reaction rates [52, 53].
3. Dynamic instabilities can cause unstable combustion and, consequently, flashback [54, 55].
4. Combustion induced vortex breakdown (CIVB), which occurs at excessive swirl numbers. The axial velocity component of combustible mixture decreases drastically at the breakdown point, which increases the potential for flashback. Also, when the fuel composition changes, the heat release rate changes, which can cause the central recirculation zone (CRZ) to expand into a rose-tulip structure that, consequently, increases the risk for flashback [56, 57].

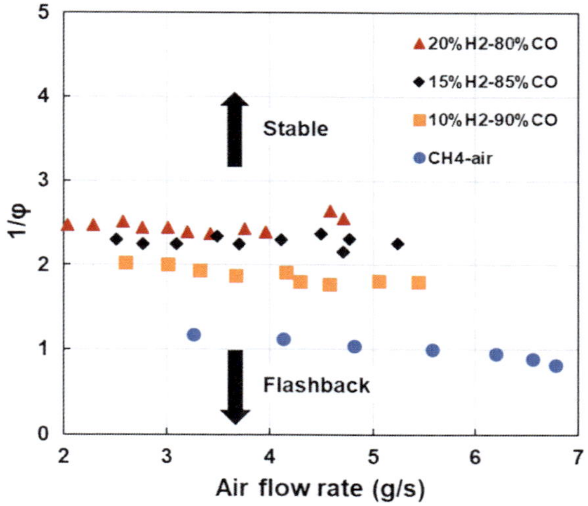


**Fig. 2.10** Flashback limits for different fuel blends at a swirl number of 0.97 [58]

Hydrogen enrichment induces significant changes in flame shape and dynamic combustion instabilities, especially at hydrogen concentrations above 60%, which makes it more difficult to accommodate/sustain the flame. These issues have encouraged researchers to study the effect of fuel composition on flashback occurrence in swirl-stabilized combustors. Dam et al. [58] shows in Fig. 2.10 the flashback limits of different fuel blends at a swirl number of 0.97. Three different fuel compositions were considered, namely 10%$H_2$ + 90%CO, 15%$H_2$ + 85%CO, and 20%$H_2$ + 80%CO, and the results are compared with air-methane combustion. It can be seen that hydrogen addition makes the flame more prone to flashback, which is attributed to faster reaction rates and, in turn, higher flame speed.

Chomiak et al. [59] conducted another study considering different coal types, as shown in Fig. 2.11. They studied different compositions of syngas generated from brown coal, bituminous coal, lignite, and coke, as listed in [59]. These syngas fuels contain different amounts of $CO_2$. The observed differences in flashback limits can be attributed to the fact that $CO_2$ slows down the reaction kinetics and, in turn, decreases the flame speed, which retards the onset of flashback.

Shelil et al. [60] performed CFD simulations of flashback in premixed swirl-stabilized air-methane flames. The flashback limit is plotted in Fig. 2.12, wherein the area under the curve denotes the unstable flashback zone, where the flow rate (and velocity) of combustible mixture is too low to balance the flame speed. It can be seen that higher flow rates are needed near stoichiometry to prevent flashback, because flame speed is maximum at stoichiometry. The effect of combustor pressure on flashback was also studied, and the results are depicted in Fig. 2.13. It can be seen that the mass flow rate needed to prevent flashback increases with combustor pressure. Even if the flame speed is assumed independent of pressure, it has to be balanced with an equal flow velocity to prevent flashback, so the pressure-induced

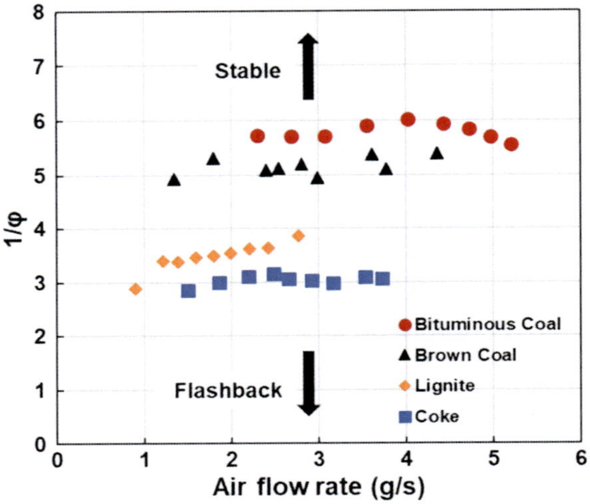


**Fig. 2.11** Flashback limits for different syngas compositions generated from different solid fuels. Swirl number = 0.97 [58]

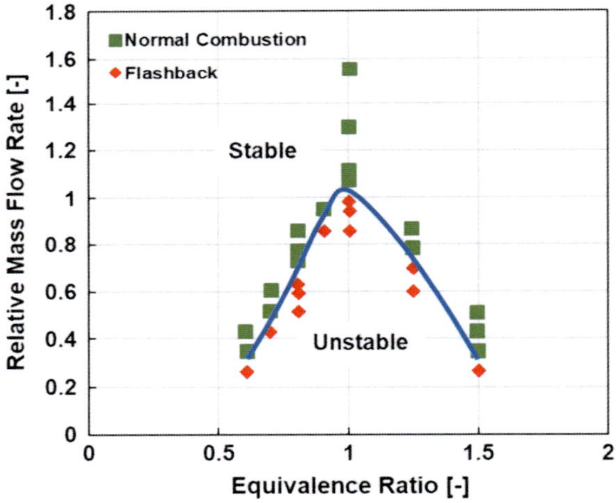


**Fig. 2.12** Flashback limit of premixed swirl-stabilized air-methane flames at atmospheric pressure for different equivalence ratios [60]

increase in density of combustible mixture will translate in higher mass flow rates at greater pressures.

Abdulsada et al. [61] studied the flashback limits for different fuel types in a swirl-stabilized combustor. They studied different $CH_4/H_2$ blends from pure $CH_4$ to pure $H_2$, in addition to the coke-oven gas (65% $H_2$, 25% $CH_4$, 6% CO, and 4% $N_2$). The study revealed that at a swirl number of 1.47 the CRZ is extended for all fuels. It was concluded that flashback occurs when the radial velocity component drops to a certain level, causing propagation of the flame front upstream.

At the end of this discussion of static instabilities, it can be concluded that hydrogen enrichment improves the resistance to blow-off but also increases the tendency for

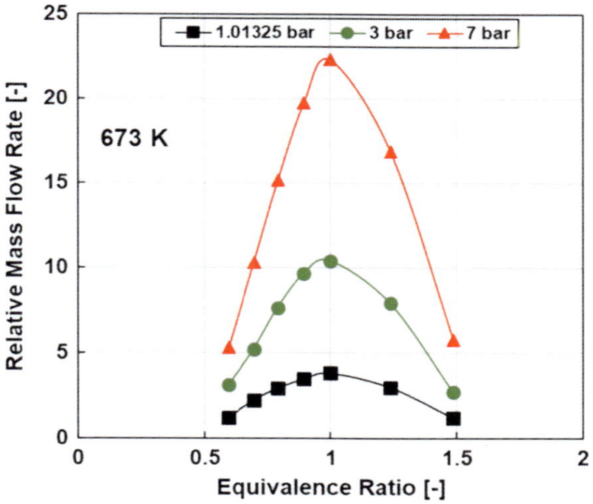


**Fig. 2.13** Effect of combustor pressure on the flashback limit of premixed swirl-stabilized air-methane flames at different equivalence ratios [60]

flashback. The net result is a narrower operability window, which has been confirmed by multiple studies for gas-turbine applications [62–65]. Hence, if the gas turbine runs on a hydrocarbon fuel that can be hydrogen-enriched on demand, the use of hydrogen should be limited to low loads only to improve the turndown. A summary of relevant studies on the flashback static instability is presented in Table 2.2.

## 2.2.2 *Dynamic Instabilities*

In this section, the main causes of dynamic combustion instabilities are discussed, and methods for suppressing them are surveyed.

### 2.2.2.1 Thermo-acoustics

The primary driving mechanism of dynamic combustion instabilities is thermo-acoustic coupling. As the term implies, two factors interplay to induce instability, namely the thermal or heat-release field and the acoustics of combustor geometry [66]. The latter is easy to comprehend and appears in plenty of examples in our daily life. Any flow in a confined passage, especially turbulent flows, generates pressure waves or noise. A simple example is the flute; blowing into it creates sound because the airflow creates broadband low-amplitude noise, but one particular frequency resonates with a higher audible amplitude. This frequency is function of the geometry of the flow passage within the flute; covering or uncovering any of the flute holes changes the geometry and thus the resonant frequency, even if the airflow remains unchanged. Gas-turbine combustors can be viewed as huge flutes; turbulent

**Table 2.2** Summary of relevant studies on the flashback static instability

| Combustible mixture | Application Experimental/Numerical | References | Remarks/ main findings |
|---|---|---|---|
| $H_2$/CO syngas blends + air and $CH_4$/air | Effect of fuel composition on flashback limit | Dam et al. [58] | Hydrogen addition makes the flame more prone to flashback, which is attributed to faster reaction rates and, in turn, higher flame speed |
| Gasified coals + air | Effect of fuel composition on flashback limit | Chomiak et al. [59] | Syngas fuels contain different amounts of $CO_2$ that slows down the reaction kinetics and, in turn, decreases the flame speed, which retards the onset of flashback |
| $CH_4$/air | Numerical CFD simulations of flashback | Shelil et al. [60] | Higher flow rates are needed near stoichiometry to prevent flashback, because flame speed is maximum at stoichiometry<br>The mass flow rate of combustible mixture needed to prevent flashback increases with combustor pressure |
| $CH_4/H_2$ blends + air and coke-oven gas + air | The flashback limits for different fuel types in a swirl stabilized combustor | Abdulsada et al. [61] | At swirl number of 1.47, the CRZ is extended for all fuels<br>Flashback occurs when the radial velocity component drops to a certain level, causing propagation of the flame front upstream |

compressor-discharge air flows through a confined passage of a particular geometry that has one or more distinct resonant frequencies. The second factor responsible for dynamic combustion instabilities is heat release within the flame. Gas-turbine flames are highly turbulent, so the map of local heat release is far from uniform. Wherever combustible mixture burns, the local released heat makes the combustion products expand at this location. Premixed combustion is thus accompanied by its own pressure fluctuations or noise. Several pressure waves coexist now within the combustor; some induced just by turbulent flow through a confined passage, and others by heat-release fluctuations. When these pressure waves interact, destructive interference occurs for most of the broadband frequencies, but one or more frequencies interfere constructively, which magnifies the amplitude of pressure fluctuations. The combustor hardware vibrates with these fluctuations, which could wear or even damage some components. In extreme cases, pieces of the combustor could break off and damage the turbine downstream [67–69]. Interestingly, all combustion systems are always associated with pressure fluctuations, even under stable operating conditions. Stable combustion is obtained, however, when the amplitude of these fluctuations is less than ~5% of the mean combustor pressure [26]. Higher amplitudes are undesired and considered an indication of dynamic combustion instability [68–70]. Flame shape and location are also altered by dynamic instabilities [71–74], which creates a serious operability concern.

Dynamic instabilities are a serious concern in LPM combustors [75–80], as they are typically operated close to their blow-off limits. The flame temperature is low enough to induce local extinction and re-ignition at multiple locations, which causes fluctuations in the heat-release map and, consequently, pressure oscillations that couple with the combustor acoustics [22, 23]. LPM combustors also feed 60–70% of the inlet air into the flame to make it lean, leaving a smaller portion for liner cooling and downstream dilution [27]. The premixed flame is free of any stoichiometric zones, i.e., the map of local temperature is almost uniform. The temperature (and density) gradients within the flame and further downstream are too small, so the resonant pressure waves can propagate freely within the quasi-uniform temperature field inside the combustor.

Recently, Speth et al. [80] investigated the dynamic instabilities for different syngas-fuel ($CO + H_2$) compositions by quantifying the overall sound-pressure levels in dB. The dB level is plotted versus equivalence ratio in Fig. 2.14 for different CO contents. It can be seen that the dB level is almost the same for all examined fuels, but

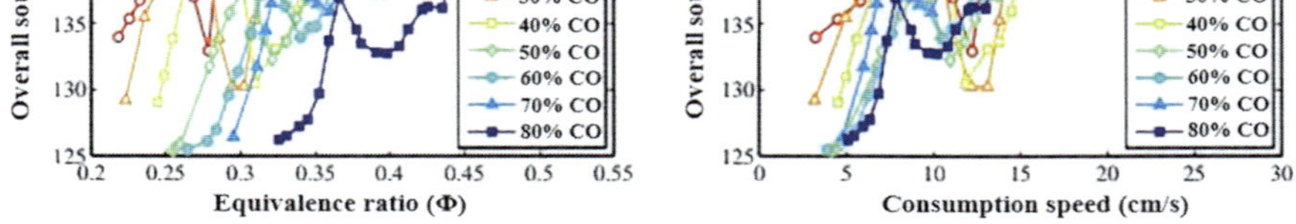


**Fig. 2.14** Overall sound pressure level versus equivalence ratio (left) and consumption speed (right) for different syngas fuel compositions [80]

the curves shift to leaner equivalence ratios as the $H_2$ content of fuel increases; i.e., the same dB level is generated at lower equivalence ratios for higher $H_2$ contents. Since the flame speed of $H_2$ is higher than that of CO, increasing the $H_2$ content of syngas increases its flame speed; nonetheless, decreasing the equivalence ratio reduces the flame temperature and, in turn, the flame speed [78]. The question to ask here is, thus, whether a constant dB level corresponds to a constant flame speed in Fig. 2.14. Three parameters are interplaying here, namely $H_2$ content, equivalence ratio (flame temperature), and flame speed. Changing the $H_2$ content of a fuel at a given equivalence ratio has a miniscule effect on adiabatic flame temperature [79], which makes the flame speed depend primarily on equivalence ratio (flame temperature). If all the dB curves are plotted against the flame speed, which is termed "consumption speed" in Fig. 2.14, all curves should collapse on each other, if the answer to the aforementioned question is true. The right plot indeed confirms that all curves collapse; hence, the dB level depends primarily on flame speed. It is worth noting here that the flame speed is a reflection of the chemical timescale, which has been found to be one of the primary parameters affecting dynamic combustion instabilities [81, 82].

#### 2.2.2.2 Suppression Methods of Dynamic Instabilities

It is mandatory to dampen the pressure oscillations in premixed combustion systems, because high amplitudes may lead to hardware damage. To mitigate the dynamic instabilities, the constructive coupling of unsteady heat release and combustor acoustics must be interrupted. This is achieved by altering either or both of them. Control methods that target the heat release are called active controls [83, 84], while methods that change the combustor acoustics are termed passive controls [85]. As the names imply, active controls monitor the amplitude-versus-frequency plot of pressure oscillations in real time, and a feedback loop intervenes to alter the heat-release map whenever the amplitude of any frequency spikes beyond a certain set threshold. Passive controls, on the other hand, do not have life monitoring; instead, they induce changes that target a certain fixed resonant frequency that is known to persist throughout the combustor operation. Both active and passive control techniques have been implemented successfully in premixed combustion systems [86, 87].

One of the most commonly implemented active techniques is to induce fluctuations in the fuel flow rate (input energy), which, in turn, alters the heat-release map. The frequency of these fluctuations is calculated and adjusted on demand by the control system to target the resonant frequencies. Another approach is catalytic combustion, which inherently has lower fluctuations in the heat-release rate; however, catalytic applications must have a stable pre-burner, which counteracts its advantage.

Passive control techniques change one or more of the following parameters: fuel type, fuel-injection devices/geometry/pressure, combustion-chamber design, as well as the installation of acoustic dampers. The latter are cavities or tubes that are attached to the liner wall to induce minor changes in the combustion-chamber geometry, which is a very simple yet very effective and widely implemented solution to target specific frequencies. Another passive approach is to split the input fuel into a main premixed stream and a smaller non-premixed pilot one, but this technique comes with a $NO_x$ penalty because of the non-premixed pilot. Some of the fuel-related techniques aim at reducing the heat-release rate or increasing the energy losses.

Currently, the passive methods are the dominant controls in combustion systems. This may be attributed to the complexity of the active methods. Richards et al. [88] presented mapping of the dynamic regimes of an engine to confirm that it can be operated well after changing the fuel type. Another very effective passive method is the perforated liner. Lei et al. [89] studied the control of combustion instabilities with a perforated liner. They measured both the pressure amplitudes and frequencies for different liners, including rigid wall as well as perforated walls with different bias volume flow rates. It can be seen from Fig. 2.15 that the pressure amplitudes decreased significantly with the use of perforated liner. They reported, however, that these dampers are not suitable for large-scale combustion systems, because the unstable energy will be very large to accommodate, and the damping effect of bias flow becomes negligible.

Lee et al. [90] studied experimentally the active control of combustion instabilities by modifying the spray characteristics of liquid fuel, which was achieved by modulating the fuel injection rate at the frequency of the instability with a proper phase shift. This technique is considered one of the most promising approaches that can dampen dynamic instabilities effectively. Other studies examined the effect of equivalence-ratio oscillation on combustion dynamics in backward-facing step combustors [91, 92]. More studies and concepts of active and passive control can be found in the literature [93–96].

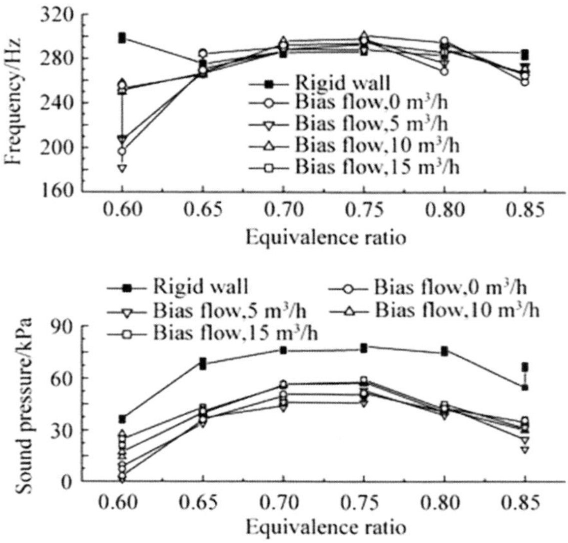


**Fig. 2.15** Frequency and sound pressure with respect to equivalence ratio and different bias flow [89]

The next section of this chapter discusses in detail the different approaches taken to overcome combustion instabilities, improve the combustion process, and reduce the exhaust emissions. Different approaches are analyzed in this section, including fuel and oxidizer flexibility (such as hydrogen enrichment and oxy-fuel combustion), variable operating conditions, and variable flame type (non-premixed, partially premixed, and premixed flames).

## 2.3 $NO_x$ Emissions in Lean Premixed Air Combustion

### *2.3.1 Prompt $NO_x$*

One of the sources of $NO_x$ formation is Prompt $NO_x$ (also known as Fenimore $NO_x$). In this mechanism, rapid reaction takes place between atmospheric nitrogen $N_2$ and intermediate hydrocarbon radicals -especially CH and $CH_2$ in the combustion chamber [97]. These reactions result in C-N species, which proceed through different reaction pathways including $O_2$ resulting in $NO_x$ formation. Prompt $NO_x$ is dominant in fuel rich mixtures. It is because of the abundance of hydrocarbon radicals available to react with $N_2$. The dependence of prompt $NO_x$ formation on the hydrocarbon fragments signifies the importance of fuel chemistry. Thermal $NO_x$, on the other hand, is insensitive to the fuel chemistry. Although it is generally considered that thermal $NO_x$ possesses the larger contribution in the total $NO_x$ formation, but changes in the fuel chemistry and combustion conditions can shift the relative prevalence of $NO_x$ formation from thermal $NO_x$ to prompt $NO_x$ [98]. Baukal et al. [99] reported that prompt $NO_x$ is weakly temperature dependent. The quantity of prompt $NO_x$ generated from overall combustion $NO_x$ emissions is minor. In the lean premixed air combustion the formation of prompt $NO_x$ is minimal. However, the contribution of this source of $NO_x$ formation significantly increases when total $NO_x$ emissions are extremely reduced. Moreover, it was described by Moreno et al. [100] that the prompt $NO_x$ mechanism becomes significant, especially when the $NO_x$ level goes below 9 ppm, which is the case of newly proposed LPM combustion systems. Generally, following reactions result in prompt $NO_x$ formation:

$$CH + N_2 \leftrightarrow HCN + N \tag{2.4}$$

$$CH_2 + N_2 \leftrightarrow HCN + NH \tag{2.5}$$

$$N + O_2 \leftrightarrow NO + O \tag{2.6}$$

$$HCN + OH \leftrightarrow CN + H_2O \tag{2.7}$$

$$CN + O_2 \leftrightarrow NO + CO \tag{2.8}$$

Miller and Bowman [101] reported the importance of prompt $NO_x$ mechanism in their combustion modeling study that total $NO_x$ formation is underestimated without considering the prompt $NO_x$. Ren and Li [102] carried out a computational study on biodiesel combustion and indicated an inverse relationship between soot emissions and prompt $NO_x$. It is possibly because of the competition for the intermediate hydrocarbon CH species; which play an active role in both soot formation and prompt $NO_x$ mechanisms. Hedley et al. [103] conducted a research study on different firing configurations in a semi-industrial scale test furnace having oxy-natural gas combustion. They reported significant dependence of temperature and $NO_x$ emissions on firing configurations and interpreted the increased prompt $NO_x$ formation as a result of increased internal recirculation. Zhilong et al. [104] investigated the effect of hydrogen addition on $NO_x$ and CO formation. They considered different mechanisms of $NO_x$ formation and indicated that hydrogen addition initially increases the prompt NO formation, but this rising trend decreases significantly when hydrogen addition is greater than 20%. They interpreted this phenomenon as the outcome of improved premixed combustion and reduced methane proportion. Moreover, they reported that hydrogen addition decreases NO formation through prompt $NO_x$ mechanism only, and the overall NO formation increases steadily.

Prompt $NO_x$ mechanism is reported to be more significant for hydrocarbons with higher carbon number. It is stated in the work of Naha and Aggarwal [105] that the prompt $NO_x$ route is prevalent in n-heptane flames while thermal $NO_x$ route becomes significant for methane flames. Fiskum et al. [106] carried out an investigation on propane and methane flames and indicated similar results, where prompt $NO_x$ was dominant for propane flames. They explained that $NO_x$ emissions increase through prompt route at higher carbon number of hydrocarbons. Costa et al. [107] investigated the turbulent propane flames and indicated the dominance of prompt route of $NO_x$ formation, thereby validating the similar results with the laminar propane flames.

### 2.3.2 *Fuel $NO_x$*

When nitrogen containing fuels are oxidized during the combustion process, the resulting NO formation is known as fuel $NO_x$ [108]. The amount of fuel $NO_x$ is dependent on the amount of fuel-bound nitrogen. Other influencing factors are hydrocarbon's oxidation and chemical kinetics of the reaction. Ammonia ($NH_3$) is being considered as a promising carbon-free fuel. Liquid fuels usually possess fuel bound nitrogen as compared to their gaseous counterparts. Several biomass derived fuels also possess larger compositions of ammonia [109]. The combustion of ammonia results in significant amount of fuel $NO_x$ emissions as ammonia is a nitrogen-containing combustion species.

Ammonia doped flames were numerically investigated using the eddy dissipation concept for predicting $NO_x$ emissions [110]. The reaction and resulting NO emissions were modeled using the GRI 2.11 mechanism [111] and good agreement with the experimental data was found. The results indicated that the rate of NO formation was higher in fuel $NO_x$ route as compared to thermal $NO_x$. Some other experimental studies have shown that the producer gas generated from seed corn and wood possesses substantial amount of ammonia and their combustion result in higher proportions of fuel $NO_x$ [112, 113]. Biomass derived gas containing ammonia was numerically investigated using a laminar flamelet model in a gas turbine combustor [114]. A simplified global reaction mechanism was used for NO formation. It was found that the fuel $NO_x$ route was dominant in the resulting overall $NO_x$ emissions.

Coming towards the combustion chemistry of ammonia, nitrogen is converted to either $NO_x$ or $N_2$, subject to the conditions of reaction. Several studies have been conducted for investigating the fuel $NO_x$ formation pathways of ammonia combustion [115–122]. Sullivan et al. [115] conducted a numerical investigation and stated that ammonia to NO conversion efficiency reduces with decreasing the concentration of ammonia. Zabetta et al. [116] conducted numerical investigation on a series of plug flow reactors and performed combustion modeling on syngas containing ammonia. They stated that increasing the pressure and staging result in effective conversion of fuel-bound nitrogen compounds to $N_2$ instead of fuel $NO_x$. For low temperature reactors, it was reported in another study that combined air staging, fuel staging, and selective non-catalytic reduction method can be an efficient approach of fuel $NO_x$ reduction [121]. Lower $NO_x$ emissions are resulted from deeper staged combustion. Li et al. [122] experimentally studied the reduced fuel $NO_x$ emissions under lean and stoichiometric conditions for fluidized bed biofuel combustion. They investigated the effect of excess oxygen (or excess air) of a range of 20–80% on fuel $NO_x$ formation, as shown in Fig. 2.16. Several biofuels are considered for comparison under fuel lean and stoichiometric conditions. For all the fuels, increasing the excess air concentration results in increased fuel-$NO_x$ formation. The slopes for all the fuels except for leather waste follow the same trend, i.e. every 10% increment in the excess

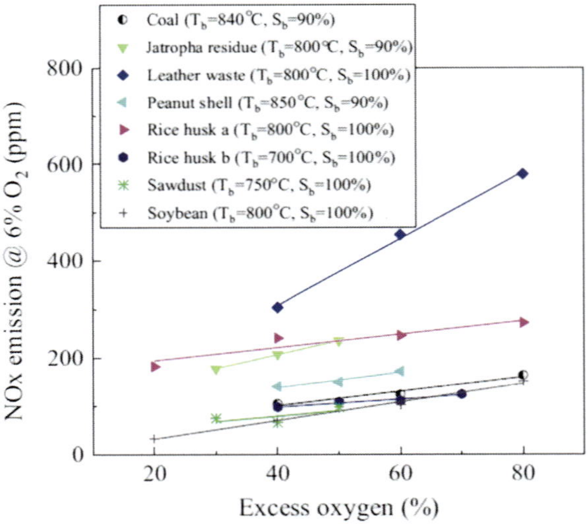


**Fig. 2.16** The effect of excess oxygen on $NO_x$ emission, where $S_b$ indicates the stoichiometry in the bed while $T_b$ represents the bed temperature [122]

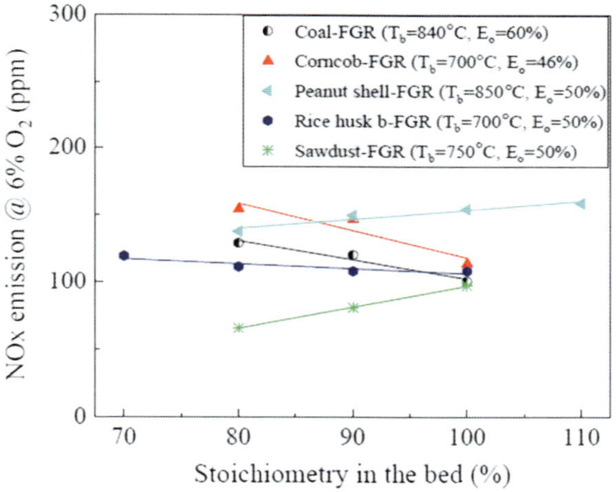


**Fig. 2.17** The influence of bed's stoichiometry on $NO_x$ emission for a range of stoichiometry from 70 to 110%, where $E_O$ represents the excess oxygen percentage and $T_b$ represents the bed temperature [122]

oxygen results in 8–29 ppm increase in $NO_x$ emissions. $NO_x$ formation increases significantly for leather waste combustion with a reported increments of 69 ppm per 10% increase in excess oxygen. Fuel NO formation is subject to the availability of higher volatile-N components in the fuel, and this exception may be due to leather waste composition. The presence of volatile-N species in the fuel result in increased fuel NO formation, especially under lean conditions [122]. The effect of stoichiometry in the bed on $NO_x$ emissions was investigated for five different biofuels, as represented in Fig. 2.17. The flow of all the gases was maintained as constant in flue gas recirculation (FGR) for eliminating the effects of superficial velocity. An inconsistency in the $NO_x$ emissions is observed from the figure. For peanut shell and sawdust, the reduction of stoichiometry results in decreasing $NO_x$ emissions because of reducing atmosphere. However, a reverse trend is observed for coal, corncob, and rice husk b. The results presented in this figure, therefore, imply that the reduction in stoichiometry does not always result in the reduction of $NO_x$ emissions for every fuel, at least over the investigated range of stoichiometry, from 70 to 110%.

## 2.4 Approaches for Efficient Combustion

Robustness, low emissions, and high efficiency are the key parameters motivating the development of gas-turbine combustors. Many approaches have been introduced and developed to achieve these goals including: (1) fuel/oxidizer-flexibility approach, (2) variable operating conditions approach, and (3) variable flame type approach.

### 2.4.1 Fuel/Oxidizer-Flexibility Approach

The fuel flexibility of a gas-turbine combustor describes how much the flame characteristics and combustor operability are affected by changing the fuel composition. This is currently an area of active research because of the interest in using different fuel types and blends. The aim of this chapter section is to analyze the effect of fuel composition on combustion characteristics, namely static stability limits (blow-off and flashback), dynamic stability, and exhaust emissions.

Natural gas is the primary gaseous fuel for gas turbines; yet, there is a worldwide interest in substitute fuels because of the escalating natural-gas prices and the depleting fossil-fuel resources. Alternative fuels include syngas, biomass, and liquefied petroleum gas. These fuels are different from natural gas not just in the physical and chemical transport properties but also in the energy content. The change in chemical composition of the fuel results in substantial change in the chemical timescale, which, consequently, affects the operability characteristics in terms of blow-off, flashback, and ignition time [123–126].

The Wobbe Index (WI) is a common indicator for fuel interchangeability in the gas-turbine industry. It applies to gaseous fuels and is defined as

$$WI = \frac{LHV_{fuel}}{\sqrt{\frac{\rho_{fuel}}{\rho_{air}}}} \tag{2.9}$$

where $LHV_{fuel}$ is the lower heating value of fuel per unit mass, and $\rho_{fuel}$ and $\rho_{air}$ are the densities of fuel and air, respectively. The density ratio is the fuel specific gravity. The Modified Wobbe Index (MWI) was introduced to account for the fuel temperature ($T_{fuel}$) as well, namely

$$MWI = \frac{LHV_{fuel}}{\sqrt{T_{fuel} \times \frac{\rho_{fuel}}{\rho_{air}}}} \tag{2.10}$$

Figure 2.18 shows the MWI map of all known gaseous fuels. Standard natural gas (mostly $CH_4$) lies at the center of the map. The larger the content of higher hydrocarbons ($C_{2+}$), the higher the MWI is at constant LHV, e.g., liquefied petroleum gas (LPG). Hydrogen enrichment increases LHV at constant MWI. Increasing the inert content ($N_2$, $CO_2$, $H_2O$) decreases both LHV and MWI, e.g., syngas and coke-oven gas (COG).

Richards et al. [127] comprehensively reviewed the fuel variability and stability in premixed hydrogen-enriched combustion. They illustrated that there is a rising interest in switching from natural gas to hydrogen-enriched methane and syngas in combined-cycle power plants to enhance the stability limits and reduce $NO_x$ emissions. $H_2$ addition was found to enhance the operability of gas-turbine combustion

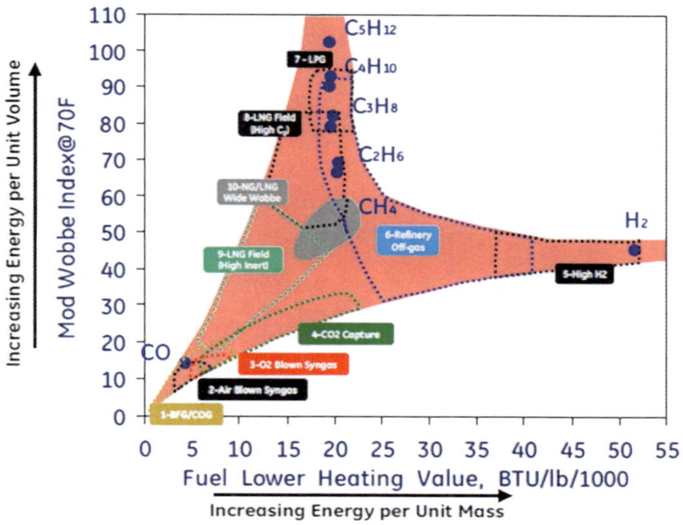


**Fig. 2.18** MWI map of all known gaseous fuels

systems significantly. Recent technologies were also discussed that affect the combustor design and development to avoid flashback, auto-ignition, and dynamic instabilities associated with $H_2$ enrichment [127]. It was reported that $H_2$ addition results in leaner blow-off limit. However, the potential for flashback increased because of the increase in flame speed, which is an adverse effect of $H_2$ enrichment [128]. Nonetheless, in order to maintain the same power output from the combustor, the total flow rate of air through the combustor needs to be increased. The extra air decreases the equivalence ratio and flame temperature, which reduces the potential for flashback. Richards et al. revealed that swirl is the most commonly used stabilization technique, and despite the abundance of past studies, further research is needed to minimize combustion instabilities while varying the fuel composition.

Lieuwen et al. [42] also discussed the impact of varying the fuel composition on the performance of LPM combustors. They quantified the laminar and turbulent flame velocities and chemical timescale. One of the main findings is that the combustion behavior of fuel mixtures, such as laminar and turbulent flame speed, can be substantially different from that of a fuel with a single-species composition. In the same context, Kido et al. [129] studied numerically the reliance of turbulent flame speed on the turbulence intensity and equivalence ratio using different fuel types. Three different fuels ($CH_4$, $C_3H_8$, and $H_2$) were examined at the same laminar flame speed, and the study revealed that they have significantly different turbulent flame speeds, as shown in Fig. 2.19. This can be attributed to differences in the transport diffusion properties.

The combined effects of fuel blending and heat-release rate on combustion instabilities have been investigated by Ferguson et al. [124]. The effect of fuel Wobbe

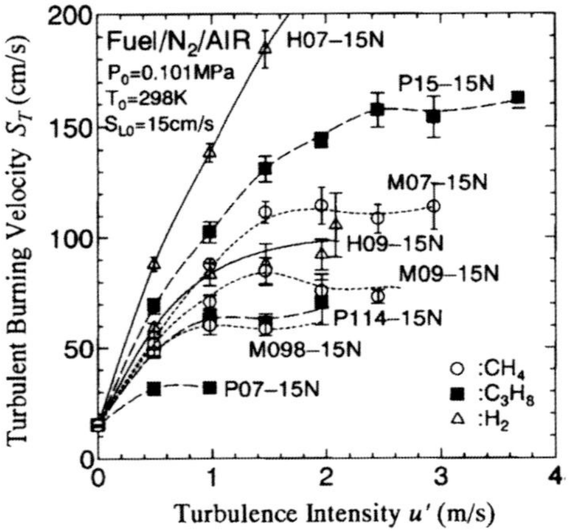


**Fig. 2.19** Reliance of turbulent burning velocity on turbulence intensity for different fuel types [129]

index (WI) was studied by mixing natural gas with different concentrations of $C_2H_6$. The study revealed that, at a fixed Wobbe index, the produced dynamic response is not constant.

#### 2.4.1.1 Hydrogen-Enrichment Approach

Hydrogen ($H_2$) fuel is pivotal for green and clean energy production, because it is a renewable and sustainable source of energy. It can be produced via steam reforming of bio-methane [130–132]. $H_2$ is one of the most powerful and highly efficient fuels because of its high energy content per unit mass [133–135]. It is well known that many countries around the world depend mainly on hydrocarbon fossil fuels for energy production, which results in enormous greenhouse-gas emissions ($CO_2$). Therefore, the $H_2$-enrichment approach is highly recommended for cleaner energy production in the coming decades, because $H_2$ is non-toxic and produces no harmful emissions (just $H_2O$). Besides, the concentration of effluent $CO_2$ decreases as that of $H_2O$ increases. $H_2$ enrichment can be implemented in a variety of power-generation systems, including internal combustion engines, steam power cycles, jet engines, and utility gas turbines [136]. The integrated gasification combined cycle (IGCC) is a widely implemented example of power-generation systems that have low carbon footprints and run on $H_2$-rich fuels generated from the gasification of solid fuels and wastes. $H_2$ enrichment is also one of the promising techniques used to enhance the stability of LPM flames, retard their blow-off, and reduce their $NO_x$ emissions. However, the effect of $H_2$ addition still requires extensive research to understand the complicated physics and reaction kinetics of $H_2$-enriched flames.

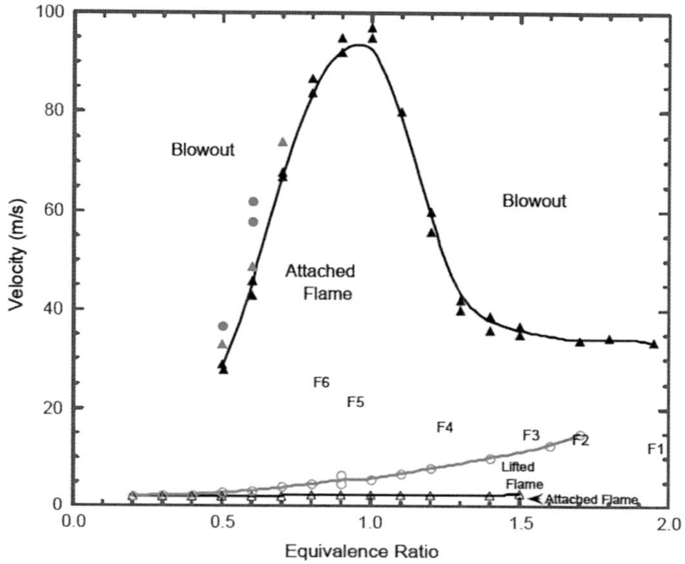


**Fig. 2.20** Blow-off velocity versus equivalence ratio of premixed $CH_4$/air flame without $H_2$ enrichment [137]

## Effect of Hydrogen on Combustor Stability Map

Schefer [137] conducted an experimental study to investigate the stability characteristics of a LPM swirl-stabilized flame with $H_2$ enrichment. The study revealed that the maximum blow-off velocity occurs at stoichiometric conditions, where the flame speed is greatest, see Fig. 2.20. Rich flames were lifted away from the burner surface, so more focus was made on lean flames. Two $H_2$ enrichment levels (10 and 20% $H_2$) were examined and compared to simple methane combustion at lean equivalence ratios, as seen in Fig. 2.21. It can be observed that $H_2$ enrichment increases the blow-off velocity, i.e., enhances the blow-off limit. This can be attributed to the saturation of the reaction zone by free OH and H radicals that boost the reaction rates, thus resulting in higher burning velocity and lower Lewis number with $H_2$ addition [138]. The Lewis number is defined as the ratio of the mass and thermal diffusion rates. The Prandtl, Schmidt, and Lewis numbers are very important dimensionless fluid-property groups that give an indication of the relative rates of diffusion. The thermo-physical transport properties play a critical role in determining the flame characteristics and stability.

Zhen et al. [139] studied the static stability limits of stoichiometric mixtures of biogas and air. Biogas (BG) was synthesized by mixing $CH_4$, $H_2$, and $CO_2$ in different proportions. Three primary compositions were examined, namely $BG_{60}$, $BG_{50}$, and $BG_{40}$, where the subscript refers to the %$CH_4$ in the BG mixture. The effect of %$H_2$ on flame stability was investigated at different Reynolds numbers of the combustible

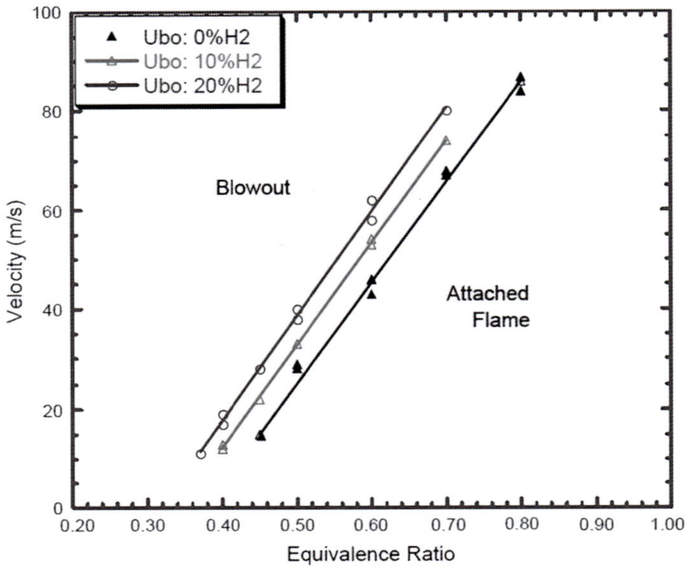


**Fig. 2.21** Blow-off velocity versus equivalence ratio of $CH_4$/air flame with $H_2$ enrichment at 0, 10, and 20 $H_2$ [137]

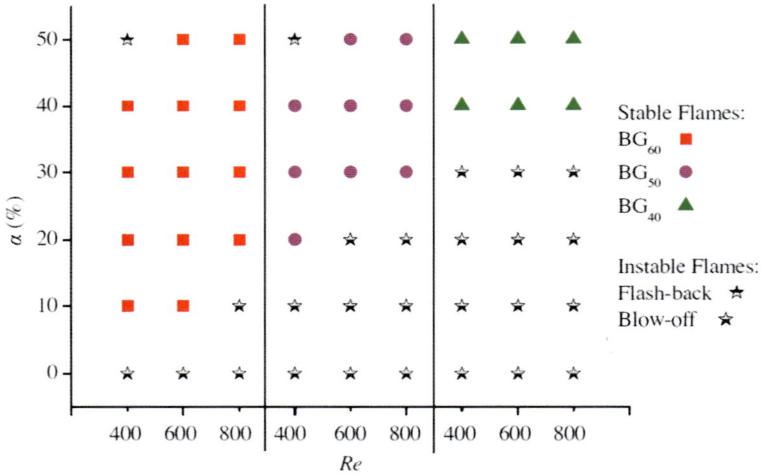


**Fig. 2.22** Stability map of stoichiometric biogas/air flames at different levels of $CH_4$ in the $CH_4/H_2/CO_2$ biogas fuel [139]

mixture. Figure 2.22 shows the upper and lower flammability limits of those three fuel compositions at different Reynolds numbers and hydrogen-enrichment levels;

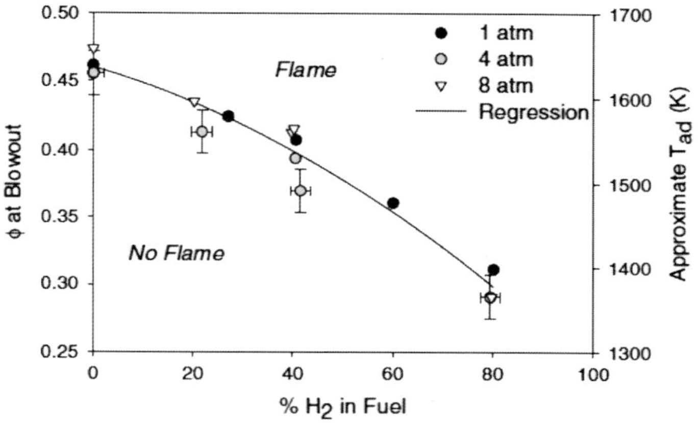


**Fig. 2.23** Lean blow-off limit of a swirl combustor at different $H_2$-enrichment levels and combustor pressures. Inlet temperature and velocity are 580 K and 40 m/s, respectively [143]

the parameter $\alpha$ denotes the $\%H_2$ in the BG mixture. It can be observed that $H_2$-free BG ($CH_4 + CO_2$ only) blows off at all examined Reynolds numbers and $CH_4$ contents, i.e., the flame is stabilized when a certain amount of $H_2$ is introduced into the combustible mixture. More $H_2$ is needed at lower $CH_4$ contents to prevent blow-off, which highlights the adverse effect of $CO_2$ on flame stability. Rashwan et al. [140, 141] confirmed this fact by showing that increased levels of $CO_2$ in the combustible mixture result in deteriorated flame stability and ultimately blow-off. It can also be observed in Fig. 2.22 that increasing both the $H_2$ and $CH_4$ contents triggers flashback, especially at low Reynolds numbers, i.e., low velocities of combustible mixture. This is attributed to greater flame speeds at higher $H_2$ levels.

Other studies conducted by Lieuwen et al. [42, 142] targeted the effect of $H_2$ enrichment on flame blow-off at different combustor pressures. It was found that $H_2$ enrichment is more effective at higher pressures, because the lean blow-off limit shifts to widen the stable operating range. In similar investigations by Strakey et al. [41, 143], the effect of $H_2$ enrichment on the lean blow-off limit of a swirl combustor was quantified at different pressures ranging from 1 to 8 atm, see Fig. 2.23. The $H_2$ content of the fuel ranged from 0% to 80%. The results reveal that increasing the $H_2$ content retards flame blow-off. However, the combustor pressure shows no discernable effect on the blow-off limit.

### Effect of Hydrogen on Laminar Burning Velocity

Tuncer [144] carried out measurements of the laminar flame speed at blow-off in a gas-turbine combustor for different $H_2$-enrichment levels. $H_2$ addition boosted the laminar flame speed over the entire examined equivalence-ratio range of 0.55–1.4 (lean to rich), which considerably extended the blow-off limits; see Fig. 2.24. Halter

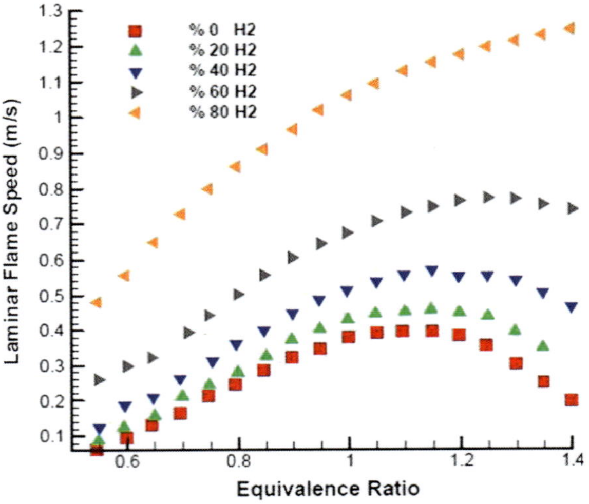


**Fig. 2.24** Blow-off laminar flame speed versus equivalence ratio for different $H_2$-enrichment levels [144]

et al. [145] explained this behavior by showing that the effect of stretch on laminar flame speed weakens when $H_2$ is added to the combustible mixture, which also explains why the blow-off velocity increases with $H_2$ enrichment, as was observed in Fig. 2.21. A small percentage of $H_2$ considerably changes the instantaneous and average flame characteristics [145].

Ilbas et al. [146–148] conducted experiments to compare the laminar burning velocities of $CH_4/H_2$ mixtures with air, ranging from 100% $CH_4$ to 100% $H_2$, over the equivalence-ratio range of 0.8–3.2. The study revealed that increasing the fuel hydrogen fraction brought about an increase in the laminar flame speed; see Fig. 2.25. The stoichiometric flame speed as function of hydrogen fraction is plotted in Fig. 2.26. The authors recommended the mixture of 30% $H_2$ and 70% $CH_4$ as an alternative fuel for power-generation plants, as it offers good flame stability, wide flammability range, and a relatively high burning velocity.

### Effect of Hydrogen on $NO_x$ and CO Emissions

Ghoniem et al. [149] studied the stability and emissions of a LPM $H_2$-enriched propane-air flame stabilized on a backward-facing step. $H_2$ addition was observed to increase the NO emissions, as shown in Fig. 2.27, because of the slight temperature rise associating $H_2$ addition; however, performance is not compromised because leaner operation with lower NO emissions becomes possible thanks to $H_2$ enrichment. Operation near the lean blow-off limit was associated with strong pressure oscillations, which couple with the heat-release fluctuations to create a strong thermo-acoustic field. Griebel et al. [150], studied the lean blow-off limits and $NO_x$ emissions of $H_2$-enriched $CH_4$/air flames at different combustor pressures. They found that $NO_x$ was reduced significantly by ~35% because $H_2$ enrichment allowed

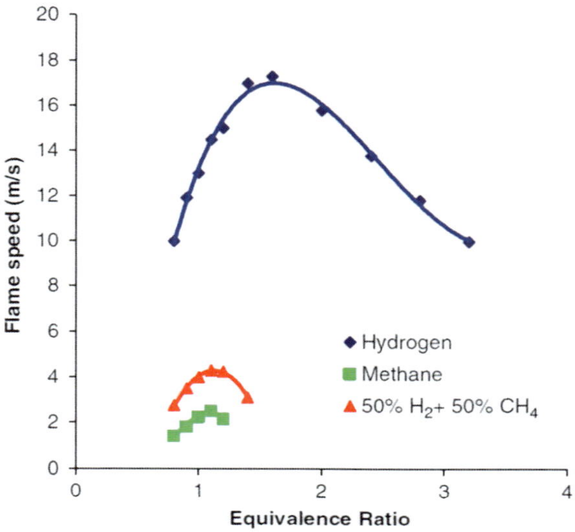


**Fig. 2.25** Flame speed of $CH_4/H_2$ fuel blends at different equivalence ratios [146]

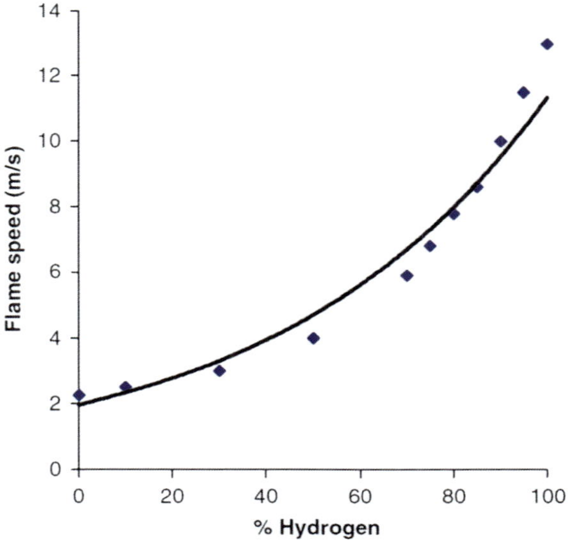


**Fig. 2.26** Stoichiometric flame speed at different hydrogen fractions of $CH_4/H_2$ fuel blend [146]

for reducing the flame temperature by 60 K. Hydrogen enrichment of natural gas is thus an approach for reducing the $NO_x$ emissions of natural-gas combustion in gas turbines. However, this approach is successful only when operating near the lean blow-off region. Lounici et al. [151] studied improving the performance of a dual-fueled gas turbine by $H_2$ enrichment. They raised the problem that the efficiency of dual-fuel engines decreases when operated at low loads, and suggested $H_2$ addition

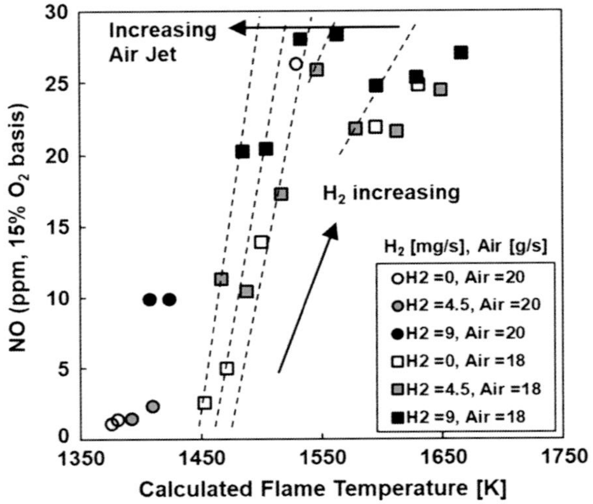


**Fig. 2.27** NO emissions at different flame temperatures in a LPM $C_3H_8/H_2$/air flame stabilized on a backward-facing step for different air and $H_2$ flow rates at a constant $C_3H_8$ flow rate of 0.66 kg/s [149]

as an effective method to overcome this problem and increase the burning velocity. It was reported that the break specific fuel consumption and the emissions of hydrocarbons, CO, and $CO_2$ were decreased with $H_2$ addition.

Schefer et al. [152] investigated the effect of $H_2$ enrichment in a LPM swirl combustor fueled by natural gas. They studied the flame stability and blow-off mechanisms for different hydrogen fractions at different fuel-to-air ratios. The study revealed that adding an amount of $H_2$ to the combustion process results in significant reduction in CO concentration but an increase in the OH radical concentration, as shown in Fig. 2.28. Thus, flame stability improves, and flame length decreases. NO concentration, however, increases slightly in the early flame sections and remains constant further downstream. This is attributed to the slight increase in flame temperature with $H_2$ enrichment, i.e., the authors did not benefit from the ability to operate at leaner equivalence ratios with $H_2$ addition. Hu et al. [153–155] conducted a numerical study of the NO concentrations in premixed $CH_4/H_2$/air flames. Figure 2.29 describes the effect of hydrogen fraction on NO concentration at different axial locations from the burner; substantial reduction in NO can be obtained with $H_2$ enrichment.

A study conducted by York et al. [156] on a 10-MW LPM gas turbine showed the possibility of operation at low $NO_x$ emissions. They examined different fuels, including natural gas and two hydrogen-enriched fuel mixtures, $60\%H_2/40\%N_2$ and $60\%H_2/30\%N_2/10\%CO$, in addition to pure $H_2$, at elevated pressure of 17 atm. Figure 2.30 shows a comparison between these fuels in terms of $NO_x$ concentration and flame temperature. All fuels containing $H_2$ generate lower $NO_x$ compared to natural gas.

Daniele et al. [157] conducted another high-pressure experimental investigation of the characteristics of LPM syngas combustion in gas turbines. The lean blow-off characteristics, exhaust emissions, and turbulent flame speed were discussed. Three syngas fuel mixtures were considered and compared with pure natural gas, namely

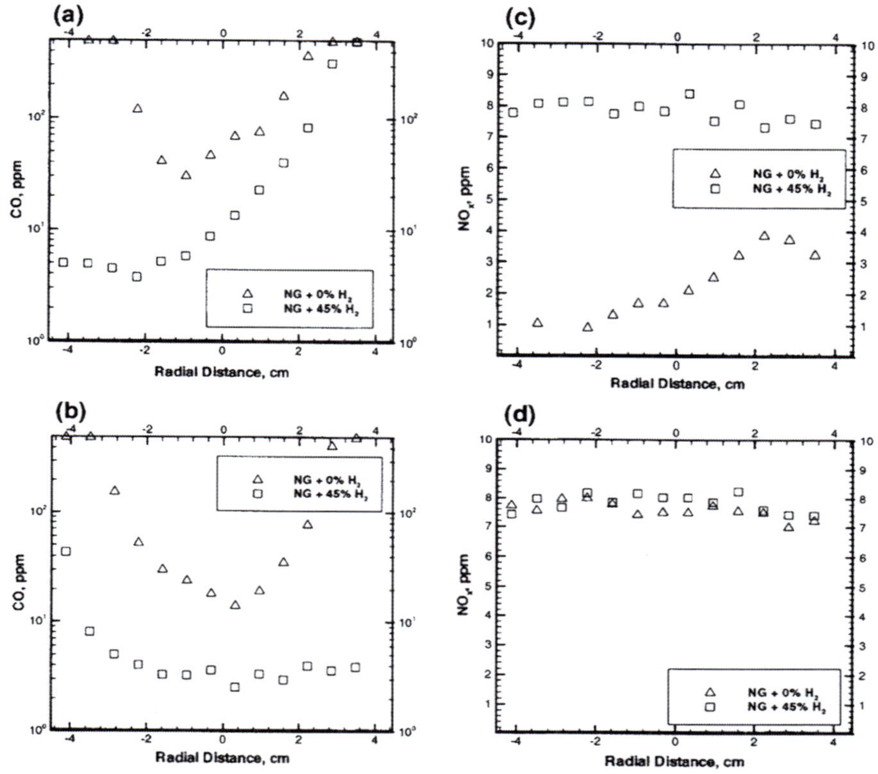


**Fig. 2.28** Effect of hydrogen enrichment on the CO and NO emissions of a LPM swirl combustor fueled by natural gas at different axial positions: **a**, **c** at Z = 5.1 cm; **b**, **d** at Z = 20.3 cm [152]

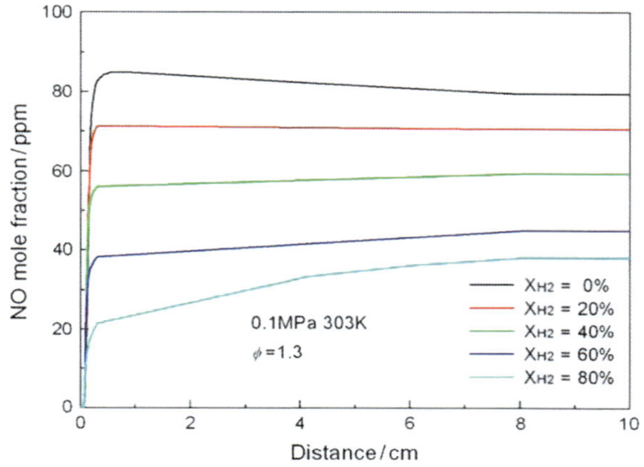


**Fig. 2.29** Axial distributions of NO concentration at different hydrogen fractions in $CH_4/H_2$/air flames [153]

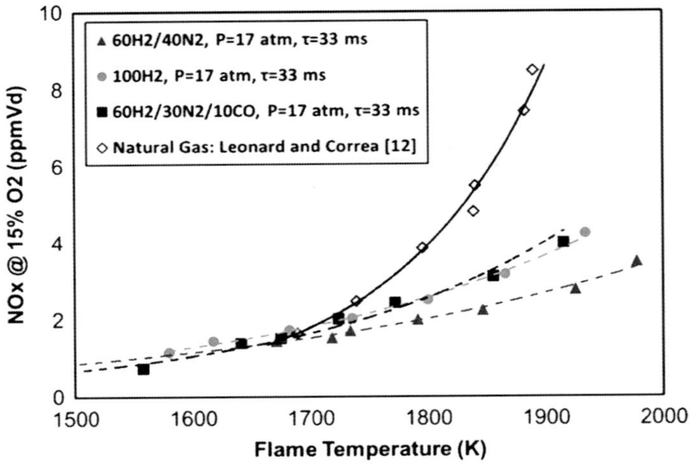


**Fig. 2.30** $NO_x$ emissions for different hydrogen-based fuels in a 10-MW LPM gas turbine [156]

20%$H_2$/20%CO/60%$CH_4$, 50%$H_2$/50%CO, and 33%$H_2$/67%CO. The experiments were performed at combustor pressures up to 30 bar with a maximum airflow rate of 0.3 kg/s. Flames were stabilized by means of recirculation zones established by a sudden enlargement in the combustor. The equivalence ratio ranged from 0.24 to 0.5. The experiments were carried out starting from a stable flame at a certain air/fuel ratio, and then the fuel mass flow rate was gradually reduced down to lean blow-off. Unlike other syngas studies, the results of Daniele et al. revealed that achieving low $NO_x$ in syngas combustion is very challenging because of the difficulty of premixing syngas effectively with air. Furthermore, the risk of flashback increases, which should be taken into consideration; precautions should thus be taken prior to the use of syngas. Ter-Maatha et al. [158] performed a feasibility study to investigate the cost effectiveness of adding $H_2$ to the combustion process. They reported that this method is very competitive to the other new methods/techniques of $NO_x$ reduction.

Ghenai [159] performed a numerical investigation of syngas-combustion in a gas turbine combustor. The effects of fuel composition and heating value were examined to quantify the emissions. Syngas fuel has a $H_2$/CO ratio in the range of 0.63–2.35 and a heating value that ranges from 8224 to 12,492 kJ/m$^3$. The syngas was obtained from a gasification process of solid fuel, such as coal. The author also studied the effect of shifting natural gas to syngas on the gas-turbine performance. It was concluded that gas turbines could burn syngas more efficiently than methane with lower NO emissions and wall temperatures. This can be attributed to the lower flame temperature of syngas flame because of the lower heating value of syngas. The results also showed that $CO_2$ emissions were reduced by 30–49% when methane was shifted to syngas, but at the expense of reducing the power by 20–55%. NO emissions were also significantly reduced by 11.5–97.5% at the same mass flow rate of fuel, but again with a power-loss penalty due to the lower heating value of syngas.

**Table 2.3** Summary the hydrogen-enrichment section for efficient combustion

| Combustion technique/approaches | Application<br>Experimental/Numerical | References | Remarks/major findings |
|---|---|---|---|
| Hydrogen enrichment | Experimental study to investigate the stability characteristics of a LPM swirl-stabilized flame with $H_2$ enrichment | Schefer [137] | Adding more than 20% of $H_2$ to $CH_4$/air combustion results in significant increase in the OH concentration and extends the lean blow-off limit of the combustor |
| | Experimental study of the static stability limits of stoichiometric mixtures of biogas and air | Zhen et al. [139] | $H_2$-free BG ($CH_4$ + $CO_2$ only) blows off at all examined Reynolds numbers and $CH_4$ contents, i.e., the flame is stabilized when a certain amount of $H_2$ is introduced into the combustible mixture<br>More $H_2$ is needed at lower $CH_4$ contents to prevent blow-off, which highlights the adverse effect of $CO_2$ on flame stability |
| | Experimental study of the effect of $H_2$ enrichment on flame blow-off at different combustor pressures | Lieuwen et al. [42, 142] | $H_2$ enrichment is more effective at higher pressures, because the lean blow-off limit shifts to widen the stable operating range |
| | Experimental study of the effect of $H_2$ enrichment on the lean blow-off limit of a swirl combustor at different pressures ranging from 1 to 8 atm | Strakey et al. [41, 143] | Increasing the $H_2$ content retards flame blow-off. However, the combustor pressure shows no discernable effect on the blow-off limit |

(continued)

**Table 2.3** (continued)

| Combustion technique/approaches | Application Experimental/Numerical | References | Remarks/major findings |
|---|---|---|---|
| Effect of hydrogen on laminar burning velocity | Measurements of the laminar flame speed at blow-off in a gas-turbine combustor for different $H_2$-enrichment levels | Tuncer [144] | $H_2$ addition boosted the laminar flame speed over the entire examined equivalence-ratio range of 0.55 to 1.4 (lean to rich), which considerably extended the blow-off limits |
| | Effect of hydrogen addition on air-methane combustion | Halter et al. [145] | The effect of stretch on laminar flame speed weakens when $H_2$ is added to the combustible mixture, which also explains why the blow-off velocity increases with $H_2$ enrichment |
| | Experiments to compare the laminar burning velocities of $CH_4/H_2$ mixtures with air | Ilbas et al. [146–148] | Increasing the fuel hydrogen fraction brought about an increase in the laminar flame speed<br>The mixture of 30% $H_2$ and 70% $CH_4$ is recommended as an alternative fuel for power-generation plants, as it offers good flame stability, wide flammability range, and a relatively high burning velocity |
| Effect of hydrogen on $NO_x$ and CO emissions | Experimental study of the stability and emissions of a LPM $H_2$-enriched propane-air flame stabilized on a backward-facing step | Ghoniem et al. [149] | $H_2$ addition increases NO emissions because of the slight temperature rise associating $H_2$ addition; however, performance is not compromised because leaner operation with lower NO emissions becomes possible thanks to $H_2$ enrichment |

(continued)

**Table 2.3** (continued)

| Combustion technique/approaches | Application Experimental/Numerical | References | Remarks/major findings |
|---|---|---|---|
| | Experimental study of the lean blow-off limits and $NO_x$ emissions of $H_2$-enriched $CH_4$/air flames at different combustor pressures | Griebel et al. [150] | $NO_x$ was reduced significantly by ~35% because $H_2$ enrichment allowed for reducing the flame temperature by 60 K. Hydrogen enrichment of natural gas is thus an approach for reducing the $NO_x$ emissions of natural-gas combustion in gas turbines. However, this approach is successful only when operating near the lean blow-off region |
| | Improving the performance of a dual-fueled gas turbine by $H_2$ enrichment | Lounici et al. [151] | The efficiency of dual-fuel engines decreases when operated at low loads $H_2$ addition suggested as an effective method to overcome this problem and increase the burning velocity<br>The break specific fuel consumption and the emissions of hydrocarbons, CO, and $CO_2$ were decreased with $H_2$ addition |
| | Effect of $H_2$ enrichment in a LPM swirl combustor fueled by natural gas | Schefer et al. [152] | Adding an amount of $H_2$ to the combustion process results in significant reduction in CO concentration but an increase in the OH radical concentration. Thus, flame stability improves, and flame length decreases |
| | Numerical study of NO concentrations in premixed $CH_4$/$H_2$/air flames | Hu et al. [153–155] | Substantial reduction in NO can be obtained with $H_2$ enrichment |

(continued)

**Table 2.3** (continued)

| Combustion technique/approaches | Application Experimental/Numerical | References | Remarks/major findings |
|---|---|---|---|
| | Experimental study on a 10-MW LPM gas turbine | York et al. [156] | All fuels containing $H_2$ generate lower $NO_x$ compared to natural gas |
| | High-pressure experimental investigation of the characteristics of LPM syngas combustion in gas turbines | Daniele et al. [157] | Achieving low $NO_x$ in syngas combustion is very challenging because of the difficulty of premixing syngas effectively with air<br>The risk of flashback increases, which should be taken into consideration; precautions should thus be taken prior to the use of syngas |
| | Feasibility study to investigate the cost effectiveness of adding $H_2$ to the combustion process | TerMaatha et al. [158] | This method is very competitive to the other new methods/techniques of $NO_x$ reduction |
| | Numerical investigation of syngas-combustion in a gas turbine combustor | Ghenai [159] | Gas turbines could burn syngas more efficiently than methane with lower NO emissions and wall temperatures. This can be attributed to the lower flame temperature of syngas flame because of the lower heating value of syngas<br>$CO_2$ emissions were reduced by 30% to 49% when methane was shifted to syngas, but at the expense of reducing the power by 20% to 55%<br>NO emissions were also significantly reduced by 11.5% to 97.5% at the same mass flow rate of fuel, but again with a power-loss penalty due to the lower heating value of syngas |

The above literature review of the hydrogen-enrichment approach shows that $H_2$ addition is beneficial for extending the lean blow-off limit by boosting the reaction rates, increasing the flame speed, and making the flame less sensitive to stretch. In terms of emissions, CO and $CO_2$ emissions are reduced. $NO_x$ emissions, however, increase if the combustor is operated at the same equivalence ratio; however, if benefit is made from the extended blow-off limit by operating the combustor leaner, then $NO_x$ emissions ultimately decrease. Flashback remains a concern, and the associated risk depends on the utilized burner technology. Alternative and synthetic fuels show promising potential to substitute natural gas, but only if the fuel is premixed properly with air prior to combustion. The reviewed studies are summarized in Table 2.3, which highlights their major findings.

#### 2.4.1.2 Oxy-fuel Combustion Approach

Unlike the aforementioned approaches that target fuel flexibility, oxy-fuel combustion is a key approach that utilizes oxidizer variability to reduce almost all combustion emissions. In this section, oxy-fuel combustion will be thoroughly described and comprehensively reviewed.

The most harmful pollutants emitted from fuel combustion in air are nitrogen oxides, $NO_x$, which are formed primarily because of the reaction of air-based nitrogen with oxygen at high flame temperatures. Interestingly, if fuel is burned in pure oxygen instead of air, which is termed oxy-fuel combustion, there will be no nitrogen to form $NO_x$; in other words, $NO_x$ emissions will be inherently eliminated [160, 161]. Other pollutants that should seriously be taken into consideration are $CO_2$ and CO; the former is a greenhouse gas, and the latter is toxic. Since the complete stoichiometric combustion of a hydrocarbon fuel in pure oxygen produces $CO_2$ and $H_2O$ only as combustion products, separating $CO_2$ for underground sequestration becomes readily possible after simple condensation of $H_2O$. By comparison, separating $CO_2$ from the exhaust of an air-based combustion process is costly and energy-intensive because of the presence of nitrogen in the exhaust. Oxy-fuel combustion thus provides a very promising solution for eliminating $CO_2$ emissions as well. This solution also reduces the volume and mass flow rates of the flue-gas stream significantly, with consequent benefits of reduced heat losses and reduced size of flue-gas treatment equipment [162].

CO now becomes the single challenging pollutant to tackle, especially that the oxy-fuel combustion process is stoichiometric with hardly any excess oxygen. It should be noted that oxygen is made available by cryogenic or membrane-based separation of air, which consumes energy. Oxygen in oxy-fuel combustion is thus a valuable resource that must be used wisely, unlike air in air-fuel combustion, which is abundant at almost no cost. Another problem in oxy-fuel combustion is the excessive flame temperature when fuel is burned in pure oxygen [163], which is beyond the metallurgical limits of the most resilient materials known. The flame must thus be diluted to regulate its temperature. The single possible solution to this problem is exhaust-gas recirculation (EGR), aka flue-gas recirculation (FGR), i.e., the flame is diluted by large amounts of recirculated effluent $CO_2$ or ($CO_2 + H_2O$) [164]. This is to avoid introducing any foreign constituents into the exhaust, other than $CO_2$ and $H_2O$, which facilitates the easy separation of $CO_2$. Unfortunately, this dilution solution makes the flame $CO_2$-rich, and $CO_2$ has an adverse effect on reaction kinetics, burning velocity, and flame stability [165]. Therefore, the combination of stoichiometric operation (lack of excess oxygen) with retarded reaction kinetics hinders the complete combustion of CO, i.e., an issue with CO emissions is created. If this issue is resolved, oxy-fuel combustion will be the long-pursued technology for zero-emission power generation.

Another challenge to face in oxy-fuel combustion is the fact that the combustion and radiative heat-transfer characteristics differ substantially from those in air-fuel combustion. This may be attributed to the significant differences in the physical properties of $CO_2$ and $N_2$ [166–169]. The replacement of $N_2$ by $CO_2$ in the oxidizer mixture affects the flame from multiple perspectives because of changes in mixture density, volumetric heat capacity, and transport properties, including thermal conductivity, mass diffusivity, and dynamic viscosity. Rashwan et al. [140, 141] illustrated the effect of $CO_2$ addition in terms of thermal conductivity, kinematic viscosity, and mass diffusivity. $CO_2$ has higher density than $N_2$, which affects gas volume, pressure drop, and flame shape. The density of air is 0.433 kg/m$^3$ at 1 atm and 800 K, while the corresponding density of $O_2/CO_2$ mixture is 0.623 kg/m$^3$ at an oxygen fraction of 21%. The higher density of $CO_2$ also leads to higher heat capacities of $O_2/CO_2$ mixtures with respect to air. At 800 K, the volumetric heat capacity of air is 480 J/m$^3$/K, while that of $O_2/CO_2$ mixture is 717 J/m$^3$/K at an oxygen fraction of 21%. The higher heat capacity of $O_2/CO_2$ mixture with respect to air directly reduces the flame temperature and causes lower flame speeds and inferior flame stability. In addition to the changes in density and volumetric heat capacity, flame speed is also affected by the gas transport properties. The thermal conductivity and dynamic viscosity of $N_2$ and $CO_2$ at different temperatures, as well as the mass diffusivities of $O_2$ in $N_2$ and $CO_2$, are available in [141].

#### Operability Degrees of Freedom of Oxy-fuel Combustion

Oxy-fuel combustion has different degrees of freedom that confine the operational space from those of air-fuel combustion. Table 2.4 compares the two side by side.

Based on Table 2.4, the operational space of oxy-fuel combustion can be represented schematically as seen in Fig. 2.31. Operation at rich equivalence ratios renders the combustion incomplete with excessive CO emissions, while the operation at lean equivalence ratios results in excess $O_2$ in the exhaust stream. This effluent $O_2$ is problematic from two aspects; first, oxygen is made available for oxy-fuel combustion by cryogenic or membrane-based separation of air, which consumes energy, so the excess $O_2 \equiv$ waste energy. Second, the presence of $O_2$ in the $CO_2 + H_2O$ exhaust complicates the separation process of $CO_2$, because $O_2$ will not condense out with $H_2O$. The operational space of oxy-fuel combustion is also bounded by a limitation

**Table 2.4** Comparison of air- and oxy-fuel combustion based on the operability degrees of freedom

| Aspect | Air-fuel combustion | Oxy-fuel combustion |
|---|---|---|
| Oxidizer | Air | $O_2/CO_2$ oxy-mixture[a] |
| Oxidizer composition | Fixed (21% $O_2$ + 79% $N_2$) | Variable %$O_2$ in $O_2/CO_2$ oxy-mixture |
| Equivalence ratio | Variable (lean) | Fixed (stoichiometric)[b] |
| Control of flame temperature | Through equivalence ratio (i.e., fuel flow rate) | Flow rate of recirculated $CO_2$, i.e., %$O_2$ in $O_2/CO_2$ oxy-mixture[c] |
| Main parameter controlling mass flow through turbine (i.e., engine load) | Airflow rate | (Fuel flow rate + $O_2$ flow rate) and/or recirculated-$CO_2$ flow rate |
| Separate control of flame temperature and engine load | Yes; the two parameters are almost independent of each other[d] | Very difficult; both parameters are simultaneously affected by the flow rate of recirculated $CO_2$ |
| Fuel type | All fuels | All fuels |

[a]$O_2/CO_2$ oxy-mixture consists of the $O_2$ separated from air + the $CO_2$ recirculated from exhaust
[b]Fixed equivalence ratio means that fuel and $O_2$ are always in fixed proportions (stoichiometric) with respect to one another
[c]Fuel and $O_2$ always yield stoichiometric flame temperature, so the flow rate of recirculated $CO_2$ is the sole control parameter of flame temperature
[d]Changing fuel flow to change the equivalence ratio can be done with almost no change in airflow rate

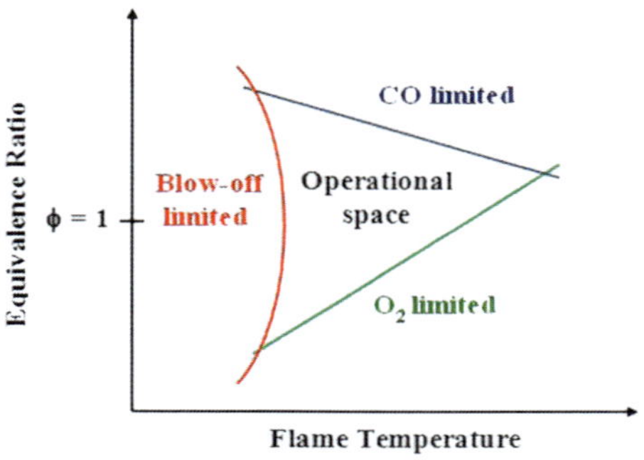


**Fig. 2.31** The operational space of oxy-fuel combustion based on equivalence ratio and flame temperature [19]

on minimum flame temperature, i.e., maximum $CO_2$ dilution. In all, the above discussion shows how limited the operability of an oxy-fuel combustor is, with respect to an air-fuel one.

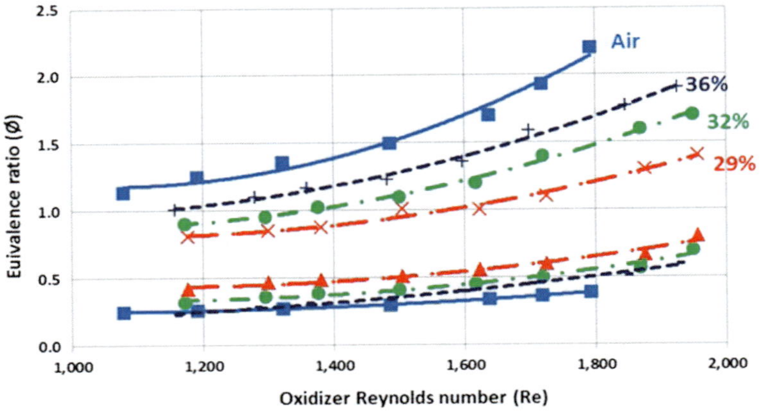


**Fig. 2.32** Flammability limits of oxy-fuel combustion at different oxygen fractions and comparison with air-fuel combustion [141]

## Effect of Oxy-fuel Combustion on Combustor Stability Map

Ditaranto et al. [170, 171] reported that the process of oxy-fuel combustion requires at least 30% oxygen fraction (≡ vol.% of $O_2$ in the $O_2/CO_2$ oxy-mixture), in order to perform in a stable manner comparable to air-fuel combustion (21% oxygen fraction). This 9% additional oxygen compensates for the higher heat capacity of $CO_2$ compared to $N_2$. Liu et al. [172] studied numerically the oxy-methane combustion characteristics in a gas-turbine model combustor. They concluded that lower oxygen fractions result in deteriorated blow-off limit. Nemitallah and Habib [173] studied experimentally and numerically the oxy-fuel combustion characteristics of non-premixed flames in a co-axial gas-turbine combustor. They reported that the stability of oxy-fuel combustion is adversely affected when the oxygen fraction is reduced below 25%. Amato et al. [45] studied the extinction mechanisms of premixed oxy-fuel flames. They showed that using $CO_2$ as diluent worsens the operability limits, due to the slower chemical reactions of this system relative to the case of burning in air. Rashwan et al. [140, 141] studied partially premixed oxy-fuel combustion and compared the results with air-fuel combustion. They focused primarily on three oxygen fractions, namely 29, 32, and 36%, but reported that the range of oxygen fraction needed to obtain a stable flame is 29–40%. In addition, they reported that air-fuel combustion has a larger stable range than oxy-fuel combustion, due to the adverse effect of introducing $CO_2$ into the combustion zone, as shown in Fig. 2.32.

Kutne et al. [174] studied the effect of oxygen fraction on oxy-fuel combustion characteristics and reported that the operation with oxygen fractions lower than 22% was not possible even at stoichiometric conditions. Baolu et al. [175] investigated experimentally the effect of $CO_2$ dilution on oxy-methane combustion in a mixed tubular burner. They reported that a stable flame was obtained for a reasonable range

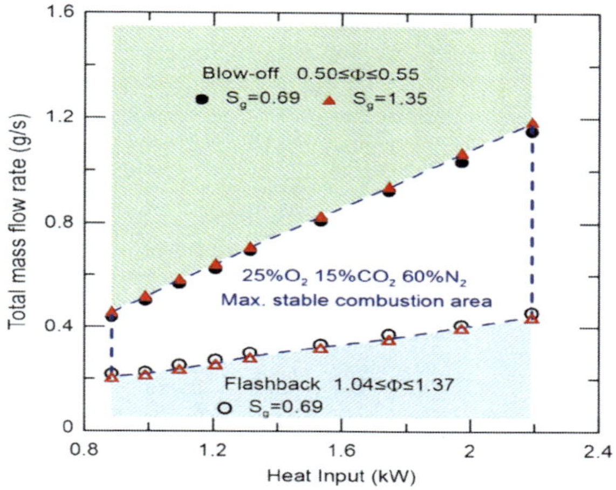


**Fig. 2.33** Stability map of natural-gas combustion in an oxidizer mixture of 25% $O_2$, 15% $CO_2$, and 60% $N_2$ [178]

of equivalence ratio within the oxygen-fraction range of 21–50%. Hu et al. [176] studied the effects of equivalence ratio (0.8–1.2), oxygen fraction (25–35%), and diluents ($N_2$ vs. $CO_2$) on the laminar flame speed of $CH_4$ under atmospheric conditions. They reported that the laminar flame speed of the premixed oxy-methane mixture peaks at stoichiometric conditions. Li et al. [177] investigated numerically the effects of exhaust-gas recirculation on the characteristics of premixed oxy-methane flames under atmospheric pressure. The results showed a shift in the flame location as well as changes in the species concentrations for the cases of oxy-combustion, as compared to the case of air-combustion. A recent study conducted by Jerzak and Kuznia [178] investigated the upper and lower flammability limits of the swirl-stabilized combustion of natural gas in different oxidizers, including air, oxygen-enriched air (25%$O_2$/75%$N_2$), and a 25%$O_2$/15%$CO_2$/60%$N_2$ mixture. The experiments were conducted at two different swirl numbers, namely 0.69 and 1.35. Figure 2.33 represents the stability map of the $O_2$/$CO_2$/$N_2$ case; the swirl number has negligible effect on the flammability limits. Interestingly, when the breadth of stable combustion zone of the 25%$O_2$/75%$N_2$ case is compared to that of the 25%$O_2$/15%$CO_2$/60%$N_2$ one, see Fig. 2.34, the latter case shows inferior stability, although both cases have 25% $O_2$. The $O_2$/$CO_2$/$N_2$ case even showed deteriorated stability compared to the simple air case, although the former has 25% $O_2$. This clearly highlights the adverse effects of $CO_2$ dilution.

Shi et al. [179] studied experimentally oxy-methane combustion at different oxygen fractions. The study confirmed some well-known observations regarding oxy-fuel flames. It was reported that the stable flame range is very small, confined by flashback at higher oxygen fractions and by blow-off at lower oxygen fractions. The dependence of flame temperature and burning velocity on oxygen fraction was also reported, as seen in Fig. 2.35. As the oxygen fraction increases, both burning velocity and flame temperature increase. The oxygen-fraction range for stable operation was reported as ~0.20 to ~0.86.

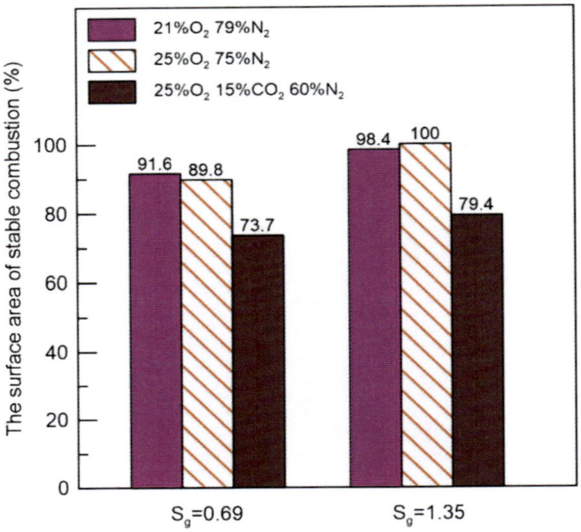


**Fig. 2.34** Comparison of flammability limits of natural-gas combustion in air, oxygen-enriched air, and an $O_2/CO_2/N_2$ mixture [178]

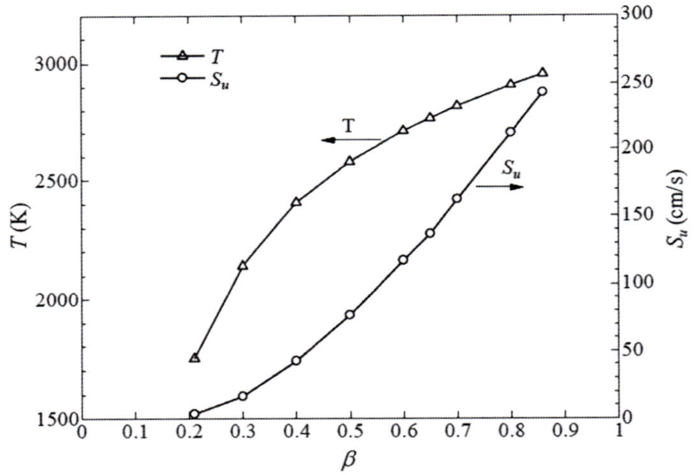


**Fig. 2.35** Adiabatic flame temperature and laminar flame speed verses the oxygen fraction in oxy-methane combustion [179]

## Effect of Oxy-fuel Combustion on NO Emissions

Kim et al. [180] studied flame stability, NO emissions, and visual flame appearance in a 0.03–0.20 MW oxy-fuel combustor using flue-gas recirculation (FGR). They

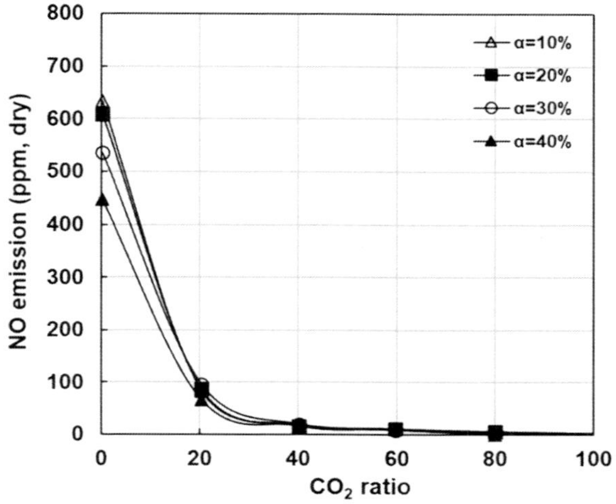


**Fig. 2.36** NO emission versus $CO_2$ ratio at output power of 0.03 MW in an oxy-fuel combustor using flue-gas recirculation (FGR) [180]

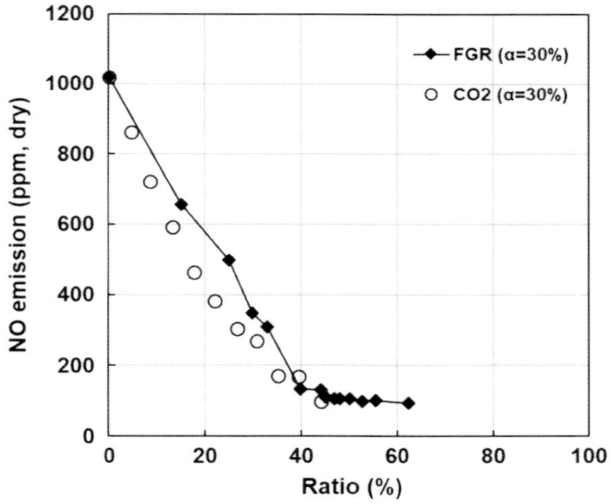


**Fig. 2.37** NO emissions versus FGR ratio and $CO_2$ ratio at output power of 0.2 MW in an oxy-fuel combustor using flue-gas recirculation (FGR) [180]

reported that, while operating at 40% FGR ratio and output power of 0.03 MW, a reduction of 93% was recorded in NO emissions. Meanwhile, while operating at output power of 0.20 MW at the same FGR ratio, the $NO_x$ reduction level was ~85% only. Figure 2.36 presents the NO emissions in ppm at different oxygen fractions in the 0.03-MW combustor. As shown in the figure, the lower the oxygen fraction the higher the NO emissions. This can be attributed to the lower turbulence intensities and higher resident time associated with the lower oxygen contents. They also reported that NO emissions are significantly reduced if the FGR ratio is increased. This can be imputed to greater dilution effects, as shown in Fig. 2.37. The study recommended that NO emissions in the FGR oxy-fuel combustor could be quite effectively reduced

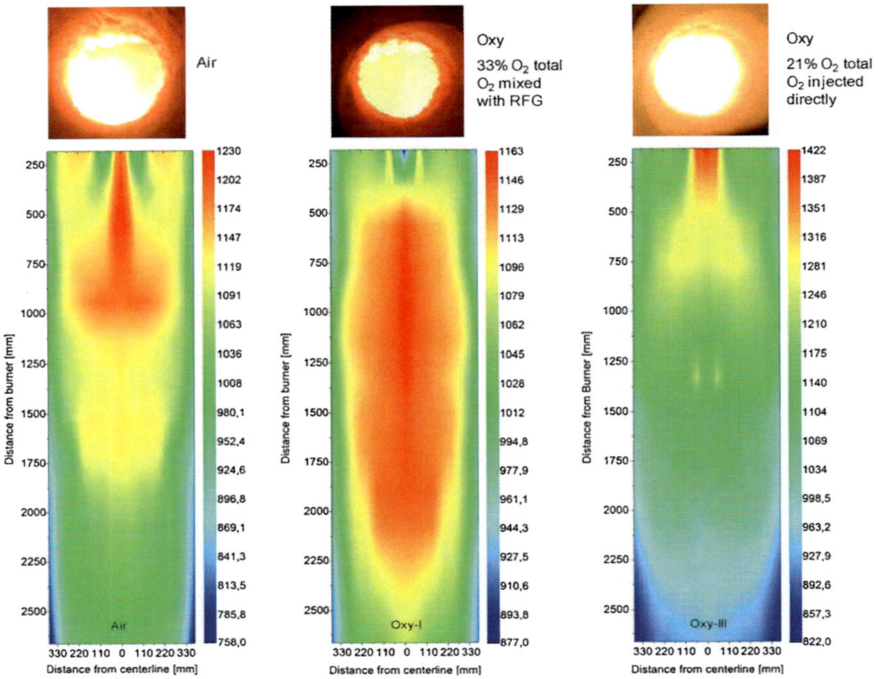


**Fig. 2.38** Comparison of air and oxy-fuel combustion using predicted temperature distributions of air-fuel (left), oxy-fuel with 33% $O_2$ mixed with RFG (middle), and oxy-fuel with 21% $O_2$ directly injected (right) [181]

at FGR ratios higher than 40%. It should, however, be taken into consideration that the higher the FGR ratio, the lower the energy and combustion efficiencies.

An additional challenge in the pathway of implementing oxy-fuel combustion for zero-emission operation is the need to determine how existing air-fuel combustion hardware will behave if switched to oxy-fuel combustion. The notion here is to undergo the switch with minimum hardware changes possible. This will convince the end users of air-fuel combustion applications to switch, especially if they do not have to scrap their existing devices and replace them with brand-new, costly oxy-fuel ones. CFD models have been established to investigate the ability of air-fuel burners to accommodate the new oxidizers. A new burner was designed by Grathwohl et al. [181] to perform a CFD investigation of oxy-fuel combustion at different oxygen fractions and compare the results with air-fuel combustion. This burner was designed to accommodate a wide range of direct-injected oxygen up to 100%. Figure 2.38 compares the temperature distributions of three cases, namely air-fuel, oxy-fuel with 33% $O_2$ mixed with RFG, and oxy-fuel with 21% $O_2$ directly injected; RFG $\equiv$ recirculated flue gas. The study revealed that the maximum temperature was recorded in the 33% $O_2$ case. Furthermore, the burner with direct $O_2$ injection provided more stable flames. The reviewed studies are summarized in Table 2.5, which highlights their major findings.

### 2.4.2 Variable Operating Conditions Approach

The operating conditions of gas-turbine combustors, such as air inlet temperature, combustion-chamber pressure, equivalence ratio, and swirl strength are the main influencers affecting combustion stability. Many recent studies were performed on the variability of the operating conditions in gas-turbine combustion systems [183–186]. Lee et al. [187] illustrated the effect of operating parameters on LPM methane-air combustion in a swirl-stabilized combustor. The study revealed that combustion instabilities were observed only when the inlet air temperature was in the range of 650–670 K. Similarly, operation in the equivalence-ratio range of 0.5–0.7 led to combustion instability as well. In addition, it was found that as the swirler angle increases, the combustion instability subsides. This behavior was attributed to the modification of stream dynamics at the surfaces of the swirler.

A parametric study was performed by Venkataraman et al. [188] on the effects of equivalence ratio, inlet velocity, and swirler angle on the combustion instabilities in a bluff-body stabilized burner fueled by natural gas. The study revealed that the combustion instability increased with inlet velocity. Figure 2.39 (left) shows the peak-to-peak pressure amplitude versus equivalence ratio at different inlet velocities in the range of 2–6 m/s. At the 6-m/s velocity, as the equivalence ratio increases above 0.48, the flame becomes unstable, but it regains stability briefly at equivalence ratios of 0.80–0.85, before losing stability again beyond 0.85. The study also revealed that greater swirler angles are beneficial for combustor stability, especially at higher equivalence ratios in the range of 0.6–0.9, as shown in Fig. 2.39 (right).

### 2.4.3 Variable Flame Characteristics

When fuel and oxidizer are introduced separately into the reaction zone, the resulting flame is non-premixed. An example is shown in Fig. 2.40 with central fuel injection and coaxial oxidizer flow. A shear layer is formed, where both flows interact and diffuse towards each other. At a particular intermediate location, the proportions of fuel and oxidizer become stoichiometric, and a hot primary reaction sheet exists at this stoichiometric locus, see Fig. 2.40. Logically, the fuel concentration is highest at the centerline within the internal diffusion-convection zone and decreases to reach zero at the reaction sheet itself. The oxidizer concentration, on the other hand, is highest at the combustor walls and reduces to zero at the reaction sheet.

Non-premixed flames are superiorly stable in different burner geometries and over wide ranges of operating conditions, because the hot stoichiometric reaction sheet is

**Table 2.5** Summary of the relevant studies on the oxy-fuel combustion approach

| Combustion technique/approaches | Application<br>Experimental/Numerical | References | Remarks/Major findings |
|---|---|---|---|
| Oxy-fuel combustion | Experimental investigation of oxy-fuel combustion as compared to air-fuel combustion | Ditaranto et al. [170, 171] | Process of oxy-fuel combustion requires at least 30% oxygen fraction to perform in a stable manner comparable to air-fuel combustion |
| | Numerical study of the oxy-methane combustion characteristics in a gas-turbine model combustor | Liu et al. [172] | Lower oxygen fractions result in deteriorated blow-off limit |
| | Experimental and numerical investigation of the oxy-fuel combustion characteristics of non-premixed flames in a co-axial gas-turbine combustor | Nemitallah and Habib [173] | The stability of oxy-fuel combustion is adversely affected when the oxygen fraction is reduced below 25% |
| | Experimental investigation of partially premixed oxy-fuel combustion | Rashwan et al. [140, 141] | The range of oxygen fraction needed to obtain a stable flame is 29–40%<br>Air-fuel combustion has a larger stable range than oxy-fuel combustion, due to the adverse effect of introducing $CO_2$ into the combustion zone |
| | Effect of oxygen fraction on oxy-fuel combustion characteristics | Kutne et al. [174] | Operation with oxygen fractions lower than 22% is not possible even at stoichiometric conditions |
| | Effects of equivalence ratio, oxygen fraction, and diluents ($N_2$ vs. $CO_2$) on the laminar flame speed of $CH_4$ under atmospheric conditions | Hu et al. [182] | Laminar flame speed of the premixed oxy-methane mixture peaks at stoichiometric conditions |

(continued)

**Table 2.5** (continued)

| Combustion technique/approaches | Application Experimental/Numerical | References | Remarks/Major findings |
|---|---|---|---|
| | Numerically investigation of the effects of exhaust-gas recirculation on the characteristics of premixed oxy-methane flames under atmospheric pressure | Li et al. [177] | The flame location shifts and the species concentrations change in the cases of oxy-combustion, as compared to the case of air-combustion |
| | Experimental Investigation of the upper and lower flammability limits of the swirl-stabilized combustion of natural gas in different oxidizers | Jerzak and Kuznia [178] | $CO_2$ dilution results in inferior stability despite oxygen enrichment |
| | Experimental study of oxy-methane combustion at different oxygen fractions | Shi et al. [179] | The stable flame range is very small, confined by flashback at higher oxygen fractions and by blow-off at lower oxygen fractions. The oxygen-fraction range for stable operation is ~0.20 to ~0.86<br>As the oxygen fraction increases, both burning velocity and flame temperature increase |
| | Experimental study of flame stability, NO emissions, and visual flame appearance in a 0.03–0.20 MW oxy-fuel combustor using flue-gas recirculation (FGR) | Kim et al. [180] | Substantial NO reductions with FGR<br>The study recommended that NO emissions in the FGR oxy-fuel combustor could be quite effectively reduced at FGR ratios higher than 40%. It should, however, be taken into consideration that the higher the FGR ratio, the lower the energy and combustion efficiencies |
| | CFD investigation of oxy-fuel combustion at different oxygen fractions (33% $O_2$ mixed with RFG, and 21% $O_2$ directly injected) | Grathwohl et al. [181] | The maximum temperature was recorded in the 33% $O_2$ case<br>The burner with direct $O_2$ injection provided more stable flames |

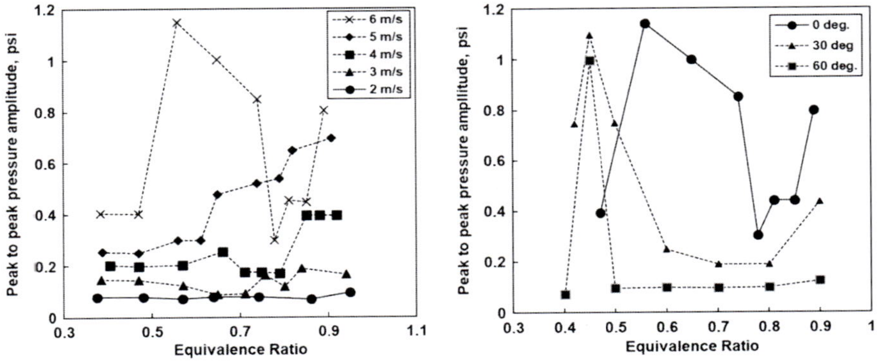


**Fig. 2.39** The effects of inlet velocity (left) and swirler angle (right) on the peak-to-peak pressure amplitude at different equivalence ratios in a bluff-body stabilized burner fueled by natural gas [188]

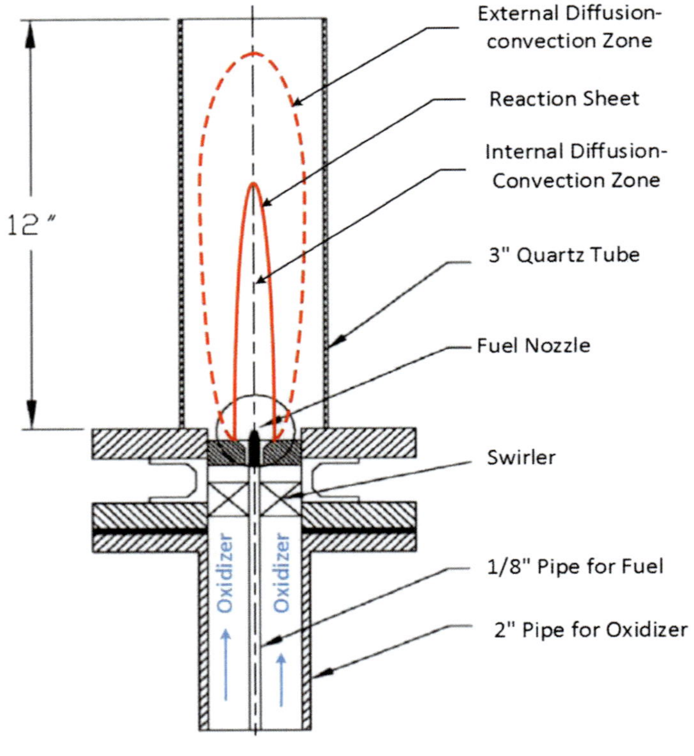


**Fig. 2.40** Reaction zones of a non-premixed jet flame in a co-axial burner [173]

always created. However, the high local temperature at this sheet triggers the formation of excessive $NO_x$ emissions (in air-fuel combustion), which is a major disadvantage of non-premixed flames. Changing the fuel type can affect the $NO_x$ emissions and the length of primary reaction zone. For instance, adding CO or $H_2$ to the fuel shortens the flame, because it requires less oxygen to complete the combustion.

Premixed flames are created when fuel and oxidizer are mixed upstream of the reaction zone under non-reacting conditions, and the combustible mixture is introduced into the reaction zone. Depending on the effectiveness of mixing, the flame could be premixed or partially premixed. In the case of premixed air-fuel flames, the flame temperature is uniform and controlled by the global equivalence ratio of combustible mixture (typically lean); i.e., no stoichiometric reaction sheet is created, contrary to non-premixed flames. The lack of the stoichiometric sheet makes the stability of premixed flames significantly inferior to that of non-premixed ones; i.e., premixed flames are more sensitive to burner geometry and operating conditions. If the global equivalence ratio is kept lean enough in a premixed flame, the flame temperature can be low enough to avoid the generation of excessive $NO_x$ emissions, which is the fundamental concept in LPM combustion.

Currently, LPM combustion is the most effective air-fuel technique for low-emission power generation. Many recent studies reported that partially or fully premixed combustion is one of the most promising solutions that can significantly decrease $NO_x$ emissions [189–191]. However, the conversion of the flame from non-premixed to premixed results in greater combustion instabilities. The degree of premixing affects the characteristics (amplitudes and frequencies) of instabilities, which brings its own challenges to burner designers and manufacturers. The instability of laminar premixed flame is one of the classic well-quantified subjects in flame dynamics, which describes the dynamic response of a perfectly premixed jet flame to weak fluctuations/perturbations [93, 136]. Practical combustion systems are, however, not laminar and never perfectly premixed, which motivated many research studies to investigate the effect of the degree of air/fuel premixing on combustion instabilities [192, 193]. Rashwan et al. [140, 141] investigated experimentally the effect of the degree of premixing on flammability limits, visual flame appearance, extinction mechanisms, and $NO_x$ and CO emissions. The flame was anchored over a perforated plate burner. Five different degrees of premixing were examined, as seen in Fig. 2.41, wherein the degree of premixing is characterized by the ratio of axial mixing length to burner pipe diameter, i.e., $L/D = 7, 25, 45, 67$, and 128. It was concluded that, as the degree of premixing increases, the zone of stable operation becomes narrower, as shown in Fig. 2.41. However, increasing the degree of premixing resulted in reduction in $NO_x$ and CO emissions by approximately 48%. It was reported that the flame color changed from pure blue at the higher degrees of premixing to reddish at the lower degrees. This was attributed to the presence of soot in the partially premixed flames, as compared to the soot-free premixed ones.

A similar study was conducted by Shih et al. [193], who investigated experimentally the influence of the degree of premixing of air-fuel combustion on flame stability and $NO_x$ emissions. They reported, contrary to Rashwan et al. [140, 141], that incomplete mixing of the reactants led to significant reduction in the stable operation range,

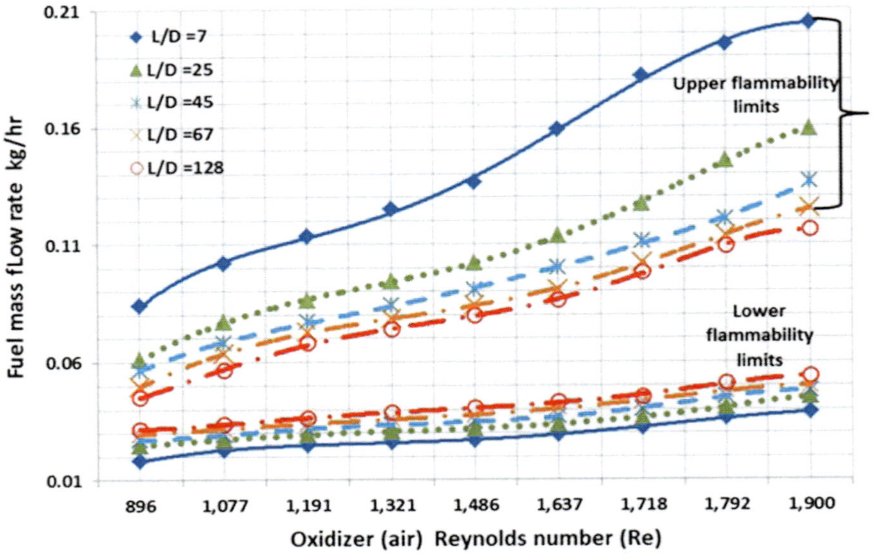


**Fig. 2.41** Flammability limits at different degrees of premixing of air-fuel combustion [140, 141]

as a result of increased dynamic pressure oscillations. Combustion instabilities and flame behavior changed significantly, when the main operability parameters, such as equivalence ratio and degree of premixing, were varied. Venkataraman et al. [188] studied the effect of degree of premixing on the stability of a bluff-body stabilized flame in a co-axial burner. The results are presented in Fig. 2.42 in terms of peak-to-peak pressure amplitude versus equivalence ratio at three degrees of premixing, namely 50, 75, and 100%. It can be seen that the 75% degree is most unstable, whereas the fully premixed (100%) flame enjoys best stability, which highlights that the correlation between dynamic instabilities and degree of premixing is intricate and not straightforward.

Another important consideration in premixed combustion is auto-ignition and ignition delay. It is known that when an ignition source is triggered to initiate the combustion of a fuel/air mixture, combustion does not initiate right away; instead, the mixture undergoes a finite ignition delay before the onset of heat release. For instance, in a gas turbine running on liquid fuel, the temperature of air entering the combustor is typically higher than the auto-ignition temperature of fuel, so combustion is initiated without the aid of an ignition source (like spark plug). However, to be able to premix this fuel with air prior to combustion, there must be a premixing length, where fuel evaporates, and the vapor mixes with air without combustion. This is carefully accounted for in the design of premixing section by ensuring that the flow residence time is shorter than the ignition-delay time of the fuel/air mixture, so combustion initiates outside the premixing hardware. The problem requires extensive investigation, because the ignition-delay time is function of fuel type, equivalence ratio, air inlet

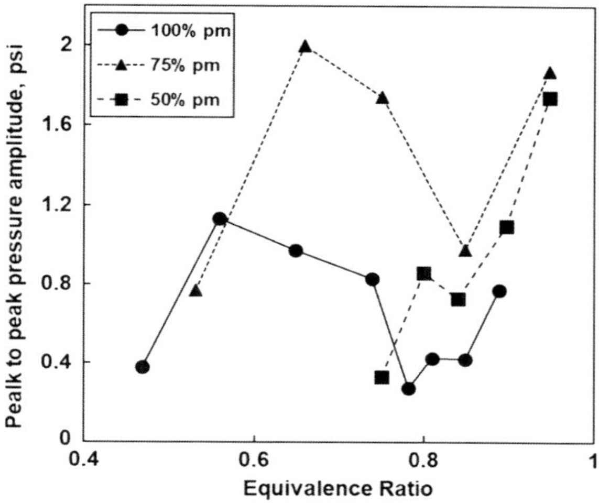


**Fig. 2.42** Peak-to-peak pressure amplitude versus equivalence ratio at different degrees of premixing in a bluff-body stabilized co-axial burner [188]

temperature, combustor pressure, and geometry of injection and premixing sections. For instance, fuels with higher hydrogen content have shorter ignition-delay times [194, 195].

Since this chapter focuses on the implementation of premixed combustion in swirl-stabilized combustors, mainly for gas-turbine applications, the next section of this chapter, Sect. 2.4, is dedicated to swirl-stabilized combustors. The effect of swirl on flame stability and emissions is investigated, as one of the well-known and effective approaches to overcome the instabilities and operability issues associated with the application of LPM combustion in gas turbines.

## 2.5 Swirl Stabilization for Stability Enhancement and Emission Reduction

### *2.5.1 Stabilization Mechanism and Swirl Number*

Swirl stabilization is one of the earliest, easiest, and most efficient techniques to keep the flame attached to the burner (i.e., prevent its blow-off) and maintain adequate flame strength, which, in turn, enhances the combustor turndown (i.e., part-load performance) [196–200]. There are many alternative ways to generate swirl, such as supplying the combustible mixture tangentially into the burner [201, 202], axial vane swirlers [203], radial vane swirlers [204], and mechanical spinners [205, 206].

The principle beyond swirl stabilization is to introduce a tangential velocity component into the flow. Consider a swirler inserted inside a constant-diameter pipe; flow entering the swirler axially will exit with axial and tangential velocity components, i.e., helical motion. Swirl induces miniscule changes in flow density, so it can be considered constant. From the continuity equation,

$$\dot{m} = \rho A_n v \tag{2.11}$$

where, $A_n$ is the cross-sectional area taken normal to the considered velocity, v. Therefore,

$$\dot{m}_{in} = \dot{m}_{out}$$

$$\rho A_{n,in} v_{in} = \rho A_{n,out} v_{out} \tag{2.12}$$

$A_{n,in}$ is the pipe cross-sectional area, and $v_{in}$ is purely axial. If $A_{n,out}$ is also taken as the pipe cross-sectional area, then $v_{out}$ must be the axial component of the exiting swirling flow, $v_{out,axial}$. Equation 2.12 thus simplifies to

$$v_{in} = v_{out,axial} \tag{2.13}$$

The exiting swirling flow, however, has another velocity component, namely $v_{out,tangential}$. This means that the total velocity, $v_{out}$, must be greater than the inlet velocity, $v_{in}$. In other words, the swirler has increased the kinetic energy of the flow, and this increase is deducted from the flow pressure, i.e., the flow expands. This expansion creates a central low-pressure zone that recirculates flow back upstream, i.e., central recirculation zone (CRZ), see Fig. 2.43. If the swirling flow comes across a sudden area enlargement while entering the combustor, a corner or outer recirculation zone (ORZ) is also created between the expanding swirling flow and the combustor corner, as seen in Fig. 2.43.

The benefit from creating recirculation zones in reacting flow is that they recirculate hot combustion products to act as an effective ignition source for the fresh incoming reactants. This increases the blow-off velocity, i.e., retards the blow-off limit, which allows for operating at higher inlet flow velocities. Since blow-off is already a critical limitation in LPM flames, as reviewed earlier, swirl offers the desired advantage of extending the operability window on the blow-off side. The relative strengths and sizes of those recirculation zones depend on many isothermal and combustion-related factors, such as swirl strength, inlet velocity, turbulence intensity, burner-exit and combustor-inlet geometries, composition of combustible mixture, equivalence ratio, inlet temperature, and heat loss from the flame.

The swirl strength is quantified by a dimensionless parameter, which is well known as the swirl number $S_n$. Different definitions have been established for the swirl number [207, 208]; a widely used definition is the ratio of the axial flux of angular momentum to the axial thrust [209], i.e.,

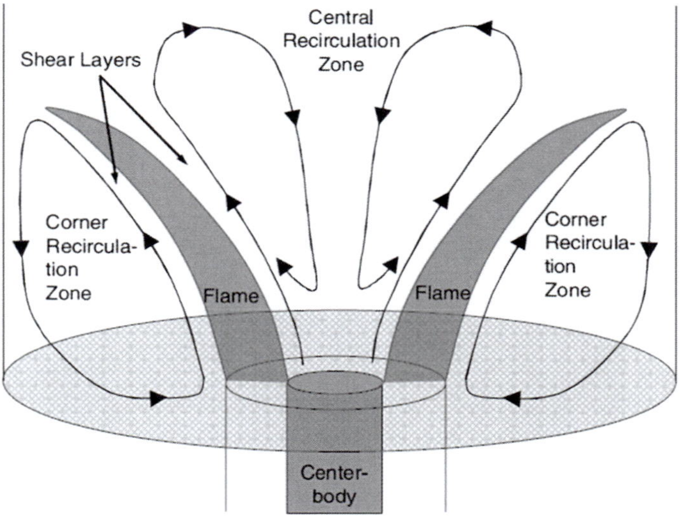


**Fig. 2.43** Schematic description of the flow field inside a swirl-stabilized combustor [41]

$$S_n = \frac{\int_0^R \rho U W r^2 dr}{R \int_0^R \rho U^2 r dr} \tag{2.14}$$

where, ρ is the density of swirling flow, $U$ and $W$ are the axial and tangential velocity components, respectively, and $R$ is the outer radius of swirler. Figures 2.44 and 2.45 depict examples of axial and radial swirlers, respectively [204]. Axial swirlers receive

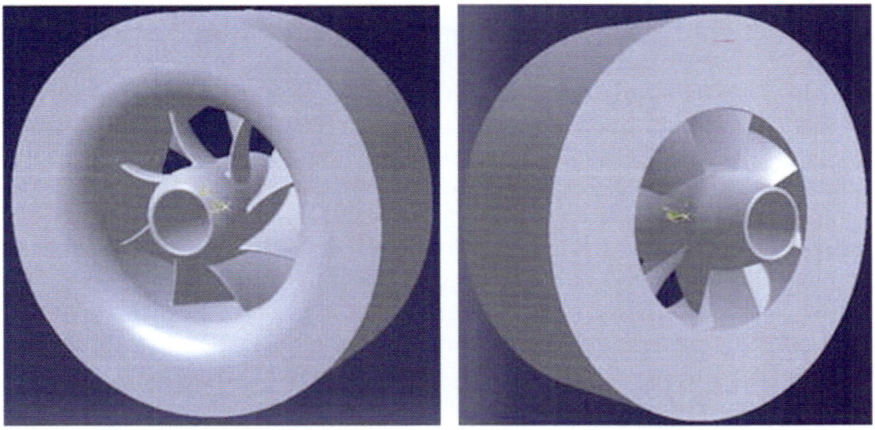

(a) (b)

**Fig. 2.44** Example of axial swirler: **a** upstream view and **b** downstream view [204]

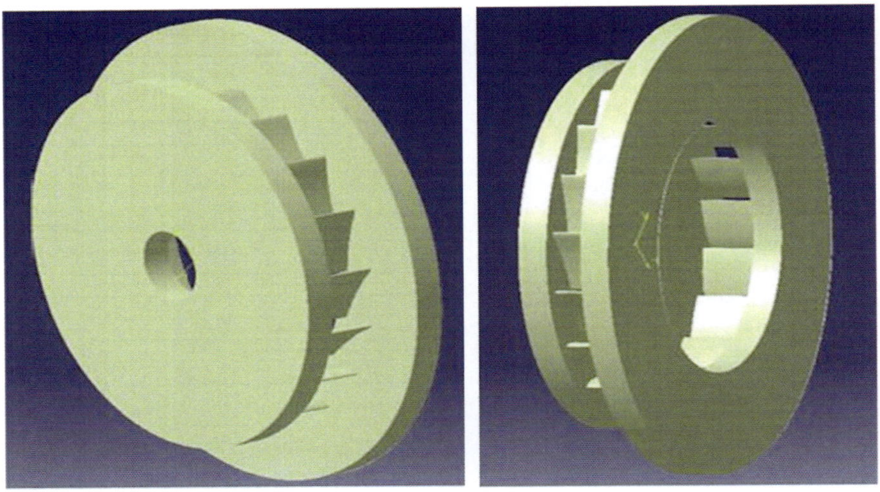

(a) (b)

**Fig. 2.45** Example of radial swirler: **a** upstream view and **b** downstream view [204]

purely axial flow and apply swirl to it. On the other hand, the flow entering a radial swirler has tangential and radially inward velocity components; the tangential one is preserved, while the radial one is converted to axial. Axial swirlers are suitable for low to moderate swirl numbers with acceptable pressure drop, whereas radial swirlers are suitable for high swirl numbers, also with acceptable pressure drop. However, it should be noted that excessive swirl strengths induce vortex breakdown that eliminates the CRZ and significantly alters the flow structure. Radial swirlers have a slight advantage over axial ones in terms of fabrication simplicity, because the vanes of radial swirlers can be simple flat sheets, whereas the vanes of axial swirlers are complicated three-dimensional airfoils, as seen in Figs. 2.44 and 2.45. The depicted axial swirler has eight equally spaced vanes; each is a NASA 8411 airfoil with inner and outer vane angles of 30° and 58°, respectively. On the other hand, the radial swirler is made of 18 equally spaced vanes, NASA manufactured as well. A swirler should generally be mounted at a certain distance upstream of the combustor inlet to allow the flow to dissipate the wakes created by the trailing edges of the swirler vanes.

Another simple technique to create swirl is to split the inlet flow into two components, one enters axially, while the other enters tangentially. The two components interact to create a turbulent swirl region. The key advantage of this aerodynamic technique is the ability to vary the swirl number simply by changing the flow split, which is not applicable for vane-type swirlers (axial and radial). El-Drainy et al. [209] developed this simple aerodynamic technique further by augmenting an axial vane swirler with a radial vane one to create a hybrid vane swirler that receives two flow components, an axial component passing through the axial side, and a tangential component passing through the radial side. Figure 2.46 shows the streamlines

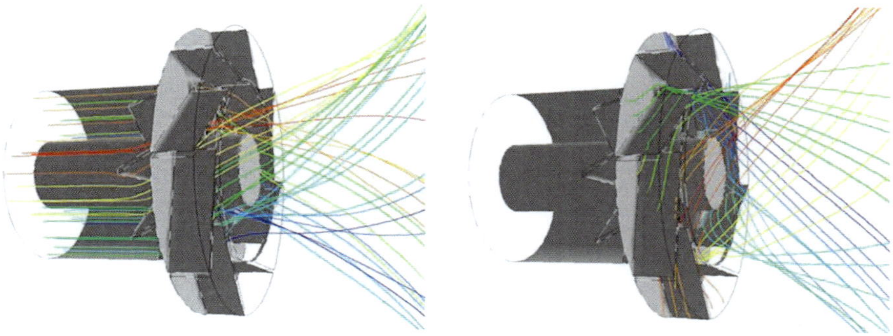

**Fig. 2.46** Axial (left) and tangential (right) flow components passing through the axial and radial vanes, respectively, of a hybrid axial/radial swirler [209]

of each flow component. This study revealed that this new hybrid swirler concept ought to enhance the combustion efficiency, because of its ability to produce higher swirl numbers at lower engine loads, which is beneficial for maintaining LPM flame stability at low part-loads.

### 2.5.2 *Effect of Swirl on Flame Stability*

It has already been discussed how swirl stabilization boosts the flame resistance to blow-off by increasing the blow-off velocity. Hence, swirl offers the desired advantage of extending the operability window of LPM combustor on the blow-off side. Swirl, however, complicates the reacting flowfield, see Fig. 2.43, with central and outer recirculation zones that confine the primary reaction zone. Any oscillations in location or size of either recirculation zone will thus induce dynamic instability. Moreover, the flowfield is highly turbulent, and the heat release within the primary reaction zone is strongly affected by the shear layers adjacent to the recirculation zones and by their effectiveness as ignition sources. The relative sizes and strengths of the recirculation zones also affect flame shape and stability. All these factors can create multiple modes of dynamic instability. It is well known that the central recalculation zone becomes strong or well established for swirl number exceeding 0.6 [210].

Jerzak and Kuznia [178] studied experimentally the effect of swirl number on the flashback and blow-off limits of premixed flames. They used a tangential type swirler [211–214] and examined three different swirl numbers, namely 0.69, 1.16, and 1.35. Figures 2.47 and 2.48 show the flashback and blow-off limits, respectively, in terms of equivalence ratio and total mass flow rate. The study revealed that higher swirl numbers provide wider operability windows.

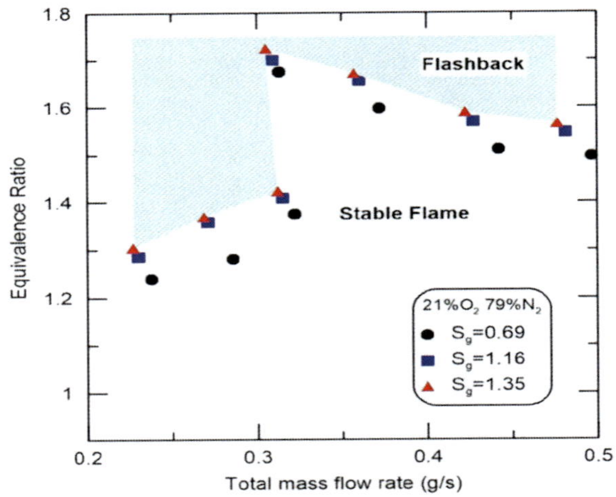


**Fig. 2.47** Flashback limit of premixed flames for different swirl numbers [178]

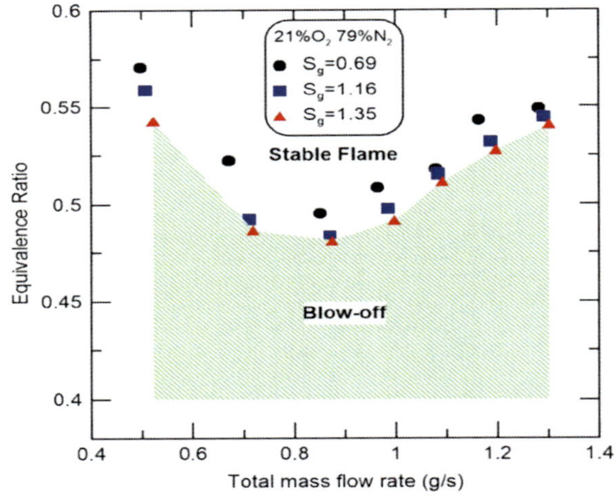


**Fig. 2.48** Blow-off limit of premixed flames for different swirl numbers [178]

Syred et al. [201] studied the effect of swirl number on the flashback and blow-off limits of coke-oven-gas (65%$H_2$/25%$CH_4$) flames. They reported that increasing the swirl number from 0.8 to 1.46 worsens the blow-off limit slightly, whereas the flashback limit is improved; see Fig. 2.49. Therefore, using higher swirl numbers is preferable for protection against flashback in hydrogen- or oxygen-enriched combustion processes.

Dam et al. [58] studied the effect of swirl number on the flashback limit of syngas fuel. The presented results in Fig. 2.50 show the flashback limits for two swirl numbers of 0.71 and 0.97. The higher swirl number demonstrates a wider stable combustion zone, which agrees with the findings of Syred et al. [201] for hydrogen-enriched fuels.

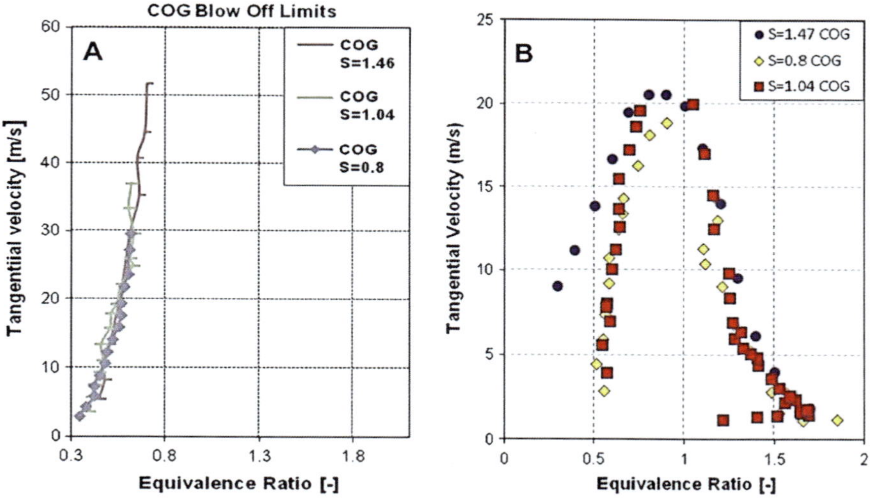


**Fig. 2.49** Blow-off (left) and flashback (right) limits of coke oven gas (65%$H_2$/25%$CH_4$) at different swirl numbers [201]

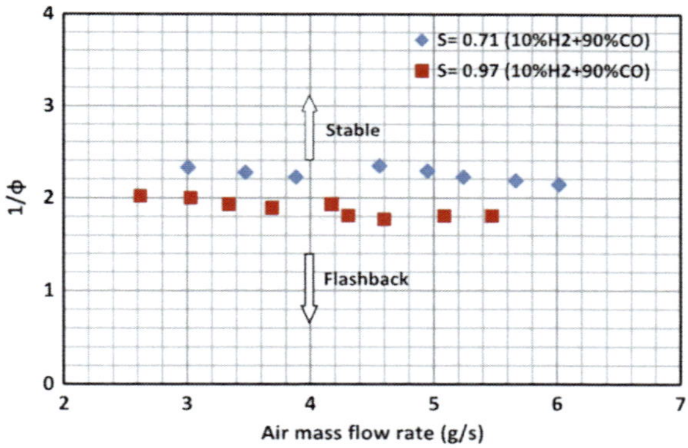


**Fig. 2.50** Flashback limits of syngas fuel (10%$H_2$/90%CO) at different swirl numbers [58]

## 2.5.3 *Effect of Swirl on $NO_x$ and CO Emissions*

One of the features of swirl is that it induces a helical motion that prolongs the residence time of reacting flow inside the combustor. This gives more time for the CO oxidation reactions to complete, thus reducing the ultimate CO emissions. The longer residence time, however, could be disadvantageous from a $NO_x$ perspective, because $NO_x$ generation depends on both flame temperature and residence time spent at this temperature. However, if leaner operation (lower flame temperature) can be implemented, benefiting from the extended operability window, $NO_x$ emissions can be kept low. The turbulence of swirl also improves the degree of fuel/air premixing, which minimizes the presence of hotter richer zones within the flame, and this helps keeping $NO_x$ low as well.

Lim et al. [215] performed an experimental study on a small-scale gas-turbine combustor to investigate the effect of combustor design on the exhaust emissions, especially $NO_x$. They used two different burner designs, namely block and cone types, and found that the cone type yields better stability and lower $NO_x$, as compared to the block type. Along with the optimization of combustor design, many devices have been developed to boost the combustion efficiency by enhancing the flow mixing. A vortex generator is a piece of equipment that generates vortices in the flow field, and it has been used in several applications. Kim et al. [216] conducted an experimental study to investigate the effect of using a vortex generator on premixed flame behavior in a gas-turbine combustor. For the sake of comparison, they measured the $NO_x$ and CO emissions with and without the use of vortex generator. The study revealed that $NO_x$ and CO could be reduced by 21.2 and 13.3%, respectively, with the use of vortex generator. Figure 2.51 compares the results of $NO_x$ and CO emissions with and without the vortex generator. Claypole and Syred [217] studied the effect of swirl aerodynamics on $NO_x$ formation at different swirl numbers in the range of 0.63–3.04. They concluded that $NO_x$ is formed in the flame front. Furthermore, the recirculation zone does not affect $NO_x$ formation despite the long residence times.

The above literature review of the swirl-stabilization approach is summarized in Table 2.6. Toward better understanding of the characteristics of premixed combustion in gas turbines, the next section of this chapter (Sect. 2.5) presents a numerical case study of turbulent premixed combustion in a gas-turbine model combustor. Temperature, species concentrations, and the flowfield are investigated over a range of operating conditions.

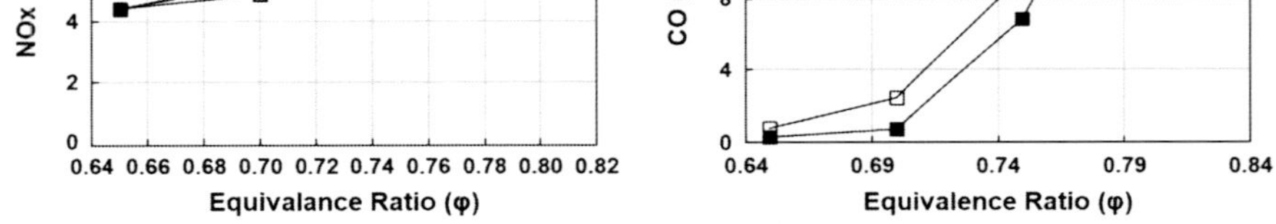


**Fig. 2.51** $NO_x$ (left) and CO (right) emissions at different equivalence ratios with and without the use of vortex generator [217]

**Table 2.6** Summary the swirl-stabilization section for efficient combustion

| Combustion technique/approaches | Application<br>Experimental/Numerical | References | Remarks/Major findings |
|---|---|---|---|
| Swirl stabilizer approach and its effects on flame stability | Developed a hybrid vane swirler | El-Drainy et al. [209] | The hybrid swirler receives two flow components (axial & tangential) and has the ability to produce higher swirl numbers at lower engine loads, which is beneficial for maintaining LPM flame stability at low part-loads |
| | Review of swirl combustion systems | Syred [210] | The central recalculation zone becomes strong or well established for swirl number exceeding 0.6 |
| | Experimental study of the effect of swirl number on the flashback and blow-off limits of premixed flames | Jerzak and Kuznia [178] | Higher swirl numbers provide wider operability windows |
| | Experimental study of the effect of swirl number on the flashback and blow-off limits of coke-oven-gas | Syred et al. [201] | Increasing the swirl number from 0.8 to 1.46 worsens the blow-off limit slightly, whereas the flashback limit is improved Using higher swirl numbers is preferable for protection against flashback in hydrogen- or oxygen-enriched combustion processe. |
| | Effect of swirl number on the flashback limit of syngas fuel | Dam et al. [58] | Higher swirl number demonstrates a wider stable combustion zone, which agrees with the findings of Syred et al. [218] for hydrogen-enriched fuels. |

(continued)

**Table 2.6** (continued)

| Combustion technique/approaches | Application Experimental/Numerical | References | Remarks/Major findings |
|---|---|---|---|
| Effect of swirl on $NO_x$ and CO emissions | Experimental study on a small-scale gas-turbine combustor to investigate the effect of combustor design on the exhaust emissions | Lim et al. [215] | The cone type burner yields better stability and lower $NO_x$, as compared to the block type |
| | Experimental study to investigate the effect of using a vortex generator on premixed flame behavior in a gas-turbine combustor | Kim et al. [216] | $NO_x$ and CO could be reduced by 21.2 and 13.3%, respectively, with the use of vortex generator |
| | Effect of swirl aerodynamics on $NO_x$ formation at different swirl numbers | Claypole and Syred [217] | $NO_x$ is formed in the flame front<br>The recirculation zone does not affect $NO_x$ formation despite the long residence times |

## 2.6 Numerical Modeling of Premixed Combustion

### 2.6.1 Turbulent Premixed Combustion

The use of premixed combustion for $NO_x$-emissions reduction has been extensively discussed in the past sections of this chapter. In most combustion applications, the flow is usually turbulent. Hence, turbulent premixed combustion involves the interaction between turbulent flow, chemical reactions, heat and mass transfer, as well as radiation. Understanding of the physics of turbulent premixed combustion is thus very challenging.

### 2.6.2 Turbulent Combustion Modeling Schemes

Numerical tools, such as CFD, are very important to predict the combustor operability under a variety of operating conditions. These tools provide very important recommendations towards the development of new burner designs. In addition, CFD tools are cost effective in the investigation of complex geometries and applications, as they (a) provide performance data before fabricating the hardware and (b) provide data that can be difficult to measure. One important factor affecting the computational efficiency is the accuracy of the numerical models used for discretizing the set of governing equations associated with combustion modeling. The aim of this section is thus to discuss intensively the available modeling techniques for turbulent premixed combustion. Special emphasis is made on the CFD studies performed on hydrogen-fuel blends ($CH_4/H_2$) using large eddy simulation (LES).

The CFD approaches for solving turbulent reacting flows are: (1) K-epsilon model, (2) direct numerical simulation (DNS), (3) Reynolds-averaged Navier-Stokes (RANS), and (4) large-eddy simulation (LES) [219, 220]. DNS is the most accurate approach in solving the turbulent flowfield. In this approach, all the scales of the flowfield are completely resolved; however, DNS is highly expensive from the perspective of computational time needed to obtain accurate solutions for complex engineering configuration. The RANS approach is the most commonly used one for engineering CFD applications. The principle beyond this approach is that the governing equations are averaged, and the flow is solved neither for the large nor the small eddies. The accuracy of this approach is mediocre; however, it can give average values of the flowfield [221]. Veynante and Egolfopoulos [222] claimed that the RANS model is not reasonable enough for solving the reacting flows accurately. This is attributed to the unsteady nature of these flows, which renders the RANS approach inaccurate. The LES approach resolves all large scales but filters and models the small ones. It is an effective approach with smaller computational cost as compared to DNS. As per the open literature, LES is considered a promising tool with many advantages for modeling industrial applications. Figure 2.52 summarizes how these computational models solve the problems. DNS is the most accurate because it computes all scales

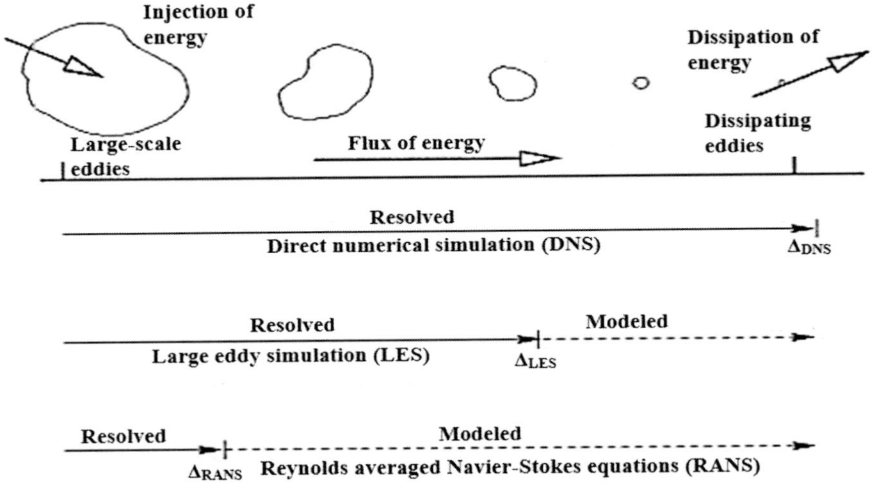


**Fig. 2.52** Common turbulent schemes and prediction methods for turbulent flows

of energy. On the contrary, RANS resolves the large-scale energy eddies only and models the rest. LES resolves the large and intermediate energy scales and models the small ones.

## 2.6.3 LES Governing Equations

As mentioned above, the LES approach is based on the separation of scales. The large scales are directly computed, while the effects of small scales are modeled. The conservation equations can be Favre-filtered as follows [223–228]:

*Continuity equation*:

$$\frac{\partial \bar{\rho}}{\partial t} + \frac{\partial \left(\bar{\rho} \tilde{u}_j\right)}{\partial x_j} = 0 \tag{2.15}$$

*Momentum equation*:

$$\frac{\partial (\bar{\rho}\, \tilde{u}_i)}{\partial t} + \frac{\partial \left(\bar{\rho}\, \tilde{u}_j \tilde{u}_i + \delta_{ij}\, \bar{P}\right)}{\partial x_j} - \frac{\partial \check{\tau}_{ij} \partial x_j}{=} \bar{\rho} g_i + \left\{\frac{\partial \sigma_{ij}}{\partial x_i}\right\} I + \left\{\frac{\partial \left(\check{\tau}_{ij} - \check{\tau}_{ij}\right)}{\partial x_j}\right\} II \tag{2.16}$$

Modeling of the SFS terms is required to close the above system of equations. Term II is neglected under the assumption that the filtered viscous stresses, $\check{\tau}_{ij}$, can be approximated to the viscous stresses evaluated in terms of the Favre-filtered velocity, $\check{\tau}_{ij}$.

*Total energy equation*:

$$\frac{\partial\left(\bar{\rho}\,\tilde{\mathrm{E}}\right)}{\partial \mathrm{t}}+\frac{\partial\left[\left(\bar{\rho}\,\tilde{\mathrm{E}}+\bar{\mathrm{P}}\right)\tilde{\mathrm{u}_{\mathrm{j}}}\right]}{\partial \mathrm{x}_{\mathrm{j}}}-\frac{\partial(\check{\tau}_{\mathrm{ij}}\tilde{\mathrm{u}_{\mathrm{i}}})}{\partial \mathrm{x}_{\mathrm{j}}}+\frac{\partial \check{q}_{\mathrm{j}}}{\partial \mathrm{x}_{\mathrm{j}}}$$
$$=\bar{\rho}\tilde{\mathrm{u}_{\mathrm{i}}}\mathrm{g}_{\mathrm{i}}-\left\{\frac{\partial\left[\bar{\rho}\left(\mathrm{h}_{\mathrm{s}}\tilde{\mathrm{u}}_{\mathrm{j}}-\check{\mathrm{h}}_{\mathrm{s}}\tilde{\mathrm{u}_{\mathrm{j}}}\right)\right]}{\partial \mathrm{x}_{\mathrm{j}}}\right\}III$$
$$+\left\{\frac{\partial\left(\overline{\tau_{ij}u_i}-\check{\tau}_{ij}\tilde{u_i}\right)}{\partial x_j}\right\}IV-\left\{\frac{\partial\left(\overline{q_j}-\check{q}_j\right)}{\partial x_j}\right\}V$$
$$-\frac{1}{2}\left\{\frac{\partial[\bar{\rho}(\,u_j\tilde{u_i}u_i-\tilde{u_j}\,u_i\tilde{u}_i)]}{\partial x_j}\right\}VI \tag{2.17}$$

where, σij represent the SFS stresses and $\left(\mathrm{h}_{\mathrm{s}}\tilde{\mathrm{u}}_{\mathrm{j}}-\check{\mathrm{h}}_{\mathrm{s}}\check{\mathrm{u}}_{\mathrm{j}}\right)$ represent the SFS enthalpy flux.

*Species transport equation*:

$$\frac{\partial\left(\bar{\rho}\tilde{Y_\alpha}\right)}{\partial t}+\frac{\partial\left(\bar{\rho}\tilde{Y_\alpha u_j}\right)}{\partial x_j}+\frac{\partial\check{\Gamma}_{j.\alpha}}{\partial x_j}=-\left\{\frac{\partial\left[\bar{\rho}\left(Y_\alpha\tilde{u}_j-\check{Y}_\alpha\check{u}_j\right)\right]}{\partial x_j}\right\}VII$$
$$-\left\{\frac{\partial\left[\bar{\rho}\left(\overline{\Gamma_{j.\alpha}}-\check{\Gamma}_{j.\alpha}\right)\right]}{\partial x_j}\right\}VIII+\{\overline{\omega_\alpha}\}IX \tag{2.18}$$

where, $\left(\mathrm{Y}_\alpha\tilde{\mathrm{u}}_{\mathrm{j}}-\check{\mathrm{Y}}_\alpha\check{\mathrm{u}}_{\mathrm{j}}\right)$ is the SFS species fluxes, and $(\omega_\alpha)$ is the filtered reaction rates.

Following similar assumptions of neglecting the term (II) for the total heat and species mass diffusion fluxes, terms V and VIII may be neglected.

*Ideal gas equation*:

$$\bar{P}=\bar{\rho}\,\check{R}\check{T} \tag{2.19}$$

$$\check{\tau}_{ij}=2\check{\mu}\left(\check{S}_{ij}-\frac{1}{3}\delta_{ij}\check{S}_u\right) \tag{2.20}$$

$$\check{q}_j = -\check{\lambda}\frac{\partial \tilde{T}}{\partial x_j} - \bar{\rho}\sum_{\alpha=1}^{N} \tilde{h}_\alpha \check{D}_\alpha \frac{\partial\left(\tilde{Y}_\alpha\right)}{\partial x_j} \tag{2.21}$$

$$\check{\Gamma}_{j.\alpha} = -\bar{\rho}\check{D}_\alpha \frac{\partial\left(\tilde{Y}_\alpha\right)}{\partial x_j} \tag{2.22}$$

where, $S_{ij} = \frac{1}{2}\left(\frac{\partial \tilde{u}_i}{\partial x_j} + \frac{\partial \tilde{u}_j}{\partial x_i}\right)$ is defined the strain rate corresponding to the Favre-filtered velocity. The molecular transport coefficients $\check{\mu}$, $\check{\lambda}$ and $\check{D}_\alpha$ are all evaluated at temperature, $\tilde{T}$.

### 2.6.4 LES for Turbulent Premixed Combustion

LES of turbulent reacting flows has been studied extensively for both premixed and non-premixed flames in gas-turbine engines [229–231] and reciprocating internal combustion engines [232]. LES can provide better understanding of many combustion phenomena, including blow-off, flashback and dynamic instabilities [222, 233–237]. Huang et al. [238] conducted a numerical LES study to simulate the combustion dynamics in a LPM swirl-stabilized backward-facing step combustor. The backward-facing step at the combustor inlet section enhances the conditions of the flow to create outer recirculation zone for better stabilization of the flame. The combustor is established with a center body fixed at the combustor inlet to help creating a central flow recirculation zone, where the hot products are mixed again with the incoming mixture of air and fuel. Figure 2.53 shows the mean temperature distribution and streamline patterns inside the combustor. Two recirculation zones were established within the combustor. The central recirculation zone was established in the wake of the center body of the combustor under the effect of swirling flow. The outer recirculation zone was established downstream of the backward-facing step

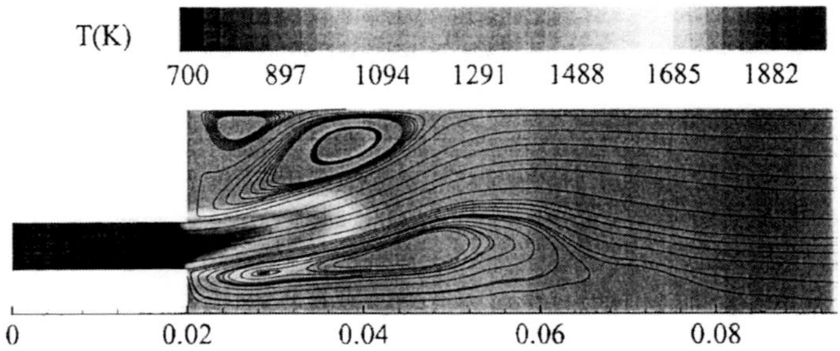


**Fig. 2.53** Predicted mean temperature field and streamlines pattern in a LPM swirl-stabilized combustor [238]

of the combustor. The sizes of both recirculation zones depend mainly on the swirl intensity, equivalence ratio, and combustor geometry.

The burning velocity is a key parameter in premixed flames for different combustible mixtures, and it is considered of crucial importance in designing and developing burner systems. This has motivated recent studies to develop high-accuracy numerical methods to simulate the reaction zone and heat-transfer processes in detail, aiming at predicting accurate values of the burning velocity. Measurements of the burning velocity in premixed hydrogen-air flames suffer from large uncertainties. This can be attributed to the high diffusivity of hydrogen, which causes combustion instabilities. In LES models, it is reasonable to select the perfectly premixed model for stable flows, but it is not suitable for predicting unstable cases.

Franzelli et al. [239] studied the combustion instabilities of lean partially premixed swirl-stabilized flames using LES. The study was performed over a range of equivalence ratios from 0.6 to 1.5 at different pressures (1, 3, 10 atm) and different temperatures (300, 500, 700 K). The results revealed that the predicted values of flame speed at ambient pressure and temperature are the closest to the values by the GRI mechanism, as shown in Fig. 2.54.

Bastiaans et al. [240] studied numerically the heat transfer and stabilization of laminar premixed $CH_4/O_2/CO_2$ and $H_2/O_2/N_2$ flames stabilized over a heat-flux burner. Higher-order numerical method with detailed chemical kinetic mechanisms and detailed transport properties was used. The aim of their work was to study the effect of the flow and temperature generated by the burner holes on flame structure and burning velocity. Furthermore, the heat flux between burner plate and the surrounding gaseous mixture was investigated at various distances and burner-plate temperatures.

The coupling between the turbulent swirl flames and combustor acoustics is not fully understood yet and needs further investigations to predict the stability of new combustion systems. Komarek and Polifke [241] studied experimentally and numerically the impact of swirl fluctuations on the flame response in a premixed swirl burner. The study revealed that the position of the swirl generator has a strong impact on the combustion dynamics but a weak effect on the heat-release rate. A summary of the available LES models and major findings is presented in Table 2.7.

## 2.7 Premixed Combustion in a Gas-Turbine Model Combustor: Numerical Case Study

CFD became an important tool for troubleshooting, optimizing, and predicting the performance of many combustion systems. In the following case study, a simplified numerical model is presented to give a brief explanation of how premixed oxy-fuel combustion is modeled. In this section, some of the main characteristics of premixed combustion, which were discussed in the previous sections, are presented based on numerical modeling, seeking better understanding of premixed combustion. The

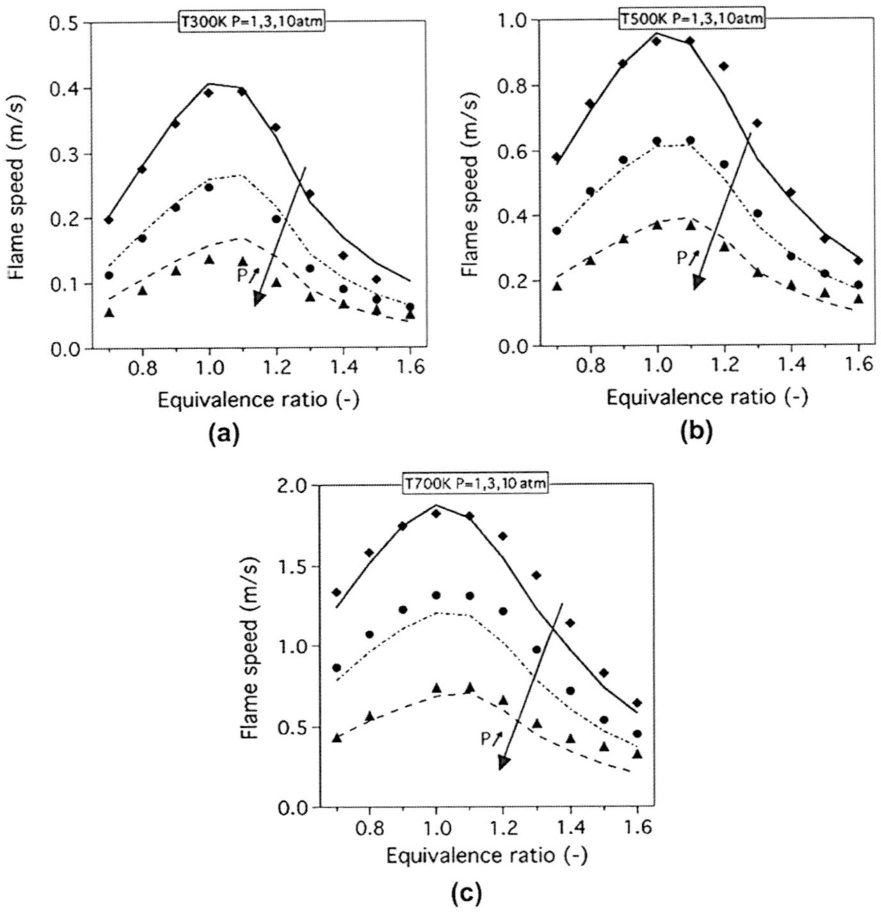


**Fig. 2.54** Predicted flame speed versus equivalence ratio at different pressures and temperatures in comparison to the GRI 3.0 detailed mechanism [239]

simulations were performed in a two-dimensional (2-D) domain for a gas-turbine model combustor burning methane under premixed oxy-fuel conditions.

### *2.7.1 Model Validation*

The work of Wilkes et al. [245] is used to validate the numerical model of the case study under investigation in this section. A 2-D representation of the combustor is used to conduct the validation study. Fuel is introduced at the centerline and air is introduced through the annulus. The combustor has a length of 0.9 m and a radius of

**Table 2.7** Summary of LES models

| Combustion technique/approaches | Description of application | References | Remarks/ Major findings |
|---|---|---|---|
| Eddy breakup type | Triangular-shaped bluff body | Fureby and moller [228] | Filtered Reaction rate determined by sub grid- scale mixing and chemical kinetics |
| Turbulent flame speed and thickened flame | Backward facing step combustor | Colin et al. [227]; Ghoniem et al. [225]; Altay et al. [91, 226]; Nemitallah et al. [92] | Turbulent flame speed used to model filtered reaction terms |
| Linear eddy | Turbulent stagnation point premixed flame | Chakravarthy and Menson [242, 243] | Small scale turbulence and chemical process treated using one-dimensional model |
| Combustion dynamics and instability | Lean premixed Swirl stabilized gas turbine | Huang et al. [238] Schmitt et al. [244] | Taking into account finite rate of chemical reactions and variable thermos-physical properties |
| Turbulent premixed flame | Hydrogen lean premixed combustion | Perez et al. [223] | Sub-filter scale model coupled with the flame prolongation of intrinsic low dimensional manifold |

0.15 m. More details about the combustor can be found in the work done by Wilkes et al. [245]. The grid generation/meshing of the 2-D axisymmetric combustor was performed using the Gambit code, then the numerical investigation of turbulent combustion of natural gas in the combustor was performed using the FLUENT 14.5 code. The simple algorithm is applied for the pressure-velocity coupling. The second-order upwind scheme is used to discretize the convection terms in all transport equations, and the second-order method is used for pressure discretization. The solution convergence is assumed when all of the residuals parameters fall below 10e-06. The numerical predictions were carried out based on the assumptions of steady flow and two-dimensional, axisymmetric, and turbulent flow regimes.

The predicted temperatures in the axial and radial directions using the numerical model are compared to the corresponding measured experimental data of Wilkes et al. [245], as presented in Figs. 2.55 and 2.56. As shown in the figures, the results of the numerical model are in good agreement with the experimental data for both the axial and radial temperature distributions.

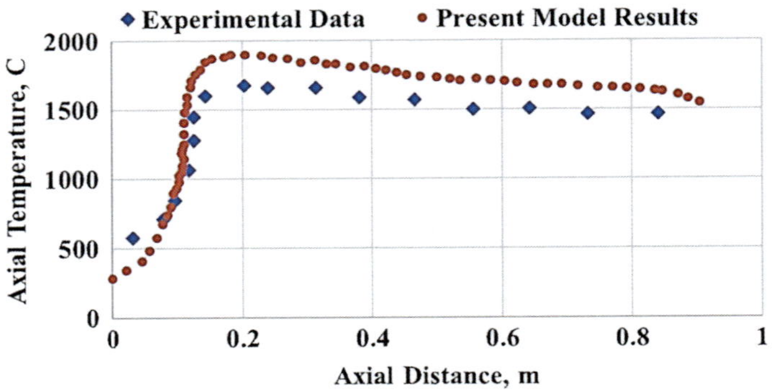


**Fig. 2.55** Comparison of the numerical model results and the experimental data by Wilkes et al. [245] in terms of axial temperature distribution

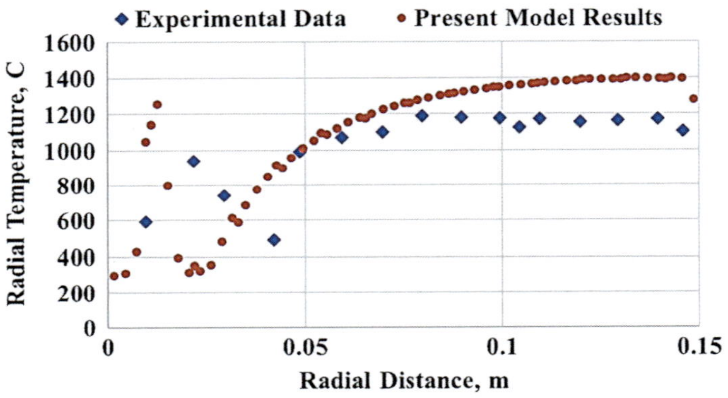


**Fig. 2.56** Comparison of the numerical model results and the experimental data by Wilkes et al. [245] in terms of radial temperature distribution

## 2.7.2 *Case Study of Premixed Oxy-fuel Combustion*

In the current case study, a premixed gas-turbine model combustor is utilized, as presented in Fig. 2.57. The combustor has one inlet section to introduce the combustible mixture ($CH_4 + O_2 + CO_2$) and one outlet section. The diameters of inlet and exit

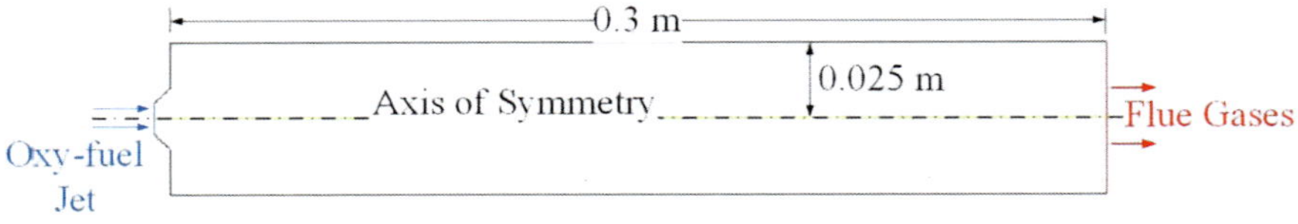


**Fig. 2.57** Physical domain of the premixed combustor

sections are 10 and 50 mm, respectively. The combustion chamber is of cylindrical shape with a length of 300 mm. The walls are quartz with a wall thickness of 6 mm. "Velocity inlet with swirl" is the inlet boundary condition with a value of 15 m/s for both the axial and tangential components. The heat-transfer coefficient through the combustor wall is set to 10 $W/m^2/K$ with a free-stream temperature of 300 K and internal and external emissivities of 1.0 and 0.93, respectively.

The Jones Lindstedt (JL) four-step mechanism was implemented for the modeling of turbulent premixed methane combustion under oxy-fuel conditions, similar to what was reported by Anderson et al. [246]. The computational geometry is two dimensional and axisymmetric. The grid used has about 81,000 nodes after performing a grid-independency study. Numerical simulations were carried out using the commercial FLUENT code. The combustion-reaction scheme in the flame region was modeled using the species-transport model with the eddy-dissipation concept as turbulence-chemistry interaction. The mixture properties were imported to the FLUENT software using a CHEMKIN file. The realizable K-epsilon turbulence model was implemented along with the discrete ordinates (DO) radiation model.

### 2.7.3 *Results and Discussion of the Case Study*

The effect of oxygen fraction in the oxidizer mixture ($O_2 + CO_2$) on the combustion characteristics of premixed methane flame are investigated numerically. Figures 2.58 and 2.59 show the effect of oxygen fraction on the axial distributions of temperature and axial velocity, respectively, at very rich conditions (equivalence ratio = 2.4) and a Reynolds number of 15,000. It can be seen from Fig. 2.58 that the higher the oxygen fraction, the higher the flame temperature. This can be attributed to reduced $CO_2$ dilution effect at higher oxygen concentrations in the oxidizer mixture. The

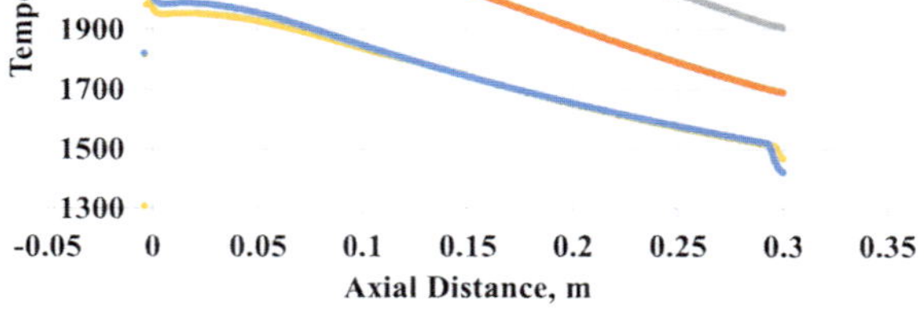


**Fig. 2.58** Axial temperature distributions at different oxygen fractions

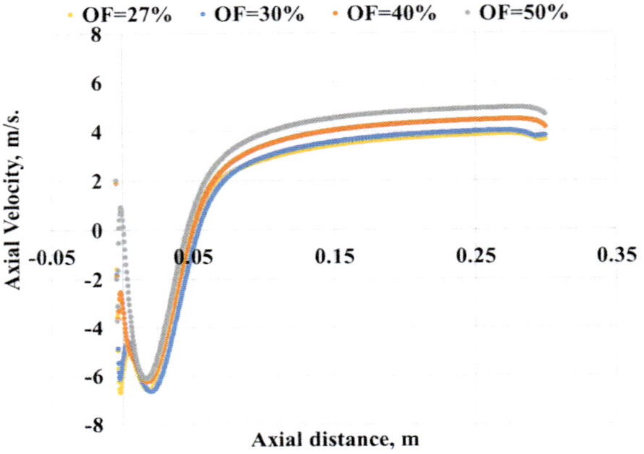


**Fig. 2.59** Axial distributions of axial velocity at different oxygen fractions

higher flame temperature results in flow expansion, which explains why the axial velocity increases with oxygen fraction, as shown in Fig. 2.59. The negative axial velocity in the early flow sections infers the presence of a center recirculation zone. Figures 2.60 and 2.61 present the CO and $CO_2$ exhaust emissions, respectively, at an equivalence ratio of 2.4. Faster reaction rates at higher oxygen fractions result in higher CO concentrations in the flame zone, as shown in Fig. 2.60. By comparison, the axial distribution of CO mole fraction is uniform at lower oxygen fractions, which infers lower reaction intensities. Furthermore, at higher oxygen fractions, fast conversion rates of CO to $CO_2$ are observed due to the abundance of oxygen and the faster reaction rates. As a result, CO concentrations decline (Fig. 2.60), and $CO_2$ concentrations increase (Fig. 2.61) in the axial direction at higher oxygen fractions.

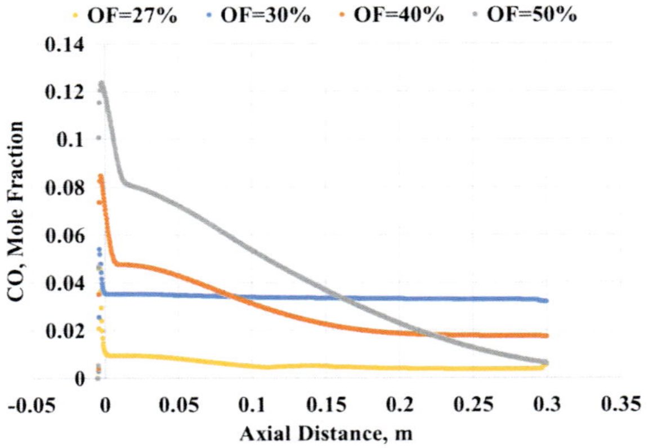


**Fig. 2.60** Axial distributions of CO mole fraction at different oxygen fractions

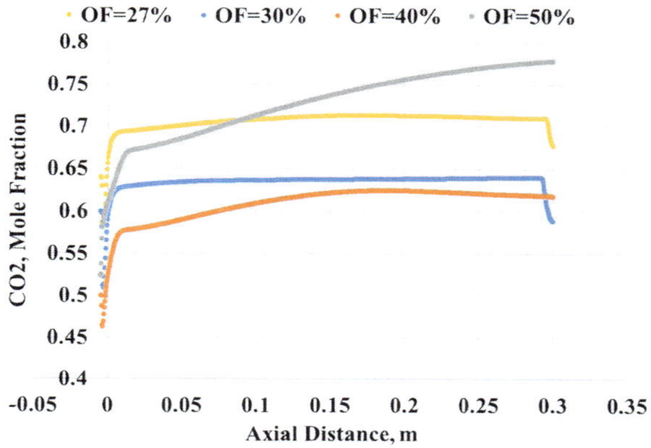


**Fig. 2.61** Axial distributions of $CO_2$ mol fraction at different oxygen fractions

The contours of temperature, stream function, and axial velocity are presented in Fig. 2.62 at an oxygen fraction of 50%, stoichiometric equivalence ratio, and a Reynolds number of 15,000. The flame zone can be identified from the temperature contours by the hot spot colored in red near the inlet section. The stream function is presented to show the center recirculation zone, which is one of the main stabilizing mechanisms of the flame. As shown in the middle contours in Fig. 2.62, an annular recirculation zone is created around the combustor centerline. The velocity contours show deflection of the flow in the tangential direction to create a swirl zone to stabilize the flame. The velocity is maximum at the inlet and decreases in the axial direction, as shown in Fig. 2.62.

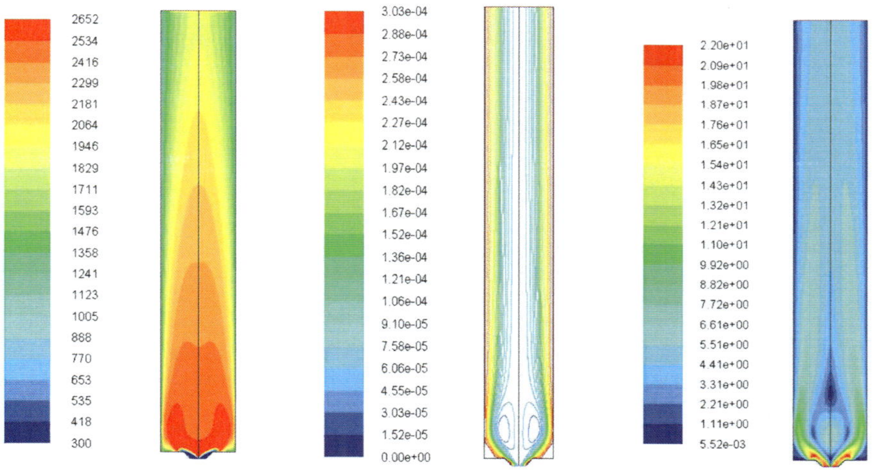


**Fig. 2.62** Contour plots of temperature, stream function, and axial velocity

## 2.8 Concluding Remarks

This chapter provides an overview of the operability and stability issues associating premixed combustion in general and lean premixed (LPM) combustion in some specific cases. First, the operability issues are discussed in terms of static and dynamic instabilities. Under the static instability, the characteristics and correlations governing the blow-off mechanism are analyzed. The main finding from this section is that the chemical timescale dominates the blow-off mechanism, and it depends mainly on fuel composition and burning velocity. Dynamic instabilities are mainly driven by thermo-acoustics, when the turbulent flow fluctuations couple with the oscillations in heat-release rate from the combustion process. Combustion instabilities are well discussed, and the methods to control them are presented. The fuel/oxidizer-flexibility approaches are analyzed with the main conclusions that fuel blending is a key parameter controlling the overall combustion behavior and that oxy-fuel combustion could be the future zero-emission technology. When hydrogen is introduced in fuel blends, it provides wider stability limits and significant reductions in $NO_x$ and CO emissions. Furthermore, the swirl-stabilization of premixed combustion is discussed in terms of the stabilization mechanism and the effect of swirl number on stability and emissions. The discussion reveals that increasing the swirl number enhances the stability limits and reduces exhaust emissions. Finally, recent developments in numerical simulations of turbulent premixed combustion are surveyed in terms of recent available numerical models, with focus on the LES technique and its corresponding sub-models and applications. A case study is presented considering the numerical modeling of premixed turbulent reacting flow in a gas-turbine combustor, and the results are validated with relevant experimental data.

**Acknowledgements** The authors wish to acknowledge the support received from King Fahd University of Petroleum & Minerals under Grant number BW191002 for the preparation of this book chapter.

## References

1. Taamallah S, Vogiatzaki K, Alzahrani FM, Mokheimer EMA, Habib MA, Ghoniem AF (2015) Fuel flexibility, stability and emissions in premixed hydrogen-rich gas turbine combustion: technology, fundamentals, and numerical simulations. Appl Energy 154:1020–1047. https://doi.org/10.1016/j.apenergy.2015.04.044
2. Escudero M, Jimnezngel, Gonzalez C, Lopez I (2013) Quantitative analysis of potential power production and environmental benefits of biomass integrated gasification combined cycles in the European Union. Energy Policy 53:63–75. https://doi.org/10.1016/j.enpol.2012.10.004
3. Rovira A, Montes MJ, Varela F, Gil M (2013) Comparison of heat transfer fluid and direct steam generation technologies for integrated solar combined cycles. Appl Therm Eng 52:264–274. https://doi.org/10.1016/j.applthermaleng.2012.12.008
4. Bolland O, Stadaas JF (1995) Comparative evaluation of combined cycles and gas-turbine systems with water injection, steam injection, and recuperation. J Eng Gas Turbines Power-Trans ASME 117:138–145. https://doi.org/10.1115/1.2812762

5. Bahr DW (1996) Aircraft turbine engine $NO_x$ emission abatement. Chapter 10. Kluwer Academic Publisher, The Netherlands
6. Correa SM (1993) A review of $NO_x$ formation under gas-turbine combustion conditions. Combust Sci Technol 87:329–362. https://doi.org/10.1080/00102209208947221
7. Lefebvre AH (1995) The role of fuel preparation in low-emission combustion. J Eng Gas Turbines Power 117:617–654. https://doi.org/10.1115/1.2815449
8. Correa SM (1998) Power generation and aeropropulsion gas turbines: From combustion science to combustion technology. Symp Combust 27:1793–1807. https://doi.org/10.1016/S0082-0784(98)80021-0
9. Gazzani M, Chiesa P, Martelli E, Sigali S, Brunetti I (2014) Using hydrogen as gas turbine fuel: premixed versus diffusive flame combustors. J Eng Gas Turbines Power 136:51504. https://doi.org/10.1115/1.4026085
10. Sreenivasan KR, Raghu S (2000) The control of combustion instability: a perspective. Curr Sci 79:867–883
11. Sheu EJ, Ghoniem AF (2014) Redox reforming based, integrated solar-natural gas plants: Reforming and thermodynamic cycle efficiency. Int J Hydrogen Energy 39:14817–14833. https://doi.org/10.1016/j.ijhydene.2014.07.086
12. Lieuwen TC, Yang V (2006) Combustion instabilities in gas turbine engines. https://doi.org/10.2514/4.866807
13. Goy CJ, James SRRS (2005) Combustion instabilities in gas turbine engines: operational experience, fundamental mechanisms, and modeling. Prog Astronaut Aeronaut 210:163–175
14. Harrje DT RFH (1972) Liquid rocket combustion instability
15. Yang V, Anderson W (1995) Liquid rocket engine combustion instability. Prog Aeronaut Astronaut 169
16. De Luca L, Price EW, Summerfield M (1992) Non-steady burning and combustion stability of solid propellants. AIAA 143
17. Culik FEC (1988) Combustion instabilities in liquid-fueled propulsion systems—an overview. AGARD Conf Combust Instab Liq Propuls Syst 450:1
18. Ma FH, Li J, Yang V, Lin KCJTA (2005) Thermo-acoustic flow instability in a scramjet combustor. AIAA, AIAA, p 3824
19. Mugridge BD (1980) Combustion driven oscillations. J Sound Vib 70:437–452. https://doi.org/10.1016/0022-460X(80)90311-9
20. Culick FEC, Yang V (1992) Prediction of the stability of unsteady motions in solid-propellant rocket motors. Prog Astronaut Aeronaut 143:719–779. https://doi.org/10.2514/5.9781600866159.0719.0779/r10.2514/5.9781600866159.0719.0779
21. Mongia HC, Held TJ, Hsiao GC, Pandalai RP (2003) Challenges and progress in controlling dynamics in gas turbine combustors. J Propuls Power 19:822–829. https://doi.org/10.2514/2.6197
22. Wicksall DM, Agrawal AK (2007) Acoustics measurements in a lean premixed combustor operated on hydrogen/hydrocarbon fuel mixtures. Int J Hydrogen Energy 32:1103–1112. https://doi.org/10.1016/j.ijhydene.2006.07.008
23. Li S, Zheng Y, Martinez DM, Zhu M, Jiang X (2015) Large-eddy simulation of flow and combustion dynamics in a lean partially-premixed swirling combustor. Phys Procedia 66:333–336. https://doi.org/10.1016/j.egypro.2015.02.080
24. Li S, Zheng Y, Zhu M, Martinez DM Jiang X (2015) Large-eddy simulation of flow and combustion dynamics in lean premixed swirling combustor. In: 12th International conference on combustion & energy utilisation, energy procedia, p 333–336
25. O'Connor J, Acharya V, Lieuwen T (2015) Transverse combustion instabilities: acoustic, fluid mechanic, and flame processes. Prog Energy Combust Sci 49:1–39. https://doi.org/10.1016/j.pecs.2015.01.001
26. Huang Y, Yang V (2009) Dynamics and stability of lean-premixed swirl-stabilized combustion. Prog Energy Combust Sci 35:293–364. https://doi.org/10.1016/j.pecs.2009.01.002
27. Darwish MA, Abdulrahim HK, Mabrouk AA, Hassan AS (2015) Cogeneration power-desalting plants using gas turbine combined cycle, desalination updates. In: Ning RY

(ed) IntechOpen. https://doi.org/10.5772/60209. Available from https://www.intechopen.com/books/desalination-updates/cogeneration-power-desalting-plants-using-gas-turbine-combined-cycle
28. Mattingly JD, Ohain Von H (2006) Elements of propulsion : gas turbines and rockets. https://doi.org/10.2514/4.861789
29. Verhoek FH (1969) Thermodynamics and rocket propulsion. J Chem Educ 46:140. https://doi.org/10.1021/ed046p140
30. Chen YC, Chang CC, Pan KL, Yang JT (1998) Flame lift-off and stabilization mechanisms of nonpremixed jet flames on a bluff-body burner. Combust Flame 115:51–65. https://doi.org/10.1016/S0010-2180(97)00336-2
31. Fan A, Wan J, Liu Y, Pi B, Yao H, Liu W (2014) Effect of bluff body shape on the blow-off limit of hydrogen/air flame in a planar micro-combustor. Appl Therm Eng 62:13–19. https://doi.org/10.1016/j.applthermaleng.2013.09.010
32. Kedia KS, Ghoniem AF (2015) The blow-off mechanism of a bluff-body stabilized laminar premixed flame. Combust Flame 162:1304–1315. https://doi.org/10.1016/j.combustflame.2014.10.017
33. Tuttle SG, Chaudhuri S, Kostka S, Kopp-Vaughan KM, Jensen TR, Cetegen BM et al (2012) Time-resolved blowoff transition measurements for two-dimensional bluff body-stabilized flames in vitiated flow. Combust Flame 159:291–305. https://doi.org/10.1016/j.combustflame.2011.06.001
34. Plee SL, Mellor AM (1979) Characteristic time correlation for lean blowoff of bluff-body-stabilized flames. Combust Flame 35:61–80. https://doi.org/10.1016/0010-2180(79)90007-5
35. Wang Q, Hu L, Yoon SH, Lu S, Delichatsios M, Chung SH (2015) Blow-out limits of nonpremixed turbulent jet flames in a cross flow at atmospheric and sub-atmospheric pressures. Combust Flame 162:3562–3568. https://doi.org/10.1016/j.combustflame.2015.06.012
36. Dawson JR, Gordon RL, Kariuki J, Mastorakos E, Masri AR, Juddoo M (2011) Visualization of blow-off events in bluff-body stabilized turbulent premixed flames. Proc Combust Inst 33:1559–1566. https://doi.org/10.1016/j.proci.2010.05.044
37. Radhakrishnan K, Heywood JB, Tabaczynski RJ (1981) Premixed turbulent flame blowoff velocity correlation based on coherent structures in turbulent flows. Combust Flame 42:19–33. https://doi.org/10.1016/0010-2180(81)90139-5
38. Figura L, Lee JG, Quay BD, Santavicca DA (2007) The effects of fuel composition on flame structure and combustion dynamics in a lean premixed combustor. In: Turbo Expo 2007, vol 2, pp 181–187. https://doi.org/10.1115/gt2007-27298
39. Wicksall D, Agarwal A (2001) Effects of fuel composition on flammability limits of a lean premixed combustor. In: ASME/IGTI Turbo Expo, New Orleans. GT2001-0007
40. Williams TC, Shaddix* CR, Schefer RW (2007) Diluted oxygen on performance of a premixed swirl-stabilized combustor. Combust Sci Technol 180:64–88. https://doi.org/10.1080/00102200701487061
41. Strakey P, Sidwell T, Ontko J (2007) Investigation of the effects of hydrogen addition on lean extinction in a swirl stabilized combustor. Proc Combust Inst 31(II):3173–3180. https://doi.org/10.1016/j.proci.2006.07.077
42. Lieuwen T, McDonell V, Petersen E, Santavicca D (2008) Fuel flexibility influences on premixed combustor blowout, flashback, autoignition, and stability. J Eng Gas Turbines Power 130:11506. https://doi.org/10.1115/1.2771243
43. Zhang Q, Noble DR, Meyers A, Xu K, Lieuwen T (2005) Characterization of fuel composition effects in $H_2$/CO/$CH_4$ mixtures upon lean blowout. ASME/IGTI Turbo Expo 129:1–13. https://doi.org/10.1115/1.2718566
44. Hudak R, Noble DR, Scarborough D (210) Carlo PAD for GT2010-22302 2010:1–11
45. Amato A, Hudak B, D'Souza P, D'Carlo P, Noble D, Scarborough D et al (2011) Measurements and analysis of CO and $O_2$ emissions in CH4/$CO_2$/$O_2$ flames. Proc Combust Inst 33:3399–3405. https://doi.org/10.1016/j.proci.2010.07.015
46. Shanbhogue SJ, Husain S, Lieuwen T (2009) Lean blowoff of bluff body stabilized flames: scaling and dynamics. Prog Energy Combust Sci 35:98–120. https://doi.org/10.1016/j.pecs.2008.07.003

47. Horlock JH, Bathie WW (2004) Advanced gas turbine cycles, vol 126. https://doi.org/10.1115/1.1789994
48. Bagdanavicius A, Bowen PJ, Syred N, Kay P, Crayford A., Sims G et al (2010) Burning velocities of alternative gaseous fuels at elevated temperature and pressure. AIAA J 48:317–329. https://doi.org/10.2514/1.43225
49. Fermoso J, Arias B, Gil MV, Plaza MG, Pevida C, Pis JJ et al (2010) Co-gasification of different rank coals with biomass and petroleum coke in a high-pressure reactor for $H_2$-rich gas production. Bioresour Technol 101:3230–3235. https://doi.org/10.1016/j.biortech.2009.12.035
50. Chiesa P, Lozza G, Mazzocchi L (2005) Using hydrogen as gas turbine fuel. J Eng Gas Turbines Power 127:73. https://doi.org/10.1115/1.1787513
51. Fritz J, Kröner M, Sattelmayer T (2004) Flashback in a swirl burner with cylindrical premixing zone. J Eng Gas Turbines Power 126:276. https://doi.org/10.1115/1.1473155
52. Checkel MD, Thomas A (1994) Turbulent combustion of premixed flames in closed vessels. Combust Flame 96:351–370. https://doi.org/10.1016/0010-2180(94)90104-X
53. Lewis B, Von Elbe G (1939) Physics of flames and explosions of gases. J Appl Phys 10:344–359. https://doi.org/10.1063/1.1707315
54. Subramanya M, Choudhuri A (2007) Investigation of combustion instability effects on the flame characteristics of fuel blends. In: 5th International energy conversion engineering conference, 25 June 2007–28 June 2007, vol 2, pp 817–30. https://doi.org/10.2514/6.2007-4796
55. Morrone B, Unich A (2009) Numerical investigation on the effects of natural gas and hydrogen blends on engine combustion. Int J Hydrogen Energy 34:4626–4634. https://doi.org/10.1016/j.ijhydene.2009.01.005
56. Kiesewetter F, Konle M, Sattelmayer T (2007) Analysis of combustion induced vortex breakdown driven flame flashback in a premix burner with cylindrical mixing zone. J Eng Gas Turbines Power 129:929. https://doi.org/10.1115/1.2747259
57. Kröner M, Fritz J, Sattelmayer T (2002) Flashback limits for combustion induced vortex breakdown in a swirl burner. J Eng Gas Turbines Power 125:413–422. https://doi.org/10.1115/1.1582498
58. Dam B, Corona G, Hayder M, Choudhuri A (2011) Effects of syngas composition on combustion induced vortex breakdown (CIVB) flashback in a swirl stabilized combustor. Fuel 90:3274–3284. https://doi.org/10.1016/j.fuel.2011.06.024
59. Chomiak J, Longwell JP, Sarofim AF (1989) Combustion of low calorific value gases; problems and prospects. Prog Energy Combust Sci 15:109–129. https://doi.org/10.1016/0360-1285(89)90012-9
60. Shelil N, Bagdanavicius A, Griffiths A, Syred N (2009) Flashback analysis for hydrogen/methane mixtures for premixed swirl combustion
61. Abdulsada M, Syred N, Griffiths A, Bowen P (2011) Effect of swirl number and fuel type upon the flashback in swirl combustors. In: 49th AIAA aerospace sciences meeting including the new horizons forum and aerospace exposition, vol 62, pp 1–16. https://doi.org/10.2514/6.2011-62
62. Noble DDR, Zhang Q, Shareef A, Tootle J, Meyers A, Lieuwen T (2006) Syngas mixture composition effects upon flashback and blowout. ASME Conf Proc 2006:357–368. https://doi.org/10.1115/gt2006-90470
63. Noble DDR, Zhang Q, Lieuwen T (2006) Syngas fuel composition sensitivities of combustor flashback and blowout. In: 23rd Annual international Pittsburgh Coal conference PCC—Coal-energy, environment and sustainable development, 25 Sept 2006–28 Sept 2006. University of Pittsburgh, School of Engineering
64. Chen Z, Qin C, Zhang Y (2014) Flame stability of partially premixed combustion for PNG/LNG interchangeability. J Nat Gas Sci Eng 21:467–473. https://doi.org/10.1016/j.jngse.2014.09.018
65. Straub DL, Ferguson DH, Casleton KH, Richards GA (2007) Effects of propane/natural gas blended fuels on gas turbine pollutant emissions. 5th U.S. national combustion meeting, pp 1–13

66. Ducruix S, Schuller T, Durox D, Candel S (2003) Combustion dynamics and instabilities: elementary coupling and driving mechanisms. J Propuls Power 19:722–734. https://doi.org/10.2514/2.6182
67. Taamallah S, Labry ZA, Shanbhogue SJ, Ghoniem AF (2015) Thermo-acoustic instabilities in lean premixed swirl-stabilized combustion and their link to acoustically coupled and decoupled flame macrostructures. Proc Combust Inst 35:3273–3282. https://doi.org/10.1016/j.proci.2014.07.002
68. Lieuwen T, Torres H, Johnson C, Zinn BT (2001) A mechanism of combustion instability in lean premixed gas turbine combustors. Trans ASME 123:182–189. https://doi.org/10.1115/1.1339002
69. Crocco L (1969) Research on combustion instability in liquid propellant rockets. Symp Combust 12:85–99. https://doi.org/10.1016/S0082-0784(69)80394-2
70. Annaswamy AM, Ghoniem AF (2002) Active control of combustion instability: theory and practice. Control Syst IEEE 22:37–54. https://doi.org/10.1109/mcs.2002.1077784
71. Selim MYE (2004) Sensitivity of dual fuel engine combustion and knocking limits to gaseous fuel composition. Energy Convers Manag 45:411–425. https://doi.org/10.1016/S0196-8904(03)00150-X
72. Krischer K, Varela H (2003) Oscillations and other dynamic instabilities. Handbook of fuel cells: fundamentals, technology, applications, vol 2, pp 679–701. https://doi.org/10.1002/9780470974001.f206052
73. Robey EH, Janus MC, Richards GA, Yip MJ, Robey E (1997) Effects of ambient conditions and fuel composition on combustion stability. In: ASME conference proceedings. 97-GT-266
74. Hendricks AG, Vandsburger U (2007) The effect of fuel composition on flame dynamics. Exp Therm Fluid Sci 32:126–132. https://doi.org/10.1016/j.expthermflusci.2007.02.007
75. Sutton G, Biblarz O (2001) Rocket propulsion elements. https://doi.org/10.1017/CBO9781107415324.004
76. Lefebvre A, Ballal D, Bahr D (2001) Gas turbine combustion, vol 40. https://doi.org/10.1002/1521-3773(20010316)40:6%3c9823::aid-anie9823%3e3.3.co;2-c
77. Lefebvre A, Ballal D, Bahr D (2010) Gas turbine combustion: alternative fuels and emissions. https://doi.org/10.1002/1521-3773(20010316)40:6%3c9823::AID-ANIE9823%3e3.3.CO;2-C
78. Van Maaren A, Thung DS, De Goey LPH (1994) Measurement of flame temperature and adiabatic burning velocity of methane/air mixtures. Combust Sci Technol 96:327–344
79. Imteyaz BA, Nemitallah MA, Abdelhafez A, Habib MA (2018) Combustion behavior and stability map of hydrogen-enriched oxy-methane premixed flames in a model gas turbine combustor. Int J Hydrogen Energy 43:16652–16666
80. Speth RL, Ghoniem AF (2009) Using a strained flame model to collapse dynamic mode data in a swirl-stabilized syngas combustor. Proc Combust Inst 32(I)I:2993–3000. https://doi.org/10.1016/j.proci.2008.05.072
81. Kushari A, Rosen LJ, Jagoda JI, Inn BT (1996) The effect of heat content and composition of fuel on pulse combustor performance. Symp Combust 26:3363–3368. https://doi.org/10.1016/S0082-0784(96)80184-6
82. Lee MC, Seo SB, Chung JH, Kim SM, Joo YJ, Ahn DH (2010) Gas turbine combustion performance test of hydrogen and carbon monoxide synthetic gas. Fuel 89:1485–1491. https://doi.org/10.1016/j.fuel.2009.10.004
83. Sattinger SS, Neumeier Y, Nabi A, Zinn BT, Amos DJ, Darling DD (2000) Sub-scale demonstration of the active feedback control of gas-turbine combustion instabilities. J Eng Gas Turbines Power 122:262. https://doi.org/10.1115/1.483204
84. Seume JR, Vortmeyer N, Krause W, Hermann J, Hantschk C-C, Zangl P et al (1998) Application of active combustion instability control to a heavy duty gas turbine. J Eng Gas Turbines Power 120:721–726. https://doi.org/10.1115/1.2818459
85. Straub DL, Richards GA (1998) Effect of fuel nozzle configuration on premix combustion dynamics. Am Soc Mech Eng 98-NaN-492. https://doi.org/10.1115/98-gt-492

86. Mcmanus KR, Poinsoti T, Candeli SM (1993) A review of active control of combustion instabilities. Prog Energy Combust Sci 19:1–29. https://doi.org/10.1016/0360-1285(93)90020-F
87. Atlantic N, Organisation T, Organisation T (2003) Active control of engine dynamics, vol 323
88. Richards GA, Janus MC, Richards GAJM (1998) Characterization of oscillations during premix gas turbine combustion. ASME J Eng Gas Turbines Power 120:294–301
89. Li L, Guo Z, Zhang C, Sun X (2010) A passive method to control combustion instabilities with perforated liner. Chin J Aeronaut 23:623–630. https://doi.org/10.1016/S1000-9361(09)60263-6
90. Lee J-Y, Lubarsky E, Zinn BT (2005) "Slow" active control of combustion instabilities by modification of liquid fuel spray properties. Proc Combust Inst 30:1757–1764. https://doi.org/10.1016/j.proci.2004.08.206
91. Altay HM, Speth RL, Hudgins DE, Ghoniem AF (2009) Flame-vortex interaction driven combustion dynamics in a backward-facing step combustor. Combust Flame 156:1111–1125. https://doi.org/10.1016/j.combustflame.2009.02.003
92. Nemitallah MA, Kewlani G, Hong S, Shanbhogue SJ, Habib MA, Ghoniem AF (2016) Investigation of a turbulent premixed combustion flame in a backward-facing step combustor; effect of equivalence ratio. Energy 95:211–222. https://doi.org/10.1016/j.energy.2015.12.010
93. Candel SM (1992) Combustion instabilities coupled by pressure waves and their active control. Symp Combust 24:1277–1296. https://doi.org/10.1016/S0082-0784(06)80150-5
94. Noiray N, Durox D, Schuller T, Candel S (2009) Dynamic phase converter for passive control of combustion instabilities. Proc Combust Inst 32(II):3163–3170. https://doi.org/10.1016/j.proci.2008.05.051
95. Jin W, Wang J, Nie Y, Yu S, Huang Z (2015) Experimental study on flame instabilities of laminar premixed $CH_4H_2$ air non-adiabatic flat flames. Fuel 159:599–606. https://doi.org/10.1016/j.fuel.2015.07.022
96. Lieuwen T (2003) Modeling premixed combustion-acoustic wave interactions: a review. J Propuls Power 19:765–781. https://doi.org/10.2514/2.6193
97. Fenimore CP (1971) Formation of nitric oxide in premixed hydrocarbon flames. In: Thirteenth symposium (international) on combustion, pp 373–380
98. Cheng AS, Upatnieks A, Mueller CJ (2006) Investigation of the impact of biodiesel fuelling on $NO_x$ emissions using an optical direct injection diesel engine. Int J Engine Res 7:297–318
99. Baukal C (2013) Oxygen-enhanced combustion. Industrial combustion, 2nd edn. CRC Press. ISBN 978-1-4398-6228-5 978-1-4398-6230-8
100. Moreno FE, Pam RL, Chirolo DS (1992) International gas research conference. Institute of Technology, Canada, vol 5, pp 402–412
101. Miller JA, Bowman CT (1989) Mechanism and modeling of nitrogen chemistry in combustion. Prog Energy Combust Sci 15:287–338
102. Ren Y, Li X (2011) Numerical simulation of the soot and $NO_x$ formations in a biodieselfuelled engine. SAE International, SAE Technical Papers Series
103. Hedley JT, Pourkashanian M, Williams A, Yap LT (1995) $NO_x$ formation in large-scale oxy-fuel flames. Combust Sci Technol 108:311–322. https://doi.org/10.1080/00102209508960404 ISSN 0010-2202, 1563-521X
104. Wei Z, Haisheng Z, Leung C, Cheung C, Huang Z (2020) Effects of $H_2$ addition on the formation and emissions of $CO/NO_2/NO_x$ in the laminar premixed biogas-hydrogen flame undergoing the flame-wall interaction. Fuel 259
105. Naha S, Aggarwal SK (2004) Fuel effects on $NO_x$ emissions in partially premixed flames. Combust Flame 139:90–105
106. Fiskum A (2008) Calculation of $NO_x$ formation in a swirl burner. M.sc. thesis, Norwegian University of Science and Technology
107. Costa, Meunier M, Carvalho MG (1998) On $NO_x$ emissions from turbulent propane diffusion flames. Ph Combust Flame 112:221–30
108. Galbiati MA, Cavigiolo A, Effuggi A, Gelosa D, Rota R (2004) Mild combustion for fuel-$NO_x$ reduction. Combust Sci Technol 176(7):1035–1054

109. Sujith Sukumaran S-CK (2013) Modeling fuel $NO_x$ formation from combustion of biomass-derived producer gas in a large-scale burner. Combust Flame 160:2159–2168
110. Weydahl T, Bugge M, Gran I, Ertesvåg I (2001) Computational modeling of nitric oxide formation in biomass combustion. Int J Appl Mech Eng 7:125–141
111. Bowman CT, Hanson RK, Davidson DF, Gardiner WC Jr, Lissianski V, Smith GP, Golden DM, Frenklach M, Goldenberg M (2013) GRI 2.11 mechanism. http://www.me.berkeley.edu/gri_mech/
112. Sethuraman S, Van Huynh C, Kong S-C (2011) Producer gas composition and $NO_x$ emissions from a pilot-scale biomass gasification and combustion system using feedstock with controlled nitrogen content. Energy Fuels 25:813–822
113. Van Huynh C, Kong S-C (2013) Characterization and challenge of development of producer gas fuel combustor. Fuel 103:987–996
114. Adouane B, Hoppesteyn P, de Jong W, van der Wel M, Hein KRG, Spliethof H (2002) Gas turbine combustor for biomass derived LCV gas, a first approach towards fuel-$NO_x$ modelling and experimental validation. Appl Therm Eng 22:959–970
115. Sullivan N, Jensen A, Glarborg P, Day MS, Grcar JF, Bell JB (2002) Ammonia conversion and $NO_x$ formation in laminar coflowing nonpremixed methane-air flames. Combust Flame 131:285–298
116. Zabetta EC, Kilpinen P, Hupa M, Stahl K, Leppalahti J, Cannon M, Nieminen J (2000) Kinetic modeling study on the potential of staged combustion in gas turbines for the reduction of nitrogen oxide emissions from biomass IGCC plants. Energy & Fuels 14:751–761
117. Drake MC, Pitz RW, Correa SM, Lapp M (1985) Nitric oxide formation from thermal and fuel-bound nitrogen sources in a turbulent nonpremixed syngas flame. In: Symposium (international) on combustion, vol 20, pp 1983–1990
118. Azuhata S, Kaji R, Akimoto H, Hishinuma Y (1981) A study of the kinetics of the $NH_3$–NO–$O_2$–$H_2O_2$ reaction. Symposium (international) on combustion, vol 18, pp 845–852
119. Sullivan N, Jensen A, Glarborg P, Day MS, Grcar JF, Bell JB, Pope CJ, Kee RJ (2002) Ammonia conversion and $NO_x$ formation in laminar coflowing nonpremixed methane-air flames. Combust Flame 131(3):285–298
120. Otomo J, Koshi M, Mitsumori T, Iwasaki H, Yamada K (2018) Chemical kinetic modeling of ammonia oxidation with improved reaction mechanism for ammonia/air and ammonia/hydrogen/air combustion. Int J Hydrogen Energy 43, 3004–3014
121. Zabetta EC, Hupa M, Reducing KS (2005) $NO_x$ emissions using fuel staging air staging and selective noncatalytic reduction in synergy. Ind Eng Chem Res 44:4552–4561
122. Li Pin-Wei, Chyang Chien-Song (2020) A comprehensive study on $NO_x$ emission and fuel nitrogen conversion of solid biomass in bubbling fluidized beds under staged combustion. J Energy Inst 93:324–334
123. Mordaunt C (2002) Dual fuel issues related to performance, emissions and combustion instability in gas turbine systems. Department of Mechanical and Nuclear Engineering, Pennsylvania State University, University Park, PA 16802, pp 1–34
124. Ferguson D, Straub D, Richards G, Robey E (2007) Impact of fuel variability on dynamic instabilities in gas turbine combustion. Fifth U.S. combustion meeting
125. Mordaunt C, Brassard C, Lee SY, Santoro RJ (2001) Combustion instability studies in a high-pressure lean-premixed model combustor under liquid fuel operation. In: International joint power generation conference, vol 1, pp 249–56
126. Nerd LOAHG (2004) A study of parameters affecting the combustion stability and emissions behavior of alstom heavy-duty gas turbines. ASME Turbo Expo 2004:93–100. https://doi.org/10.1115/GT2004-53228
127. Richards GA, McMillian MM, Gemmen RS, Rogers WA, Cully SR (2001) Issues for low-emission, fuel-flexible power systems. Prog Energy Combust Sci 27:141–169. https://doi.org/10.1016/S0360-1285(00)00019-8
128. Shelil N, Bagdanavicius A, Syred N, Griffiths A, Bowen PJ (2010) Premixed swirl combustion and flashback analysis with hydrogen/methane mixtures. In: 48th AIAA aerospace sciences meeting including the new horizons forum and aerospace exposition, 4 Jan 2010–7 Jan 2010

129. Kido H, Nakahara M, Nakashima K, Hashimoto J (2002) Influence of local flame displacement velocity on turbulent burning velocity. Proc Combust Inst 29:1855–1861. https://doi.org/10.1016/S1540-7489(02)80225-5
130. Said SAM, Simakov DSA, Mokheimer EMA, Habib MA, Ahmed S, Waseeuddin M et al (2015) Computational fluid dynamics study of hydrogen generation by low temperature methane reforming in a membrane reactor. Int J Hydrogen Energy 40:3158–3169. https://doi.org/10.1016/j.ijhydene.2015.01.024
131. Sheu EJ, Mokheimer EMA, Ghoniem AF (2015) A review of solar methane reforming systems. Int J Hydrogen Energy 40:12929–12955. https://doi.org/10.1016/j.ijhydene.2015.08.005
132. Li Y, Zhang N, Cai R (2013) Low $CO_2$-emissions hybrid solar combined-cycle power system with methane membrane reforming. Energy 58:36–44. https://doi.org/10.1016/j.energy.2013.02.005
133. Edwards PP, Kuznetsov VL, David WIF, Brandon NP (2008) Hydrogen and fuel cells: towards a sustainable energy future. Energy Policy 36:4356–4362. https://doi.org/10.1016/j.enpol.2008.09.036
134. Ball M, Wietschel M (2009) The future of hydrogen—opportunities and challenges. Int J Hydrogen Energy 34:615–627. https://doi.org/10.1016/j.ijhydene.2008.11.014
135. HyWays (2007) The European hydrogen energy roadmap. HyWays
136. Wicksall DM, Agrawal AK, Schefer RW, Keller JO (2005) The interaction of flame and flow field in a lean premixed swirl-stabilized combustor operated on $H_2/CH_4$/air. Proc Combust Inst 30(I)I:2875–2883. https://doi.org/10.1016/j.proci.2004.07.021
137. Schefer RW (2003) Hydrogen enrichment for improved lean flame stability. Int J Hydrogen Energy 28:1131–1141. https://doi.org/10.1016/S0360-3199(02)00199-4
138. Jackson GS, Sai R, Plaia JM, Boggs CM, Kiger KT (2003) Influence of $H_2$ on the response of lean premixed $CH_4$ flames to high strained flows. Combust Flame 132:503–511. https://doi.org/10.1016/S0010-2180(02)00496-0
139. Zhen HS, Leung CW, Cheung CS, Huang ZH (2016) Combustion characteristic and heating performance of stoichiometric biogas-hydrogen-air flame. Int J Heat Mass Transf 92:807–814. https://doi.org/10.1016/j.ijheatmasstransfer.2015.09.040
140. Rashwan SS, Ibrahim AH, Abou-arab TW (2015) Experimental investigation of oxy-fuel combustion of CNG flames stabilized over perforated-plate burner. In: 18th International flame research foundation, Friesing, Munich, pp 1–11
141. Rashwan SS, Ibrahim AH, Abou-Arab TW, Nemitallah MA, Habib MA (2016) Experimental investigation of partially premixed methane-air and methane-oxygen flames stabilized over a perforated-plate burner. Appl Energy 169:126–137. https://doi.org/10.1016/j.apenergy.2016.02.047
142. Zhang Q, Noble DR, Lieuwen T (2005) Blowout measurements in a syngas-fired gas turbine combustor. In: 22nd Annual international Pittsburgh Coal conference PCC 2005, vol 1, pp 313–318
143. Cheng RK, Littlejohn D, Strakey PA, Sidwell T (2009) Laboratory investigations of a low-swirl injector with $H_2$ and $CH_4$ at gas turbine conditions. Proc Combust Inst 32(II):3001–3009. https://doi.org/10.1016/j.proci.2008.06.141
144. Tuncer O (2013) Premixed combustion of hydrogen and syngas fuels in gas turbine combustors. Formatex 946–957
145. Halter F, Chauveau C, Gökalp I (2007) Characterization of the effects of hydrogen addition in premixed methane/air flames. Int J Hydrogen Energy 32:2585–2592. https://doi.org/10.1016/j.ijhydene.2006.11.033
146. Ilbas M, Crayford AP, Yilmaz I, Bowen PJ, Syred N (2006) Laminar-burning velocities of hydrogen-air and hydrogen-methane-air mixtures: an experimental study. Int J Hydrogen Energy 31:1768–1779. https://doi.org/10.1016/j.ijhydene.2005.12.007
147. Ilbas M, Yilmaz I, Kaplan Y (2005) Investigations of hydrogen and hydrogen-hydrocarbon composite fuel combustion and $NO_x$ emission characteristics in a model combustor. Int J Hydrogen Energy 30:1139–1147. https://doi.org/10.1016/j.ijhydene.2004.10.016

148. Ilbas M, Yilmaz I, Veziroglu TN, Kaplan Y (2005) Hydrogen as burner fuel: modelling of hydrogen-hydrocarbon composite fuel combustion and $NO_x$ formation in a small burner. Int J Energy Res 29:973–990. https://doi.org/10.1002/er.1104
149. Ghoniem AF, Annaswamy A, Park S, Sobhani ZC (2005) Stability and emissions control using air injection and $H_2$ addition in premixed combustion. Proc Combust Inst 30(I)I:1765–1773. https://doi.org/10.1016/j.proci.2004.08.175
150. Griebel P, Boschek E, Jansohn P (2007) Lean blowout limits and $NO_x$ emissions of turbulent, lean premixed, hydrogen-enriched methane/air flames at high pressure. J Eng Gas Turbines Power 129:404. https://doi.org/10.1115/1.2436568
151. Lounici MS, Boussadi A, Loubar K, Tazerout M (2014) Experimental investigation on NG dual fuel engine improvement by hydrogen enrichment. Int J Hydrogen Energy 39:21297–21306. https://doi.org/10.1016/j.ijhydene.2014.10.068
152. Schefer RW, Wicksall DM, Agrawal AK (2002) Combustion of hydrogen-enriched methane in a lean premixed swirl-stabilized burner. Proc Combust Inst 29:843–851. https://doi.org/10.1016/S1540-7489(02)80108-0
153. Hu E, Huang Z, Zheng J, Li Q, He J, Liu B et al (2009) Numerical study on laminar burning velocity and NO formation of premixed methane-hydrogen-air flames. Int J Hydrogen Energy 34:6545–6557. https://doi.org/10.1016/j.ijhydene.2009.05.080
154. Hu E, Huang Z, Liu B, Zheng J, Gu X (2009) Experimental study on combustion characteristics of a spark-ignition engine fueled with natural gas-hydrogen blends combining with EGR. Int J Hydrogen Energy 34:1035–1044. https://doi.org/10.1016/j.ijhydene.2008.11.030
155. Hu E, Huang Z, He J, Miao H (2009) Experimental and numerical study on lean premixed methane-hydrogen-air flames at elevated pressures and temperatures. Int J Hydrogen Energy 34:6951–6960. https://doi.org/10.1016/j.ijhydene.2009.06.072
156. York WD, Ziminsky WS, Yilmaz E (2013) Development and testing of a low $NO_x$ hydrogen combustion system for heavy-duty gas turbines. J Eng Gas Turbines Power 135:22001. https://doi.org/10.1115/1.4007733
157. Daniele S, Jansohn P, Boulouchos K (2009) Experimental investigation of lean premixed syngas combustion at gas turbine relevant conditions : lean blow out limits, emissions and turbulent flame speed. Ital Sect Combust Inst Combust Colloquai II-4:1–11
158. TerMaath CY, Skolnik EG, Schefer RW, Keller JO (2006) Emissions reduction benefits from hydrogen addition to midsize gas turbine feedstocks. Int J Hydrogen Energy 31:1147–1158. https://doi.org/10.1016/j.ijhydene.2005.10.002
159. Ghenai C (2010) Combustion of syngas fuel in gas turbine can combustor. Adv Mech Eng. https://doi.org/10.1155/2010/342357
160. Habib MA, Nemitallah MA (2015) Design of an ion transport membrane reactor for application in fire tube boilers. Energy 81:787–801. https://doi.org/10.1016/j.energy.2015.01.029
161. Scheffknecht G, Al-Makhadmeh L, Schnell U, Maier J (2011) Oxy-fuel coal combustion—A review of the current state-of-the-art. Int J Greenh Gas Control 5:16–35. https://doi.org/10.1016/j.ijggc.2011.05.020
162. Aneke M, Wang M (2015) Process analysis of pressurized oxy-coal power cycle for carbon capture application integrated with liquid air power generation and binary cycle engines. Appl Energy 154:556–566. https://doi.org/10.1016/j.apenergy.2015.05.030
163. Habib MA, Nemitallah MA, Ahmed P, Sharqawy MH, Badr HM, Muhammad I et al (2015) Experimental analysis of oxygen-methane combustion inside a gas turbine reactor under various operating conditions. Energy 86:105–114. https://doi.org/10.1016/j.energy.2015.03.120
164. Granados DA, Chejne F, Mejía JM (2015) Oxy-fuel combustion as an alternative for increasing lime production in rotary kilns. Appl Energy 158:107–117. https://doi.org/10.1016/j.apenergy.2015.07.075
165. Fu Q, Kansha Y, Song C, Liu Y, Ishizuka M, Tsutsumi A (2014) A cryogenic air separation process based on self-heat recuperation for oxy-combustion plants. Appl Energy. https://doi.org/10.1016/j.apenergy.2015.03.039

166. Nemitallah MA (2016) A study of methane oxy-combustion characteristics inside a modified design button-cell membrane reactor utilizing a modified oxygen permeation model for reacting flows. J Nat Gas Sci Eng 28:61–73. https://doi.org/10.1016/j.jngse.2015.11.041
167. Nemitallah MA, Habib MA, Mezghani K (2015) Experimental and numerical study of oxygen separation and oxy-combustion characteristics inside a button-cell LNO-ITM reactor. Energy 84:600–611. https://doi.org/10.1016/j.energy.2015.03.022
168. Habib MA, Salaudeen SA, Nemitallah MA, Ben-Mansour R, Mokheimer EMA (2016) Numerical investigation of syngas oxy-combustion inside a LSCF-6428 oxygen transport membrane reactor. Energy 96:654–665. https://doi.org/10.1016/j.energy.2015.12.043
169. Ben Mansour R, Nemitallah MA, Habib MA (2013) Numerical investigation of oxygen permeation and methane oxy-combustion in a stagnation flow ion transport membrane reactor. Energy 54:322–332. https://doi.org/10.1016/j.energy.2013.03.027
170. Ditaranto M, Hals J (2006) Combustion instabilities in sudden expansion oxy-fuel flames. Combust Flame 146:493–512. https://doi.org/10.1016/j.combustflame.2006.04.015
171. Ditaranto M, Anantharaman R, Weydahl T (2013) Performance and $NO_x$ emissions of refinery fired heaters retrofitted to hydrogen combustion. Energy Procedia 37:7214–7220. https://doi.org/10.1016/j.egypro.2013.06.659
172. Liu CY, Chen G, Sipöcz N, Assadi M, Bai XS (2012) Characteristics of oxy-fuel combustion in gas turbines. Appl Energy 89:387–394. https://doi.org/10.1016/j.apenergy.2011.08.004
173. Nemitallah MA, Habib MA (2013) Experimental and numerical investigations of an atmospheric diffusion oxy-combustion flame in a gas turbine model combustor. Appl Energy 111:401–415. https://doi.org/10.1016/j.apenergy.2013.05.027
174. Kutne P, Kapadia BK, Meier W, Aigner M (2011) Experimental analysis of the combustion behaviour of oxyfuel flames in a gas turbine model combustor. Proc Combust Inst 33:3383–3390. https://doi.org/10.1016/j.proci.2010.07.008
175. Shi B, Hu J, Ishizuka S (2015) Carbon dioxide diluted methane/oxygen combustion in a rapidly mixed tubular flame burner. Combust Flame 162:420–430. https://doi.org/10.1016/j.combustflame.2014.07.022
176. Hu X, Yu Q, Liu J, Sun N (2014) Investigation of laminar flame speeds of $CH_4/O_2/CO_2$ mixtures at ordinary pressure and kinetic simulation. Energy 70:626–634. https://doi.org/10.1016/j.energy.2014.04.029
177. Li YH, Chen GB, Chao YC (2015) Effects of flue gas addition on the premixed oxy-methane flames in atmospheric condition. Energy Procedia 75:3054–3059. https://doi.org/10.1016/j.egypro.2015.07.623
178. Jerzak W, Kuźnia M (2016) Experimental study of impact of swirl number as well as oxygen and carbon dioxide content in natural gas combustion air on flame flashback and blow-off. J Nat Gas Sci Eng 29:46–54. https://doi.org/10.1016/j.jngse.2015.12.054
179. Shia B, Zhua Z, Wanga N, Lub P, Ishizukac S (2015) An experimental study on oxy-fuel combustion of methane under various oxygen mole fractions an experimental study on oxy-fuel combustion of methane under various oxygen mole fractions. In: 8th International symposium on coal combustion Beijing, China
180. Kim HK, Kim Y, Lee SM, Ahn KY (2007) NO reduction in 0.03–0.2 MW oxy-fuel combustor using flue gas recirculation technology. Proc Combust Inst 31(I)I:3377–3384. https://doi.org/10.1016/j.proci.2006.08.083
181. Grathwohl S, Lemp O, Maier J, Schnell U, Scheffknecht G (2009) Ein Brennerkonzept zum Einsatz unter Oxyfuel-Bedingungen. VDI Ber 2056:301–306
182. Hu E, Huang Z, He J, Jin C, Zheng J (2009) Experimental and numerical study on laminar burning characteristics of premixed methane-hydrogen-air flames. Int J Hydrogen Energy 34:4876–4888. https://doi.org/10.1016/j.ijhydene.2009.03.058
183. Lieuwen T (2002) Experimental investigation of limit-cycle oscillations in an unstable gas turbine combusto. J Propuls Power 18:61–67. https://doi.org/10.2514/2.5898
184. Straub DL, Richards GA (1999) Effect of axial swirl vane location on combustion dynamics. ASME Paper 99-GT-109. https://doi.org/10.1115/99-gt-109

185. Li G, Gutmark EJ (2005) Effect of exhaust nozzle geometry on combustor flow field and combustion characteristics. Proc Combust Inst 30(II):2893–2901. https://doi.org/10.1016/j.proci.2004.08.189
186. Acharya S, Uhm JH (2005) Characteristics and control of combustion instabilities in a swirl-stabilized spray combustor. Combust Process Propuls 157–168. https://doi.org/10.1016/b978-012369394-5/50019-6
187. Lee SY, Seo S, Broda JC, Pal S, Santoro RJ (2000) An experimental estimation of mean reaction rate and flame structure during combustion instability in a lean premixed gas turbine combustor. Proc Combust Inst 28:775–782. https://doi.org/10.1016/S0082-0784(00)80280-5
188. Venkataraman KK, Preston LH, Simons DW, Lee BJ, Lee JG, Santavicca DA (1999) Mechanism of combustion instability in a lean premixed dump combustor. J Propuls Power 15:909–918. https://doi.org/10.2514/2.5515
189. Barlow RS, Karpetis AN, Frank JH, Chen JY (2001) Scalar profiles and NO formation in laminar opposed-flow partially premixed methane/air flames. Combust Flame 127:2102–2118. https://doi.org/10.1016/S0010-2180(01)00313-3
190. Xue H (2003) $NO_x$ emissions in n-heptane/air partially premixed flames. Combust Flame 132:723–741. https://doi.org/10.1016/S0010-2180(02)00534-5
191. Chen L, Yong SZ, Ghoniem AF (2012) Oxy-fuel combustion of pulverized coal: characterization, fundamentals, stabilization and CFD modeling. Prog Energy Combust Sci 38:156–214. https://doi.org/10.1016/j.pecs.2011.09.003
192. Broda JC, Seo S, Santoro RJ, Shirhattikar G, Yang V (1998) An experimental study of combustion dynamics of a premixed swirl injector. Symp Combust 27:1849–1856. https://doi.org/10.1016/S0082-0784(98)80027-1
193. Shih WP, Lee JG, Santavicca DA (1996) Stability and emissions characteristics of a lean premixed gas turbine combustor. Symp Combust 26:2771–2778. https://doi.org/10.1016/S0082-0784(96)80115-9
194. Wang B, Wang Z, Shuai S, Xu H (2014) Combustion and emission characteristics of Multiple Premixed Compression Ignition (MPCI) mode fuelled with different low octane gasolines. Appl Energy 160:769–776. https://doi.org/10.1016/j.apenergy.2015.01.115
195. Benajes J, Martín J, Novella R, Thein K (2016) Understanding the performance of the multiple injection gasoline partially premixed combustion concept implemented in a 2-Stroke high speed direct injection compression ignition engine. Appl Energy 161:465–475. https://doi.org/10.1016/j.apenergy.2015.10.034
196. Durox D, Moeck JP, Bourgouin JF, Morenton P, Viallon M, Schuller T et al (2013) Flame dynamics of a variable swirl number system and instability control. Combust Flame 160:1729–1742. https://doi.org/10.1016/j.combustflame.2013.03.004
197. Srodowiska RO, Jerzak W (2015) Formation of CO in the oxy-fuel premixed flame. Annu Set Environ Prot
198. Merlo N, Boushaki T, Chauveau C, De Persis S, Pillier L, Sarh B et al (2013) Experimental study of oxygen enrichment effects on turbulent non-premixed swirling flames. Energy Fuels 27:6191–6197. https://doi.org/10.1021/ef400843c
199. Palies P, Durox D, Schuller T, Candel S (2011) Experimental study on the effect of swirler geometry and swirl number on flame describing functions. Combust Sci Technol 183:704–717. https://doi.org/10.1080/00102202.2010.538103
200. De A, Acharya S (2012) Parametric study of upstream flame propagation in hydrogen-enriched premixed combustion: effects of swirl, geometry and premixedness. Int J Hydrogen Energy 37:14649–14668. https://doi.org/10.1016/j.ijhydene.2012.07.008
201. Syred N, Giles A, Lewis J, Abdulsada M, Valera Medina A, Marsh R et al (2014) Effect of inlet and outlet configurations on blow-off and flashback with premixed combustion for methane and a high hydrogen content fuel in a generic swirl burner. Appl Energy 116:288–296. https://doi.org/10.1016/j.apenergy.2013.11.071
202. Syred N, Abdulsada M, Griffiths A, O'Doherty T, Bowen P (2012) The effect of hydrogen containing fuel blends upon flashback in swirl burners. Appl Energy 89:106–110. https://doi.org/10.1016/j.apenergy.2011.01.057

203. Ishak MSA, Jaafar MNM, Omar WZW (2015) Experimental analysis on the formation of CO–NO–HC in swirling flow. J Teknol 72:21–26. https://doi.org/10.11113/jt.v72.3909
204. Palies P, Durox D, Schuller T, Candel S (2010) The combined dynamics of swirler and turbulent premixed swirling flames. Combust Flame 157:1698–1717. https://doi.org/10.1016/j.combustflame.2010.02.011
205. Hoffmann S, Habisreuther P, Lenze B (1994) Development and assessment of correlations for predicting stability limits of swirling flames. Chem Eng Process 33:393–400. https://doi.org/10.1016/0255-2701(94)02011-6
206. Freitag M, Janicka J (2007) Investigation of a strongly swirled unconfined premixed flame using les. Proc Combust Inst 31(I):1477–1485. https://doi.org/10.1016/j.proci.2006.07.225
207. Flohr P, Stuttaford P (2013) Modern gas turbine systems. https://doi.org/10.1533/9780857096067.2.151
208. Mellor AM (1990) Design of modern turbine combustors
209. Eldrainy YA, Aly HS, Saqr KM, Jaafar MNM A multiple inlet swirler for gas turbine combustors. In: Proceedings of international conference on fluid mechanics, heat transfer and thermodynamics. World Academic Science and Engineering Technology, Tokyo, Japan, pp 1260–1263
210. Syred N (2006) A review of oscillation mechanisms and the role of the precessing vortex core (PVC) in swirl combustion systems. Prog Energy Combust Sci 32:93–161. https://doi.org/10.1016/j.pecs.2005.10.002
211. Syred N, Beér JM (1974) Combustion in swirling flows: a review. Combust Flame 23:143–201. https://doi.org/10.1016/0010-2180(74)90057-1
212. Shelil N (2009) Flashback studies with premixed swirl combustion. Cardiff University
213. Cavaliere DE, Kariuki J, Mastorakos E (2013) A comparison of the blow-off behaviour of swirl-stabilized premixed, non-premixed and spray flames. Flow Turbul Combust 91:347–372. https://doi.org/10.1007/s10494-013-9470-z
214. Johnson MR, Littlejohn D, Nazeer WA, Smith KO, Cheng RK (2005) A comparison of the flowfields and emissions of high-swirl injectors and low-swirl injectors for lean premixed gas turbines. Proc Combust Inst 30(II):2867–2874. https://doi.org/10.1016/j.proci.2004.07.040
215. Lim AH, Kim HS, Ahn KY, Kim YM (2003) The study for designs of lean-premixed low $NO_x$ combustor. In: Korean society of combustion symposium, vol 27, pp 83–88
216. Kim G, Lee YD, Sohn CH, Choi KW, Kim HS (2015) Experimental investigation on combustion and emission characteristics of a premixed flame in a gas-turbine combustor with a vortex generator. Appl Therm Eng 77:57–64. https://doi.org/10.1016/j.applthermaleng.2014.11.073
217. Claypole TC, Syred N (1981) The effect of swirl burner aerodynamics on $NO_x$ formation. Symp Combust 18:81–89. https://doi.org/10.1016/S0082-0784(81)80013-6
218. Ramadan IA, Ibrahim AH, Abou-Arab TW, Rashwan SS, Nemitallah MA, Habib MA (2016) Effects of oxidizer flexibility and bluff-body blockage ratio on flammability limits of diffusion flames. Appl Energy 178:19–28. https://doi.org/10.1016/j.apenergy.2016.06.009
219. Peters N (2001) Turbulent combustion. Meas Sci Technol 12:2022. https://doi.org/10.1088/0957-0233/12/11/708
220. Veynante D, Vervisch L (2002) Turbulent combustion modeling. Prog Energy Combust Sci 28:193–266. https://doi.org/10.1016/S0360-1285(01)00017-X
221. Poinsot T, Veynante D (2005) Theoretical and numerical combustion. https://doi.org/10.1016/j.combustflame.2007.07.002S0010-2180(07)00176-9
222. Angelberger C, Veynante D, Egolfopoulos F (2001) LES of chemical and acoustic forcing of a premixed dump combustor. Flow Turbul Combust 65:205–222. https://doi.org/10.1023/A:1011477030619
223. Perez FE, Groth CPT, Gülder ÖL (2014) Large-eddy simulation of lean hydrogen-methane turbulent premixed flames in the methane-dominated regime. Int J Hydrogen Energy 39:7147–7157. https://doi.org/10.1016/j.ijhydene.2014.02.028
224. Hernández-Pérez FE, Yuen FTC, Groth CPT, Gülder ÖL (2011) LES of a laboratory-scale turbulent premixed Bunsen flame using FSD, PCM-FPI and thickened flame models. Proc Combust Inst 33:1365–1371. https://doi.org/10.1016/j.proci.2010.06.010

225. Ghoniem AF, Park S, Wachsman A, Annaswamy A, Wee D, Murat Altay H (2005) Mechanism of combustion dynamics in a backward-facing step stabilized premixed flame. Proc Combust Inst 30(I)I:1783–1790. https://doi.org/10.1016/j.proci.2004.08.201
226. Murat Altay H, Speth RL, Hudgins DE, Ghoniem AF (2009) The impact of equivalence ratio oscillations on combustion dynamics in a backward-facing step combustor. Combust Flame 156:2106–2116. https://doi.org/10.1016/j.combustflame.2009.07.024
227. Colin O, Ducros F, Veynante D, Poinsot T (2000) A thickened flame model for large eddy simulations of turbulent premixed combustion. Phys Fluids 12:1843–1863. https://doi.org/10.1063/1.870436
228. Fureby C, Moller S-I (1995) Large eddy simulation of reacting flows applied to bluff body stabilized flames. AIAA J 33:2339–2347
229. di Mare F, Jones WP, Menzies KR (2004) Large eddy simulation of a model gas turbine combustor. Combust Flame 137:278–294. https://doi.org/10.1016/j.combustflame.2004.01.008
230. Kim W-W, Menon S, Mongia HC (1999) Large-eddy simulation of a gas turbine combustor flow. Combust Sci Technol 143:25–62. https://doi.org/10.1080/00102209908924192
231. Moin P (2002) Advances in large eddy simulation methodology for complex flows. Int J Heat Fluid Flow 23:710–720. https://doi.org/10.1016/S0142-727X(02)00167-4
232. Haworth DC, Jansen K (2000) Large-eddy simulation on unstructured deforming meshes: towards reciprocating IC engines. Comput Fluids 29:493–524. https://doi.org/10.1016/S0045-7930(99)00015-8
233. Selle L, Lartigue G, Poinsot T, Koch R, Schildmacher KU, Krebs W et al (2004) Compressible large eddy simulation of turbulent combustion in complex geometry on unstructured meshes. Combust Flame 137:489–505. https://doi.org/10.1016/j.combustflame.2004.03.008
234. Sommerer Y, Galley D, Poinsot T, Ducruix S, Lacas F, Veynante D (2004) Large eddy simulation and experimental study of flashback and blow-off in a lean partially premixed swirled burner. J Turbul 5:37–41. https://doi.org/10.1088/1468-5248/5/1/037
235. Stone C, Menon S (2003) Large-eddy simulation of combustion dynamics in swirling flows 19
236. Stone C, Menon S (2003) Open-loop control of combustion instabilities in a model gas turbine combustor. J Turbul 4
237. Shinjo J, Mizobuchi Y, Ogawa S (2003) LES of unstable combustion in a gas turbine combustor. Lecture notes in computer science (including subseries lecture notes in artificial intelligence and lecture notes in bioinformatics), vol 2858, pp 234–244.
238. Huang Y, Sung H-G, Hsieh S-Y, Yang V (2003) Large-eddy simulation of combustion dynamics of lean-premixed swirl-stabilized combustor. J Propuls Power 19:782–794. https://doi.org/10.2514/2.6194
239. Franzelli B, Riber E, Gicquel LYM, Poinsot T (2012) Large eddy simulation of combustion instabilities in a lean partially premixed swirled flame. Combust Flame 159:621–637. https://doi.org/10.1016/j.combustflame.2011.08.004
240. Yu JF, Yu R, Bai XS, Bastiaans RJM, Van Oijen JA, De Goey LPH (2016) Heat transfer and flame stabilization of laminar premixed flames anchored to a heat-flux burner. Int J Hydrogen Energy 41:2037–2051. https://doi.org/10.1016/j.ijhydene.2015.11.105
241. Komarek T, Polifke W (2010) Impact of swirl fluctuations on the flame response of a perfectly premixed swirl burner. J Eng Gas Turbines Power 132:61503. https://doi.org/10.1115/1.4000127
242. Chakravarthy VK, Menon S (2001) Subgrid modeling of turbulent premixed flames in the flamelet regime. Flow Turbul Combust 65:133–161. https://doi.org/10.1023/A:1011456218761
243. Chakravarthy VK, Menon S (2001) Large-eddy simulation of turbulent premixed flames in the flamelet regime. Combust Sci Technol 162:175–+. https://doi.org/10.1080/00102200108952141
244. Schmitt P, Poinsot T, Schuermans B, Geigle KP (2007) Large-eddy simulation and experimental study of heat transfer, nitric oxide emissions and combustion instability in a swirled turbulent high-pressure burner. J Fluid Mech 570:17. https://doi.org/10.1017/S0022112006003156

245. Wilkes NS, Guilbert PW, Shepherd CM, Simcox S (1989) The application of HARWELL-FLOW3D to combustion problems
246. Andersen J, Rasmussen CL, Giselsson T, Glarborg P (2009) Global combustion mechanisms for use in CFD modeling under oxy-fuel conditions. Energy Fuels 23:1379–1389. https://doi.org/10.1021/ef8003619

# Chapter 3
# Burner Designs for Clean Power Generation in Gas Turbines

## 3.1 Introduction

The increasingly stricter environmental regulations encouraged researchers to develop combustion systems that can meet such restrictions. Older gas turbine engines for power generation used non-premixed flame combustors thanks to their superior stability characteristics [1–3]. However, these combustors are no longer used because they generate unacceptably high concentrations of $NO_x$ pollutant emissions [4]. An approach for controlling $NO_x$ emissions is flue gas recirculation (FGR), where reactants are diluted with inert (cooled) exhaust gases. However, mixing the reactants with inert exhaust gases results in lower flame burning velocity and, consequently, lower combustion efficiency, and the combustor is prone to loss of stability [5]. Modern technologies and concepts have been introduced in the last five decades to the gas-turbine power-generation industry, such as lean premixed combustion (LPM) and catalytic combustion [6]. The latter was found to be expensive, unsafe, and less durable [7]. In LPM combustors, fuel and air are homogenously mixed in lean proportions upstream of the combustor to control the flame temperature and, as a result, thermal $NO_x$ emissions are significantly reduced. Nowadays, most manufacturers adopt the LPM technology in novel gas turbine combustors to replace the non-premixed technology for the control of combustion temperature and $NO_x$ emissions. However, switching from the non-premixed combustion mode to the lean premixed one causes undesired combustion characteristics, including increased levels of flame instabilities, auto-ignition, and flashback. Furthermore, operation under extra lean conditions at low loads adversely affects flame stability through excessive CO emissions and increased tendency to blowout [8]. These challenges necessitate the development of novel combustor designs that can operate stably under extra lean conditions while keeping emissions at low levels.



M. A. Nemitallah et al., *Approaches for Clean Combustion in Gas Turbines*,
Fluid Mechanics and Its Applications 122,
https://doi.org/10.1007/978-3-030-44077-0_3

In this chapter, novel burner designs adopting premixed combustion are introduced for the sake of improving the combustor stability and performance in gas turbine applications under different operating conditions. Such burner designs include Stagnation Point Reverse Flow (SPRF) with internal flue gas recirculation (IFGR) [9], dry low $NO_x$ (DLN) or dry low emission (DLE) combustors [10], EV-burners, and micromixers (MMs) [11]. In addition to these novel burner designs, perforated plate (PP) burner is considered as one of the oldest and important techniques used in many combustion applications. The technology of PP burner is discussed carefully in this chapter as a basis for the development of the novel MM combustion technology. The development of syngas-fuelled micro-mixing gas-turbine burner is also discussed in this chapter.

Despite reaching very low levels of $NO_x$ and CO emissions considering the abovementioned techniques, the complete control of all emissions has not been achieved yet considering the lean premixed air combustion technique. Additional technologies should also be applied, alongside with premixed combustion, for the capture of $CO_2$ emissions. $CO_2$ is one of the primary greenhouse gases and contributes by over 60% to the global warming problem, because anthropogenic activities release over 30 Gt $CO_2$ per year into the atmosphere. $CO_2$ capture technologies have thus been introduced to capture these vast amounts of $CO_2$ emissions [12]. There are three main categories for $CO_2$ capture technologies, namely pre-combustion, oxy-fuel combustion, and post-combustion. Among these technologies, oxy-fuel combustion is very promising for the capture of $CO_2$ in gas turbine applications. In this technology, fuel is burned with pure oxygen to result in $CO_2$ and $H_2O$ as main combustion products, which makes the process of $CO_2$ separation simple and economic after $H_2O$ condensation [13]. However, oxy-fuel combustion brings its own challenges in terms of combustion stability and combustor efficiency [14]. Different fuels have different combustion and emission characteristics, which reflects on combustor stability at the different loading conditions. This necessitates the development of fuel-flexible burners that can handle different fuels based on the loading conditions to sustain stable flame. The fuel flexibility approach, considering mainly hydrogen-enriched combustion of hydrocarbons, is addressed in detail in this chapter.

Oxy-fuel combustion is a very promising technology for the successful deployment of zero emissions power plants (ZEPPs). The pure oxygen required for the oxy-fuel combustion process is currently obtained via cryogenic distillation in air-separation units (ASU), which consumes copious amounts of energy and hinders the economic application of oxy-fuel combustion technology. The thermodynamic and economic penalties incurred in cryogenic air separation can easily offset any advantage gained by oxy-fuel combustion. Such shortcomings prompted many researchers to develop and explore alternative membrane-based air-separation systems for oxy-fuel combustion applications. The application of membranes for power production in high-temperature membrane reactors (HTMRs) is discussed in this chapter.

The use of fossil fuel is expected to increase significantly by mid-century due to the large rise in the world energy demand despite the effective integration of renewable energies in the energy production sector. This increase, alongside with the development of stricter emission regulations, forced the manufacturers of combustion systems, especially gas turbines, to develop novel combustion techniques for the control of $NO_x$ and $CO_2$ emissions, the latter being a greenhouse gas responsible for more than 60% to the global warming problem. The main objective of this chapter is to perform an extensive review of the state-of-the-art advances in combustion techniques and burner designs for clean combustion in gas turbines. The chapter is extended to link such environment-friendly burners with carbon-capture technologies, targeting full control of emissions out of gas turbines. In particular, oxy-fuel combustion, as a promising carbon-capture technology, is investigated for clean energy production while capturing the effluent $CO_2$. The performance of modern gas turbine combustors is evaluated under oxy-combustion conditions and compared with air-combustion in terms of flame stability and the associated emissions. This chapter addresses different burner designs and combustion techniques for clean power production in gas turbines. Combustion and emission characteristics, flame instabilities, and solution techniques are presented, such as lean premixed air-fuel (LPM) and premixed oxy-fuel combustion techniques, and the combustor performance is compared for both cases. The fuel flexibility approach is also discussed, as one of the combustion techniques for controlling emissions and reducing flame instabilities, focusing on the hydrogen-enrichment and the integrated fuel-flexible premixed oxy-combustion approaches. State of the art burner designs for gas turbine combustion applications are discussed in this chapter, including stagnation point reverse flow (SPRF) burner, dry low-$NO_x$ (DLN) and dry low-emission (DLE) burners, EnVironmental burners (including EV, AEV, and SEV burners), perforated plate (PP) burner, and micromixer (MM) burner. Special emphasis is made on the MM combustor technology, as one of the most recent advances in gas turbines for stable premixed flame operation with wide turndown and effective control of $NO_x$ emissions. Since the generation of pure oxygen is prerequisite to oxy-combustion, oxygen-separation membranes became of immense importance either for air separation for clean oxy-combustion applications or for conversion/splitting of the effluent $CO_2$ into useful chemical and energy products. The different carbon-capture technologies, along with the most recent carbon-utilization approaches towards $CO_2$ emissions control, are also discussed.

## 3.2 Lean Premixed Air Combustion

### 3.2.1 *Combustion and Emissions Characteristics*

$NO_x$ is a generic term for the nitrogen oxides, primarily NO and $NO_2$, which contribute to the formation of smog and acid rain, as well as depletion of the tropospheric ozone. CO, as a product in the cases of incomplete combustion, is a toxic, colorless, odorless, and tasteless gas. $NO_x$ is formed within the combustor at elevated firing temperatures. The dominant mechanism for $NO_x$ formation is the thermal (Zeldovich) mechanism [10]. Controlling the temperature within the combustor can result in significant control of $NO_x$ emissions. For example, when the temperature is raised from 1800 to 2100 K, the amount of $NO_x$ produced in milliseconds at 2100 K is of the same order as the amount produced in few seconds at 1800 K for the same oxygen fraction [15]. $NO_x$ formation is thus highly dependent on flame temperature. Moreover, the equivalence ratio plays a key role in forming both $NO_x$ and CO as illustrated in Fig. 3.1. $NO_x$ increases exponentially near stoichiometric firing, whereas CO emissions become excessive near the lean blowout limit. The typical window of operation for methane fuel, for example, is 0.45–0.70 in premixed combustors. Beyond this window, special flame stabilization means are necessary, such as pilot flames or sub-chambers [15, 16].

Developers are seeking technologies to expand the operation range with low $NO_x$ emissions. One solution is to manipulate the geometry of the combustor. Staged combustion, is another solution to reduce the $NO_x$ emissions while maintaining the design requirements [17]. Rich-Quick-Lean (RQL) or Over-Fire-Air (OFA) implement staging of combustion air, where the primary flame zone is maintained fuel-rich to enjoy excellent stability, then the reacting flow is diluted rapidly with the remaining combustion air to jump from rich to lean with minimum residence time near stoichiometry [18, 19]. This method is more reliable and economical way than changing

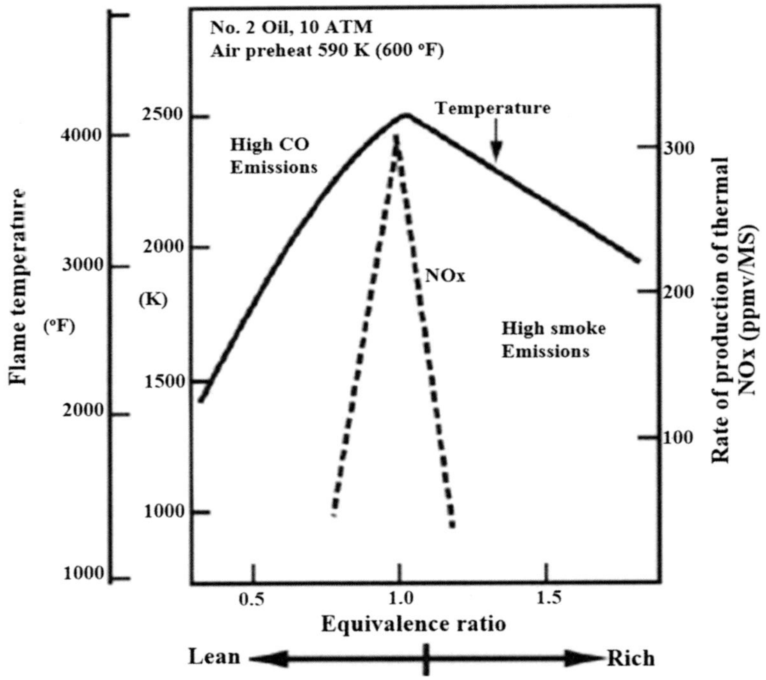


**Fig. 3.1** $NO_x$ and CO formation as function of equivalence ratio [15]

the geometry [20]. OFA has been extensively employed in air staging technique [19]. Zhou et al. [21] studied the $NO_x$ emissions and overall performance of a staged-air combustion system using a large-scale horizontal laboratory furnace, which was equipped with a swirl burner and fired with heavy oil. They investigated the effects of conditions of air staging (i.e., OFA ratio and position) on the emissions of $NO_x$ and CO both experimentally and numerically. It was reported from their experimental results that in heavy-oil combustion both the location and ratio of OFA affected the emissions of $NO_x$; the larger the distance between burner outlet and OFA location, the lower the $NO_x$ emissions with a negligible rise in the burnout loss. Furthermore, a 22.22% reduction in $NO_x$ emissions was reported when a ratio of 20% OFA was implemented in an industrial furnace. Liu et al. [22] numerically investigated the impact of OFA location and ratio on the combustion of coal and, particularly, on the $NO_x$ emissions. They reported that by increasing the OFA ratio the burner-zone temperature and fly-ash carbon content increased while $NO_x$ emissions decreased. Similarly, when the OFA location was increased, reduction in $NO_x$ emissions and an increase in fly-ash carbon content were also observed. Hodzic et al. [23] also reported that both the ratio and distance between the OFA nozzle and burner outlet considerably reduced the $NO_x$ emissions. Yang et al. [24] performed a numerical study on a fuel-oil industrial-scale combustor to establish the impact of staged combustion on $NO_x$ emissions. They reported that with greater staging there is a decrease in total $NO_x$ emissions, since higher OFA ratio causes lower flame temperature and oxygen concentration, which leads to a lower rate of thermal $NO_x$ formation. They also reported that deviation from stoichiometric primary flame zone decreases $NO_x$ emissions, and that the rate of reduction is proportional to the residence time. They also reported that air staging has no significant influence on the rate of burnout.

Staging can also be done on the fuel side by varying the equivalence ratio of burners [25]. This approach enhances the annular combustor efficiency [26]. Yang et al. [27] focused on fuel staging to reduce the $NO_x$ emissions from low ranked coal. The influence of several parameters (such as reburn fuel fraction, residence time, stoichiometric ratio etc.) on $NO_x$ emissions was investigated. They reported reduction in emissions $NO_x$ when reducing the stoichiometric ratio within the combustion zone down to 0.8. Zhao et al. [28] investigated the effect of equivalence ratio and other parameters on fuel staging in wood-powder cyclone gasification. They reported that there is no obvious effect of fuel staging on the reduction of the tar content in the produced gas. The range of optimum equivalence ratio was 0.26–0.29, and the ratio of staged fuel was not greater than 20%. Samarasinghe et al. [29] investigated the impact of fuel staging on flame instability and structure in a swirl-stabilized can combustor. They reported that fuel staging effectively curbs the instability when the overall equivalence ratio was either increased by staging or remained constant during staging. Lin et al. [30] studied a pre-mixer and pre-vaporizer multi-hole tube burner with a radially in-line dual-stages fuel arrangement. They reported that the pre-mixer and pre-vaporizer multi-hole had a significant impact on the reduction of $NO_x$ emissions.

### 3.2.2 $NO_x$ Emissions Under Lean Premixed Combustion

$NO_x$ usually refers to nitrogen dioxide ($NO_2$) and nitric oxide (NO) [31]. In majority of the high temperature application for heating/burning, the $NO_x$ mainly contains NO. Based on the mechanism for $NO_x$ formation, the $NO_x$ can be divided into three categories: (a) thermal $NO_x$, (b) prompt $NO_x$, and (c) fuel $NO_x$ [32, 33]. However, there are other mechanisms like, $N_2O$ pathway [34], $NO_2$ pathway and NNH pathway [35] for $NO_x$ formation in combustion [36, 37]. Figure 3.2 illustrates the overall mechanism of the formation of $NO_x$ and its process of reduction, where Vol-N represents the volatile nitrogen compounds ($NH_3$ or HCN).

The thermal $NO_x$ forms when the temperature inside the combustor becomes higher than the stoichiometric conditions. $NO_x$ formation of this type increases rapidly with increasing temperature. The formation of thermal $NO_x$ is a result of Zeldovich mechanism [38]:

$$N_2 + O_2 \rightarrow NO, NO_2 \tag{3.1}$$

As the name prompt implies, the prompt $NO_x$ formation occurs by fast reaction of $O_2$, $N_2$ with the hydrocarbon (HC) radicals:

$$CH_4 + O_2 + N_2 \rightarrow NO, NO_2, CO_2, H_2O \text{ trace species.} \tag{3.2}$$

Also, the fuel $NO_x$ can form due to the direct oxidation of organo-nitrogen compounds present in the fuel:

$$RxN + O_2 \rightarrow NO, NO_2, CO_2, H_2O \text{ trace species.} \tag{3.3}$$

Under premixed combustion conditions, the combustion temperature can be reduced to levels at which the emissions of thermal $NO_x$ can be controlled [13]. However, $NO_x$ can be formed under such combustion conditions through prompt and fuel mechanisms. In this section, focus is made on both the prompt $NO_x$ and fuel $NO_x$ [39, 40].

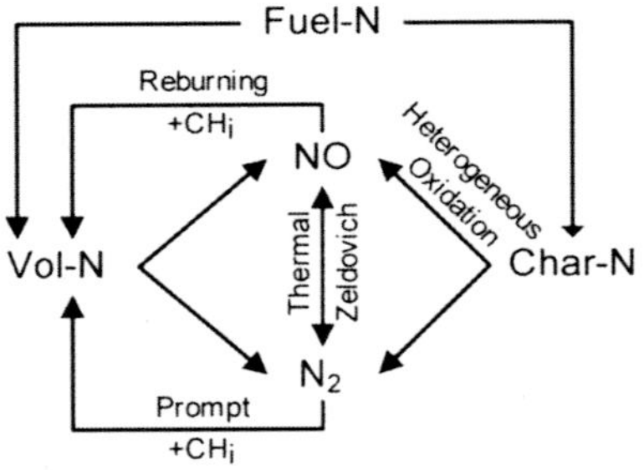


**Fig. 3.2** Mechanisms for $NO_x$ formation and its reduction from combustion process [41]

#### 3.2.2.1 Prompt $NO_x$

Prompt $NO_x$ forms as a result of free radicals from high-density fuels (e.g. natural gas) reacting with oxygen and nitrogen in a region of fuel-rich flame inside the combustion zone [42, 43]. This mechanism was first identified by Fenimore [44], as per the mechanism reactions provided in Table 3.1. Furthermore, besides the fuels with higher energy density, there are other factors that can lead to prompt $NO_x$ formation like, shorter residence time, combustion in low temperature, fuel-rich environment generated by gas turbines, surface burners, and some multistage combustion systems [45]. It was revealed in a study using gaseous fuel that some of the prompt $NO_x$ reacts with HC radicals and, as a result, it gets converted into $N_2$ when the excess air factor is less than 0.65 [46]. In fact, the higher number of carbon atoms promote the formation of prompt $NO_x$. Several complex chain reactions occur during the generation of this type of $NO_x$ producing various intermediary species in-between. The amount of HCN formed is a function of HC radicals. CH, $CH_2$, C, $C_2H$ was found as the most common fragments HC fuels, with CH having the highest influence and $C_2H$ having the least, which participates in producing prompt $NO_x$ [47]. Schefer et al. [48, 49] reported that most of the $NO_x$ formed at the bottom part of the flame is prompt $NO_x$. They claimed that CH is the main reason behind the formation of prompt $NO_x$ in this region.

Fackler et al. [51] investigated experimentally and numerically the formation of $NO_x$ in single-jet stirred reactor of high intensity operating on LPM combustion technique. The numerical part of their study was performed using chemical reactor network (CRN) of 3.0 elements and detailed chemistry. They reported for their setup that most of the formed $NO_x$ is prompt $NO_x$, see Fig. 3.3. The higher prompt $NO_x$ formation was attributed to the operation under higher equivalence ratios than the usual range for $CH_4$ (0.45–0.65). Most of the $NO_x$ was formed in the turbulent flame

**Table 3.1** Reactions involving Fenimore prompt mechanism [50]

| S/L No. | Reactants | | Direction | Products | | |
|---|---|---|---|---|---|---|
| 1 | CH | $N_2$ | → | HCN | N | |
| 2 | HCN | O | → | NCO | H | |
| 3 | NCO | O | → | NO | CO | |
| 4 | NCO | OH | → | NO | H | CO |
| 5 | NCO | $O_2$ | → | NO | $CO_2$ | |
| 6 | HCN | O | → | NH | CO | |
| 7 | NCO | H | → | NH | CO | |
| 8 | NH | OH | → | N | $H_2O$ | |
| 9 | NH | O | → | NO | H | |
| 10 | NH | $O_2$ | → | NO | OH | |
| 11 | N | $O_2$ | → | NO | O | |
| 12 | N | OH | → | NO | H | |

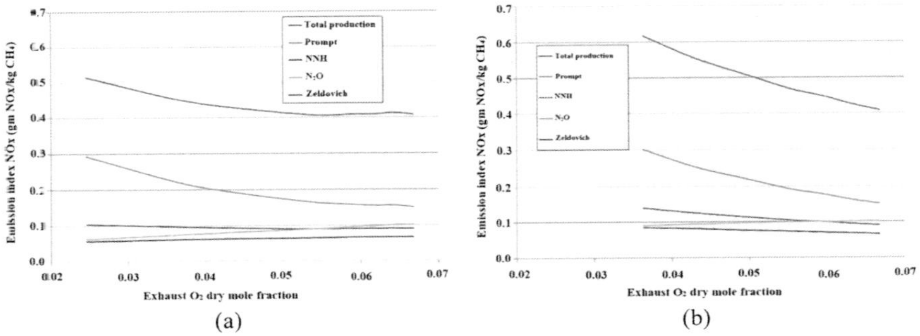


**Fig. 3.3** Emission index for different $NO_x$ formation mechanism from CRN model with $CH_4$ diluted with [51]: **a** $N_2$ and **b** $CO_2$

zone where the flame brush in the numerical model was considered as a perfectly stirred reactor operating close to its blowout limit. The prompt $NO_x$ contributed in the recirculating zone and increased with the increase of dilution level. The concentration of $N_2$ used for dilution also played an important role in $NO_x$ formation. In a study of flame stability improvement in LPM combustion, Cho and Chung [52] observed $NO_x$ formation in different level of hydrogen addition. They concluded that adding more hydrogen can lead to higher $NO_x$ formation for the same equivalence ratio (Ø) and stretch rate (κ). However, this $NO_x$ can be reduced with increasing hydrogen while operating close to extinction condition in LPM combustion. Hydrogen addition increased the OH and H radicals consequently increasing prompt $NO_x$, whose formation was observed more than the thermal $NO_x$, see Fig. 3.4a. Prompt and thermal $NO_x$ formation reduced at higher hydrogen mole fraction close to extinction limit as shown in Fig. 3.4b. More than 50% reduction of NO emission was achieved

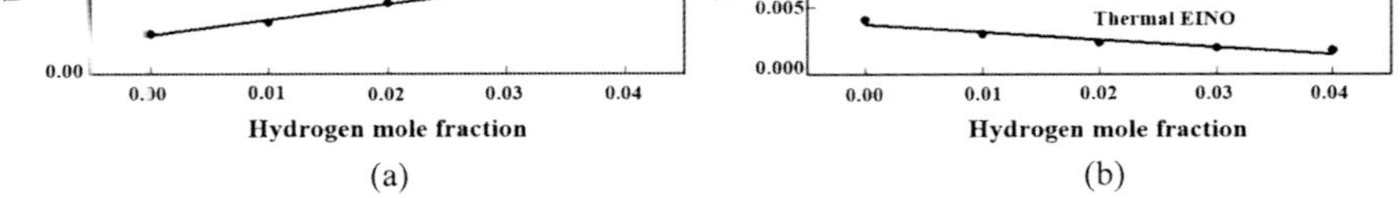


**Fig. 3.4** Prompt and thermal $NO_x$ emission with hydrogen addition expressed in terms of the emission index of NO (EINO) for $CH_4$–$H_2$/air premixed combustion under 298.15 K and 1.0 atm in [52]: **a** Ø = 0.6 at stretch rate near extinction limit and **b** κ = 200 $s^{-1}$ at Ø near extinction limit

in identical flame temperature at the same heat capacity condition even with the varying hydrogen addition.

#### 3.2.2.2 Fuel $NO_x$

Formation of fuel $NO_x$ is a complicated phenomenon that occurs as a result of changing the structure of chemical bonding of nitrogen and HC molecules. Nitrogen species in fuel reacts with air particles in combustion, consequently producing fuel $NO_x$ or fuel-bound $NO_x$. Literature indicates that this type of $NO_x$ emission is not very significant using natural gas. However, fuel $NO_x$ could be an important source of $NO_x$ emission in case of syngas and distillates of lower quality [39, 53, 54]. In syngas combustion, ammonia is the nitrogen species that reacts with air and converts into $NO_x$. Ideally, this ammonia should be separated from the syngas after being formed during coal gasification process. In reality, few of this ammonia remains with the syngas before entering the combustion chamber and generates fuel $NO_x$. Rich-lean strategy can be incorporated to reduce the fuel $NO_x$ formation [55]. Glarborg et al. [56] revealed that this type of $NO_x$ is formed in reaction of HCN and $NH_3$ produced from the conversion of fuel-bounded nitrogen with the other existing combustion radicals. The bonding of nitrogen in fuel is weaker than the diatomic nitrogen present in air. This weaker nitrogen-carbon bonds can be broken easily during the combustions consequently producing higher fuel $NO_x$ than the thermal $NO_x$ [57]. Zhang et al. [58] studied the influence of various functional forms of nitrogen on the emission of fuel-$NO_x$ at 900 °C and concluded that the emission varies in the order of Glycine, pyrrole, pyridine, and methylenedi-p-phenylene diisocyanate (MDI).

Nakata [59] investigated the fuel $NO_x$ formation and its reduction in an integrated coal gasification combined cycle (IGCC) system. They reported that fuel-$NO_x$ formation characteristics can be heavily influenced by the air mixing condition in the secondary zone and with the gradual mixing of air at the secondary inlet. The conversion ratio of $NO_x$ from $NH_3$ reduced with the increase of combustor pressure. They observed that the fuel-NO formation rises with increasing the reaction temperature. The rapid formation of fuel-NO occurred when the reaction temperature reached 2000 K and above as shown in Fig. 3.5.

### *3.2.3 Combustion Instabilities and Solution Techniques*

The conversion of the flame from the non-premixed type to the premixed one results in increased levels of flame instabilities, which brings its own challenges to the gas turbine manufacturers. In premixed combustion systems, premixing of fuel and oxidizer as well as flame stabilization is usually achieved through swirl. Creation of a swirl-induced recirculation zone downstream of the burner is an efficient way of stabilizing the flame when a physical flame holder is not available [60]. However,

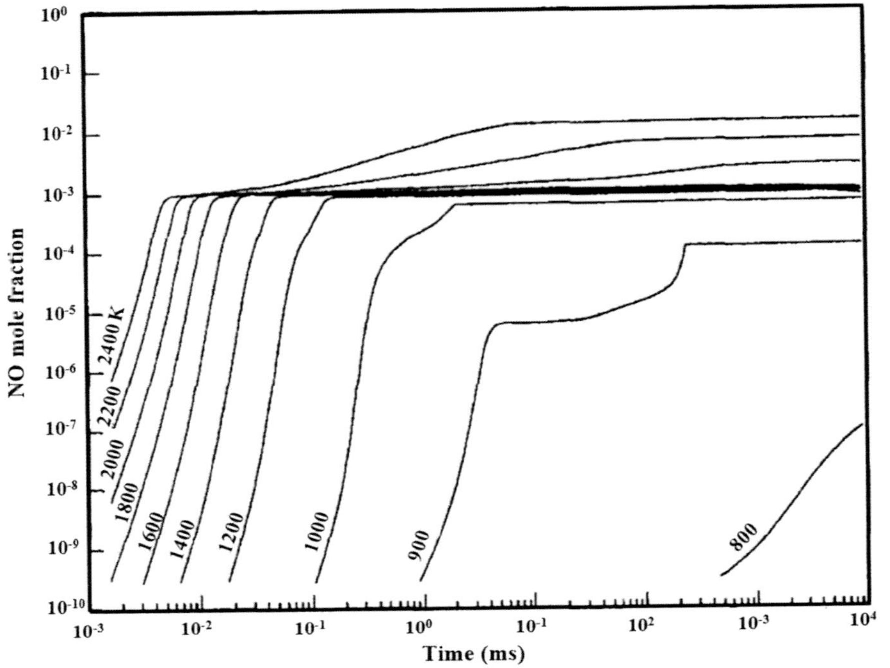


**Fig. 3.5** NO formation with different reaction temperature in an integrated coal gasification combined cycle (IGCC) system under stoichiometric conditions [59]

premixing of the reactants increases the homogeneity of the temperature and species-concentration fields, which makes the combustor more responsive to swirl-induced oscillations in equivalence ratio and/or pressure [61–63]. Permanent damages may occur when the frequencies of oscillations interfere constructively with the cold-flow frequencies and/or natural frequencies of the combustion-system geometry, a situation that is known as resonance and characterized by high amplitude pressure pulsations [61].

Combustion instabilities can be categorized into static and dynamic instabilities. The mechanisms for static instabilities are flame blowout and flashback. Such mechanisms are mainly function of the operating conditions. The flame blows out when the flow speed exceeds the flame speed, whereas flashback occurs when the flame speed exceeds the flow speed. On the other hand, dynamic instabilities, or thermos-acoustics, are generated due to coupling between the heat release and the acoustic fluctuations of pressure resulting in unwanted pressure waves [64, 65]. At a certain level, such pressure waves could be destructive and may result in damage of the combustor hardware [66, 67]. Premixed combustion is considered stable when the amplitude of pressure fluctuation is less than 5% of the mean chamber pressure. Beyond 5% is unstable or oscillatory combustion [66–69].

Combustion instability can be developed because of two reasons. The first reason is that instability can be initiated by any artificial perturbation outside the combustion chamber. The second reason is that instability can be generated within the system. For the first reason, the instabilities are initiated by an external perturbation which is stable to small disturbances [70, 71]. For the second reason, the instabilities arise from the oscillations and it is mainly caused by a small fluctuation that is grows out of due to the system noise [6, 72, 73]. The pressure fluctuations lead to instability depending on the excitation mechanism and the driving forces of these fluctuations and their corresponding damping process [6]. Combustion instabilities in gas turbines can be classified into the following categories: low, intermediate, and high frequency instabilities [74–76]. Many researchers have developed methods and techniques for controlling these instabilities [25, 77, 78]. Rashwan et al. [77] defined two methods for damping combustion oscillations in gas turbines, namely active and passive control. In the passive control technique, some changes are made to limit the rate of heat release or to increase the rate of energy loss through changing the fuel/oxidizer compositions, fuel-injection device, design of combustion chamber, or through installing acoustic dampers [78]. In the active control technique, on the other hand, an energy source is introduced into the combustion system to interrupt the coupling between heat release and the acoustic waves [25].

Recently, several techniques are being proposed to avert the circumstances leading to oscillations in combustion systems while maintaining wide ranges of operability. These techniques include strategies to control the mixing profile. Two active control strategies stand out, namely pulsing injection of pilot fuel at the frequency of instability [79] and fuel staging [80]. The latter technique allows for decreasing $NO_x$ emissions and suppressing instabilities simultaneously without the sophistication of high-frequency actuation of the pulsing-injection technique. Lacarelle et al. [81] studied the effect of different fuel injection strategies on the residence time delays and quality of mixing in a premixed system. They reported that for a given distribution of fuel, an increasing quality of mixing is achieved when the fuel injection line is formed of two rotating jets to cross-flow angles (25° and 55°) as compared to a 45° single jet angle. Lacarelle and Paschereit [82] designed a fluidics oscillator to avert the drawbacks of mechanically triggered flows. The mixing efficiency of four of such injectors, applied in a generic jet-in-crossflow arrangement, was compared with that of a typical non-oscillating jet. It was reported that the fluidics injectors performed better than the normal jets and that, for a given quality of mixing, about 50% decrease in mixing length could be achieved by using fluidics injectors. Lee et al. [83] investigated the concepts of both axial and radial flow micromixing using experimental and numerical (i.e. CFD) analyses. They reported that axial micromixing increases the tendency for flashback, while radial micromixing offers lower emissions performance. The utilization of MM method, therefore, seems promising in achieving low emissions and eliminating flashback. Estefanos [84] investigated the influences of air-fuel mixing on $NO_x$ emissions and combustion instabilities in a lean premixed system. He reported that injecting fuel axially into a region of strong unsteady mixing tends to eliminate the combustion dynamics by damping the unsteady mixing.

## 3.3 Oxy-combustion for Carbon Capture

### *3.3.1 Oxy-fuel Combustion Technology*

Burning fuel in pure oxygen instead of air, i.e., oxy-fuel combustion, serves primarily to facilitate simple and economic capture of the effluent $CO_2$ emissions. Since $CO_2$ and $H_2O$ are the main combustion products, the former can be separated and captured in high-purity form after condensing the latter. The process of oxy-fuel combustion, however, is characterized by high flame temperatures, which necessitates recycling part of the flue gases (mainly $CO_2$) back into the combustor to dilute the reactants and control the flame temperature [85]. There has been considerable progress in the development of oxy-combustion technology after the International Panel for Climate Change (IPCC, 2005) published its special report on $CO_2$ capture and storage [86, 87]. However, for stable operation under oxy-combustion conditions, the oxygen fraction in the oxidizer mixture ($O_2$ plus $CO_2$) should not fall below a certain level [88–92]. $CO_2$ addition affects the combustion characteristics, in terms of adiabatic flame temperature, radiative heat transfer, transport properties, chemical kinetics, and flame macrostructure. Therefore, oxy-combustion in highly $CO_2$ concentrated environments has its own challenges and opportunities [93–97].

### *3.3.2 Comparison of Air-Fuel Versus Oxy-fuel Combustion Concepts*

The concepts of air-fuel and oxy-fuel combustion differ significantly in terms of combustor operability as well as flame characteristics. The main control parameter in the operation of air-fuel combustors in gas turbines is the equivalence ratio, which is used to control the flame temperature and prevent/mitigate combustion instabilities. The overall equivalence ratio is always kept lean, especially in premixed combustors, to reduce $NO_x$ emissions and tune the combustor-exit temperature to the cycle requirements and the metallurgical limits of the turbine first stage. The fact that air is an abundant low-cost resource allows for lean operation. The composition of air as an $O_2/N_2$ oxidizer remains constant throughout all modes of gas-turbine operation. Flue-gas recirculation (FGR) is optional and serves primarily to reduce $NO_x$ emissions and mitigate combustion instabilities. Oxy-fuel combustors are, however, quite different. The generation of pure oxygen for combustion is an energy-consuming process, which makes this oxygen a limited precious resource (as compared to air). Oxy-fuel combustors are, thus, typically operated near stoichiometry to conserve the oxygen. FGR becomes a must in this case to replace the equivalence ratio as the main control parameter. Part of the effluent $CO_2$ is recirculated in precise proportions to dilute the primary reactants for controlling the flame temperature [98] and tuning the combustor-exit temperature to the cycle requirements and the metallurgical limits of the turbine first stage [99]. The composition of the $O_2/CO_2$ oxidizer is, thus, variable based on the FGR ratio. Air and oxy-combustors are thus significantly different in design and operability. Nonetheless, current gas-turbine users will not be easily motivated to switch to oxy-combustion, unless their current lean premixed (LPM) air-based gas turbines are retrofitted to implement oxy-combustion with minimal hardware changes. Converting a traditional air-combustion system to utilize oxy-combustion thus requires careful investigation of combustor operability and flexibility.

From the perspective of combustion physics and reaction kinetics, air-fuel and oxy-fuel combustion are again quite different because of replacing the air-based $N_2$ with $CO_2$. The difference in physical properties of $N_2$ and $CO_2$ significantly affects the physical properties of the reacting mixture [100–102], i.e., density, volumetric heat capacity, and transport properties [103]. The transport properties comprise of mass diffusivity, dynamic viscosity, and thermal conductivity. $CO_2$ has higher density than $N_2$, which affects flame shape, bulk velocity, and the pressure drop across combustor headend, as well as the total mass flowing through the combustor and turbine. The $CO_2/O_2$ oxidizer also has higher volumetric heat capacity compared to air, which in turn reduces the flame temperature, thereby reducing flame speed and stability. Since the rate of chemical kinetics is directly proportional to the laminar flame speed and combustion efficiency, the presence of $CO_2$ in increased concentrations in oxy-fuel combustion, results in reduction in both laminar flame speed and combustion efficiency, because retarded chemical kinetic rates have been observed in oxy-fuel flames [104, 105]. An additional challenge in oxy-fuel flames is $CO_2$ dissociation, which is an endothermic reaction that consumes energy and affects the reaction kinetics [106–109], especially at high $CO_2$ concentrations. It can thus be deduced that if an air-fuel combustor is converted to oxy-fuel operation at the same equivalence ratio and without any geometry changes, a 21% oxygen fraction (by vol.) in the $O_2/CO_2$ oxidizer will not be enough to sustain stable flame [94, 110], i.e., the minimum oxygen fraction is greater than 21%. Ditaranto et al. [111, 112] have shown that the oxy-combustion process requires at least 30% oxygen fraction to achieve stable combustion as compared to air-based combustion. The higher oxygen concentration is required to compensate for the higher volumetric heat capacity of $CO_2$ compared to $N_2$. Such challenges necessitate the modification of burner designs to be able to hold stable flames under oxy-fuel combustion conditions.

### 3.3.3 Premixed Oxy-fuel Combustion

Speth et al. [113] investigated the dynamic steadiness of premixed stoichiometric $CH_4/CO_2/O_2$ flames. The adiabatic flame temperature ($T_{ad}$) of the reactant mixture was adjusted through the oxygen fraction. Although air and oxy-flames have different flame speeds at the same $T_{ad}$, similarity was found between the two flames in the fact that $T_{ad}$ primarily rules the transformation from one stability mode to another. The above finding was supported by high speed images that showed similar vortex collapse modes. Since, $T_{ad}$ changes the turbulent flame geometries so these modes, which are symptomatic of related turbulent flame geometries, are thereby representing a change in $T_{ad}$. Kutne et al. [114] also studied and compared oxy and air-methane flames but under partially premixed conditions. They considered equivalence ratios of 0.5–1.0 and oxygen fractions of 20–40%. OH*-chemiluminescence imaging and Raman spectroscopy were used to quantify the flame structure. It was reported that the oxygen fraction has a major effect on flame characteristics, whereas the equivalence ratio has a little effect. It was also observed in air-flames that the inner recirculation zone is located further upstream, as compared to oxy-flames, indicating inferior stability of the latter. The authors also confirmed that the 21% oxygen fraction of air is unattainable in oxy-flames, which have a higher minimum threshold. The inferior combustion temperatures and flame speeds of oxy-flames were attributed to the higher volumetric heat capacity of $CO_2$ compared to $N_2$. Numerous other researches also reinforced these observations [95, 107, 115–118]. Temperature profiles were found to be matched in $CH_4$/air and $CH_4/36\%O_2/64\%CO_2$ flames, according to Song et al. [95]. Comparable values were reported in the literature [88, 119–121]. Rashwan et al. [122] reported similar findings; an oxygen fraction of 29–40% is needed for stable oxy-combustion with FGR. Moreover, oxy-combustion limits the combustor stability window, because of the negative impact of adding $CO_2$ on the combustion kinetics. Abdelhafez et al. [123] showed that no stable oxy-flames could be obtained with oxygen fractions below 22%, even at stoichiometric conditions, in a swirl-stabilized premixed gas-turbine model combustor. They added that the effect of oxygen fraction is significantly greater than that of equivalence ratio, i.e., stable flames can be obtained at much leaner conditions with higher oxygen fractions. Jerzak and Kuznia [124] compared the blowout mechanisms of premixed $CH_4/O_2/N_2$ and $CH_4/O_2/CO_2$ flames in a swirl combustor. They reported that oxy-flames with FGR have significantly narrower operability windows compared to air-flames. Hu et al. [125] examined the laminar flame speeds of $CH_4/O_2/N_2$ and $CH_4/O_2/CO_2$ flames at atmospheric conditions, targeting the effects of oxygen fraction, equivalence ratio, and dilution ($CO_2$ vs. $N_2$). The laminar flame speed was reported as a quadratic function of oxygen fraction. Moreover, the measured flame speed of an $N_2$-diluted flame was about five times that of a $CO_2$-diluted flame, keeping other parameters constant. Oh et al. [115] supplemented carbon monoxide (CO) as a subordinate fuel along with $CH_4$ in a non-premixed oxy-fuel gas turbine combustor; as the CO mole fraction was increased, the experiments indicated longer flame lengths and inferior stability.

Saanum and Ditaranto [126] studied a new high-pressure oxy-fuel burner designed at SINTEF (The Foundation for Scientific and Industrial Research in Norway) with a 100-kW input load and pressure of 10 bar. The swirl-stabilized oxy-burner uses a mixture of $O_2$ and $CO_2$ as oxidizer and generates a flame in a cylindrical double-wall combustor. It was observed that as the pressure increases, the CO level decreases radically. The flame also becomes more compact at higher pressures. It was found that stable flames are obtained at oxygen concentrations above 36.5% at 20 kW and above 40% at 40 kW. An experimental study was performed by Rashwan et al. [118] on partially premixed methane-oxygen and methane-air flames over a perforated-plate burner. The results showed that as the premixing ratio increases, the emissions of CO and $NO_x$ decrease. The flame color changed to pure blue at higher premixing ratios from reddish color at lower ratios; high levels of soot formation were observed in non-premixed flames. The results also indicated that as the oxygen fraction increases in the oxidizer mixture, the stability of the oxy-fuel flames increases due to the associated increase in flame speed. Ramadan et al. [107] experimentally investigated different non-premixed flames involving air-fuel, $O_2$-enriched-air fuel and oxy-fuel flames stabilized over a bluff-body of various blockage ratios. The results showed that flame stability increases with increasing fuel momentum for the same blockage ratio, caused by the corresponding increase in Reynolds number, turbulence, and mixing.

## 3.4 Fuel-Flexible Combustion Approach

The non-renewable nature of fossil fuels and the stringent emissions regulations have pushed combustion researchers to look for efficient utilization of all available fossil fuels and develop alterations in the existing gas turbine technologies for utilizing non-conventional fuels. Original gas turbine manufacturers (OEMs) have been facing this challenge of fuel flexibility because most of the traditional gas turbines have been designed and manufactured for standard natural gas only. Non-conventional fuels pose quite a different set of chemical and transport properties, as compared to standard fuels, which causes operability issues in traditional gas turbines. This created a wider need for research in fuel-flexible combustors.

### *3.4.1 Fuel Flexibility*

The Wobbe Index (WI) is a common indicator for interchangeability and fuel characteristics in the gas turbine industry. It applies to gaseous fuels and is defined as the ratio of lower heating value (LHV) per unit mass of fuel to the square root of fuel specific gravity, where the specific gravity is the ratio of fuel density to air density; see Eq. (3.1). The Modified Wobbe Index (MWI) was introduced to account also for the fuel temperature. Figure 3.6 shows the MWI map of all known gaseous fuels. Standard natural gas (mostly $CH_4$) lies at the center of the map. The larger the content of higher hydrocarbons ($C_{2+}$), the higher the MWI is at constant LHV, e.g., liquefied petroleum gas (LPG). Hydrogen enrichment increases LHV at constant MWI. Increasing the inert content ($N_2$, $CO_2$, $H_2O$) decreases both LHV and MWI, e.g., syngas and coke-oven gas (COG). Table 3.2 complements Fig. 3.6 by categorizing the different fuels.

$$\mathrm{WI} = \frac{LHV_{fuel}}{\sqrt{\frac{\rho_{fuel}}{\rho_{air}}}} \quad \text{and} \quad \mathrm{MWI} = \frac{LHV_{fuel}}{\sqrt{T_{fuel} \times \frac{\rho_{fuel}}{\rho_{air}}}} \tag{3.4}$$

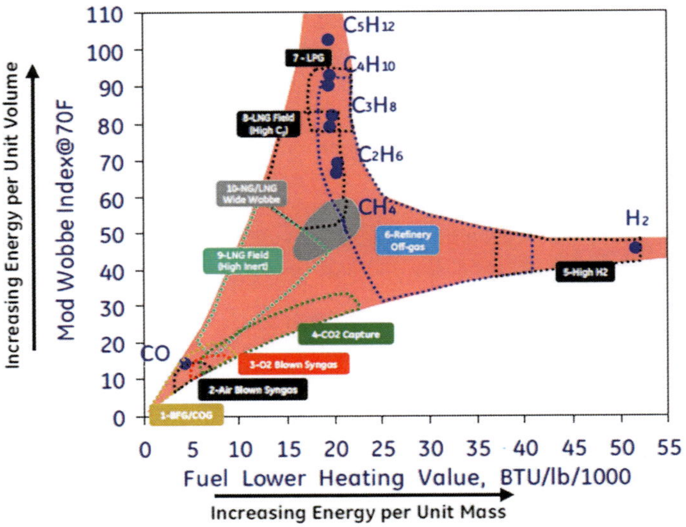


**Fig. 3.6** MWI map of all known gaseous fuels (from www.ge.com)

**Table 3.2** Wobbe Index ranges of different fuels [127]

| Fuel category | Wobbe Index range (MJ/Nm$^3$) |
|---|---|
| Syngas type | 24–29 |
| Natural gas type | 48–53 |
| LPG type | 72–87 |
| Methane | 47–53 |
| Hydrogen | 40–48 |
| Carbon monoxide | 13 |

Justification and validation of using the Wobbe index has been widely discussed in the literature [128]. Different gaseous fuels exist for potential use in gas turbines. Even for natural gas itself, different compositions are found from different geographical locations. Some applications dictate running the gas turbine on high-$H_2$ fuels or low-quality ones (such as COG). The MWI was thus developed to characterize these different fuels and baseline them to standard natural gas. The power input to a combustor is given as the product of LHV per unit volume of fuel and the volumetric fuel flow rate. This implies higher volumetric fuel flow rate for lesser LHV per unit volume, for providing the same input power to the combustor. The fuel density is the link between LHVs per unit volume and per unit mass. The MWI thus accounts for the commonly used LHV per unit mass and the fuel density.

An empirical relation was developed to link the input power to combustor pressure drop and fuel WI, as follows

$$P_{input} \propto \sqrt{\Delta P} \cdot WI \tag{3.5}$$

The combustor pressure drop represents the air flow rate through the engine and, thus, its load. For a characteristic equivalence ratio (flame temperature), the fuel flow rate can also be linked to the combustor pressure drop. Hence, the input power is proportional to the product of pressure drop and WI, as the latter involves heating value.

Gas-turbine manufacturers have established standards for combustor design on natural gas and use the WI or MWI for gaseous fuel interchangeability and characteristics with fixed pressure drops and geometry of the combustor [129, 130]. Combustors are typically designed to operate within narrow ranges of Wobbe Index.

## 3.4.2 *Fuel-Flexible Combustion Approaches*

### 3.4.2.1 Hydrogen Enrichment

Hydrogen ($H_2$) is one of the simplest chemical elements available and is a promising fuel for carbon-emissions-free operation. As $H_2$ has a higher heating value per unit mass compared to other hydrocarbon fuels, it has been used as rocket propellant since 1950s for space-exploration uses. It is an abundant, clean energy carrier, confined in water, different hydrocarbons, and various organic compounds. Its efficient extraction in bulk quantities from these resources is a challenging task. Meanwhile, natural-gas reforming with steam is currently being used to produce hydrogen. It is also the by-product of many industrial and chemical processes. In pre-combustion carbon-capture techniques, it is viable to use this by-product $H_2$ as fuel, either along with natural gas or, in some potential cases, alone. Since traditional gas turbines have been historically designed for standard natural gas, the variations in fuel composition and/or switching to a different fuel are challenges being faced by the gas turbine designers [131, 132]. Even hydrogen enrichment poses a challenge, especially in premixed combustors, because the chemical and transport properties of hydrogen are quite different from standard fuels. Moreover, the storage and cost of hydrogen are also major concerns. The use of hydrogen as pure fuel in gas turbines is rather limited. There are quite a few of such power plants that use pure hydrogen as fuel. A 12-MW power plant (non-premixed combustion system), opened in Fusina, Italy, in 2010, is the first of its kind to operate completely on pure hydrogen [133].

Hydrogen is typically used to enrich hydrocarbon fuels, especially natural gas ($CH_4$), in gas turbines. Since, the adiabatic flame temperature, heating value, and laminar flame speed, are quite larger for hydrogen as compared to methane, hydrogen enrichment caters for better combustion quality by increasing both the combustion efficiency and fuel economy [134]. Patil et al. [135] reported that the power of a LPM engine fueled by compressed natural gas increased by 11% and the fuel consumption decreased by 8% with hydrogen enrichment, as compared to operation of the same engine with compressed natural gas only [135]. Biogas combustion is also enhanced with hydrogen enrichment, especially the lean blowout limit [136].

Switching from standard natural gas to $C_{2+}$ fuels (higher WI) dictates the use of smaller fuel-injection orifices, whereas switching to low-quality fuels (lower WI) has to be accompanied by enlargement in fuel orifices. Hydrogen enrichment, on the other hand, induces minimal changes in WI, see Fig. 3.6, so a gas turbine can be hydrogen-enriched to a certain extent without geometry changes. If WI is within the range of 30–50 MJ/Nm$^3$, it has been reported that hydrogen can be used with methane in combustors without any considerable geometry modification [137]. The heating value of hydrogen is far greater than that of methane, which implies that a little amount of hydrogen can replace a bigger amount of methane, thereby posing no significant geometry-change requirements in a combustor designed for methane. It should be noted, however, that despite its common use in the gas-turbine industry, the WI does not account for changes in flame speed and ignition and chemical delay times [132]; therefore, high-$H_2$ require special burner designs to create a safe margin to flashback.

One of the primary reasons for using hydrogen with methane is its capability to reduce $CO_2$ emissions; pure hydrogen results in no $CO_2$ emissions, just $H_2O$. The concentration of $H_2O$ in the flue gases increases with hydrogen enrichment. However, since the adiabatic flame temperature of hydrogen combustion is considerably higher than that of methane, the generation of thermal $NO_x$ increases [134]. It has been reported that with 28–36% hydrogen enrichment, very low $NO_x$ emissions can be achieved [138] at the expense of moderate increase in the concentration of unburned hydrocarbons. To combine the benefits of increased power output, fuel economy, and reduced $NO_x$ emissions, an optimal hydrogen fraction of 20% has been reported for compressed natural gas engines [139–143].

#### 3.4.2.2 Syngas Combustion

Syngas is a fuel gas mixture consisting primarily of $H_2$, CO, and some $CO_2$. It is produced from the gasification of virtually any hydrocarbon fuel, feedstock, or waste. Some industries also generate $H_2$-rich gas mixtures as byproducts, e.g., coke-oven gas, which can fuel gas turbines for power generation. Syngas (including $H_2$) can also be intentionally produced by reforming fossil fuels. Integrated Gasification Combined Cycle (IGCC) plants have been developed over a number of decades to utilize this principle, and they are gaining significant interest in recent years for pre-combustion capture of $CO_2$ to meet the stringent environmental regulations governing greenhouse-gas emissions. Implementing the water-gas shift reaction allows for converting most CO to $CO_2$; the latter is captured again prior to combustion, so the gas turbine combustor runs on very $H_2$-rich syngas fuel.

The concept of pre-combustion carbon capture has triggered fuel reforming to generate syngas, which motivated research to evaluate syngas combustion under LPM conditions for further control of emissions without affecting the efficiency.

However, the conversion of current natural-gas fueled combustors to work with high-$H_2$ syngas is not feasible yet. This may be attributed to the higher tendency of syngas for flashback and to the higher syngas flow rates required to maintain the same power obtained from natural gas [144], since syngas has lower WI. Without any enlargement of fuel orifices, this increased flow rate will induce higher velocities, which might also cause early blowout of the flame as well. In addition, it might result in instable operation of the compressor and ineffective cooling of the turbine blades [144, 145]. This necessitates more research to evaluate properly the combustion and emissions characteristics of syngas for safe application in gas turbines.

As per the open literature, a great number of studies have been performed on syngas combustion considering mixtures of $H_2$ and CO with different concentrations. On the other hand, there is a limited number of studies on the combustion and emission characteristics of syngas fuel mixtures containing other gases such as $H_2O$ and $CO_2$ as well. There is also lack of experimental data on the laminar burning velocity of syngas diluted with $NH_3$ or mixed with $CH_4$. Actually, the composition of syngas mixture depends on the adopted gasification method and the utilized feedstock. Typical compositions of syngas mixtures, considering different gasification approaches, ranges from 0–9.3% $CH_4$, 0–20.4% $H_2O$, 1.3–29.6% $CO_2$, 8.1–60.5% CO, and 6.8–50.4% $H_2$, in addition to other impurities [146]. Both CO and $H_2$ gases have different burning characteristics than conventional hydrocarbons. A number of recent investigations on syngas combustion of different hydrogen concentrations reported adverse effects on flame structure and stability as well as flashback and blowout limits [147–150]. Fu et al. [150] reported significant changes in flame structure using syngas fuels of different hydrogen concentrations, even for the same operating pressure and equivalence ratio. The presence of $CO_2$ and $H_2O$ in syngas also affects its combustion characteristics significantly. As both are highly radiative species, high concentrations of $CO_2$ and/or $H_2O$ in syngas may have drastic effects on the temperature and radiation characteristics of the flame, in addition to undesired flame dynamics [151]. Lee et al. [152] showed that $CO_2$ has insignificant effect on the ignition delay time in syngas combustion, based on experiments on syngas mixtures of different hydrogen contents with $CO_2$ concentrations above 14% under higher operating pressure.

#### 3.4.2.3 Ammonia Combustion

The use of ammonia ($NH_3$) either for carbon-free combustion or for the reduction of $NO_x$ emissions in conventional combustion systems has recently received lot of interest despite the safety concerns associated with $NH_3$ combustion. In this section, a condensed review is presented of the performance and combustion characteristics of $NH_3$ as a fuel for gas turbine applications. $NH_3$ enrichment of hydrocarbon fuels can result in better control of combustion temperature and emissions. Although the idea is not new, there is a limited number of studies on the performance of $NH_3$ as a fuel for power production in gas turbines. Actually, the idea was initiated during the 1960s [153–155]. Such studies reported higher ignition energy for $NH_3$ due to its low reactivity as compared to hydrocarbon fuels. $NH_3$-air combustion under stoichiometric conditions has longer quenching distance, 3.5 times that of propane fuel. Furthermore, the combustion of $NH_3$ showed limited operability range based on equivalence ratio as compared to propane fuel [154]. Moreover, the amount of air supplied to a combustor for burning $NH_3$ has to be lower to increase the residence time for complete combustion. This reduces the inlet Reynolds number and, hence, flow mixing and turbulence are adversely affected, resulting in reduced combustion efficiency [155]. $NH_3$ flames also have low laminar burning velocity in the range of 0.06–0.08 m/s [156]. However, $NH_3$ combustion showed some good characteristics in terms of faster dissociation rate, which may result in faster flames of comparable characteristics to some hydrocarbon fuels. It was reported that 28% dissociation of $NH_3$ could be utilized as enrichment fuel for gas turbine applications designed to work on hydrocarbons. Other studies considered $NH_3$ as an engine fuel and showed limited success in terms of reduced $NO_x$ emissions [157, 158].

Recently, a number of studies have been performed considering $NH_3$ a flexible fuel for gas turbine applications [159, 160]. They reported that the associated emissions with the combustion of ammonia limits the wide application of such fuel technology. The upper and lower concentrations of $NH_3$ by volume in air, corresponding to the upper and lower flammability limits under atmospheric conditions, were found to be 33.5% and 14.8%, respectively [161]. Preheating of the reactants resulted in wider operability limits, whereas the presence of water vapor restricts the operability limits. SPG Advanced Propulsion and Energy [159, 160] is one of few companies that tried to produce commercial systems for $NH_3$ combustion. They presented a series of reports summarizing the challenges involved in the development of a reliable $NH_3$ combustion technology, including: (1) slower reactivity and, accordingly, lower combustion temperature, (2) combustion instabilities, (3) vaporization approach of $NH_3$ for improved combustion, and (4) cracking of $NH_3$ upstream of the burner for enhanced burning velocity. The research is ongoing, trying to improve the performance of $NH_3$ as fuel for gas turbine applications, supported by the global trend of decarbonizing fuels for reduced carbon emissions out of combustion systems. Focus is made on understanding the slower reactivity of $NH_3$, particularly in fuel blends with hydrogen and methane. Over-production of OH radicals and enhanced reactivity were reported at operating equivalence ratios in the range of 1.05–1.25 [161]. Such results are promising if $NH_3$ is blended with other light hydrocarbons for gas turbine applications. More research is required in this field to reduce the time required for the development of reliable $NH_3$ combustion systems on the industrial scale.

### 3.4.3 *Fuel-Flexible Premixed Oxy-fuel Combustion*

The use of low-carbon or carbon-free fuels (e.g., $H_2$) seems to be a satisfactory solution for controlling $CO_2$ emissions. However, hydrogen production and storage on the industrial scale is expensive, and more research work is required on this subject. Moreover, the world still depends on fossil fuels for meeting the increasing energy demand because of their competitive price. Therefore, combining oxy-combustion of low-carbon fuels with $H_2$-enrichment provides a good solution for controlling both $NO_x$ and $CO_2$ emissions at reasonable cost. $H_2$-enriched fuels have better combustion and emissions characteristics compared to non-enriched fuels [162–164]. The use of premixed combustion technique also maintains the flame temperature inside the gas turbine combustor within safe limits, especially under $H_2$-enriched oxy-fuel conditions. However, the application of premixed combustion with $H_2$-enrichment brings its own challenges in terms of flame stabilization.

Mazas et al. [165] conducted experimental and numerical analyses of the laminar flame velocity of premixed oxygen-enhanced methane combustion. They measured the flame velocity using the conical flame method, while the calculations were carried out using the detailed kinetic mechanism GRI-Mech. 3.0. The study was conducted at different equivalence ratios from 0.6 to 1.5 with the molar oxygen-enrichment ratio $[O_2/(O_2 + N_2)]$ ranging from 0.21 to 0.5, and the effect of adding water vapor was also analyzed with molar fraction varying from 0 to 4.5. The results indicated linear decrease of the laminar burning velocity with the addition of water vapor. For a given $O_2$-enrichment level, the laminar burning velocity was independent of the operating equivalence ratio. They concluded that the effect of adding water vapor on the laminar flame velocity is mainly a thermodynamic one and does not affect the reaction kinetics. A numerical study on the combustion characteristics of premixed oxy-methane combustion at various flue-gas dilution ratios was carried out by Li et al. [96]. As shown in Fig. 3.7, the laminar burning velocity with $CO_2$-dilution is restrained, as compared to that with $N_2$-dilution. The burning velocity with $CO_2$-dilution was found to be close to that of air/methane stoichiometric combustion at a diluent/oxygen ratio of 200%. The calculations showed small differences in adiabatic flame temperature for various diluents ($N_2/H_2O$/flue gas/$CO_2$) at lower diluent/oxygen ratios, but such differences increase as the diluent/oxygen ratio increases.

Using $H_2$-enrichment and syngas in combined cycles is attracting the interest of research community as an alternative to natural gas [166–168]. Syngas and $H_2$-enriched fuels are regarded as the prominent alternative fuels for the future energy mix [169–171]. Recycling of the industrial by-product hydrogen gas and the use of $H_2$-rich fuels after pre-combustion carbon removal can be integrated with combined cycle plants. It has been reported that the addition of $H_2$ to hydrocarbons can reduce the soot formation and CO emissions drastically [172–176]. Moreover, the addition of even a small amount of $H_2$ to the fuel mixture can affect both the instantaneous and average characteristics of a turbulent premixed flame [177], necessitating rigorous research on the combustion dynamics of such flames with varied fuel compositions.

Gersen et al. [178] investigated the ignition delay time of hydrogen/methane mixtures at various compositions in a rapid compression machine. They found that the ignition delay time decreases with the increase in $H_2$ content and can be correlated quantitatively to the hydrogen fraction of the fuel in accordance with the mixing law proposed by Cheng and Oppenheim [179]. The laminar burning velocity tends to increase linearly with the addition of hydrogen [180, 181]. The unstretched flame-propagation speed and the unstretched laminar burning velocity were reported to increase with fuel hydrogen fraction in a study conducted by Tang et al. [182]. They

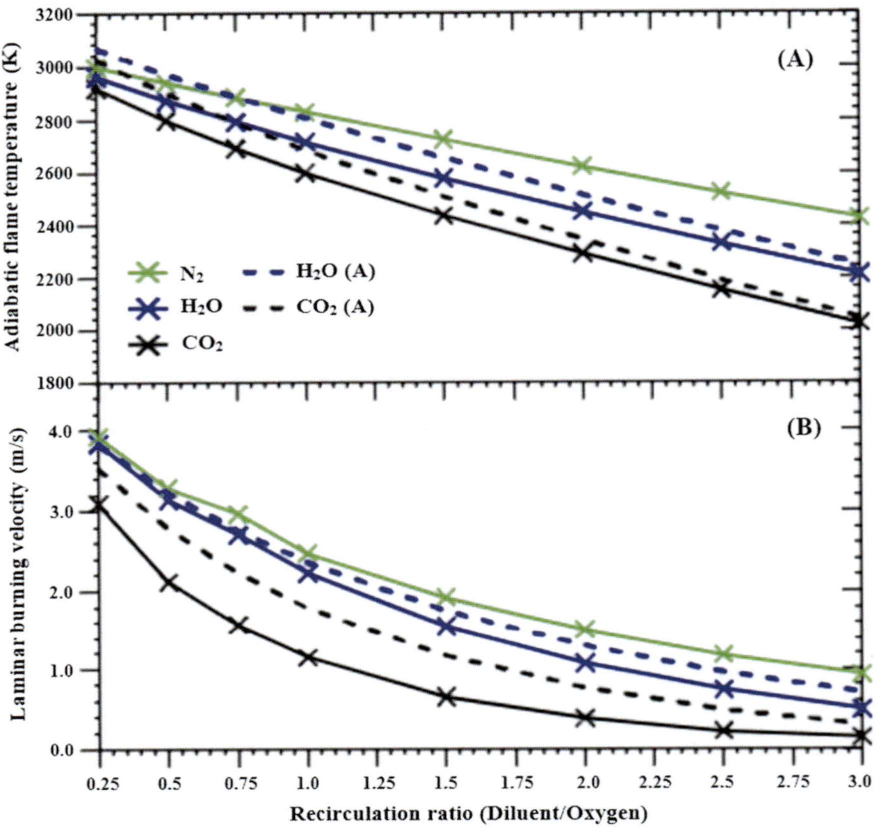


**Fig. 3.7** Effect of recirculation ratio on adiabatic flame temperature and laminar burning velocity of premixed methane flames in different oxidizer environments [96]

also found that this dependence becomes even more remarkable at higher hydrogen fractions. Sankaran and Im [183] concluded that the high laminar burning velocity of $H_2$ is attributed to its higher molecular diffusivity. Furthermore, the addition of $H_2$ to hydrocarbon fuels in lean flames has been found to increase the turbulent burning velocity monotonically [184]. In rich flames, however, $H_2$ addition had no apparent effect on the burning velocity but was reported to depend on the type of hydrocarbon fuel. Halter et al. [177] reported similar dependence of the turbulent burning velocity on hydrogen fraction in premixed $H_2$-enriched methane/air flames. They further concluded that the combustion intensity increases with hydrogen addition, implying that the turbulent burning velocity increases faster than the laminar one with hydrogen addition. This was attributed to the fact that the flame-front curvature distribution tends to enlarge slightly with hydrogen addition, which enhances the smaller-scale flame wrinkling, resulting in higher turbulent burning velocity. Similar findings were reported by Daniele et al. [185, 186].

Hydrogen enrichment enhances the reaction rate and burning velocity due to the increased H, O, and OH concentrations in the flame, as reported by Hu et al. [180]. Similar observations were reported by Wang et al. [187] from their numerical analysis of the effect of $H_2$-enrichment on methane/air combustion using the GRI-Mech 3.0 mechanism. Moreover, they suggested that $H_2$-enrichment reduces the formation of $CH_2O$ and $CH_3CHO$ species, consequently reducing the aldehyde emissions in methane combustion. An increase of ~20% in OH mole fraction was reported by Schefer [188] when blending 20% hydrogen with methane. Due to the high combustibility of hydrogen, flames with higher hydrogen fractions are characterized by higher reaction-zone temperatures, although the adiabatic flame temperature of the fuel blend may decrease with hydrogen addition. The reaction-zone expands faster because of its higher temperature, consequently inhibiting the cooler recirculation flow of combustion products into the reaction zone [189].

Schefer [188] conducted an experimental study of flame stability with hydrogen addition. He found that the lean blowout limit is maximum at stoichiometric conditions and reduces for both the lean and rich conditions. Moreover, he reported that the lean blowout limit could be reduced by around 15% by the addition of up to 20% hydrogen to the fuel. A laboratory-scale investigation of a low-swirl injector (LSI) was conducted by Cheng and Littlejohn [190] to study the air-combustion characteristics of pure hydrogen and $N_2$-diluted hydrogen with the aim of implementing this in the gas turbines of the integrated gasification combined cycle (IGCC). They found an optimum swirl number of 0.51 for $H_2$ in the various LSI configurations examined, as compared to 0.54 for hydrocarbon LSI. The lean blowout (LBO) limit was found to occur at a nearly constant equivalence ratio of 0.17 at different bulk velocities, indicating that the LBO of LSI hardly depends on Reynolds number. Another parameter that has been found to predict the LBO limit successfully is the Damkohler number. Zhang et al. [191] reported an interesting finding that the LBO limit occurs at a Damkohler number of 0.4 for various fuel blends of $H_2/CO/CH_4$. Taamallah et al. [192] presented a comprehensive review of syngas use in gas-turbine combustion. They stated that the composition of syngas is a vital parameter that determines the flame characteristics, as the physical and chemical properties of $H_2$, CO, and $CH_4$ are significantly different, which affects the combustor performance. A higher percentage of $H_2$ in the fuel blend results in minor reduction in adiabatic flame temperature. However, the addition of hydrogen has been observed to lower the LBO limit, allowing the combustor to remain lit at lower equivalence ratios. Nonetheless, the higher turbulent flame speed, characteristic of $H_2$-enriched flames, increases the risk of early flashback even at lower equivalence ratios.

Lounici et al. [193] studied the improvement of a natural-gas engine by hydrogen enrichment. They reported a reduction in break specific fuel consumption as well as the emissions of hydrocarbon, CO, and $CO_2$. Schefer et al. [188, 194] revealed that adding an amount of $H_2$ to the combustion process results in significant reduction in CO concentration and a slight increase in $NO_x$ emissions, as presented in Fig. 3.8, in addition to improved flame stability and shorter flame length. Tuncer [195] showed experimentally that hydrogen enrichment considerably extends the lean blow-off limit. However, despite the above-mentioned enhancements, the implementation of

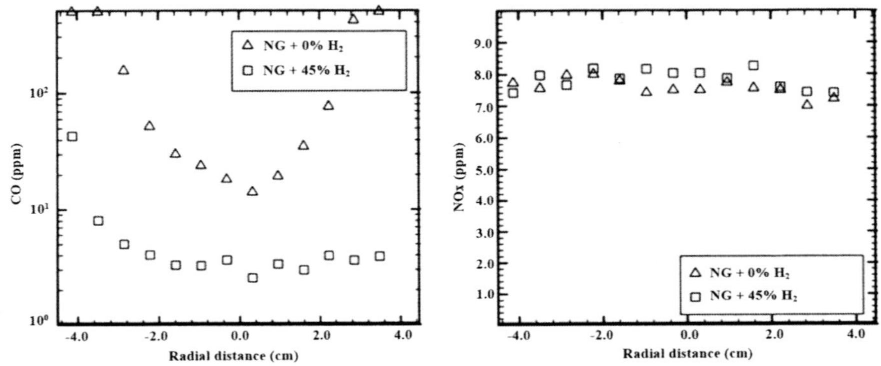


**Fig. 3.8** Effect of hydrogen enrichment on CO and $NO_X$ emissions [194]

hydrogen-enriched combustion is still limited because of the associated phenomena, including flashback, auto ignition, and combustion dynamics.

## 3.5 Gas Turbine Combustion Systems

### *3.5.1 Stagnation Point Reverse Flow (SPRF) Burners*

As previously reviewed, non-premixed flames induce excessive thermal $NO_X$ emissions because of the existence of high-temperature stoichiometric zones within the flame. However, non-premixed flames can still be utilized with reduced $NO_X$ emissions, if mixing is enhanced within the combustor to minimize the stoichiometric zones. This can be achieved through careful design of the combustor to control the mixing process and minimize the residence time of combustion products within the flame core [196]. This approach is most effective in cases of jet flames with high coaxial air velocity. The high-velocity flow improves the shear and, accordingly, enhances the mixing of air and fuel before combustion takes place, which reduces the tendency of creation of stoichiometric zones. The high-velocity jet also reduces the residence time of combustion products within flame core, resulting in lower $NO_X$ emissions.

Another approach to controlling $NO_X$ emissions is flue-gas recirculation (FGR), where the reactants are diluted with inert (cooled) exhaust gases to lower the concentrations of oxygen before the combustion process and, consequently, extend the ignition delay period [197–199]. Longer ignition delay allows air and fuel to mix within the combustor before the onset of combustion, even in the non-premixed flame mode. Besides, recirculating the exhaust gases at a reasonable rate spreads the combustion zone out throughout the whole mixing zone, rather than over a thin front. Consequently, a uniform temperature distribution is obtained within the combustion zone and, accordingly, $NO_x$ emissions are reduced to lower levels. However, mixing the reactants with inert exhaust gases results in lower flame burning velocity and, consequently, lower combustion efficiency, and the combustor is prone to loss of static stability.

Recently, the idea of a novel gas-turbine combustor design, called stagnation-point reverse-flow (SPRF) combustor, has been demonstrated to meet such conflicting requirements of stable flame and low emission levels [200]. The size of SPRF combustor is small, which promotes internal gas recirculation, because the outflow ports are on the same geometrical plane as the inflow, while the opposing end is closed [201]. This design creates a low-velocity region near the closed end, as shown in Fig. 3.9, which helps to stabilize the combustion process. As they leave the combustor, the hot products interact with the incoming reactants to form a more reactive mixture. The reactants are diluted with substantial amounts of burnt gases laden with radicals, which lowers the ignition temperature, resulting in lower $NO_x$ and CO emissions as well as enhanced blowout limit [201–203].

A few recent studies were performed to evaluate the performance of SPRF combustor considering ranges of operating parameters. Gopalakrishnan et al. [204] investigated experimentally the mechanisms for reducing $NO_x$ emissions in a non-swirling SPRF combustor in non-premixed mode. Lifted flames were obtained away from the injector, which resulted in better mixing of air and fuel before the onset of combustion and, consequently lower $NO_x$ emissions. Bobba et al. [205] studied experimentally the stabilization mechanisms and flame structure of premixed methane-air flames in

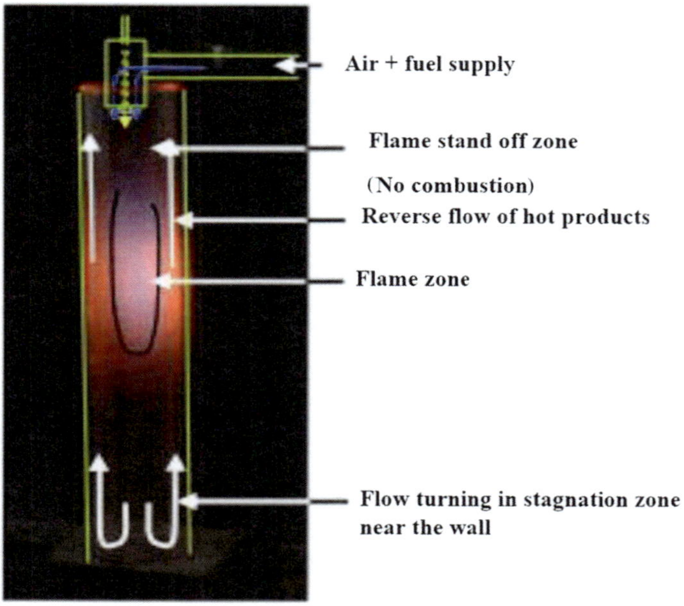


**Fig. 3.9** Schematic diagram presenting the key features of the SPRF gas turbine combustor [201]

a non-swirling SPRF combustor. The results showed two stabilization zones within the combustor, a primary zone located downstream of the inlet section, where strong turbulence and low velocities are obtained. The increased turbulence strength in the shear layer resulted in more entrainment of the recirculated exhaust gases, thus improved reaction rates, and enhanced flame stability. The secondary stabilization zone is located near the stagnation zone beside the wall. The analysis of the turbulent flame structure revealed that the flame is mainly in the thin reaction layer regime throughout the entire combustor. Castela et al. [206] conducted an experimental study to understand the combustion regimes occurring in a SPRF burner. For a given preheat inlet temperature of air, the recorded data showed a considerable effect of increasing the air inlet velocity on the reduction of $NO_x$ emissions, as presented in Fig. 3.10. Furthermore, increasing the air inlet velocity resulted in stable flame operation at leaner conditions. Undapalli et al. [207] performed large eddy simulations (LES) of a SPRF combustor, considering both non-premixed and premixed combustion modes. The premixed flame was attached to the injector, whereas the non-premixed flame was lifted away. The flame lift-off in the non-premixed mode enabled longer ignition delay for better mixing of the reactants. This may justify the similar $NO_x$ emission levels obtained in both the non-premixed and premixed combustion modes.

Moderate or intense low-oxygen dilution (MILD), aka flameless combustion, is one of the combustion techniques that also uses reverse flow burner configurations and can exhibit very low $NO_x$ emissions and wider flammability limits [208–210]. Flameless combustion can be obtained within a combustor by raising the inlet temperature of reactants above the mixture auto-ignition temperature while entraining enough inert exhaust gases to increase the ignition delay. Such conditions result in

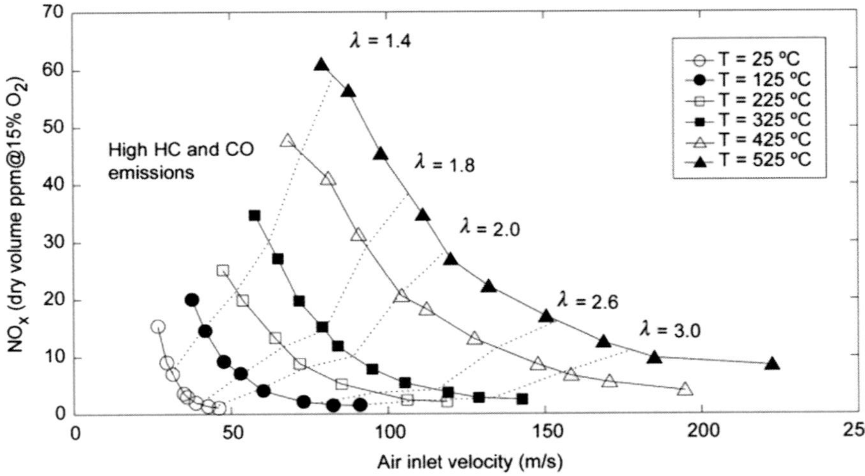


**Fig. 3.10** Effect of inlet air velocity on $NO_x$ emissions for different inlet temperatures and excess air factors [206]

no visible flame while keeping emissions at very low levels. Despite being used in a number of industrial applications, the implementation of MILD combustion in gas turbines is still in the research and development phase [211–213]. A number of studies were performed considering MILD combustion for gas turbine applications. Luckerath et al. [214] studied flameless combustion for gas turbine applications at an operating pressure 20 bar. Levy et al. [215] proposed a novel gas turbine combustor design that can adopt MILD combustion by creating a large recirculation zone within the combustor. In another study, Lammel et al. [216] were able to achieve low CO and $NO_x$ emissions using a combustor with high power density. In addition, Arghode et al. [217] discussed the concept of flameless combustion for gas turbine applications and were able to achieve ultra-low $NO_x$ emissions.

### 3.5.2 Dry Low-$NO_x$/Low-Emissions (DLN/DLE) Burners

The recent advances in the development of future gas turbine technologies largely target efficiency improvements and nitrogen oxides ($NO_x$) emission reduction. Several methods of lowering $NO_x$ emissions have been established. $NO_x$ formation during combustion is a strong function of flame temperature [218]. Lowering the temperature using diluents like water or steam apparently is the most promising solution for reducing $NO_x$ emissions, particularly in stationary power-generation systems. However, this usually results in lower overall engine efficiency, possible corrosion due to water impurities, and reduction or quenching of CO burn-out, which makes it counterproductive [219]. Solutions not involving steam or water injection include lean premixed (LPM) combustors implementing Dry Low $NO_x$ (DLN) or Dry Low Emission (DLE) technologies [220, 221]. Such combustors in gas turbine applications can reduce $NO_x$ emissions down to the single digit level [222]. Figure 3.11 shows a schematic diagram comparing the DLE concept with a conventional burner. Both burners are swirl stabilized. The size of fuel injector in the DLE burner is larger to account for the larger quantity of airflow in the premixing zone.

The major design requirements for optimum performance of DLE burners are (1) conforming to emission standards at peak load and (2) regulating the emissions across the range of engine load. Other important parameters to ensure combustion stability and wide operability range under all conditions include the system ability to respond rapidly to load changes, maintaining minimum combustions acoustics, as well as smooth switching from one fuel to another, especially in dual-engine engines [223]. Different manufacturers have different design approaches to the DLE combustor concept with the sole aim of reducing $NO_x$ emissions without utilizing steam or water injection.

Figure 3.12 shows the cross-section of a Solar Low $NO_x$ burner designed for the multiple-can combustors developed by the Solar Turbine Company. It is a fuel-injector/swirl-air assembly that contains an 18-vain air swirler with radial flow orientation to provide the required swirl to the primary air, thereby aiding air-fuel mixing by ensuring recirculation in the primary combustion zone. This burner concept allows

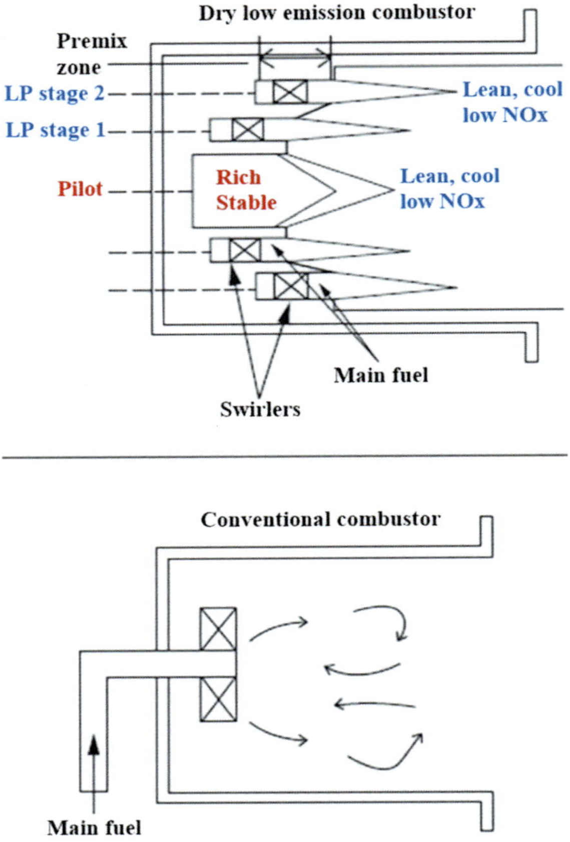


**Fig. 3.11** Schematic diagram comparing a typical DLE combustor with a conventional one [222]

three modes of fuel injection: pilot, channel, and 18-spoke swirler injection. The later mode provides better mixing. This design was found to reduce $NO_x$ emission levels to less than 10 ppmv and also lowered both CO and unburned hydrocarbon emissions when fueled by natural gas at pressures of up to 11 bar [224]. The schematic diagram shown in Fig. 3.13 represents the design concept of the premixed two-staged DLN-1 combustor that is capable of using both natural gas and liquid fuels [225]. The burner comprises of four main parts: fuel injectors, liner, venturi, and centerbody/cap assembly. These components provide three core zones: primary, secondary, and dilution zones.

The burner operates in four distinct modes, namely Primary, Lean-Lean, Secondary, and Premix, as shown in Fig. 3.14. In the primary mode, fuel is supplied through the primary nozzle only, and the primary air through the swirlers. Initial flame ignition and low loads of up to 20% are conducted in this mode. The lean-lean mode is selected at intermediate engine loads by supplying fuel to both the primary and secondary zones. The fuel mixes with air and passes through the swirler at the

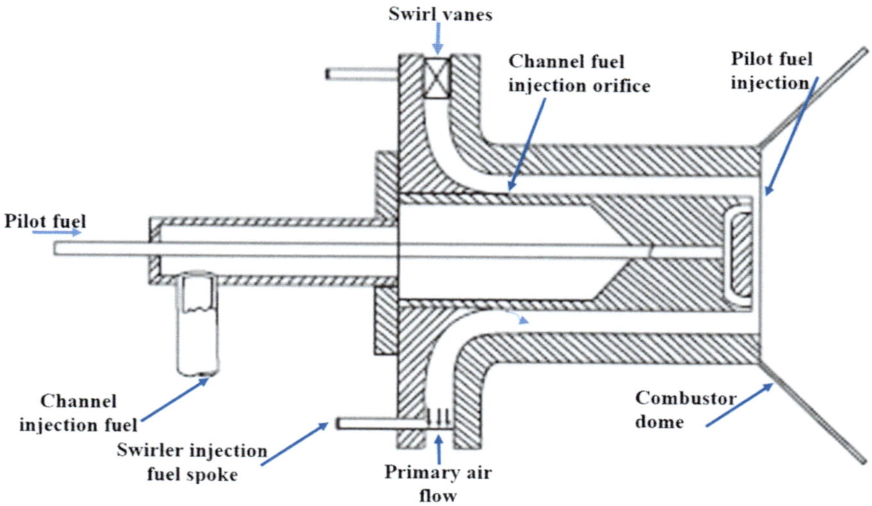


**Fig. 3.12** Solar low $NO_x$ burner system for natural gas [223]

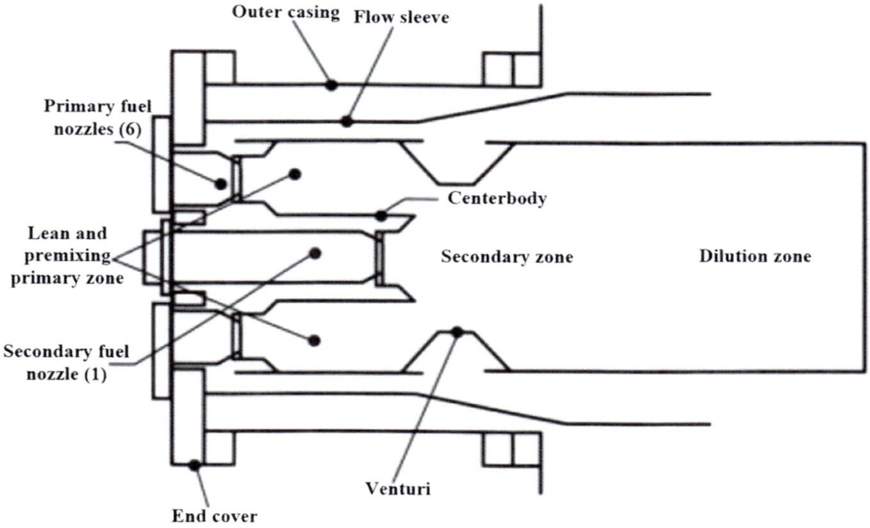


**Fig. 3.13** General electric DLN combustor [225]

exit of centerbody, thereby forming the secondary combustion zone. The secondary mode of operation is the transition mode between lean-lean and premix modes, with the flame being sustained in the secondary zone only. The fuel supplied to the primary nozzle is slowly reduced while that of the secondary nozzle is increased, which extinguishes the primary flame. In the premix mode, some fuel is still supplied to

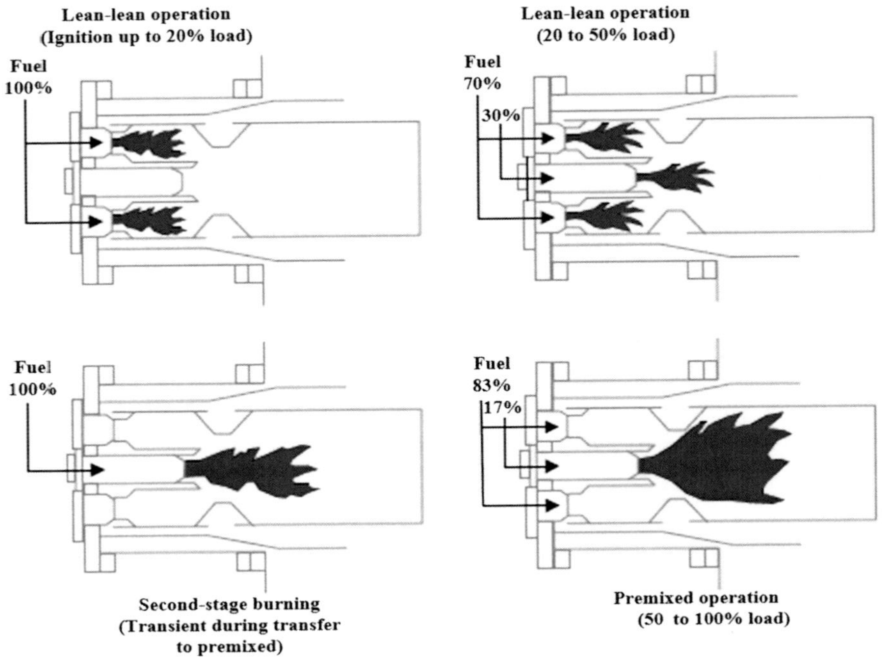


**Fig. 3.14** Fuel-staged DLN modes of operation [225]

the primary nozzle, but the flame is sustained in the secondary flame zone only. This mode provides minimum emissions at the exhaust. When this combustor operates using natural gas at base load, the levels of CO and $NO_x$ can be as low as 25 ppmv and 9.0 ppmv, respectively [223].

### *3.5.3 EnVironmental (EV/AEV/SEV) Burners*

Although the market of LPM gas turbines has many competing combustion technologies, some stand out based on their ultralow emissions and remarkable flexibility. EV-burners, for example, compete strongly with micromixers (to be discussed in the next section), especially that EV-burners offer significantly effective flame stabilization over a wide range of operation, which enhances engine turndown. Both micromixers and EV-burners have never been examined under oxy-combustion conditions; thus, the following condensed review is dedicated to EV-burners to discuss their potential for implementation in zero-emission gas turbines.

The acronym EV-burner stands for Enhanced-Vortex burner. The ability of this technology to achieve ultralow $NO_x$ emissions in LPM gas turbines has triggered the name EnVironmental (EV) burner [226], which also abbreviates to EV-burner. Its geometry consists of two half-cone shells, where one is slightly offset in the radial direction with respect to the other while maintaining parallel axes; see Fig. 3.15. Two tangential slots of constant width are thus created between the shells. Combustion air enters the burner tangentially through these slots, creating swirling core flow with vortex breakdown near burner exit [227]. The swirl strength increases in the axial direction because of the increasing diameter; the divergence angle is designed carefully to position the vortex breakdown point close to burner exit (see Fig. 3.16).

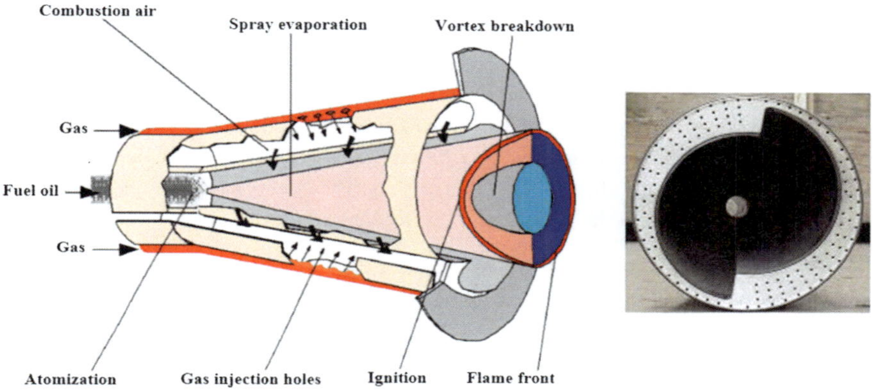


**Fig. 3.15** EV-burner principle [228]

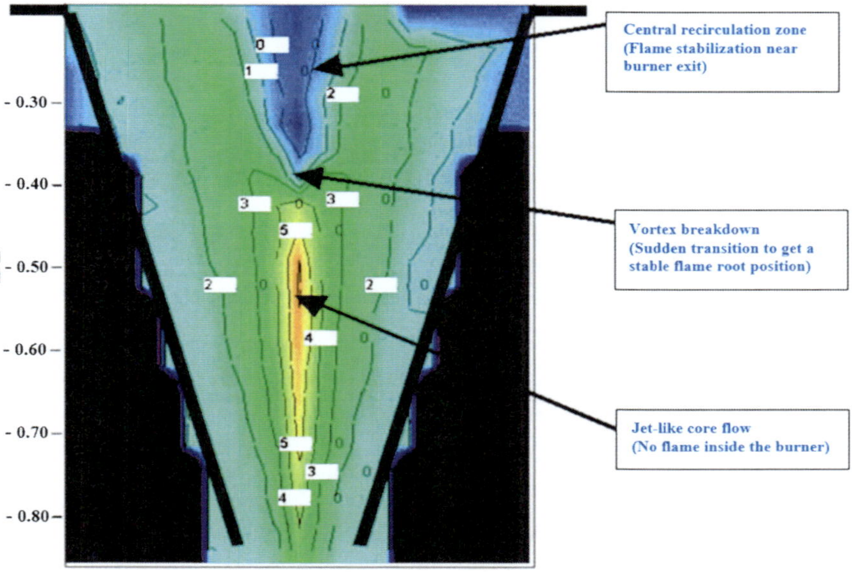


**Fig. 3.16** Axial velocity distribution inside an EV-burner [228]

Premix gaseous fuel is injected transversely into the combustion air through a row of holes just upstream of each air slot; this transverse injection boosts mixing effectiveness, which is a key prerequisite for ultralow $NO_x$ emissions. EV-burners have dual-fuel capability, i.e., gaseous fuel (DLN technology) and liquid fuel (with water injection) [226].

Vortex breakdown, which is one of the main characteristics of EV-burners, is an abrupt and drastic change in the structure of strongly swirling flows. Some studies reported that the onset of vortex breakdown is independent of the flow Reynolds number but depends mainly on the swirl number [229]. Vortex breakdown occurs in two different forms, depending on the burner design, namely the axisymmetric bubble-like and the spiral-shaped forms. The axial flow decelerates along the vortex axis, and the velocity gradients within the vortex get weaker as flow recirculation is distributed over a large area within the combustor [230]. This behavior results in perfect mixing of the reactants and higher flame stability, even at startup and under ultra-low load conditions. The perfect mixing prevents the creation of high-temperature spots within the flame, and, consequently, $NO_x$ emissions can be kept at the minimum level. The flow core accelerates downstream, preventing flame flashback. Stable operation can thus be achieved under very lean conditions, resulting in complete combustion at controlled temperature as well as lower CO and $NO_x$ emissions.

Several features [228] distinguish the EV-burner design from other DLN premix hardware [231]:

1. Strong swirl is created merely by tangential air entry; no sophisticated swirlers are needed.
2. The distribution of air entry in the axial direction strongly accelerates the core flow, which creates a natural barrier against flashback.
3. An inner or central recirculation zone exists downstream of vortex breakdown point, which effectively stabilizes the flame.
4. The vortex breakdown point is aerodynamically fixed in the free flow without the need for extra stabilization hardware (such as bluff-bodies, ramps, dump expansions, etc.) that could potentially be damaged due to flame impingement.
5. No pilot (non-premixed) fuel injection is needed to assist the stabilization of premixed flame at medium to high loads; the burner can operate in 100% premixed mode, which allows for achieving ultralow $NO_x$ emissions in the absence of non-premixed combustion.

At low loads, where the burner equivalence ratio is below the premix extinction limit, a separate pilot circuit injects fuel through a lance at burner centerline. The core flow becomes fuel-rich, and the resulting flame is again stabilized at the vortex breakdown point. Similarly, if the burner is operated with liquid fuel, it is injected through a central plain jet nozzle in the cone head. The jet disintegrates into small droplets that disperse over the entire flow field inside the burner cone. The resulting flame is stabilized with the help of inner recirculation zone.

The need to enhance the evaporation and mixing of liquid fuel for further reduction of the associated $NO_x$ emissions, led to the development of the Advanced EV (AEV) burner [232–235], see Fig. 3.17. Four inlet air slots are used instead of two, which reduces the tangential variation of radial velocity component and thus prevents the impingement of fuel droplets on cone walls. An additional mixing tube is affixed to the burner, in order to accommodate full evaporation of liquid fuel and boost mixing in gaseous phase. The axial and tangential velocity profiles inside conical section (Fig. 3.18) show a jet-like core flow with a distinct small body vortex. At cone exit, the peak axial velocity at centerline exceeds twice its average [228]. This high axial velocity of core flow persists throughout the mixing tube, thus increasing the margin to flashback. Even the lowest velocity near the walls of mixing tube can be increased by proper admission of a film of air (<10% of total airflow). This film significantly reduces the flashback risk at the wall and dilutes any local fuel-rich residues.

The EV-burner was first developed and implemented by the former Alstom Company in its GT11 gas turbine, which hosted 36 burners plus a central ignition burner in a hexagonal three-ring arrangement [236]. All burners were fired in premixed mode

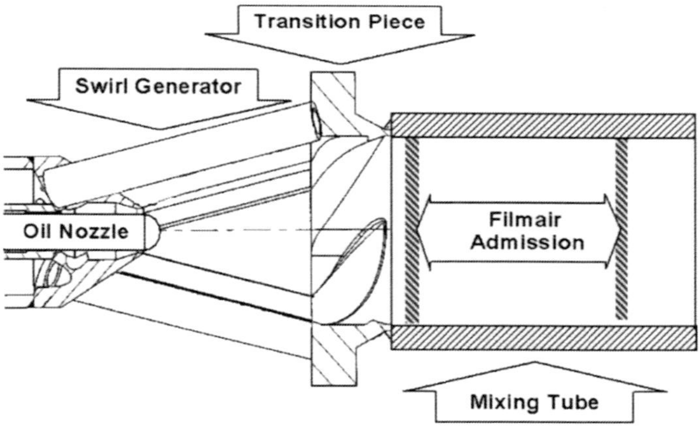


**Fig. 3.17** Design features of AEV-burner [228]

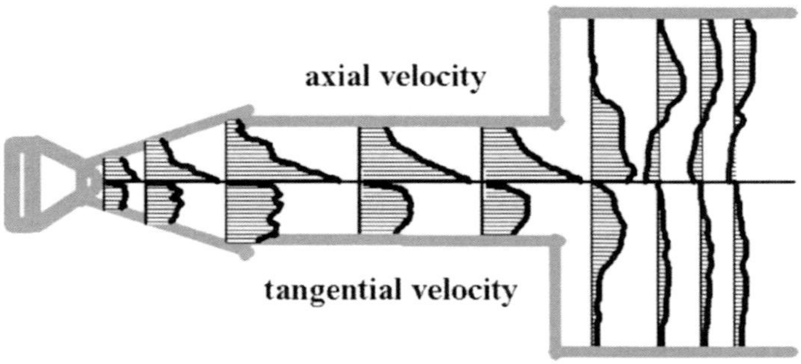


**Fig. 3.18** Axial (top) and tangential (bottom) velocity profiles in AEV-burner [228]

only because that initial design had not been equipped with central pilot injection yet. Overall combustor load and turndown were achieved by switching individual burners on and off. The same hexagonal arrangement was also used in the Alstom GT8, GT9, GT13, and GT11N2 gas turbines [237]. The aforementioned EV-burner design with pilot-fuel injection was first implemented in the annular combustors of the Alstom GT13E2 and GT10 gas turbines, and later in the GT8C2, GT24, and GT26 engines [238]. Today, all Alstom gas turbines have low-$NO_x$ combustion systems based on EV and AEV-burners. Although Alstom has been recently acquired by General Electric, the latter believes in the strong potential of EV-burner technology and invests in further developing it.

Multiple studies highlighted the wide flexibility of EV-burners. Paschereit et al. [239] examined the effect of extending the central fuel lance closer to the vortex breakdown point. Such simple design modification allowed for fixing the axial location of breakdown point over wide ranges of operating conditions and for better control of the combustion instabilities. Even without central fuel injection, extending the central lance changed the burner aerodynamics and the resulting flame. Flow visualizations were performed to quantify the effect of extended lance on combustion instabilities (pressure and heat release oscillations) and on emissions of $NO_x$, CO, and unburned hydrocarbons. The authors pursued the optimization of axial lance position and amount of central fuel at different burner equivalence ratios and output powers.

Zajadatz et al. [227] investigated controlled splits of gaseous fuel between the main premix circuit and the central pilot lance to create a staged premixed burner that generates even lower $NO_x$ emissions at lower part loads compared to the traditional operation of EV-burner. This smart fuel staging required no major design changes to the burner or its fuel-supply system, which makes the concept applicable to most existing operational EV gas turbines. In a similar study, Cho et al. [240] examined numerically the effects of fuel staging and fuel-air unmixedness on $NO_x$ emissions. The authors recommended fuel splitting between the premix circuit and the central pilot lance, in order to reduce unmixedness and the resulting $NO_x$. Another design enhancement was also proposed to boost fuel-air mixing, namely by increasing the interval between fuel holes in the premix circuit to give fuel more time to mix with air.

Güthe et al. [241] also examined fuel staging by comparing three modes of burner operation, namely perfectly premixed (fuel premixed with air in the air supply line upstream of the burner), standard (fuel injected through the premix circuit of the burner), and staged (fuel split between the premix circuit and the central pilot lance). The first mode was taken as a reference case that is free of any spatial variation in equivalence ratio, in order to decouple the effects of unmixedness and fluctuating turbulence intensity on combustion instabilities. The third mode of operation utilized the inherent flexibility of EV-burners to optimize the fuel split for controlling both instabilities and emissions. It was reported that allowing the flame anchoring point (vortex breakdown) to move further inside the burner cone increases the pressure drop across the burner and deteriorates the control of combustion instabilities. The staged mode was thus optimized to control the flame position, fluctuations, and emissions. Biagioli et al. [242] confirmed that one of the main drivers of instabilities in EV-burners is the unsteady displacement in flame anchoring point, which induces unsteady variation in flame surface and the corresponding heat release. The authors highlighted two aspects that make EV-burners more challenging than other technologies, namely variations in turbulent burning rate in such strongly swirled flows and the jump in sensible enthalpy across the flame. Both aspects were analyzed to prove a strong correlation between the tendency of the flame to be displaced and the measured amplitude of pressure pulsations. Döbbeling et al. [243] implemented closed-loop-controlled adaptive fuel staging that monitors the pulsations in flame anchoring point and adjusts the fuel split between the premix circuit and the central pilot lance to maintain flame stability. $NO_x$ was also observed to improve with the implementation of this closed-loop control.

Since the EV-burner enjoys efficient flame-stabilization (based on vortex breakdown), its use in a combustor extends the operational flexibility of high loads to lower loads and colder flame temperatures, which enhances engine turndown. Magni et al. [244] discussed an upgrade that can be retrofitted to existing AEV-based GT13E2 engines to extend the minimum environmental part load from 60% down to 30%, which allows for full operational flexibility with high efficiency within the 30–100% load range. By closing the inlet guide vanes (to reduce airflow inside combustor) and redistributing the fuel to fewer lit burners, the flame temperature can be maintained at reasonably high levels, which significantly reduces CO emissions. Reiss et al. [245] fueled the GT13E2 engine successfully with syngas (45% $H_2$, 48% CO, and 7% $CO_2$) of a lower heating value of 13.9 MJ/kg. To overcome the problems of high burning velocity and large fuel volume flux, minor design changes were implemented by concentrating the fuel injection near the burner end. Fuel dilution by nitrogen was also examined down to 6.8 MJ/kg. The recorded emissions were 20–25 ppm $NO_x$ and <5 ppm CO independent of load, which indicates remarkable flexibility. Wind et al. [246] tested fueling the Alstom GT26 engine with natural gas containing variable amounts of hydrogen and higher hydrocarbons ($C_{2+}$), targeting wider fuel flexibility. The authors examined the primary EV-burners as well as the Sequential ones (SEV-burners) of the reheat combustor. Hydrogen and higher hydrocarbons not only affect the burning velocity but also the auto-ignition delay time, which is a critical design parameter of the SEV-burners. The standard operating procedure of the engine had to be only slightly modified to accommodate those variations in fuel composition without any hardware modifications.

Based on this condensed review, EV-burners offer supreme flexibility, stability, and turndown, which makes them a promising option for oxy-fuel combustors to overcome the deteriorated performance associated with $CO_2$-diluted oxy-combustion (as compared to traditional air-fuel combustion). Since EV-burners have never been examined under oxy-combustion conditions before, extensive investigation is needed to quantify the applicability of this technology in zero-emission gas turbines.

### *3.5.4 Micromixer (MM) Combustion Technology*

Micromixer (MM) burners are a successful emerging LPM combustion technology that is already implemented on the utility scale (e.g., General Electric's F-class gas turbines) [247–249]. The MM technology is based on the concept of mixing of fuel and oxidizer at the micro-scale. A MM consists of a two-dimensional array of straight, parallel, equally-spaced tubes (each of a diameter of few millimeters) welded to two faceplates; one forms the inlet section of the burner, while the other forms the exit section. A fuel gas plenum is thus created around the tubes between the plates. The incoming fresh oxidizer stream is distributed among these tubes, and fuel is supplied from the gas plenum and injected radially (cross-flow) through side holes (each <1 mm in diameter) in the tube walls. The fuel holes are located axially near the inlet faceplate; a premixing region is created inside each tube, and the length of the tube is calculated to guarantee a fully-developed and fully-premixed flow at the exit section of each tube. Fuel is injected at the same axial location from different holes distributed equally along the circumference of the tube, in order to minimize the length required for generating fully premixed flow. The faceplate geometry, in terms of diameter and number of tubes, is optimized to minimize the emissions and widen the operability limits. The operating equivalence ratio and the pressure drop within the tubes are controlled to prevent flashback.

Fluid mixing plays a major role in combustion and the resulting emissions. As the global trend is shifting towards microsystem technology development, this vital fluid-mixing process is also in search of mixing the fuel and oxidizer at the micro level. Mixing the reactants on the macroscopic scale only gives rise to turbulent mixtures, whereas flows at the micro level are streamlined and characteristically laminar. The molecular diffusion process is also involved under certain standardized conditions, though such a mixing strategy is generally quite inefficient due to very low molecular diffusivities. For preventing this inefficiency and increasing the rate of fluid mixing, distinctive devices like MMs are used [250]. The mixing rate can be increased by any of the active or passive mixing methods [251–253]. The active ones include the use of numerous external fields like electric, magnetic, acoustic, etc. for changing the fluid flow rate. Passive approaches, on the other hand, involve changes in the tube geometry by using different types of inserts. Consideration of the fluid stream structure and mixing mechanism are required for the optimization of the mixing process. Therefore, mathematical modeling plays a major role in the investigation of these flows. The design of MMs has been broadly examined with the aid of computational fluid dynamics (see references [254–261]). Reference [253] has elaborated on the general working principles of different micromixers. Research was mostly conducted at constant bulk jet velocity of the reactant mixture, as this represents real operating conditions of LPM DLE combustors. These units are normally operated at fixed pressure drop across the headend, resulting in nearly constant inlet velocity of the combustible mixture. This assumption is applicable if flow compressibility effects are neglected, which is a common exercise among gas turbine manufacturers.

MMs are reviewed in this section, starting from basic perforated-plate burners up to the development of MM technology.

#### 3.5.4.1 Perforated Plate Burners

Perforated-plate (PP) burners are widely used in household and industrial applications such as boilers, furnaces, etc. PP burners are the precursor of MMs. A PP burner comprises of a simple perforated plate that receives a premixed or partially premixed flow of fuel and oxidizer on one side and distributes this flow among its array of holes to exit as jets on the other side. Conical-shaped flames are formed downstream of the perforated plate. These flames are susceptible to various instabilities like flashback, quenching, and blow off. Therefore, it is very important to study the stabilization mechanism of such flames. Kedia and Ghoniem [262] studied laminar, premixed, methane-air flames stabilized on a heat-conducting perforated-plate. They investigated the conditions of flame stability which led to blow off. Unsteady, 2-D simulations were performed with thorough chemical kinetics for methane-air species combustion. The joint impact of heat transfer and flame stretch is credited for flame blow off principle. Jithin et al. [263] numerically studied the characteristics of premixed propane-air flames stabilized over a perforated plate using 3-D simulations. They reported that increasing the inlet velocity increased the flame stand-off distance. For adiabatic plates, this distance was zero, i.e., the flame rested on the plate. However, by increasing the plate thermal conductivity, the stand-off distance increased again. Higher equivalence ratios made the flames move closer to the plate, which increased the heat flux transferred to the plate. Similar findings were reported by Altay et al. [264], who investigated the stability characteristics of PP-stabilized methane-air flames under various operating conditions using a 2-D numerical model. Conductive plates were considered with thermal conductivities up to 10 W/m K only. Jamal et al. [265], however, considered conductivity values as high as 385 W/m K and proved that if considerable heat is transferred through the plate to preheat the incoming combustible mixture, the flames become shorter and more stable. The authors examined the effects of plate material, thickness, and hole diameter and showed that heat transfer through the plate is the hidden primary control parameter behind all these effects. Kedia and Ghoniem [266] numerically studied the blow-off conditions and flame stabilization of PP-stabilized flames for different inlet velocities. They reported that the dynamic stability of the flame is highly dependent on the heat transfer from the flame to burner plate; this transfer was, however, considered as heat loss because the thermal conductivity was limited to 1.5 W/m K only. Rashwan et al. [267] experimentally studied partially premixed oxy-fuel flames anchored over a PP burner. At fixed equivalence ratio, stable flames were attainable only over the oxygen fraction range of 29–42% by volume in the $O_2/CO_2$ oxidizer. The premixing ratio of fuel/air mixture was also found to affect flame stabilization. Edacheri Veetil et al. [268] numerically studied the effect of hole geometry in PP burner on flame structure under rich and lean conditions. They reported that, with reduction in hole spacing, the curvature of flame base increases and the flame thickness decreases. Hindasageri et al. [269] experimentally and numerically studied the heat transfer distribution of impinging flame from PP burner for three different configurations (inline, staggered, and star) over range of Reynolds number (50–600) and at three different pitch to diameter (p/d) ratios (1.67, 2.33, 3.33) but 7 mm pitch is considered optimal for the entire study. Oh et al. [270] reported that the oscillations induced by combustion in a PP burner can be well stabilized by decreasing the hole diameter or by reducing the plate thickness. Wang and Wen [271] numerically studied laminar hydrogen flames in a PP installed channel and observed M-shaped flames. They found that the overall effect of PP was that flame speed decreases downstream of it. Li et al. [272] experimentally studied flame stabilization on a PP burner of a square cross-sectional channel and reported four phases of flame front, i.e. laminar, jet, turbulent, and secondary flame front. The flame shapes for three hole configurations of p/d = 2.33 with varying Reynolds number are illustrated in Fig. 3.19. Oh et al. [270] performed a numerical study on combustion instability in PP burners. The results

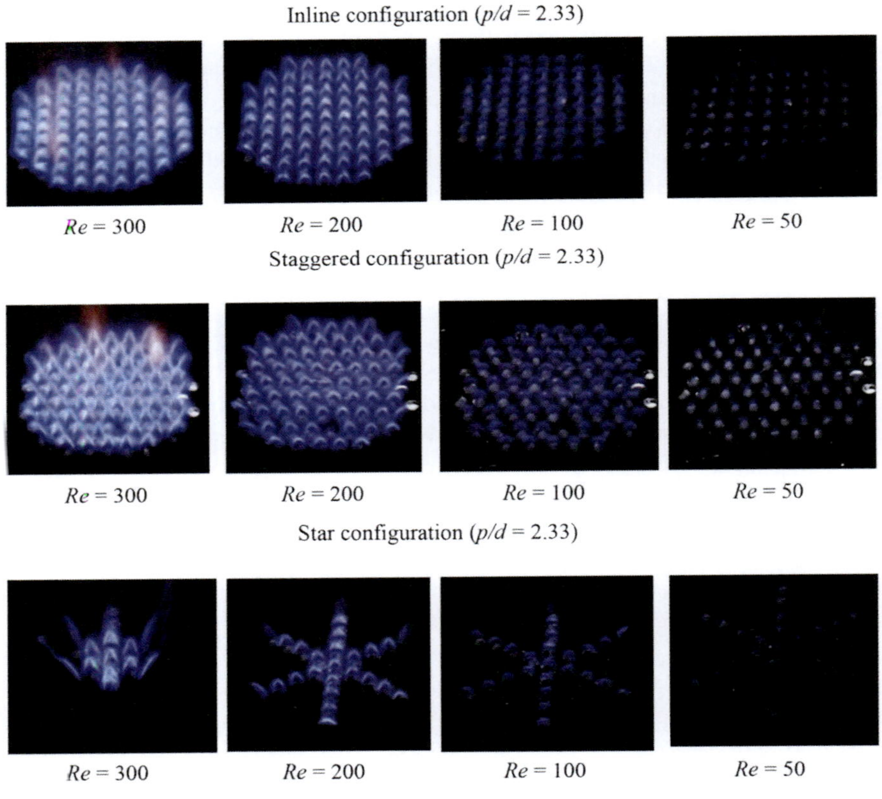


**Fig. 3.19** Flame shapes corresponding to three hole configurations for different Reynolds numbers at a pitch to diameter (p/d) ratio of 2.33 [269]

showed that plates with smaller holes are better for stabilizing the oscillations caused by combustion.

#### 3.5.4.2 Micromixer Combustion Characteristics

The main features of the micromixer (MM) technology are staging capability, scalability, and fuel flexibility. York et al. [273] conducted a full-can durability test of MM nozzles at the combustion conditions of General Electric's F-class gas-turbine at full load conditions considering air combustion of fuel blends of hydrogen, nitrogen and natural gas. Using a fuel blend of hydrogen and nitrogen (hydrogen is more than 90% by volume) and after 100 h of firing, $NO_x$ emissions were at the level of single-digit ppm. Funke et al. [274] performed a combined experimental and numerical investigation on the effect of momentum-flux ratio on $NO_x$ emissions and flame stabilization mechanism in a MM nozzle considering air combustion of hydrogen. The results showed significant effect of flame stabilization mechanism on $NO_x$ emissions. Dodo et al. [275] tested MM air-combustion of IGCC-syngas fuel simulants containing hydrogen, methane, and nitrogen with a hydrogen content of 40–65%. Three fuel blends were considered with 0, 30, and 50% carbon-capture rate. Stable combustion was recorded for all fuel blends with $NO_x$ emissions at the single-digit ppm level. Later, the same group [275] evaluated the Mitsubishi Hitachi Power Systems MM performance in an IGCC plant with focus on the effect of flame shape; slenderer flames were reported to show superior performance in terms of $NO_x$ emissions and combustor component temperatures.

Flashback, and consequent hardware failure, is the primary challenge associated with hydrogen combustion because of very high flame speeds. Lean Direct Injection (LDI) was found to drastically reduce the risk of flashback in $H_2$ combustors [276]. LDI designs are based on achieving rapid mixing through lean fuel injection from numerous locations within the combustion zone. Miniature-scale distributed premixer arrays, aka micromixers, were developed and tested in several studies, in order to reduce $NO_x$ emissions safely with $H_2$-rich fuel. The LDI approach was mimicked in these MMs but with the addition of a small-scale premixing region. If carefully designed, MMs could approach the LDI operability characteristics. Some of the examined configurations were jet-in-crossflow mixing [277], co-flow mixing [278], and swirl-induced mixing in miniature cups with axial and radial inflow [279].

The combustion of high-$H_2$ fuels using air in MMs has been examined in past studies [273–275], but never under oxy-combustion conditions. Hernandez et al. [280] indicated vast differences in the characteristics of $H_2$/air flames versus natural gas/air ones in MMs. The same group [278] later successfully developed and tested a flashback-free MM fueled by a variety of fuels ranging from 100% $H_2$ to 100% natural gas. Best operability was attained at a firing temperature below 1470 K using a fuel blend of 87% $H_2$ and 13% natural gas. It was concluded that MMs are excellent candidates for future gas turbines fueled by hydrogen or a wide variety of syngas compositions. Zhang et al. [281, 282] proved the effectiveness of MM combustion technology in reducing $NO_x$ in a syngas-fueled combustor. They combined $CO_2$

dilution with MM combustion to reduce $NO_x$ and lower the faceplate temperature at MM exit. It was reported that low CO and $NO_x$ emissions can be obtained simultaneously, which is typically a challenge in most swirl-based dry-low-emissions (DLE) combustors. By adjusting the parameters of operation, the researchers managed to achieve 2 ppm $NO_x$15 and 3 ppm CO15. The use of $CO_2$ dilution was reported to lower $NO_x$ at the expense of CO emissions. The effect of syngas composition was analyzed through the corresponding C/H ratio. Lower ratios showed promising results in terms of lower faceplate temperature and lower $NO_x$ and CO emissions.

Based on the above discussion of the different burner designs and combustion techniques for clean combustion in gas turbines, Table 3.3 summarizes the main features of the different technologies in terms of their advantages and the disadvantages.

## 3.6 High-Temperature Membrane Reactors (HTMRs)

Retrofitting an existing power plant to work under oxy-combustion conditions requires the integration of an air-separation unit (ASU) with the conventional combustion system, where the ASU is used to produce the required amount of oxygen for combustion. At the present state, ASU technology based on cryogenic distillation can meet the flow and purity requirements of large-scale boilers. In this unit, air is compressed, cooled, and cleaned before introducing it to the distillation column to make the separation process. Air is separated inside the ASU into a nitrogen-rich stream and an oxygen-rich one [283]. Cryogenic air separation consumes about 0.24 kWh/kg-$O_2$ at a purity of 95% [284, 285], which corresponds to more than 15% of the total plant output power [286–288].

In high-temperature membrane reactors (HTMR), however, both oxygen separation and oxy-fuel combustion processes occur within the same unit. If successful, this should allow for the replacement of conventional combustors by HTMRs. In such membrane reactors, as shown in Fig. 3.20, air is passed on the feed side of the membrane, and a mixture of fuel plus recirculated $CO_2$ is passed on the permeate side. The membranes are selective to oxygen and separate the feeding oxygen to the permeate side, where it reacts with the fuel within the same unit. The oxygen permeation flux depends mainly on the membrane temperature and the difference in oxygen partial pressure on both sides of the membrane. Part of the heat of combustion is used to heat-up the membrane (to temperatures above 650 °C) to be activated for oxygen separation. Numerical investigations of the HTMR concept are abundant [101, 102, 289–291], but experimental studies are very rare [292].

Tan et al. [293] fabricated hollow fiber membrane reactors (HFMRs) based on LSCF-6428 crystals to examine the catalytic combustion of methane. A simple mathematical model was developed that combines the local oxygen permeation rate with approximate catalytic reaction kinetics and can be used to predict the reactor performance for methane combustion. A mixture of $CH_4$ and He was used as sweep gas. The effects of operating temperature and methane and feed-air flow rates were studied both experimentally and theoretically. It was concluded that coating platinum on the

**Table 3.3** Advantages and disadvantages of the different burners for clean combustion in gas turbines

| Burner design | Advantages | Disadvantages |
|---|---|---|
| Stagnation Point Reverse Flow (SPRF) burners | • Small size<br>• Promotes internal gas recirculation<br>• Excellent flame stability<br>• The reactants are diluted with part of the burnt gases laden with radicals, which lowers the ignition temperature, resulting in lower $NO_x$ and CO emissions as well as enhanced blowout limit<br>• Suitable for flameless combustion | • Lower combustion efficiency<br>• Lower flame burning velocity |
| Dry Low-$NO_x$ (DLN) and Dry Low-Emission (DLE) burners | • Acceptable $NO_x$ and CO emissions | • Achieve flame stability using swirlers<br>• Less compact<br>• Flame instabilities are encountered under low load condition |
| Environmental burners (EV) (including AEV, and SEV) | • Superior flame stability over wide range of operations based on vortex breakdown<br>• Achieve ultra-low $NO_x$ emissions<br>• Dual fuel capability (gaseous and liquid)<br>• No sophisticated swirlers are needed (strong swirl is created merely by tangential air entry)<br>• Accelerated core flow that creates a natural barrier against flashback<br>• No need for extra stabilization hardware<br>• No pilot (non-premixed) fuel injection is needed to assist the stabilization of premixed flame<br>• Enhances engine turndown time<br>• Excellent candidate for oxy-fuel combustors | • Imperfect fuel-air mixing in fuel staging<br>• Unsteady variation in flame surface<br>• Variations in turbulent burning rate |

(continued)

**Table 3.3** (continued)

| Burner design | Advantages | Disadvantages |
|---|---|---|
| Perforated Plate (PP) burners | • Suitable for household and industrial applications<br>• Flame symmetry | • Susceptible to various flame instabilities like flashback, quenching, blow off, etc. |
| Micromixer (MM) burners | • Staging flexibility<br>• Fuel flexibility<br>• Can achieve single digit $NO_x$ emission level<br>• Scalability<br>• Excellent candidate for oxy-fuel combustors | • Combustion dynamics near stoichiometric condition<br>• Still under the development phase for gas turbine applications |

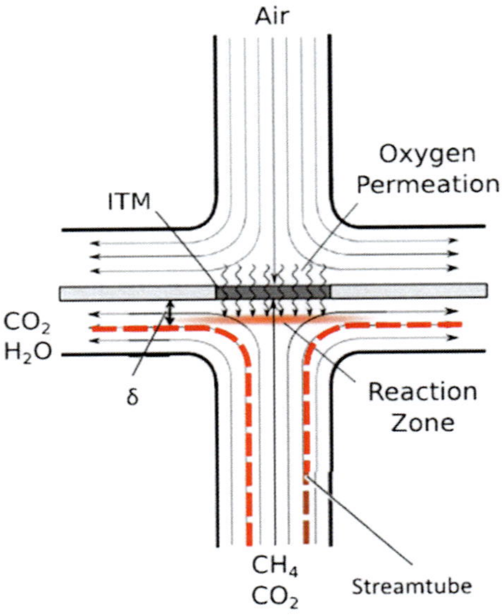


**Fig. 3.20** Schematic representation of a stagnation flow high temperature membrane reactor [292]

feed side can improve both methane conversion and oxygen permeation. Wei et al. [294] successfully investigated $CO_2$-diluted methane oxy-combustion in a U-shaped hollow fiber membrane; a study that is reviewed here in detail due to the importance of its findings. The membrane was based on $(Pr_{0.9}La_{0.1})_2(Ni_{0.74}Cu_{0.21}Ga_{0.05})O_{4-\delta}$ (PLNCG) with an inner diameter of 0.5 mm. Complete combustion was achieved at 975 °C for 450 h. The membrane was swept with a mixture of $CH_4$ and $CO_2$. For an air flow rate of 150 mL/min on the feed side and a $CH_4 + CO_2$ flow rate of 5.34 mL/min on the sweep side with 10% $CH_4$ (≡0.534 mL/min $CH_4$), the associated oxygen permeation flux was 0.7 mL $min^{-1}$ $cm^{-2}$. No details on the equivalence ratio were reported, but if at least stoichiometric combustion is assumed, the flow rate of permeated oxygen would have to be $\geq 2 \times 0.534 = 1.068$ mL/min, from which one can deduce that the inner surface area of hollow membrane was at least $1.068/0.7 = 1.53\ \text{cm}^2$; this area was not reported. The corresponding minimum membrane length would thus have to be $1.53/(\pi \times 0.05) = 9.71$ cm.

The results of Wei et al. [294] showed excellent reaction performance and good stability under oxy-fuel reaction conditions. Partial and total oxidation of methane took place simultaneously at the U-shaped PLNCG membrane. The operation showed that complete combustion can only occur at temperatures higher than 975 °C, see Fig. 3.21a. Partial oxidation of methane commenced well below 875 °C. Between 900 and 925 °C, the oxygen flux increased gradually with temperature, which increased the CO concentration due to partial oxidation. Between 925 and 950 °C, total oxidation commenced as the temperature and oxygen flux increased, which kept the CO concentration fairly constant (independent of temperature), because a balance was reached between CO generation (partial oxidation) and CO consumption (total oxidation). Beyond 950 °C, the total-oxidation reactions gradually dominated as

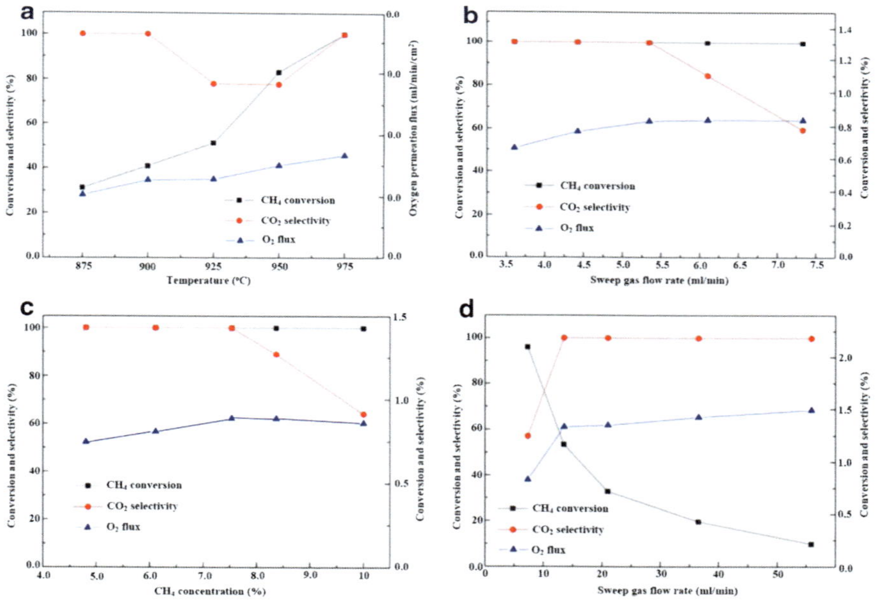


**Fig. 3.21** **a** Effect of temperature on $CH_4$ conversion, $CO_2$ selectivity, and oxygen permeation flux in the oxy-fuel combustion of methane [294]. Conditions: $F_{air} = 150$ mL/min, $F_{CH4+CO2} = 5.34$ mL/min, $C_{CH4} = 10\%$. **b** Effect of sweep-gas flow rate (low range) on $CH_4$ conversion, $CO_2$ selectivity, and oxygen permeation flux in the oxy-fuel combustion of methane at 975 °C [294]. Conditions: $F_{air} = 150$ mL/min, $C_{CH4} = 10\%$. **c** Effect of $CH_4$ concentration in the sweep gas on the $CH_4$ conversion, $CO_2$ selectivity, and oxygen permeation flux in the oxy-fuel combustion of methane at 975 °C [294]. Conditions: $F_{air} = 150$ mL/min, $F_{CH4+CO2} = 7.33$ mL/min. **d** Effect of sweep-gas flow rate (high range) on $CH_4$ conversion, $CO_2$ selectivity, and oxygen permeation flux in the oxy-fuel combustion of methane at 975 °C [294]. Conditions: $F_{air} = 150$ mL/min, $F_{CH4} = 0.733$ mL/min

the temperature and oxygen flux further increased, until complete combustion was achieved at 975 °C. Figure 3.21b explains why a flow rate of 5.34 mL/min was chosen for the $CH_4 + CO_2$ sweep gas. The oxygen permeation flux was found to increase with sweep-gas flow rate up to 0.7 at 5.34 mL/min, beyond which no significant flux enhancement was observed if the sweep-gas flow rate was further increased. Moreover, complete combustion (total oxidation) was sacrificed beyond 5.34 mL/min, which can be attributed to a Damkohler number below unity (chemical time scale exceeds flow residence time inside the membrane). Figure 3.21c explains the choice of $CH_4$ concentration. A peak in oxygen flux exists at 7.6% $CH_4$; for a $CH_4 + CO_2$ flow rate of 7.33 mL/min, the corresponding $CH_4$ flow is 0.557 mL/min, which is almost the same optimum value of 0.534 mL/min used in the temperature analysis of Fig. 3.21a. Complete combustion was not fully attained beyond 0.557 mL/min $CH_4$, which proves that this oxy-combustion process is extremely sensitive to $CH_4$ flow rate. Figure 3.21d finally analyzes the effect of increasing the sweep-gas flow rate well beyond the optimum value. The $CH_4$ flow rate was limited to 0.733 mL/min to avoid rich combustion. While the oxygen permeation flux increases to 1.5 mL $min^{-1}$ $cm^{-2}$ at 56 mL/min, the $CH_4$ conversion (partial oxidation) was sacrificed substantially (drops down to 10%). Since combustion commences by partial oxidation, it can be concluded that 90% of $CH_4$ exits the membrane as is, i.e., without reaction.

Recently, many studies investigated the integration of oxygen separation membranes in power cycles for clean energy production [290, 295–301]. Mancini and Mitsos [302] proposed a monolith-structure design of an oxygen transport reactor for power production in the range of 300–500 MWe. Based on 3-D modeling, Nemitallah et al. [290] designed an oxygen transport reactor for replacement of a gas turbine combustor to produce 5.0 MWe of power. Habib and Nemitallah [289] designed multi-separated membrane reactors for the application in fire-tube boilers, see Fig. 3.22, to produce about 5.0 MWe of power. The results are promising for future application of membrane reactors for power production; however, the size of the membrane reactor is much bigger than normal combustors for the same output power.

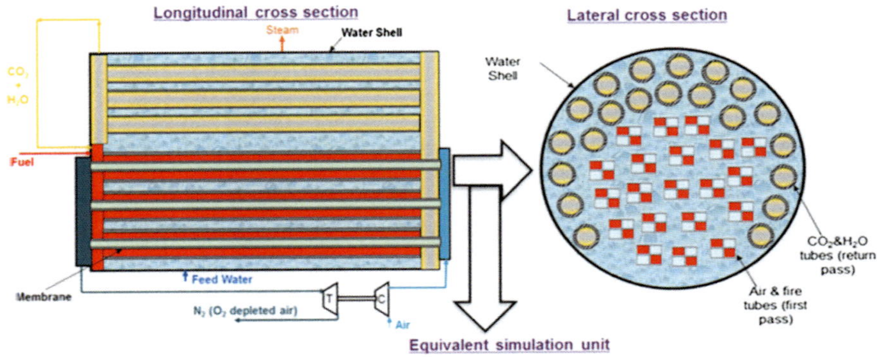


**Fig. 3.22** Design of a two-path fire tube boiler incorporating multi membrane reactors [289]

## 3.7 International Trends in CCS/CCUS Technologies

This section discusses the international trends in the application of carbon capture and sequestration/utilization (CCS/CCUS) technologies for the control of the global warming. Different routes for $CO_2$ utilization are discussed here in brief. Carbon dioxide ($CO_2$), the most important greenhouse gas, is gaining increasing interest as $C_1$ building-block for the synthesis of fuels and industrial chemicals [303]. Due to its thermodynamic stability, direct transformations of $CO_2$ are highly energy intensive processes resulting in an actual utilization of less than 1.0% of the global emission [304]. Two approaches have been adopted for $CO_2$ utilization, namely the direct utilization and the conversion of $CO_2$ into useful products. Direct $CO_2$ utilization is adopted in different industrial applications including water treatment, soft drinks, and fire extinguishers [305]. On the other hand, $CO_2$ conversion into useful products was found to be economic if linked with $CO_2$ capture technology [306]. In addition, a closed-loop carbon capture cycle can be built while combustion. But $CO_2$ has non-desirable combustion characteristics being an inert gas of low Gibbs free energy under moderate combustion condition [307]. Microalgae technique for $CO_2$ direct utilization can be promising as cultivating 1.0 ton of microalgae can fix 1.8 tons of environmental $CO_2$ [308]. Indirect use of $CO_2$ includes its use as a raw material to synthesize other useful chemical products through hydrogenation [309], cycloaddition to epoxides [310, 311], and carbonylation of alcohols or amines [312, 313]. Other chemical industries convert $CO_2$ into other products including urea and organic carbonates. $CO_2$ also can be utilized to produced syngas fuel, methanol, dimethyl ether, plastic products, and dimethyl carbonate [306, 314].

Existing $CO_2$ utilization technologies, based on the available literature, are summarized, quantified and reported in Table 3.4. These technologies make use of the captured $CO_2$ from power plants and gas turbines, as main emitters of $CO_2$. Some of these technologies have immense potential while others have limited potential to commercialization.

## 3.8 Concluding Remarks

In this chapter, the most recent combustion technologies have been discussed, including lean premixed air-fuel combustion, premixed oxy-fuel combustion, fuel-flexible combustion, and the integrated fuel-flexible premixed oxy-combustion. The latter has proven to facilitate effective control of both flame stability and emissions under the different loading conditions. Hydrogen-enriched combustion has shown better combustion efficiency, wider flame stability limits, and better fuel economy. The state-of-the-art burner designs for gas turbine applications have been addressed in detail in this chapter with their technology readiness level for the different applications. Different burner designs have been discussed including stagnation point reverse flow (SPRF), dry low-$NO_x$ (DLN) and dry low-emission (DLE), EnVironmental (EV,

**Table 3.4** Summary of $CO_2$ utilization technologies [315]

| Technology | References | Description |
|---|---|---|
| Algae cultivation | Benemann [316]<br>Wilson et al. [317]<br>Wang et al. [318] | $CO_2$ can be added to the growth process of algae, significantly increasing its productivity |
| Baking soda | Huang and Tan [319]<br>Song [320] | Another type of mineralization, $CO_2$ can be contacted with sodium rich brine, producing sodium bicarbonate or baking soda |
| Bauxite residue | Dilmore et al. [321]<br>Power et al. [322] | $CO_2$ can be injected into bauxite residue slurry to partially neutralize the slurry |
| Beverage carbonation | Descoins et al. [323]<br>Oelkers et al. [324] | High-purity $CO_2$ can be used to carbonate beverages |
| Chemical synthesis | North et al. [325]<br>Ma et al. [326] | $CO_2$ can be used as a feedstock in the production of a variety of chemicals, including acetic acid, alcohols, and sugars |
| Coffee decaffeination | Chen et al. [327]<br>Franca [328] | $CO_2$ can be used in its supercritical state as the solvent for decaffeinating coffee |
| Concrete curing | Zhan et al. [329]<br>Poon et al. [330] | At a precast concrete production facility, on-site waste $CO_2$ from flue gas can be permanently stored as limestone within the concrete |
| Electronics | Zhang and Han [331] | $CO_2$ is used, mostly as a cleaning fluid, in a small number of printed circuit board manufacturing applications |
| Enhanced Coal Bed Methane (ECBM) Recovery | Zuo-tang et al. [332]<br>Ozdemir [333] | $CO_2$ can be injected into a coal seam to displace methane, which is recovered at the surface |
| Enhanced Geothermal Systems (EGS) | Pruess [334]<br>Olasolo et al. [335] | Supercritical $CO_2$ could be used either as the circulating heat exchange fluid or as the working fluid in a supercritical $CO_2$ geothermal power cycle |

(continued)

**Table 3.4** (continued)

| Technology | References | Description |
|---|---|---|
| Enhanced oil recovery | Ferguson et al. [336] | $CO_2$, if injected into a depleted oil reservoir, decreases oil viscosity and allows the oil to flow to a production well more easily. The $CO_2$ can be permanently stored in the depleted oil reservoir when production concludes |
| Desalination | McGinnis et al. [337]<br>Al-Hallaj et al. [338]<br>EL-Naas et al. [339] | $CO_2$, mixed with $H_2O$ brine at high pressure and low temperature, forms a hydrate of $CO_2$ surrounded by $H_2O$ molecules. The hydrate is removed and rinsed, then goes through multiple stages to remove dissolved solids in the brine, resulting in an exhaust stream of potable water |
| Fire suppression technology | Guo and Fu [340]<br>Tilman et al. [341]<br>Lenihan et al. [342] | $CO_2$ can be applied to a fire to reduce the oxygen level low enough to stop combustion. It is also used in fire extinguishers and fire protection systems |
| Food processing, preservation, and packaging | Jasim and Alam [343]<br>Han [344]<br>Vaclavik and Christian [345] | $CO_2$ can be used during food processing for things such as cooling or spoilage prevention, and during packaging as an inert atmosphere to extend the shelf life of food products. Many industries, such as food packaging and wine making, use $CO_2$ as an inert gas |
| Formic acid | Schaub and Paciello [346]<br>Kortlever et al. [347] | $CO_2$ can be electro-reduced to produce formic acid and $O_2$ |
| Horticulture | Prior et al. [348]<br>Marbel et al. [349] | Plant growth in a greenhouse occurs at an optimal level of $CO_2$ concentration. $CO_2$ can be captured from on-site processes or can be provided by off-site industries |

(continued)

**Table 3.4** (continued)

| Technology | References | Description |
|---|---|---|
| Injection to conventional methanol synthesis | Huff and Sandford [350]<br>Zhang et al. [351] | $CO_2$ can be injected upstream of the methanol reformer during conventional methanol synthesis, increasing the yield of the system |
| Methanol | Wang et al. [352]<br>Kothandaraman et al. [353] | $CO_2$ can be combined with $H_2$, compressed, and reacted to produce methanol and water |
| Mineral carbonation | Huijgen et al. [354]<br>Mayoral et al. [355] | $CO_2$ can be used to produce a replacement aggregate product for the construction industry by contacting a medium-concentrate $CO_2$ with mineral-loaded alkaline brine. The $CO_2$ from the gas will precipitate out as a limestone or dolomite equivalent |
| Pharmaceutical processes | Huisman and Gray [356]<br>Kellaway [357] | In the pharmaceutical industry, $CO_2$ is used in many of the technologies listed, including inerting and supercritical fluid extraction |
| Polymer processing | Nalawade et al. [358]<br>Cooper [359] | $CO_2$ can be used as a feedstock in a variety of polymer processing routes |
| Power generation | Pipitone and Bolland [360]<br>Lombardi [361]<br>Beer [362] | Supercritical $CO_2$ could be used as the working fluid in various power cycles, regardless of the heat source |
| Pulp and paper processing | Oral et al. [363]<br>Gavrilescu [364] | $CO_2$ can be used to decrease acidity during pulp washing operations |
| Refrigerant gas | Kauf [365]<br>Colasson and Haberschill [366]<br>Agrawal and Roberts [367] | $CO_2$ can be used as a refrigerant gas replacement for more toxic gases |

(continued)

AEV and SEV), perforated plate (PP), and micromixer (MM) burners. Reverse flow burners exhibit very low $NO_x$ emission with wider flammability limits; however, this technology is still in the research and development phase for application in gas turbines. The advanced EnVironmental (AEV) burner resulted in the widest flame stability limits with the lowest level of $NO_x$ emissions among the considered burner

**Table 3.4** (continued)

| Technology | References | Description |
|---|---|---|
| Supercritical $CO_2$ as a solvent | Nalawade et al. [358]<br>Henon et al. [368] | Supercritical $CO_2$ can be used as a solvent for high-pressure extraction and to separate targeted compounds. It is also used in industries for dry cleaning for certain processes due to its low critical temperature |
| Water treatment | Espinoza and Santamarina [369] | $CO_2$ can be used to re-mineralize water after desalination process using the RO (reverse osmosis), or for the sake of reducing acidity |
| Welding | Sun et al. [370] | $CO_2$ is used to prevent oxidation of the welding metal |

designs in this chapter. Alstom invented such burner technology and implemented it in several industrial gas turbines. The novel MM burner design seems to have also wide stability limits, especially near the blowout limit. Micromixer technology is a well-developed technology with a technology-readiness level of 9.0, and it is already implemented on the utility scale (e.g., General Electric's F-class gas turbines). The application of oxy-fuel combustion technology is a very promising approach towards zero-emissions power plant (ZEPP) application while capturing $CO_2$. However, the cost of oxygen separation stands as the main barrier towards the wide application of this technology. Compact high temperature membrane reactors (HTMRs) seems to be a promising solution towards the reduction of oxygen-separation cost through the application of in situ oxy-combustion inside the membrane separation unit. However, more development is required for the membrane material to overcome the low oxygen permeation flux and membrane contamination after short period of operation.

**Acknowledgements** The authors wish to acknowledge the support received from King Fahd University of Petroleum & Minerals under Grant number BW191002 for the preparation of this book chapter.

## References

1. Pilavachi PA (2000) Power generation with gas turbine systems and combined heat and power. Appl Therm Eng 20(15):1421–1429
2. Tsoutsanis E, Meskin N, Benammar M, Khorasani K (2016) A dynamic prognosis scheme for flexible operation of gas turbines. Appl Energy 164:686–701

3. Tahan M, Tsoutsanis E, Muhammad M, Abdul Karim ZA (2017) Performance-based health monitoring, diagnostics and prognostics for condition-based maintenance of gas turbines: a review. Appl Energy 198:122–144
4. Akhtar S, Pi S, Shamim T (2018) Numerical investigation of flame structure and blowout limit for lean premixed turbulent methane-air flames under high pressure conditions. Appl Energy 228:21–32
5. Baltasar J, Carvalho MG, Coelho P, Costa M (1997) Flue gas recirculation in a gas-fired laboratory furnace: measurements and modelling. Fuel 76:919–929
6. Huang Y, Yang V (2009) Dynamics and stability of lean-premixed swirl-stabilized combustion. Prog Energy Combust Sci 35:293–364
7. Trimm DL (1983) Catalytic combustion (review). Appl Catal 7:249–282
8. Franzelli B, Riber E, Gicquel LYM, Poinsot T (2012) Large Eddy simulation of combustion instabilities in a lean partially premixed swirled flame. Combust Flame 159:621–637
9. Bobba MK, Gopalakrishnan P, Seitzman JM, Zinn BT (2006) Characteristics of combustion processes in a stagnation point reverse flow combustor. In: ASME Turbo Expo 2006: power for land, sea, and air, vol 1, GT2006-91217, pp 867–875
10. Evulet AT, ELKady AM, Branda AR, Chinn D (2009) On the performance and operability of GE's dry low $NO_x$ combustors utilizing exhaust gas recirculation for postcombustion carbon capture. Energy Proc 1:3809–3816
11. Aigner M, Müller G (1993) Second-generation low-emission combustors for ABB gas turbines: field measurements with GT11N-EV. J Eng Gas Turbines Power 115(3):533–536
12. Leung DYC, Caramanna G, Maroto-Valer MM (2014) An overview of current status of carbon dioxide capture and storage technologies. Renew Sustain Energy Rev 39:426–443
13. Nemitallah MA, Abdelhafez A, Habib MA (2019) Experimental and numerical investigations of structure and stability of premixed swirl-stabilized $CH_4/O_2/CO_2$ flames in a model gas turbine combustor. Energy Fuels 33:2526–2537
14. Kanniche M, Gros-Bonnivard R, Jaud P, Valle-Marcos J, Amann J-M, Bouallou C (2010) Pre-combustion, post-combustion and oxy-combustion in thermal power plant for $CO_2$ capture. Appl Therm Eng 30:53–62
15. Bender WR (2006) Lean pre-mixed combustion. In the gas turbine handbook. U.S. Department of Energy, Morgantown, WV 26505, pp 218–224
16. Ohkubo Y (2005) Low-$NO_x$ combustion technology, vol 41, Tech. No. 1
17. Mongia HC (1998) Aero-thermal design and analysis of gas turbine combustion systems: current status and future direction. In: 34th AIAA/ASME/SAE/ASEE joint propulsion conference and exhibit, 13–15 July 1998/Cleveland, OH
18. Hill SC, Smoot LD (2000) Modeling of nitrogen oxides formation and destruction in combustion systems. Prog Energy Combust Sci 26:417–458
19. Soo M, Jae H, Massoudi M, Hwang J (2017) Effect of staged combustion on low $NO_x$ emission from an industrial-scale fuel oil combustor in South Korea. Fuel 210:282–289. https://doi.org/10.1016/j.fuel.2017.08.065
20. Hayashi S, Yamada H (2000) $NO_x$ emissions in combustion of lean premixed mixtures injected into hot burned gas. Proc Combust Inst 28(2):2443–2449
21. Zhou H, Ren T, Yang Y (2015) Impact of OFA on combustion and NO x emissions of a large-scale laboratory furnace fired by a heavy-oil swirl burner. Appl Therm Eng 90:994–1006. https://doi.org/10.1016/j.applthermaleng.2015.07.076
22. Liu H, Liu Y, Yi G, Nie L, Che D (2013) Effects of air staging conditions on the combustion and $NO_x$ emission characteristics in a 600 MW wall fired utility boiler using lean coal. https://doi.org/10.1021/ef401354g
23. Hodzic N, Kazagic A, Smajevic I (2016) Influence of multiple air staging and reburning on $NO_x$ emissions during co-firing of low rank brown coal with woody biomass and natural gas. Appl Energy 168:38–47. https://doi.org/10.1016/j.apenergy.2016.01.081
24. Yang J, Sun R, Sun S, Zhao N, Hao N, Chen H et al (2014) Experimental study on $NO_x$ reduction from staging combustion of high volatile pulverized coals. Part 1. Air staging. Fuel Process Technol 126:266–75. https://doi.org/10.1016/j.fuproc.2014.04.034

25. Sreenivasan KR, Raghu S (2000) The control of combustion instability: a perspective. Curr Sci 79:867–883
26. Yamamoto T, Shimodaira K, Kurosawa Y, Yoshida S, Matsuura K (2017) Investigations of a staged fuel nozzle for aeroengines by multi-sector combustor test. In: Proceedings ASME Turbo Expo 2010 power land, sea air, pp 1–13
27. Yang J, Sun R, Sun S, Zhao N, Hao N, Chen H et al (2015) Experimental study on $NO_x$ reduction from staging combustion of high volatile pulverized coals. Part 2. Fuel staging. Fuel Process Technol 138:445–454. https://doi.org/10.1016/j.fuproc.2015.06.019
28. Zhao Y, Sun S, Zhang T, Zhou H (2013) Experimental research on fuel staging cyclone gasification of wood powder. Fuel 103:53–57. https://doi.org/10.1016/j.fuel.2011.08.020
29. Samarasinghe J, Culler W, Quay BD, Santavicca DA, Connor JO (2017) The Effect of fuel staging on the structure and instability characteristics of swirl-stabilized flames in a lean premixed multinozzle can combustor. J Eng Gas Turbines Power 139:1–10. https://doi.org/10.1115/1.4037461
30. Lin Y, Peng Y, Liu G (2017) Investigation on $NO_x$ of a low emission combustor design with multihole premixer-prevaporizer, pp 1–8
31. Office of Federal Register (2011) 60.2 Definitions, U.S. Code of Federal Regulations Title 40 Part 60, vol 6, pp 1–1106
32. Environmental Protection Agency (EPA) (1999) Nitrogen oxides ($NO_x$), why and how they are controlled. Epa-456/F-99-006R, November, p 48
33. Öztürk S, Eyriboyun M (2010) $NO_x$ formation in combustion of natural gases used in Turkey under different conditions. J Therm Sci Technol 30(2):95–102
34. Malte PC, Pratt DT (1974) The role of energy-releasing kinetics in $NO_x$ formation: fuel-lean, jet-stirred CO-air combustion. Combust Sci Technol 9(5–6):221–231
35. Bozzelli JW, Dean AM (1995) O + NNH: a possible new route for $NO_x$ formation in flames. Int J Chem Kinet 27(11):1097–1109
36. Carter CD, Barlow RS (1994) Simultaneous measurements of NO, OH, and the major species in turbulent flames. Opt Lett 19(4):299
37. Kumar P (2012) An experimental and numerical study of $NO_x$ formation mechanisms in NH3-H2-Air flames. Iowa State University
38. Zeldovich YB (1947) Oxidation of Nitrogen in combustion (transl by M. Shelef). Academy of Sciences of USSR, Institute of Chemical Physics, Moscow-Leningrad
39. Baukal C (2005) Everything you need to know about $NO_x$: controlling and minimizing pollutant emissions is critical for meeting air quality regulations. Met Finish 103(11):18–24
40. Zeini E, Hesham Y (2010) Q-Chem steam boilers $NO_x$ emissions reduction. In: Proceedings of the 2nd annual gas processing symposium, pp 41–49
41. Normann F (2010) Oxy-fuel combustion—the control of Nitrogen Oxides. Chalmers University of Technology, Göteborg, Sweden
42. Liss B, Wilson BR, Wilson BW (2017) System and method for producing low $NO_x$ air emissions from gasification power plants. US 20170218284A1
43. Liss B, Wilson BR, Wilson BW (2019) System and method for producing low $NO_x$ air emissions from gasification power plants. US010329500 B2
44. Fenimore CP (1971) Formation of nitric oxide in premixed hydrocarbon flames. Symp Combust 13(1):373–380
45. Barnes FJ, Bromly JH, Edwards TJ, Mandyczewsky R (1988) $NO_x$ emissions from radiant gas burners. J Inst Energy 155:184–188
46. Zhong BJ, Roslyakov PV (1996) Study on prompt $NO_x$ emission in boilers. J Therm Sci 5(2)
47. Lieuwen TC, Yang V (2013) Gas turbine emissions. Cambridge aerospace series. Cambridge University Press, Cambridge
48. Schefer RW, Namazian M, Kelly J (1991) Simultaneous Raman scattering and laser-induced fluorescence for multispecies imaging in turbulent flames. Opt Lett 16(11):858
49. Schefer RW, Namazian M, Kelly J (1991) CH, OH and $CH_4$ concentration measurements in a lifted turbulent-jet flame. Symp Combust 23(1):669–676

50. Mcdonell V et al (2014) Development and evaluation of gas fuel interchangeability criteria and methodologies, California
51. Fackler KB, Karalus MF, Novosselov IV, Kramlich JC, Malte PC (2011) Experimental and numerical study of $NO_x$ formation from the lean premixed combustion of $CH_4$ mixed with $CO_2$ and $N_2$. In: Proceedings ASME Turbo Expo, vol 2, Parts A and B, pp 55–63
52. Cho ES, Chung SH (2009) Improvement of flame stability and $NO_x$ reduction in hydrogen-added ultra lean premixed combustion. J Mech Sci Technol 23(3):650–658
53. Pavri R, Moore GD (2003) Gas turbines emissions and control. GE Power Syst 17:29–43, no. GER-4211
54. Bender WR. Lean pre-mixed combustion. In: Gas turbine handbook, pp 217–227
55. Richards G, Weiland N, Strakey P (2006) Combustion strategies for syngas and high-hydrogen fuel. In: Gas turbine handbook, pp 203–211
56. Glarborg P, Jensen AD, Johnsson JE (2003) Fuel nitrogen conversion in solid fuel fired systems. Prog Energy Combust Sci 29(2):89–113
57. Schnelle KB, Dunn RF, Ternes ME (2016) Air pollution control technology handbook, 2nd edn. CRC Press
58. Zhang L, Su D, Zhong M (2015) The effect of functional forms of nitrogen on fuel-$NO_x$ emissions. Environ Monit Assess 187(1):1–8
59. Nakata T, Sato M, Hasegawa T (1998) Reaction of fuel $NO_x$ formation for gas turbine conditions. J Eng Gas Turbines Power 120(3):474–480
60. Aliyu M, Nemitallah MA, Said SA, Habib MA (2016) Characteristics of $H_2$-enriched $CH_4$-$O_2$ diffusion flames in a swirl-stabilized gas turbine combustor: experimental and numerical study. Int J Hydrogen Energy 41:20418–20432. https://doi.org/10.1016/j.ijhydene.2016.08.144
61. Lacarelle A, Moeck J, Konle H, Vey S (2007) Effect of fuel/air mixing on $NO_x$ emissions and stability in a gas premixed combustion system. In: AIAA Aerospace Sciences Meeting and Exhibit 2007, pp 1–14. https://doi.org/10.2514/6.2007-1417
62. Bunce NA, Quay BD, Santavicca DA (2013) Interaction between swirl number fluctuations and vortex shedding in a single-nozzle turbulent swirling fully-premixed combustor. J Eng Gas Turbines Power 136:021503. https://doi.org/10.1115/1.4025361
63. Lieuwen T, Cho JH (2005) Coherent acoustic wave amplification/damping by wrinkled flames. J Sound Vib 279:669–686. https://doi.org/10.1016/j.jsv.2003.11.050
64. Huang Y, Ratner A (2009) Experimental investigation of thermoacoustic coupling for low-swirl lean premixed flames. J Propuls Power 25:365–373. https://doi.org/10.2514/1.36310
65. Emadi M, Kaufman K, Burkhalter MW, Salameh T, Gentry T, Ratner A (2015) Examination of thermo-acoustic instability in a low swirl burner. Int J Hydrogen Energy 40:13594–13603. https://doi.org/10.1016/j.ijhydene.2015.08.016
66. Hart RW, McClure FT (1965) Theory of acoustic instability in solid-propellant rocket combustion. Symp Combust 10:1047–1065. https://doi.org/10.1016/S0082-0784(65)80246-6
67. Crocco L (1969) Research on combustion instability in liquid propellant rockets. Symp Combust 12:85–99. https://doi.org/10.1016/S0082-0784(69)80394-2
68. Marshall WM, Pal S, Woodward RD, Santoro RJ (2006) Combustion instability studies using gaseous methane and liquid oxygen. In: 42nd AIAA/ASME/SAE/ASEE Joint Propulsion Conference & Exhibit 2006
69. Sutton G, Biblarz O (2001) Rocket propulsion elements. https://doi.org/10.1017/cbo9781107415324.004
70. Candel S (2002) Combustion dynamics and control: progress and challenges. Proc Combust Inst 29:1–28. https://doi.org/10.1016/S1540-7489(02)80007-4
71. Wicker JM, Greene WD, Kim SI, Yang V (1996) Triggering of longitudinal combustion instabilities in solid rocket motors: nonlinear combustion response. J Propuls Power 12:1148–1158. https://doi.org/10.2514/3.24155
72. Lieuwen TC, Yang V (2006) Combustion instabilities in gas turbine engines. https://doi.org/10.2514/4.866807
73. Sirignano WA, Liu F (1999) Performance increases for gas-turbine engines through combustion inside the turbine. J Propuls Power 15:111–118. https://doi.org/10.2514/2.5398

74. Stone C, Menon S (2002) Swirl control of combustion instabilities in a gas turbine combustor. Proc Combust Inst 29:155–160. https://doi.org/10.1016/S1540-7489(02)80024-4
75. Venkataraman KK, Preston LH, Simons DW, Lee BJ, Lee JG, Santavicca DA (1999) Mechanism of combustion instability in a lean premixed dump combustor. J Propuls Power 15:909–918. https://doi.org/10.2514/2.5515
76. Murat Altay H, Speth RL, Hudgins DE, Ghoniem AF (2009) The impact of equivalence ratio oscillations on combustion dynamics in a backward-facing step combustor. Combust Flame 156:2106–2116. https://doi.org/10.1016/j.combustflame.2009.07.024
77. Rashwan SS, Nemitallah MA, Habib MA (2016) Review on premixed combustion technology: stability, emission control, applications, and numerical case study. Energy Fuels. https://doi.org/10.1021/acs.energyfuels.6b02386
78. Li L, Guo Z, Zhang C, Sun X (2010) A passive method to control combustion instabilities with perforated liner. Chin J Aeronaut 23:623–630. https://doi.org/10.1016/S1000-9361(09)60263-6
79. Hoffmann S, Weber G, Judith H, Hermann J, Otthmann A (1998) Application of active combustion control to siemens heavy duty gas turbines. Appl Veh Technol Panel Symp Lisbon, Port, vol 14. https://doi.org/10.1115/1.2818459
80. Kokanović S, Guidati G, Torchalla S, Schuermans B (2006) Active combustion control system for reduction of $NO_x$ and pulsation levels in gas turbines. In: ASME Turbo Expo 2006 power land, sea air, pp 1–10
81. Lacarelle A, Matho L, Paschereit CO (2010) Scalar mixing enhancement in a swirl stabilized combustor trough passive and active injection control. In: 48th AIAA Aerospace Sciences Meeting Including the New Horizons Forum and Aerospace Exposition 2010. https://doi.org/10.2514/6.2010-1332
82. Lacarelle A, Paschereit CO (2012) Increasing the passive scalar mixing quality of jets in crossflow with fluidics actuators. J Eng Gas Turbines Power 134:021503. https://doi.org/10.1115/1.4004373
83. Lee H, Hernandez S, McDonell V, Steinthorsson E, Mansour A, Hollon B (2009) Development of flashback resistant low-emission micro-mixing fuel injector for 100% hydrogen and syngas fuels. In: Proceedings ASME Turbo Expo 2009 power land, sea air, pp 411–419. https://doi.org/10.1115/gt2009-59502
84. Estefanos WS (2016) Effects of the fuel-air mixing on combustion instabilities and $NO_x$ emissions in lean premixed combustion
85. Habib M et al (2011) A review of recent developments in carbon capture utilizing oxy-fuel combustion in conventional and ion transport membrane systems. Int J Energy Res 35(9):741–764
86. Edenhofer O (2014) Mitigation of climate change. Cambridge University Press, Cambridge
87. Stanger R, Wall T, Spörl R, Paneru M, Grathwohl S, Weidmann M, Scheffknecht G, McDonald D, Myöhänen K, Ritvanen J (2015) Int J Greenhouse Gas Control 40:55–125
88. Nemitallah MA, Habib MA (2013) Experimental and numerical investigations of an atmospheric diffusion oxy-combustion flame in a gas turbine model combustor. Appl Energy 111:401–415
89. Nemitallah M, Alkhaldi S, Abdelhafez A, Habib M (2018) Effect analysis on the macrostructure and static stability limits of oxy-methane flames in a premixed swirl combustor. Energy 159:86–96
90. Amato A, Hudak B, D'Carlo P, Noble D, Scarborough D, Seitzman J, Lieuwen T (2011) Methane Oxy-combustion for Low $CO_2$ cycles: blow-off measurements and analysis. J Eng Gas Turbines Power 33:061503-1
91. Kutne P, Kapadia K, Meier W, Aigner M (2011) Experimental analysis of the combustion behavior of oxyfuel flames in a gas turbine model combustor. Proc Combust Inst 33:3383–3390
92. Watanabe H, Shanbhogue S, Taamallah S, Chakroun NW, Ghoniem AF (2016) The structure of swirl-stabilized turbulent premixed $CH_4$/air and $CH_4/O_2/CO_2$ flames and mechanisms of intense burning of oxy-flames. Combust Flame 174:111–119

93. Runyon J et al (2015) Methane-Oxygen flame stability in a generic premixed gas turbine swirl combustor at varying thermal power and pressure. In: ASME Turbo Expo 2015: turbine technical conference and exposition 2015. American Society of Mechanical Engineers
94. Xie Y et al (2013) Experimental and numerical study on laminar flame characteristics of methane oxy-fuel mixtures highly diluted with $CO_2$. Energy Fuels 27(10):6231–6237
95. Song Y et al (2015) The chemical mechanism of the effect of $CO_2$ on the temperature in methane oxy-fuel combustion. Int J Heat Mass Transf 86:622–628
96. Li Y-H et al (2015) Effects of flue gas recirculation on the premixed oxy-methane flames in atmospheric condition. Energy 89:845–857
97. Kim TH et al (2016) Chemical and radiation effects on flame extinction and $NO_x$ formation in oxy-methane combustion diluted with $CO_2$. Fuel 177:235–243
98. Habib MA, Nemitallah MA, Ahmed P, Sharqawy MH, Badr HM, Muhammad I, Yaqub M (2015) Experimental analysis of oxygen-methane combustion inside a gas turbine reactor under various operating conditions. Energy 86:105–114
99. Granados DA, Chejne F, Mejía JM (2015) Oxy-fuel combustion as an alternative for increasing lime production in rotary kilns. Appl Energy 158:107–117
100. Nemitallah MA (2016) A study of methane oxy-combustion characteristics inside a modified design button-cell membrane reactor utilizing a modified oxygen permeation model for reacting flows. J Nat Gas Sci Eng 28:61–73
101. Nemitallah MA, Habib MA, Mezghani K (2015) Experimental and numerical study of oxygen separation and oxy-combustion characteristics inside a button-cell LNO-ITM reactor. Energy 84:600–611
102. Habib MA, Salaudeen SA, Nemitallah MA, Ben-Mansour R, Mokheimer EMA (2016) Numerical investigation of syngas oxy-combustion inside a LSCF-6428 oxygen transport membrane reactor. Energy 96:654–665
103. Taamallah S, Vogiatzaki K, Alzahrani FM, Mokheimer EMA, Habib MA, Ghoniem AF (2015) Fuel flexibility, stability and emissions in premixed hydrogen-rich gas turbine combustion: Technology, fundamentals, and numerical simulations. Appl Energy 154:1020–1047
104. Wicksall DM, Agrawal AK, Schefer RW, Keller JO (2005) The interaction of flame and flow field in a lean premixed swirl-stabilized combustor operated on $H_2/CH_4$/air. Proc Combust Inst 30(2):2875–2883
105. Williams TC, Shaddix CR, Schefer RW (2007) Effect of syngas composition and $CO_2$-diluted oxygen on performance of a premixed swirl-stabilized combustor. Combust Sci Technol 180(1):64–88
106. Rashwan SS, Ibrahim AH, Abou-arab TW (2015) Experimental investigation of oxy-fuel combustion of CNG flames stabilized over a perforated-plate burner. In: 18th International Flame Res Found. Friesing, Munich, pp 1–11
107. Ramadan IA, Ibrahim AH, Abou-Arab TW, Rashwan SS, Nemitallah MA, Habib MA (2016) Effects of oxidizer flexibility and bluff-body blockage ratio on flammability limits of diffusion flames. Appl Energy 178:19–28
108. Jerzak W, Kuźnia M (2016) Experimental study of impact of swirl number as well as oxygen and carbon dioxide content in natural gas combustion air on flame flashback and blow-off. J Nat Gas Sci Eng 29:46–54
109. Shi B, Zhu Z, Wang N, Lub P, Ishizukac I (2015) An experimental study on oxy-fuel combustion of methane under various oxygen mole fractions. In: 8th international symposium on coal combustion, Beijing, China
110. Joo PH, Charest MRJ, Groth CPT, Gülder ÖL (2013) Comparison of structures of laminar methane–oxygen and methane–air diffusion flames from atmospheric to 60atm. Combust Flame 160(10):1990–1998
111. Ditaranto M, Anantharaman R, Weydahl T (2013) Performance and $NO_x$ emissions of refinery fired heaters retrofitted to hydrogen combustion. Energy Proc 37:7214–7220
112. Ditaranto M, Hals J (2006) Combustion instabilities in sudden expansion oxy–fuel flames. Combust Flame 146(3):493–512

113. Speth RL, Altay HM, Ghoniem AF (2012) Dynamics and stability limits of syngas combustion, pp 1–10
114. Kutne P, Kapadia BK, Meier W, Aigner M (2011) Experimental analysis of the combustion behaviour of oxyfuel flames in a gas turbine model combustor. Proc Combust Inst 33(2):3383–3390
115. Oh J, Noh D, Ko C (2013) The effect of hydrogen addition on the flame behavior of a non-premixed oxy-methane jet in a lab-scale furnace. Energy 62:362–369
116. Liu CY, Chen G, Sipöcz N, Assadi M, Bai XS (2012) Characteristics of oxy-fuel combustion in gas turbines. Appl Energy 89(1):387–394
117. Yin C, Yan J (2016) Oxy-fuel combustion of pulverized fuels: combustion fundamentals and modeling. Appl Energy 162:742–762
118. Rashwan SS, Ibrahim AH, Abou-Arab TW, Nemitallah MA, Habib MA (2016) Experimental investigation of partially premixed methane-air and methane-oxygen flames stabilized over a perforated-plate burner. Appl Energy 169:126–137
119. Taamallah S, Chakroun NW, Watanabe H, Shanbhogue SJ, Ghoniem AF (2017) On the characteristic flow and flame times for scaling oxy and air flame stabilization modes in premixed swirl combustion. Proc Combust Inst 36(3):3799–3807
120. Watanabe H, Shanbhogue SJ, Taamallah S, Chakroun NW, Ghoniem AF (2016) The structure of swirl-stabilized turbulent premixed $CH_4$/air and $CH_4/O_2/CO_2$ flames and mechanisms of intense burning of oxy-flames. Combust Flame
121. Amato A, Hudak R, Noble DR, Scarborough D, D'Carlo PA, Seitzman JM, Lieuwen T (2010) Methane oxy-combustion for low $CO_2$ cycles: blowoff measurements and modeling. In: Proceedings Asme Turbo Expo, pp 1–11
122. Rashwan SS, Ibrahim AH, Abou-Arab TW, Nemitallah MA, Habib MA (2017) Experimental study of atmospheric partially premixed oxy-combustion flames anchored over a perforated plate burner. Energy
123. Abdelhafez A, Nemitallah MA, Rashwan SS, Habib MA (2018) Adiabatic flame temperature for controlling the macrostructures and stabilization modes of premixed methane flames in a model gas-turbine combustor. Energy Fuels 32:7868–7877
124. Jerzak W, Kuźnia M (2016) Experimental study of impact of swirl number as well as oxygen and carbon dioxide content in natural gas combustion air on flame flashback and blow-off. J Nat Gas Sci Eng 29:46–54
125. Hu X, Yu Q, Liu J, Sun N (2014) Investigation of laminar flame speeds of $CH_4/O_2/CO_2$ mixtures at ordinary pressure and kinetic simulation. Energy
126. Ditaranto M, Saanum I (2015) Operation of a swirl-stabilized oxy-fuel burner in pressurized conditions. Flexible and clean fuel conversion to industry, Freising, Germany, 1–3 June 2015, Paper n. 23
127. Treloar RD (2005) Gas installation technology. Wiley-Blackwell, February 2009, ISBN: 978-1-405-14682-1
128. Casleton KH et al (2008) System issues and tradeoffs associated with syngas production and combustion, vol 2202, May 2008
129. Klimstra J (1986) Interchangeability of gaseous fuels—the importance of the Wobbe-index. SAE Trans 95(6):962–972
130. Döbbeling K, Meeuwissen T, Zajadatz M, Flohr P (2008) Fuel flexibility of the Alstom GT13E2 medium sized gas turbine, no. 43130, pp 719–725
131. Richards GA, McMillian MM, Gemmen RS, Rogers WA, Cully SR (2001) Issues for low-emission, fuel-flexible power systems. Prog Energy Combust Sci 27(2):141–169
132. Mcdonell V, Petersen E (2018) Fuel flexibility influences on premixed combustor blowout, flashback, autoignition, and stability, vol 130, Jan 2008, pp 1–10
133. Cocchi S, Sigali S (2010) Development of a low-$NO_x$ hydrogen-fuelled combustor for 10 MW class gas turbines, no 43970, pp 1025–1035
134. Cellek MS, Pınarbas A (2017) ScienceDirect investigations on performance and emission characteristics of an industrial low swirl burner while burning natural gas, methane, hydrogen-enriched natural gas and hydrogen as fuels, vol 3

135. Patil KR, Khanwalkar PM, Thipse SS, Kavathekar KP (2009) Development of HCNG blended fuel engine with control of $NO_x$ emissions. In: Second international conference on emerging trends in engineering & technology, pp 1068–1074
136. Nagalingam B, Duebel F, Schmillen K (1983) Performance study using natural gas, hydrogen-supplemented natural gas and hydrogen in AVL research engine. Int J Hydrogen Energy 8(9):715–720
137. Lamnaouer M (2007) Flashback analysis of ULN hydrogen enriched natural gas mixtures. Tech Rep, Univ Cent, Florida
138. Hoekstra RL, Collier K, Mulligan N, Chew L (1995) Experimental study of a clean burning vehicle fuel. Int J Hydrogen Energy 20(9):737–745
139. Munshi SR, Nedelcu C, Harris J, Edwards T, Williams J, Lynch F et al (2004) Hydrogen blended natural gas operation of a heavy duty turbocharged lean burn spark ignition engine. SAE Tech Pap
140. Ma F, Liu H, Wang Y, Li Y, Wang J, Zhao S (2008) Combustion and emission characteristics of a port-injection HCNG engine under various ignition timings. Int J Hydrogen Energy 33(2):816–822
141. Ma F et al (2012) Effect of compression ratio and spark timing on the power performance and combustion characteristics of an HCNG engine. Int J Hydrogen Energy 37(23):18486–18491
142. Wang J et al (2007) Combustion behaviors of a direct-injection engine operating on various fractions of natural gas–hydrogen blends. Int J Hydrogen Energy 32(15):3555–3564
143. Ma F, Wang Y, Liu H, Li Y, Wang J, Zhao S (2007) Experimental study on thermal efficiency and emission characteristics of a lean burn hydrogen enriched natural gas engine. Int J Hydrogen Energy 32(18):5067–5075
144. Chacartegui R, Sánchez D, De Escalona JM, Muñoz A, Sánchez T (2013) Gas and steam combined cycles for low calorific syngas fuels utilization. Appl Energy 101:81e92
145. He F, Li Z, Liu P, Ma L, Pistikopoulos EN (2012) Operation window and part-load performance study of a syngas fired gas turbine. Appl Energy 89:133e41
146. Chacartegui R, Sánchez D, de Escalona JM, Jimenez-Espadafor F, Muñoz A, Sánchez T (2012) Sphera project: assessing the use of syngas fuels in gas turbines and combined cycles from a global perspective. Fuel Process Tech 103:134–145
147. Frassoldati A, Faravelli T, Ranzi E (2007) The ignition, combustion and flame structure of carbon monoxide/hydrogen mixtures. Note 1: detailed kinetic modeling of syngas combustion also in presence of nitrogen compounds. Int J Hydrogen Energy 32(15):3471–3485
148. Noble DR, Zhang QG, Lieuwen T (2006) Hydrogen effects upon flashback and blowout. In: Proceedings of ICEPAG, p 24012
149. Lieuwen T (2008) Flashback characteristics of syngas-type fuels under steady and pulsating conditions. In: School of aerospace engineering. Georgia Institute of Technology, Atlanta, Georgia. http://doi.org/10.2172/924769
150. Fu J, Tang C, Jin W, Thi LD, Huang Z, Zhang Y (2013) Study on laminar flame speed and flame structure of syngas with varied compositions using oh-plif and spectrograph. Int J Hydrogen Energy 38:1636–1643
151. Wang J, Huang Z, Kobayashi H, Ogami Y (2012) Laminar burning velocities and flame characteristics of $CO\text{-}H_2\text{-}CO_2\text{-}O_2$ mixtures. Int J Hydrogen Energy 37:19158–19167
152. Lee HC, Jiang LY, Mohamad AA (2014) A review on the laminar flame speed and ignition delay time of Syngas mixtures. Int J Hydrogen Energy 39:1105–1121
153. Verkamp FJ, Hardin M, Williams JR (1967) Ammonia combustion properties and performance in gas turbine burners. In: International symposium of combustion, vol 11, pp 985–992
154. Pratt DT (1967) Performance of ammonia fired gas turbine combustors. San Diego, USA. Report T-9-TS-67-5 Solar
155. Newhall H, Starkman ES (1966) Theoretical performance of ammonia as a gas turbine fuel. SAE, Technical paper 660768
156. Li J, Huang H, Kobayashi N, He Z, Nagai Y (2014) Study on using hydrogen and ammonia as fuels: combustion characteristics and $NO_x$ formation. Int J Energy Res 38:1214–1223

157. Ryu K, Zacharakis-Jutz GE, Kong SC (2014) Performance enhancement of ammonia-fuelled engine by using dissociation catalyst for hydrogen generation. Int J Hydrogen Energy 39:2390–2398
158. Lear B (2012) Ammonia-fuelled combustion turbines. Available online https://nh3fuel.files.wordpress.com/2012/05/lear_nh3.pdf (2017-10-29)
159. Karabeyoglu A, Brian E, Stevens J, Cantwell B (2012) Development of ammonia based fuels for environmentally friendly power generation. In: 10th international energy conversion engineering conference, p 4055
160. Karabeyoglu A, Brian E (2012) Fuel conditioning system for ammonia fired power plants. NH3 Fuel Association. Available online https://nh3fuel.files.wordpress.com/2012/10/evans-brian.pdf (2016-11-21)
161. Valera-Medina A, Xiao H, Owen-Jones M, David WIF, Bowen PJ (2018) Ammonia for power. Prog Energy Combust Sci 69:63–102
162. Zhen HS, Cheung CS, Leung CW, Choy YS (2012) Effects of hydrogen concentration on the emission and heat transfer of a premixed LPG-hydrogen flame. Int J Hydrogen Energy 37:6097–6105
163. Mansouri Z, Aouissi M, Boushaki T (2016) Numerical computations of premixed propane flame in a swirl-stabilized burner: effects of hydrogen enrichment, swirl number and equivalence ratio on flame characteristics. Int J Hydrogen Energy 41:9664–9678
164. Wu L, Kobayashi N, Li Z, Huang H (2016) Experimental study on the effects of hydrogen addition on the emission and heat transfer characteristics of laminar methane diffusion flames with oxygen-enriched air. Int J Hydrogen Energy 41:2023–2036
165. Mazas A, Fiorina B, Lacoste D, Schuller T (2011) Effects of water vapor addition on the laminar burning velocity of oxygen-enhanced methane flames. Combust Flame 158:2428–2440
166. Richards G, McMillian M, Gemmen R, Rogers W, Cully S (2001) Issues for low-emission, fuel-flexible power systems. Prog Energy Combust Sci 27:141–169
167. Lieuwen T, McDonell V, Petersen E, Santavicca D (2008) Fuel flexibility influences on premixed combustor blowout, flashback, auto ignition, and stability. J Eng Gas Turbines Power 130:11506–11516
168. Griebel P, Boshek E, Jansohn P (2007) Lean blowout limits and $NO_x$ emissions of turbulent, lean premixed, hydrogen-Enriched methane air flames at high pressure. J Eng Gas Turbines Power 129:404–410
169. Wietschel M, Ball M (2009) The future of hydrogen—opportunities and challenges. Hydrog Econ Oppor Challenges 613–639. https://doi.org/10.1017/cbo9780511635359.021
170. Edwards PP, Kuznetsov VL, David WIF, Brandon NP (2008) Hydrogen and fuel cells: towards a sustainable energy future. Energy Policy 36:4356–4362. https://doi.org/10.1016/J.ENPOL.2008.09.036
171. European Commission (2008) Hyways: the European hydrogen roadmap: contract SES6-502596, EUR-OP, Luxembourg
172. Choudhuri AR, Gollahalli SR (2000) Combustion characteristics of hydrogen-hydrocarbon hybrid fuels. Int J Hydrogen Energy 25:451–462. https://doi.org/10.1016/S0360-3199(99)00027-0
173. Gu M, Chu H, Liu F (2016) Effects of simultaneous hydrogen enrichment and carbon dioxide dilution of fuel on soot formation in an axisymmetric coflow laminar ethylene/air diffusion flame. Combust Flame 166:216–228. https://doi.org/10.1016/j.combustflame.2016.01.023
174. Park SH, Lee KM, Hwang CH (2011) Effects of hydrogen addition on soot formation and oxidation in laminar premixed $C_2H_2$/air flames. Int J Hydrogen Energy 36:9304–9311. https://doi.org/10.1016/j.ijhydene.2011.05.031
175. Pandey P, Pundir BP, Panigrahi PK (2007) Hydrogen addition to acetylene-air laminar diffusion flames: studies on soot formation under different flow arrangements. Combust Flame 148:249–262. https://doi.org/10.1016/j.combustflame.2006.09.004
176. Gülder ÖL, Snelling DR, Sawchuk RA (1996) Influence of hydrogen addition to fuel on temperature field and soot formation in diffusion flames. Symp Combust 26:2351–2358. https://doi.org/10.1016/S0082-0784(96)80064-6

177. Halter F, Chauveau C, Gökalp I (2007) Characterization of the effects of hydrogen addition in premixed methane/air flames. Int J Hydrogen Energy 32:2585–2592. https://doi.org/10.1016/j.ijhydene.2006.11.033
178. Gersen S, Anikin NB, Mokhov AV, Levinsky HB (2008) Ignition properties of methane/hydrogen mixtures in a rapid compression machine. Int J Hydrogen Energy 33:1957–1964. https://doi.org/10.1016/j.ijhydene.2008.01.017
179. Cheng RK, Oppenheim AK (1984) Autoignition in methane hydrogen mixtures. Combust Flame 58:125–139. https://doi.org/10.1016/0010-2180(84)90088-9
180. Hu E, Huang Z, He J, Jin C, Zheng J (2009) Experimental and numerical study on laminar burning characteristics of premixed methane–hydrogen–air flames. Int J Hydrogen Energy 34:4876–4888. https://doi.org/10.1016/j.ijhydene.2009.03.058
181. Di Sarli V, Di Benedetto A (2007) Laminar burning velocity of hydrogen–methane/air premixed flames 32:637–646. https://doi.org/10.1016/j.ijhydene.2006.05.016
182. Tang C, Huang Z, Jin C, He J, Wang J, Wang X et al (2008) Laminar burning velocities and combustion characteristics of propane–hydrogen–air premixed flames. Int J Hydrogen Energy 33:4906–4914. https://doi.org/10.1016/j.ijhydene.2008.06.063
183. Sankaran R, Im HG (2006) Effects of hydrogen addition on the Markstein length and flammability limit of stretched methane/air premixed flames. Combust Sci Technol 178:37–41. https://doi.org/10.1080/00102200500536217org/10.1080/00102200500536217
184. Nakahara M, Kido H (2008) Study on the turbulent burning velocity of hydrogen mixtures including hydrocarbons. AIAA J 46. https://doi.org/10.2514/1.23560
185. Daniele S, Jansohn P, Mantzaras J, Boulouchos K (2011) Turbulent flame speed for syngas at gas turbine relevant conditions. Proc Combust Inst 33:2937–2944. https://doi.org/10.1016/J.PROCI.2010.05.057
186. Daniele S, Jansohn P, Boulouchos K (2009) Experimental investigation of lean premixed syngas combustion at gas turbine relevant conditions: lean blow out limits, emissions and turbulent flame speed. Ital Sect Combust Inst
187. Wang J, Huang Z, Tang C, Miao H, Wang X (2009) Numerical study of the effect of hydrogen addition on methane-air mixtures combustion. Int J Hydrogen Energy 34:1084–1096. https://doi.org/10.1016/j.ijhydene.2008.11.010
188. Schefer RW (2003) Hydrogen enrichment for improved lean flame stability. Int J Hydrogen Energy 28:1131–1141
189. Kim HS, Arghode VK, Gupta AK (2009) Flame characteristics of hydrogen-enriched methane-air premixed swirling flames. Int J Hydrogen Energy 34:1063–1073. https://doi.org/10.1016/j.ijhydene.2008.10.035
190. Cheng RK, Littlejohn D (2008) Laboratory study of premixed $H_2$-Air and $H_2$–$N_2$-air flames in a low-swirl injector for ultralow emissions gas turbines. J Eng Gas Turbines Power 130:031503. https://doi.org/10.1115/1.2836480
191. Zhang Q, Noble D, Lieuwen T (2006) Characterization of fuel composition effects in $H_2$/CO/$CH_4$ mixtures upon lean blowout. J Eng Gas Turbines Power 129(3):688–694
192. Taamallah S, Vogiatzaki K, Alzahrani FM, Mokheimer EMA, Habib MA, Ghoniem AF (2015) Fuel flexibility, stability and emissions in premixed hydrogen-rich gas turbine combustion: technology, fundamentals, and numerical simulations. Appl Energy 154:1020–1047. https://doi.org/10.1016/j.apenergy.2015.04.044
193. Lounici MS, Boussadi A, Loubar K, Tazerout M (2014) Experimental Investigation on NG dual fuel engine improvement by hydrogen enrichment. Int J Hydrogen Energy 39:21297–21306
194. Schefer RW, Wicksall DM, Agrawal AK (2002) Combustion of hydrogen-enriched methane in a lean premixed swirl-stabilized burner. Proc Combust Inst 29:843–851
195. Tuncer O (2013) Premixed combustion of hydrogen and syngas fuels in gas turbine combustors. In: International conference on renewable energy research and applications, Madrid, Spain, pp 20–23
196. Kim S-H, Yoon Y, Jeung I-S (2000) Nitrogen oxides emissions in turbulent hydrogen jet non-premixed flames: effects of coaxial air and flame radiation. Proc Combust Inst 28:463–470

197. Arai M (2000) Flue gas recirculation for low $NO_x$ combustion system. In: Proceedings of 2000 international joint power generation conference, Miami Beach, Florida, vol 7
198. Ling Zhongqian, Zhou Hao, Ren Tao (2015) Effect of the flue gas recirculation supply location on the heavy oil combustion and $NO_x$ emission characteristics within a pilot furnace fired by a swirl burner. Energy 91:110–116
199. Tahmasebzadehbaie M, Sayyaadi H (2016) Efficiency enhancement and $NO_x$ emission reduction of a turbo-compressor gas engine by mass and heat recirculations of flue gases. Appl Therm Eng 99:661–671
200. Neumeier Y, Weksler Y, Zinn BT, Seitzman JM, Jagoda J, Kenny J (2005) Ultra-low emissions combustor with non-premixed reactants injection. In: AIAA 2005-3775 41st AIAA/ASME/SAE/ASEE joint propulsion conference & exhibit, 10–13 July 2005, Tucson, Arizona
201. Neumeier Y, Zinn B, Weksler Y, Seitzman J, Jacoga J, Kenny J (2005) Novel Combustion for ultra-low emissions with non-premixed reactants injection. AIAA-2005-3775
202. Bobba MK (2007) Flame stabilization and mixing characteristics in a stagnation point reverse flow combustor. PhD Thesis, Georgia Institute of Technology, Atlanta, GA
203. Gopalatrishnan P (2008) Effects of the reacting flowfield on combustion process in a stagnation point reverse flow combustor. PhD Thesis, Georgia Institute of Technology, Atlanta, GA
204. Gopalakrishnan P, Bobba MK, Seitzman JM (2007) Controlling mechanisms for low $NO_x$ emissions in a non-premixed stagnation point reverse flow combustor. Proc Combust Inst 31:3401–3408
205. Bobba MK, Gopalakrishnan P, Periagaram K, Seitzman JM (2008) Flame structure and stabilization mechanisms in a stagnation-point reverse-flow combustor. J Eng Gas Turbines Power 130:031505-1
206. Castela M, Veríssimo AS, Rocha AMA, Costa M (2012) Experimental study of the combustion regimes occurring in a laboratory combustor. Combust Sci Technol 184(2):243–258
207. Undapalli S, Srinivasan S, Menon S (2009) LES of premixed and non-premixed combustion in a stagnation point reverse flow combustor. Proc Combust Inst 32:1537–1544
208. Wnning JA, Wnning JG (1997) Flameless oxidation to reduce thermal no-formation. Prog Energy Combust Sci 23(1):8194
209. Perpignan AAV, Gangoli Rao A, Roekaerts DJEM (2018) Flameless combustion and its potential towards gas turbines. Prog Energy Combust Sci 69:2862
210. Cavaliere A, de Joannon M (2004) Mild combustion. Prog Energy Combust Sci 30(4):329366
211. Kruse S, Kerschgens B, Berger L, Varea E, Pitsch H (2015) Experimental and numerical study of MILD combustion for gas turbine applications. Appl Energy 148:456–465
212. Katsuki M, Hasegawa T (1998) The science and technology of combustion in highly preheated air. In: 27th symposium (International) on combustion, 27, 31353146
213. Doan NAK, Swaminathan N, Minamoto Y (2018) DNS of MILD combustion with mixture fraction variations. Combust Flame 189:173189
214. Luckerath R, Meier W, Aigner M (2008) Flox combustion at high pressure with different fuel compositions. J Eng Gas Turbines Power 130:011505
215. Levy Y, Sherbaum V, Arfi P (2004) Basic thermodynamics of Floxcom, the low-$NO_x$ gas turbines adiabatic combustor. Appl Therm Eng 24:1593–1605
216. Lammel O, Schutz H, Schmitz G, Luckerath R, Stohr M, Noll B, Aigner M, Hase M, Krebs W (2010) Flox combustion at high power density and high flame temperatures. J Eng Gas Turbines Power 132:121503
217. Arghode VK, Gupta AK, Bryden KM (2012) High intensity colorless distributed combustion for ultra low emissions and enhanced performance. Appl Energy 92:822–830
218. Alne KS (2007) Reduction of $NO_x$ emissions from the gas turbines for Skarv Idun (Master's thesis, Institutt for energi-og prosessteknikk)
219. Huang Y, Yang V (2009) Dynamics and stability of lean-premixed swirl-stabilized combustion. Prog Energy Combust Sci 35(4):293–364
220. Nemitallah MA, Rashwan SS, Mansir IB, Abdelhafez AA, Habib MA (2018) Review of novel combustion techniques for clean power production in gas turbines. Energy Fuels 32(2):979–1004

221. Boyce MP (2012) Gas turbine engineering handbook, 4th edn. Elsevier, pp 66–70. ISBN 0123838428
222. Schorr MM, Chalfin J (1999) Gas turbine $NO_x$ emissions approaching zero—is it worth the price? General Electric Power Generation, Report No. GER, 4172
223. Lefebvre AH, Ballal DR (2010) Gas turbine combustion: alternative fuels and emissions, 3rd edn. CRC press, pp 398–414
224. Lee S, Speight JG, Loyalka SK (2015) Handbook of alternative fuel technologies, 2nd edn. CRC Press, Boca Raton, FL
225. Davis LB (1996) Dry low $NO_x$ combustion systems for GE heavy-duty gas turbines. In: ASME 1996 international gas turbine and aeroengine congress and exhibition. American Society of Mechanical Engineers, pp V003T06A003-V003T06A003
226. Zajadatz M, Pennell D, Bernero S, Paikert B, Zoli R, Döbbeling K (2013) Development and implementation of the advanced environmental burner for the Alstom GT13E2. J Eng Gas Turbines Power 135(6):061503
227. Zajadatz M, Lachner R, Bernero S, Motz C, Flohr P (2007) Development and design of Alstoms's staged fuel gas injection EV burner for $NO_x$ reduction. In: Proceedings of ASME Turbo Expo 2007 power land, sea air, 14–17 May 2007, Montreal, Canada
228. Döbbeling K, Hellat J, Koch H (2007) 25 Years of BBC/ABB/Alstom lean premix combustion technologies. J Eng Gas Turbines Power
229. Lebovich S (1984) Vortex stability and breakdown, survey and extension. AIAA J 22(9):1192–1206
230. Escadier MP (1988) Vortex breakdown: observations and explanations. Prog Aero Sci 25:189–229
231. Blouch J, Li H, Mueller M, Hook R (2012) Fuel flexibility in LM2500 and LM6000 dry low emission engines. J Eng Gas Turbines Power 134(5):051503
232. Eckardt D (2014) Gas turbine powerhouse: the development of the power generation gas turbine at BBC–ABB–ALSTOM, 2nd edn
233. Bothien MR, Pennel D, Zajadatz M, Döbbeling K (2013) On key features of the AEV burner engine implementation for operational flexibility. In: ASME Turbo Expo, pp GT2013-95693
234. Marchmont C, Florjancic S, Kappis W (2013) Alstom gas turbine technology developments. In: ASME Turbo Expo, pp GT2013-95625
235. Marchmont C, Florjancic S (202) Alstom gas turbine technology trends. In: ASME Turbo Expo, pp 587–598
236. Sattelmayer T, Felchlin MP, Haumann J, Hellat J, Styner D (1992) Second generation low-emission combustors for ABB gas turbines: burner development and tests at atmospheric pressure. J Eng Gas Turbines Power 114:118–125
237. Aigner M, Mayer A, Schiessel P, Strittmatter W (1990) Second-generation low-emission combustors for ABB gas turbines: tests under full-engine conditions. In: Proceedings of the ASME Turbo Expo
238. Aigner M, Müller G (1993) Second-generation low-emission combustors for ABB gas turbines: field measurements with GT11N-EV. J Eng Gas Turbines Power
239. Paschereit CO, Flohr P, Knöpfel H, Geng W, Steinbach C, Stuber P, Bengtsson K, Gutmark E (2002) Combustion control by extended EV burner fuel lance. In: American Society of Mechanical Engineers, International Gas Turbine Institute, Turbo Expo (Publication) IGTI
240. Cho CH, Baek GM, Sohn CH, Cho JH, Kim HS (2013) A numerical approach to reduction of $NO_x$ emission from swirl premix burner in a gas turbine combustor. Appl Therm Eng
241. Güthe F, Lachner R, Schuermans B, Biagioli F (2006) Flame imaging on the ALSTOM EV-burner: thermo acoustic pulsations and CFD-validation. AIAA Pap
242. Biagioli F, Güthe F, Schuermans B (2008) Combustion dynamics linked to flame behaviour in a partially premixed swirled industrial burner. Exp Therm Fluid Sci
243. Döbbeling K, Meeuwissen T, Zajadatz M, Flohr P (2008) Fuel flexibility of the Alstom GT13E2 medium sized gas turbine. In: ASME Turbo Expo, pp 719–725
244. Magni F, Grimm F, Sorato S, Micheli M (2016) GT13E2 low part load operation: extended flexibility down to 30% load. In: ASME Turbo Expo, pp GT2016-57317

245. Reiss F, Griffin T, Reyser K (2002) The ALSTOM GT13E2 medium BTU gas turbine. In: ASME Turbo Expo, pp 705–712
246. Wind T, Güthe F, Syed K (2014) Co-firing of hydrogen and natural gases in lean premixed conventional and reheat burners (Alstom GT26). In: Volume 4A: combustion, fuels and emissions
247. Funke HHW, Beckmann N, Keinz J, Abanteriba S (2017) Numerical and experimental evaluation of a dual-fuel dry-low-$NO_x$ micromix combustor for industrial gas turbine applications. In: Proceedings of the ASME Turbo Expo
248. Haj Ayed A, Kusterer K, Funke HHW, Keinz J, Striegan C, Bohn D (2015) Experimental and numerical investigations of the dry-low-$NO_x$ hydrogen micromix combustion chamber of an industrial gas turbine. Propuls Power Res
249. Asai T, Dodo S, Karishuku M, Yagi N, Akiyama Y, Hayashi A (2015) Performance of multiple-injection dry low-$NO_x$ combustors on hydrogen-rich syngas fuel in an IGCC pilot plant. J Eng Gas Turbines Power
250. Rudyak V, Minakov A (2014) Modeling and optimization of Y-type micromixers. Micromachines 5(4):886–912
251. Karnik R. Microfluidic mixing, pp 1177–1186
252. Karnidakis G, Beskok A, Aluru N (2005) Microflows and nanoflows
253. Conlisk AT (2007) Introduction to microfluidics. By Patrick Tabeling. Oxford University Press, 312 pp. ISBN 019 856864 9. £49.95 (hardback), J Fluid Mech 570:503
254. Hoffmann M, Schluter M, Rubiger N (2006) Experimental investigation of liquid–liquid mixing in T-shaped micro-mixers using $\mu$-LIF and $\mu$-PIV. Chem Eng Sci 61:2968–2976
255. Hong CC, Choi JW (2004) A novel in-plane passive microfluidic mixer with modified Tesla structures. Lab Chip 4:109–113
256. Jiménez J (2005) The growth of a mixing layer in a laminar channel. J Fluid Mech 535:245–254
257. Aubin J, Fletcher DF, Xuereb C (2005) Design of micromixers using CFD modelling. Chem Eng Sci 60(8–9):2503–2516 (SPEC ISS)
258. Vanka SP, Luo G, Winkler CM (2004) Numerical study of scalar mixing in curved channels at low Reynolds numbers. AIChE J 50(10):2359–2368
259. Stroock AD, Dertinger SK, Whitesides GM, Ajdari A (2002) Patterning flows using grooved surfaces. Anal Chem 74(20):5306–5312
260. Wang L, Ma S, Jing W, Bi H, Han X (2014) Mixing enhancement of a passive microfluidic mixer containing triangle baffles, vol 9
261. Hong F, Cao J, Cheng P (2011) A parametric study of AC electrothermal flow in microchannels with asymmetrical interdigitated electrodes, vol 38
262. Kedia KS, Ghoniem AF (2012) Mechanisms of stabilization and blowoff of a premixed flame downstream of a heat-conducting perforated plate. Combust Flame 159(3):1055–1069
263. Jithin EV, Kishore VR, Varghese RJ (2014) Three-dimensional simulations of steady perforated-plate stabilized propane-air premixed flames. Energy Fuels 28(8):5415–5425
264. Altay HM, Park S, Wu D, Wee D, Annaswamy AM, Ghoniem AF (2009) Modeling the dynamic response of a laminar perforated-plate stabilized flame. Proc Combust Inst 32 I(1):1359–1366
265. Jamal M, Ibrahim H, Ali M, Elmahallawy M, Abdelhafez A, Nemitallah A, Rashwan S, Habib A (2017) Structure and lean extinction of premixed flames stabilized on conductive perforated plates. Energy Fuels
266. Kedia KS, Ghoniem AF (2013) An analytical model for the prediction of the dynamic response of premixed flames stabilized on a heat-conducting perforated plate. Proc Combust Inst 34(1):921–928
267. Rashwan SS, Ibrahim AH, Abou-Arab TW, Nemitallah MA, Habib MA (2017) Experimental study of atmospheric partially premixed oxy-combustion flames anchored over a perforated plate burner. Energy 122:159–167
268. Edacheri Veetil J, Aravind B, Mohammad A, Kumar S, Velamati RK (2017) Effect of hole pattern on the structure of small scale perorated plate burner flames. Fuel 216:722–733

269. Hindasageri V, Kuntikana P, Vedula RP, Prabhu SV (2015) An experimental and numerical investigation of heat transfer distribution of perforated plate burner flames impinging on a flat plate. Int J Therm Sci 94:156–169
270. Oh S, Shin Y, Kim Y (2016) Stabilization effects of perforated plates on the combustion instability in a lean premixed combustor. Appl Therm Eng 107:508–515
271. Wang CJ, Wen JX (2014) The effect of a perforated plate on the propagation of laminar hydrogen flames in a channel—a numerical study. Int J Hydrogen Energy 39(36):21335–21342
272. Li Q et al (2018) Experimental study of flame propagation across a perforated plate. Int J Hydrogen Energy 43(17):8524–8533
273. York WD, Ziminsky WS, Yilmaz E (2013) Development and testing of a low $NO_x$ hydrogen combustion system for heavy-duty gas turbines. Eng Gas Turbines Power 135:022001
274. Funke HHW, Boerner S, Keinz J, Kusterer K, Kroniger D, Kitajima J, Kazari M, Horikawa A (2012) Numerical and experimental characterization of low $NO_x$ micromix combustion principle for industrial hydrogen gas turbine applications. Proc ASME Turbo Expo 2:1069–1079
275. Dodo S, Asai T, Koizumi H, Takahashi H, Yoshida S, Inoue H (2011) Combustion characteristics of a multiple-injection combustor for dry low-$NO_x$ combustion of hydrogen-rich fuels under medium pressure. Proc ASME Turbo Expo 2:467–476
276. Weiland NT, Sidwell TG, Strakey PA (2013) Testing of a hydrogen diffusion flame array injector at gas turbine conditions. Combust Sci Technol
277. Funke HHW, Boerner S, Krebs W, Wolf E (2011) Experimental characterization of low nox micromix prototype combustors for industrial gas turbine applications. In: Proceedings of the ASME Turbo Expo
278. Dodo S, Asai T, Koizumi H, Takahashi H, Yoshida S, Inoue H (2013) Performance of a multiple-injection dry low $NO_x$ combustor with hydrogen-rich syngas fuels. J Eng Gas Turbines Power
279. Hollon B, Steinthorsson S, Mansour A, McDonell V, Lee H (2011) Ultra-low emission hydrogen/syngas combustion with a 1.3 MW injector using a micro-mixing lean-premix system. Vol. 2 Combust Fuels Emiss Parts A B
280. Hernandez SR, Wang Q, McDonell V, Mansour A, Steinthorsson E, Hollon B (2008) Micro mixing fuel injectors for low emissions hydrogen combustion. Vol. 3 Combust Fuels Emiss Parts A B
281. Zhang Y, Yang T, Liu X, Tian L, Fu Z, Zhang K (2012) Reduction of emissions from a syngas flame using micromixing and dilution with $CO_2$. Energy Fuels
282. Zhang Y, Zhang H, Tian L, Ji P, Ma S (2015) Temperature and emissions characteristics of a micro-mixing injection hydrogen-rich syngas flame diluted with $N_2$. Int J Hydrogen Energy
283. Chen L, Yong SZ, Ghoniem A (2012) Oxy-fuel combustion of pulverized coal: characterization, fundamentals, stabilization and CFD modeling. Prog Energy Combust Sci 38:156–214
284. Haslbeck J, Capicotto P, Juehn N, Lewis E, Rutkowski M, Woods M et al (2007) Bituminous coal to electricity, vol 1. Washington D.C., DOE/NETL-1291
285. BERR (2007) AEA Energy & Environment. COAL R309 BERR/Pub URN 07/1251
286. Andersson K, Johnsson F (2006) Process evaluation of an 865 MWe lignite fired $O_2/CO_2$ power plant. Energy Convers Manage 47:3487–3498
287. Okawa M, Kimura N, Kiga T, Takano S, Arai K, Kato M (1997) Trial design for a $CO_2$ recovery power plant by burning pulverized coal in $O_2/CO_2$. Energy Convers Manage 38:S123–S127
288. Varagani R, Chatel F, Pranda P, Rostam M, Lu Y, Bose A (2005) The 4th annual conference on carbon sequestration. Alexandria, VA
289. Habib MA, Nemitallah MA (2015) Design of an ion transport membrane reactor for application in fire tube boilers. Energy 81:787–801
290. Nemitallah MA, Habib MA, Ben-Mansour R, Ghoniem AF (2014) Design of an ion transport membrane reactor for gas turbine combustion application. J Memb Sci 450:60–71

291. Nemitallah MA, Habib MA, Ben-Mansour R (2013) Investigations of oxy-fuel combustion and oxygen permeation in an ITM reactor using a two-step oxy-combustion reaction kinetics model. J Memb Sci 432:1–12
292. Kirchen P, Apo DJ, Hunt A, Ghoniem AF (2013) A novel ion transport membrane reactor for fundamental investigations of oxygen permeation and oxy-combustion under reactive flow conditions. Proc Combust Inst 34(2):3463–3470
293. Tan X, Li K, Thursfield A, Metcalfe IS (2008) Oxyfuel combustion using a catalytic ceramic membrane reactor. Catal Today 131(1–4):292–304
294. Wei Y, Wang Y, Tang J, Li Z, Wang H (2013) Oxy-fuel combustion for $CO_2$ capture using a $CO_2$-tolerant oxygen transporting membrane. AIChE J 59(10):3856–3862
295. Puig-Arnavat M, Søgaard M, Hjuler K, Ahrenfeldt J, Henriksen UB, Hendriksen PV (2015) Integration of oxygen membranes for oxygen production in cement plants. Energy 91:852–865
296. Duan L, Yue L, Qu W, Yang Y (2015) Study on $CO_2$ capture from molten carbonate fuel cell hybrid system integrated with oxygen ion transfer membrane. Energy 93:20–30
297. Hwang KR, Park JW, Lee SW, Hong S, Lee C-B, Oh DK et al (2015) Catalytic combustion of the retentate gas from a $CO_2/H_2$ separation membrane reactor for further $CO_2$ enrichment and energy recovery. Energy 90:1192–1198
298. Chiesa P, Romano MC, Spallina V, Turi DM, Mancuso L (2013) Efficient low $CO_2$ emissions power generation by mixed conducting membranes. Energy Proc 37:905–913
299. Manzolini G, Gazzani M, Turi DM, Macchi E (2013) Application of hydrogen selective membranes to IGCC. Energy Proc 37:2274–2283
300. Voleno A, Romano MC, Turi DM, Chiesa P, Ho MT, Wiley DE (2014) Post-combustion $CO_2$ capture from natural gas combined cycles by solvent supported membranes. Energy Proc 63:7389–7397
301. Romano MC (2013) Ultra-high $CO_2$ capture efficiency in CFB oxyfuel power plants by calcium looping process for $CO_2$ recovery from purification units vent gas. Int J Greenh Gas Control 18:57–67
302. Mancini ND, Mitsos A (2011) Ion transport membrane reactors for oxy-combustion part II: analysis and comparison of alternatives. Energy 36:4721e39
303. Aresta M, Dibenedetto A, Quaranta E (2016) State of the art and perspectives in catalytic processes for $CO_2$ conversion into chemicals and fuels: the distinctive contribution of chemical catalysis and biotechnology. J Catal 343:2–45
304. An R, Yu B, Li R, Wei Y-M (2018) Potential of energy savings and $CO_2$ emission reduction in China's iron and steel industry. Appl Energy 226:862–880
305. Belizón M, Fernández-Ponce MT, Casas L, Mantell C, Martínez de la Ossa-Fernández EJ (2018) Supercritical impregnation of antioxidant mango polyphenols into a multilayer PET/PP food-grade film. J $CO_2$ Utilization 25:56–67
306. Rafiee A, Khalilpour KR, Milani D, Panahi M (2018) Trends in $CO_2$ conversion and utilization: a review from process systems perspective. J Environ Chem Eng 6:5771–5794
307. Wang L, Alsaker N, Skreiberg Q, Hovd B (2017) Effect of carbonization conditions on $CO_2$ gasification reactivity of biocarbon. Energy Proc 142:932–937
308. Durán I, Rubiera F, Pevida C (2018) Microalgae: potential precursors of $CO_2$ adsorbents. J $CO_2$ Utilization 26:454–464
309. Jiang X, Jiao Y, Moran C, Nie X, Gong Y, Guo X, Walton KS, Song C (2018) $CO_2$ hydrogenation to methanol on PdCu bimetallic catalysts with lower metal loadings. Catal Commun, in press. https://doi.org/10.1016/j.catcom.2018.09.006
310. Zhang S, Liu X, Li M, Wei Y, Zhang G, Han J, Zhu X, Ge Q, Wang H (2018) Metal-free amino-incorporated organosilica nanotubes for cooperative catalysis in the cycloaddition of $CO_2$ to epoxides. Catal Today, in press. https://doi.org/10.1016/j.cattod.2018.07.004
311. Maya EM, Rangel-Rangel E, Díaz U, Iglesias M (2018) Efficient cycloaddition of CO2 to epoxides using novel heterogeneous organocatalysts based on tetramethylguanidine-functionalized porous polyphenylenes. J $CO_2$ Utilization 25:170–179
312. Marchegiani M, Nodari M, Tansini F, Massera C, Della Ca N (2017) Urea derivatives from carbon dioxide and amines by guanidine catalysis: easy access to imidazolidin-2-ones under solvent-free conditions. J $CO_2$ Utilization 21:553–561

313. Liu P, Yang Y, White MG (2013) Theoretical perspective of alcohol decomposition and synthesis from $CO_2$ hydrogenation. Surf Sci Rep 68:233–272
314. Leonzio G (2018) State of art and perspectives about the production of methanol, dimethyl ether and syngas by carbon dioxide hydrogenation. J $CO_2$ Utilization 27:326–354
315. Davis T, Bucher T, Jesse Goellner RM (2013) Novel $CO_2$ utilization concepts. https://doi.org/10.1016/j.enconman.2007.12.029
316. Benemann JR (1997) $CO_2$ mitigation with microalgae systems. Energy Convers Manage 38:S475–S479. https://doi.org/10.1016/S0196-8904(96)00313-5
317. Wilson MHH, Groppo J, Placido A, Graham S, Morton SA, Santillan-Jimenez E et al (2014) $CO_2$ recycling using microalgae for the production of fuels. Appl Petrochemical Res 1–13. https://doi.org/10.1007/s13203-014-0052-3
318. Wang B, Li Y, Wu N, Lan CQ (2008) $CO_2$ bio-mitigation using microalgae. Appl Microbiol Biotechnol 79:707–718. https://doi.org/10.1007/s00253-008-1518-y
319. Huang CH, Tan CS (2014) A review: $CO_2$ utilization. Aerosol Air Qual Res 14:480–499. https://doi.org/10.4209/aaqr.2013.10.0326
320. Song C (2002) $CO_2$ Conversion and utilization: an overview. $CO_2$ Convers Util 809:1–2. https://doi.org/10.1021/bk-2002-0809.ch001
321. Dilmore R, Lu P, Allen D, Soong Y, Hedges S, Fu JK et al (2008) Sequestration of $CO_2$ in mixtures of bauxite residue and saline wastewater. Energy Fuels 22:343–353. https://doi.org/10.1021/ef7003943
322. Power G, Gräfe M, Klauber C (2011) Bauxite residue issues: I. Current management, disposal and storage practices. Hydrometallurgy 108:33–45. https://doi.org/10.1016/j.hydromet.2011.02.006
323. Descoins C, Mathlouthi M, Le Moual M, Hennequin J (2006) Carbonation monitoring of beverage in a laboratory scale unit with on-line measurement of dissolved $CO_2$. Food Chem 95:541–553. https://doi.org/10.1016/j.foodchem.2004.11.031
324. Oelkers EH, Gislason SR, Matter J (2008) Mineral carbonation of $CO_2$. Elements 4:333–337. https://doi.org/10.2113/gselements.4.5.333
325. North M, Pasquale R, Young C (2010) Synthesis of cyclic carbonates from epoxides and $CO_2$. Green Chem 12:1514. https://doi.org/10.1039/c0gc00065e
326. Ma J, Sun N, Zhang X, Zhao N, Xiao F, Wei W et al (2009) A short review of catalysis for $CO_2$ conversion. Catal Today 148:221–231. https://doi.org/10.1016/j.cattod.2009.08.015
327. Chen Y, Brown PH, Hu K, Black RM, Prior RL, Ou B et al (2011) Supercritical $CO_2$ decaffeination of unroasted coffee beans produces melanoidins with distinct NF-κB inhibitory activity. J Food Sci 76. https://doi.org/10.1111/j.1750-3841.2011.02304.x
328 Franca AS (2016) Coffee: decaffeination. Encycl Food Heal 232–236. http://dx.doi.org/10.1016/B978-0-12-384947-2.00183-5
329. Zhan BJ, Xuan DX, Poon CS, Shi CJ (2016) Effect of curing parameters on $CO_2$ curing of concrete blocks containing recycled aggregates. Cem Concr Compos 71:122–130. https://doi.org/10.1016/j.cemconcomp.2016.05.002
330. Zhan B, Poon C, Shi C (2013) $CO_2$ curing for improving the properties of concrete blocks containing recycled aggregates. Cem Concr Compos 42:1–8. https://doi.org/10.1016/j.cemconcomp.2013.04.013
331. Zhang X, Han B (2007) Cleaning using $CO_2$-based solvents. Clean Soil Air Water 35:223–229. https://doi.org/10.1002/clen.200700007
332. Zuo-tang W, Guo-xiong W, Rudolph V, Diniz da Costa JC, Pei-ming H, Lin X (2009) Simulation of $CO_2$-geosequestration enhanced coal bed methane recovery with a deformation-flow coupled model. Proc Earth Planet Sci 1:81–89. https://doi.org/10.1016/j.proeps.2009.09.015
333. Ozdemir E (2009) Modeling of coal bed methane (CBM) production and $CO_2$ sequestration in coal seams. Int J Coal Geol 77:145–152. https://doi.org/10.1016/j.coal.2008.09.003
334. Pruess K (2006) Enhanced geothermal systems (EGS) using $CO_2$ as working fluid-A novel approach for generating renewable energy with simultaneous sequestration of carbon. Geothermics 35:351–367. https://doi.org/10.1016/j.geothermics.2006.08.002

335. Olasolo P, Juárez MC, Morales MP, Damico S, Liarte IA (2016) Enhanced geothermal systems (EGS): a review. Renew Sustain Energy Rev 56:133–144. https://doi.org/10.1016/j.rser.2015.11.031
336. Ferguson RC, Nichols C, Van Leeuwen T, Kuuskraa VA (2009) Storing $CO_2$ with enhanced oil recovery. Energy Proc 1:1989–1996. https://doi.org/10.1016/j.egypro.2009.01.259
337. McGinnis RL, Hancock NT, Nowosielski-Slepowron MS, McGurgan GD (2013) Pilot demonstration of the $NH_3/CO_2$ forward osmosis desalination process on high salinity brines. Desalination 312:67–74. https://doi.org/10.1016/j.desal.2012.11.032
338. Al-Hallaj S, Parekh S, Farid MM, Selman JR (2006) Solar desalination with humidification-dehumidification cycle: review of economics. Desalination 195:169–186. https://doi.org/10.1016/j.desal.2005.09.033
339. El-Naas MH, Al-Marzouqi AH, Chaalal O (2010) A combined approach for the management of desalination reject brine and capture of CO2. Desalination 251:70–74. https://doi.org/10.1016/j.desal.2009.09.141
340. Guo TN, Fu ZM (2007) The fire situation and progress in fire safety science and technology in China. Fire Saf J 42:171–182. https://doi.org/10.1016/j.firesaf.2006.10.005
341. Tilman D, Reich P, Phillips H, Menton M, Patel A, Vos E et al (2000) Fire suppression and ecosystem carbon storage. Ecology 81:2680–2685. https://doi.org/10.1890/0012-9658(2000)081%5b2680:fsaecs%5d2.0.co;2
342. Lenihan JM, Bachelet D, Neilson RP, Drapek R (2008) Simulated response of conterminous United States ecosystems to climate change at different levels of fire suppression, $CO_2$ emission rate, and growth response to $CO_2$. Glob Planet Change 64:16–25. https://doi.org/10.1016/j.gloplacha.2008.01.006
343. Ahmed J, Alam T (2012) An overview of food packaging: material selection and the future of packaging. Handb Food Process Des 1237–1283. https://doi.org/10.1002/9781444398274.ch41
344. Han JH (2005) Innovations in food packaging. Innov Food Packag 517. https://doi.org/10.1016/b978-012311632-1/50046-2
345. Vaclavik V, Christian E (2008) Food preservation and processing. Essentials Food Sci 425–446. https://doi.org/10.1007/978-0-387-69940-0_17
346. Schaub T, Paciello RA (2011) A process for the synthesis of formic acid by $CO_2$ hydrogenation: thermodynamic aspects and the role of CO. Angew Chemie Int Ed 50:7278–7282. https://doi.org/10.1002/anie.201101292
347. Kortlever R, Balemans C, Kwon Y, Koper MTM (2015) Electrochemical $CO_2$ reduction to formic acid on a Pd-based formic acid oxidation catalyst. Catal Today 244:58–62. https://doi.org/10.1016/j.cattod.2014.08.001
348. Prior SA, Brett Runion G, Christopher Marble S, Rogers HH, Gilliam CH, Allen Torbert H (2011) A review of elevated atmospheric CO2 effects on plant growth and water relations: implications for horticulture. HortScience 46:158–162
349. Christopher Marble S, Prior SA, Brett Runion G, Allen Torbert H, Gilliam CH, Fain GB (2011) The importance of determining carbon sequestration and greenhouse gas mitigation potential in ornamental horticulture. HortScience 46:240–244
350. Huff CA, Sanford MS (2011) Cascade catalysis for the homogeneous hydrogenation of $CO_2$ to methanol. J Am Chem Soc 133:18122–18125. https://doi.org/10.1021/ja208760j
351. Zhang C, Jun KW, Kwak G, Lee YJ, Park HG (2016) Efficient utilization of carbon dioxide in a gas-to-methanol process composed of CO2/steam-mixed reforming and methanol synthesis. J CO2 Util 16:1–7. https://doi.org/10.1016/j.jcou.2016.05.005
352. Wang WH, Himeda Y, Muckerman JT, Manbeck GF, Fujita E (2015) $CO_2$ hydrogenation to formate and methanol as an alternative to photo- and electrochemical $CO_2$ reduction. Chem Rev 115:12936–12973. https://doi.org/10.1021/acs.chemrev.5b00197
353. Kothandaraman J, Goeppert A, Czaun M, Olah GA, Prakash GKS (2016) Conversion of $CO_2$ from air into methanol using a polyamine and a homogeneous ruthenium catalyst. J Am Chem Soc 138:778–781. https://doi.org/10.1021/jacs.5b12354

354. Huijgen WJJ, Comans RNJ, Witkamp GJ (2007) Cost evaluation of $CO_2$ sequestration by aqueous mineral carbonation. Energy Convers Manag 48:1923–1935. https://doi.org/10.1016/j.enconman.2007.01.035
355. Mayoral MC, Andrés JM, Gimeno MP (2013) Optimization of mineral carbonation process for $CO_2$ sequestration by lime-rich coal ashes. Fuel 106:448–454. https://doi.org/10.1016/j.fuel.2012.11.042
356. Huisman GW, Gray D (2002) Towards novel processes for the fine-chemical and pharmaceutical industries. Curr Opin Biotechnol 13:352–358. https://doi.org/10.1016/S0958-1669(02)00335-X
357. Kellaway I (2001) Transport processes in pharmaceutical systems, vol 228. https://doi.org/10.1016/s0378-5173(01)00823-7
358. Nalawade SP, Picchioni F, Janssen LPBM (2006) Supercritical carbon dioxide as a green solvent for processing polymer melts: processing aspects and applications. Prog Polym Sci 31:19–43. https://doi.org/10.1016/j.progpolymsci.2005.08.002
359. Cooper AI (2000) Polymer synthesis and processing using supercritical carbon dioxide. J Mater Chem 10:207–234. https://doi.org/10.1039/a906486i
360. Pipitone G, Bolland O (2009) Power generation with $CO_2$ capture: technology for $CO_2$ purification. Int J Greenh Gas Control 3:528–534. https://doi.org/10.1016/j.ijggc.2009.03.001
361. Lombardi L (2003) Life cycle assessment comparison of technical solutions for $CO_2$ emissions reduction in power generation. Energy Convers Manag 44:93–108. https://doi.org/10.1016/S0196-8904(02)00049-3
362. Beér JM (2007) High efficiency electric power generation: the environmental role. Prog Energy Combust Sci 33:107–134. https://doi.org/10.1016/j.pecs.2006.08.002
363. Oral J, Sikula J, Puchyr R, Hajny Z, Stehlik P, Bebar L (2005) Processing of waste from pulp and paper plant. J Clean Prod 13:509–515. https://doi.org/10.1016/j.jclepro.2003.09.005
364. Gavrilescu D (2008) Energy from biomass in pulp and paper mills. Environ Eng Manag J 7:537–546
365. Kauf F (1999) Determination of the optimum high pressure for transcritical $CO_2$-refrigeration cycles. Int J Therm Sci 38:325–330. https://doi.org/10.1016/S1290-0729(99)80098-2
366. Colasson S, Haberschill P (2010) Effect of refrigerant charge on global performances of a transcritical $CO_2$ heat pump. Sustain Refrig Heat Pump Technol Conf Stock, Sweden, pp 1–7. https://doi.org/10.3969/j.issn.0258-2724.2010.05.005
367. Agrawal R, Roberts M (2000) Dual mixed refrigerant cycle for gas liquefaction
368. Hénon FE, Camaiti M, Burke ALC, Carbonell RG, DeSimone JM, Piacenti F (1999) Supercritical $CO_2$ as a solvent for polymeric stone protective materials. J Supercrit Fluids 15:173–179. https://doi.org/10.1016/s0896-8446(99)00005-4
369. Lee SY, Seo S, Broda JC, Pal S, Santoro RJ (2000) An experimental estimation of mean reaction rate and flame structure during combustion instability in a lean premixed gas turbine combustor. Proc Combust Inst 28:775–782. https://doi.org/10.1016/S0082-0784(00)80280-5
370. Sun J, Liu X, Tong Y, Deng D (2014) A comparative study on welding temperature fields, residual stress distributions and deformations induced by laser beam welding and $CO_2$ gas arc welding. Mater Des 63:519–530. https://doi.org/10.1016/j.matdes.2014.06.057

# Chapter 4
# Gas Turbine Performance for Different Burner Technologies

## 4.1 Introduction

Currently, fossil fuel combustion is the chief source of energy required for producing power. However, combustion of fossil fuel has its high hazardous effects on the environment in relationships of nitrogen oxides ($NO_x$), sulfur oxides ($SO_x$) and carbon emissions [1]. Carbon emission is being considered as the main cause for the greenhouse effect. This lead to increasing fear of global warming and raising required efforts to reduce carbon emissions. Many technologies have, thus, been invented to reduce those effects [2]. Among the three main technologies including oxyfuel, pre- and post-combustion, oxyfuel can reduce $CO_2$ emissions up to 20% [3]. Oxy-fuel combustion utilizes oxygen instead of air and hence eliminates nitrogen from the oxidant gas stream. This reduces the volume and mass of the flue gas stream. Accordingly, the size of the flue gas treatment equipment is greatly reduced [4]. The application of oxy-fuel combustion also eliminates $NO_x$ formation due to absence of nitrogen in the combustion process. However, fuel combustion in pure oxygen makes the adiabatic flame temperature higher than that of burning hydrocarbon fuel in air [5]. The excessive rise in combustion temperature may not be suitable for applications like gas turbines [6]. To avoid these complications, some of the exhaust $CO_2$ is recirculated back into the combustor to dilute the reactants and control the flame temperature. Premixing of combustion reactants prevents the creation of hot stoichiometric combustion zones.

The technology of Lean premixed (LPM) combustion is used in numerous combustion systems (such as large scale and aviation gas turbine systems and automotive internal combustion engines). This is primarily because of its benefits such as more efficient combustion and lower pollutant emissions when compared to non-premixed systems and configurations [7, 8]. The LPM combustion technique has the advantage of less emission due to the reduction in flame temperature, which reduces the thermal nitric oxides emissions for air combustion [8]. It should be noted here that fuel and



M. A. Nemitallah et al., *Approaches for Clean Combustion in Gas Turbines*,
Fluid Mechanics and Its Applications 122,
https://doi.org/10.1007/978-3-030-44077-0_4

prompt $NO_x$ are normally having insignificant proportions particularly in gaseous fuels. However, a weakness of pre-mixed combustion is that the flame is more prone to thermo-acoustic instability and/or flashback [9]. Combustion instability has been considered as one of the most critical phenomena encountered during the development of LPM combustor systems [10, 11]. Moreover, if not anchored properly, LPM flames may blow off leading to what is referred to as lean flammability limit or static instability [12–14]. Combining oxy-combustion with LPM combustion can result in almost full control of emissions and flame temperature while capturing $CO_2$ efficiently. However, the replacement of $N_2$ by $CO_2$ may tighten the stable operation range of a given burner. The physical properties of $N_2$ and $CO_2$ are dissimilar and necessitate the modification of the burner design to adapt for such differences. Under the elevated combustion temperature conditions, $CO_2$ has a higher thermal capacity than that of $N_2$ (about 1.6 times) [15]. This higher thermal capacity acts as a thermal energy reservoir causing a reduction in the burning rate. High concentrations of $CO_2$, therefore, may result in flame blowout. Moreover, $CO_2$ is an active molecule in the infrared region, so the radiation heat transfer will be fully different from that in the case of air combustion. In addition, under high-temperature operation, $CO_2$ becomes active and participates in the reactions and, accordingly, affects the rate of reaction kinetics within the combustor [16]. All such changes while adapting oxy-combustion technology should result in distinct behavior of the generated oxy-flame as compared to air-flame. Accordingly, many questions arise when considering such changes. These include the reasonable range of operation for the equivalence ratio, the $CO_2$ concentration in the oxidant and the hydrogen content if enrichment is required for stability improvement. As well, the performance of swirl flames in comparison to micromixer technology is questionable. The effect of combustor design and flow configuration on flame stability are investigated in a series of numerical studies by Peng et al. [17, 18] and Zuo et al. [19, 20]. Peng et al. [17] performed a detailed numerical study considering premixed hydrogen/air flames in a micro combustor with and without front-cavity. The results showed that the front-cavity widens the flame operability range. Zuo et al. [19] studied numerically the effect of flow configuration on the distribution of the wall temperature in a micro combustor, and the results showed that the counter flow configuration resulted in more uniform wall temperature than the co-flow configuration.

Solutions that do not involve steam or water injection include lean premixed (LPM) combustors implementing Dry Low $NO_x$ (DLN) or Dry Low Emission (DLE) technologies. Such combustors in gas turbine applications can reduce $NO_x$ emissions down to the single digit level. Different manufacturers have different design approaches to the DLE combustor concept with the sole aim of reducing $NO_x$ emissions without utilizing steam or water injection. Although the market of LPM gas turbines has many competing combustion technologies, some stand out based on their ultralow emissions and remarkable flexibility. Environmental (EV)-burners, for example, compete strongly with micromixers, especially that EV-burners offer significantly effective flame stabilization over a wide range of operation, which enhances engine turndown. The main features of the micromixer (MM) technology are staging capability, scalability, and fuel flexibility. Both micromixers and EV-burners have never been examined under oxy-combustion conditions; thus, the following condensed review is dedicated to EV-burners to discuss their potential for implementation in zero-emission gas turbines. The focus is made in this chapter on evaluating the performance of gas turbines considering the different burner designs. The performance of the different burners is evaluated in this chapter based on flame operability/stability over fuel-flexible/oxidizer-flexible operating conditions. In addition, the ability of the different burners to reduce emissions is also evaluated. The talk will be extended to include the replacement of current gas turbine combustors with what is called high-temperature membrane reactors (HTMR) for zero emissions power plant applications. In summary, the future research requires further investigations of oxyfuel combustion applications to achieve flame stability under $CO_2$ combustion chamber environment for low $NO_x$ applications. Thus, establishment of combustor stability maps based on operable ranges of equivalence ratio and $CO_2$ concentration. The future industry may require use of different fuels, therefore, the influence of the adiabatic flame temperature of the input fuel mixture is of high concern. The effect of all of these parameters on the flame macrostructure and the flame temperature is of high importance to gas turbine manufacturers.

## 4.2 Dry Low-$NO_x$/Low-Emissions (DLN/DLE) Burners for Gas Turbines

The clear majority of today's operational gas turbines and combustion applications utilizes air for combustion. The use of air as oxidizer generates an exhaust stream that consists mainly of $CO_2$, $H_2O$, $O_2$, and $N_2$. $CO_2$ is a greenhouse gas that is primarily responsible for the escalating global warming problem. The world is still largely dependent on fossil fuels for energy and will be for many decades to come [21], but $CO_2$ is inevitably created as a product of burning fossil fuels. The Paris Agreement [22] has declared that limiting future temperature rise to 2 °C will be very difficult to achieve without the implementation of carbon capture and sequestration. Separating $CO_2$ from the other products of air-combustion is a costly and complicated process [23]. If fuel is burned, however, in pure oxygen instead of air, i.e., oxy-fuel combustion, the products will consist mainly of $CO_2$ and $H_2O$. Carbon capture can thus be implemented at the lowest cost after simple condensation of the water vapor. The volume of exhaust stream is also significantly smaller in the absence of air-based nitrogen, which reduces the size and increases the efficiency of the treatment equipment. One of the other benefits of employing oxy-combustion is the inherent drastic reduction of nitrogen oxides emissions, which result from the reaction of air-based oxygen and nitrogen.

Air and oxy-combustion, however, differ greatly in the degrees of freedom that bound the combustor operability. Air is an abundant low-cost resource, so air-fuel gas turbines are operated under lean conditions to keep the turbine-inlet temperature within hardware-safe limits. The generation of pure oxygen either through cryogenic techniques or other air separation units, on the other hand, is power consuming, so an oxy-fuel gas turbine will have to be operated close to stoichiometry for efficient utilization of both oxygen and fuel. Burning fuel in pure oxygen, however, results in extreme flame temperatures that exceed the hardware limits. Some of the effluent $CO_2$, thus, has to be recirculated to dilute the reactants and control the adiabatic flame temperature, which is termed exhaust-gas recirculation. The reactants are thus fuel plus an $O_2/CO_2$ diluted oxidizer.

The ability to retrofit existing lean premixed (LPM) air-based gas turbines and other combustion applications to implement oxy-combustion with no significant hardware changes will offer one of the essential motivations for most customers to switch to oxy-combustion. This, however, requires better understanding of the stability and characteristics of oxy-fuel flames. The study presented in the next section (Sect. 4.3), thus, aims at investigating premixed, swirl-stabilized $CH_4/O_2/CO_2$ flames in a model gas-turbine combustor that mimics the geometry of operational LPM DLE DACRS air-fuel combustors, where DLE stands for Dry Low Emissions and DACRS stands for Dual Annular Counter-Rotating Swirlers [24]. The reader is referred elsewhere for the history and development of this technology (e.g., Ref. [25]). Intentional deviation from stoichiometric operation (typical of oxy-combustion) is made in this study to widen the test scope and explore a two-dimensional space of equivalence ratio and oxygen fraction in the $O_2/CO_2$ diluted oxidizer. The stability map of the examined combustor geometry was identified by quantifying the flashback and blowout limits within this 2-D space. An analysis of flame shape was also performed by imaging selected flames. No direct measurements of flame speed were made, but the findings of flame stability and shape are used to infer key information on flame speed.

Studies of flame speed or burning velocity in the literature are abundant and will not be fully reviewed here. Some pertinent studies targeted oxy-methane flames [26, 27], while the majority focused on air-fuel conditions. Other studies considered heavier hydrocarbons like ethane [28] and propane [29], and even liquid fuels like ethanol [30], butanol [31], and heptane/octane [32]. It was unanimously agreed in all studies that for lean mixtures both adiabatic flame temperature ($T_{ad}$) and flame speed ($FS$) increase with equivalence ratio. Kaskan [33] and Van Maaren et al. [34] correlated $FS$ to $T_{ad}$ directly as $\log(FS) \propto 1/T_{ad}$ for various fuels. $FS$ was also found to increase if $T_{ad}$ is increased by reactant preheating for gaseous [35] as well as liquid fuels [36]. Reducing $T_{ad}$ by dilution (with $CO_2$, $H_2O$, $N_2$, helium, or argon) resulted in lower $FS$ [37, 38]. In oxy-combustion, this corresponds to reducing the oxygen fraction in $O_2/CO_2$ diluted oxidizer. The conclusion to be made from this survey, thus, is that $FS$ is a single function of $T_{ad}$; any change of the adiabatic flame temperature by preheating or dilution affects the flame speed.

The majority of past oxy-combustion studies examined coal as a carbon-rich fuel to boost the concentration of effluent $CO_2$ and, thus, its capture efficiency [39, 40]. Non-premixed and staged partially premixed conditions were considered. The premixed swirl-stabilized $CH_4/O_2/CO_2$ flames of the study presented in the next section (Sect. 4.3), on the other hand, were rarely investigated [41]. Taamallah et al. [9] compared lean premixed swirl-stabilized air and oxy-methane flames based on the conditions that induce flame stabilization in the outer recirculation zone. The oxygen fraction (dilution ratio) of oxy-flame was adjusted to match the adiabatic flame temperatures of the lean air and oxy-flames. A Strouhal number was introduced based on the inlet flow velocity and the azimuthal spinning frequency of outer recirculation zone to characterize the flow within this zone. They reported that the value of this number is constant at ~0.12 for both the air and oxy-flames, irrespective of inlet Reynolds number. This finding was used to conclude that the inlet bulk velocity should be the parameter of choice, rather than Reynolds number, if similar flow conditions are to be maintained in the outer recirculation zone—a conclusion that motivated the study in Sect. 4.3 to investigate keeping the inlet bulk velocity constant for all flames examined here. Taamallah et al. [9] also measured the temperature within the outer recirculation zone and used it to compute and compare the local extinction strain rates. They reported that a single correlation exists between a characteristic flow time (inverse of the spinning frequency) and a characteristic flame time (inverse of extinction strain rate) for both air and oxy-flames. An analysis of similar flow and flame time scales was thus be performed in the study presented in Sect. 4.3 based on the findings obtained here.

Shroll et al. [42] examined the dynamic stability characteristics of stoichiometric premixed swirl-stabilized $CH_4/O_2/CO_2$ flames. The oxygen fraction (dilution ratio) was used to control $T_{ad}$. Oxy-methane flames were found to mimic air-methane ones in the fact that the transition from one stability mode to another is governed primarily by $T_{ad}$ of reactant mixture, although oxy and air-flames have very different burning velocities at a given $T_{ad}$. This finding was supported by high-speed images that showed comparable vortex breakdown modes indicative of analogous turbulent flame geometries that change with $T_{ad}$. Swirl-stabilized oxy and air-methane flames were also compared by Kutne et al. [43] in a model gas-turbine combustor but under partially premixed conditions. Oxygen fractions of 20–40% and equivalence ratios of 0.5–1.0 were considered. Flame structure was quantified by OH*-chemiluminescence imaging and Raman measurements. It was reported that oxygen fraction strongly affects the flame characteristics, whereas equivalence ratio has a minor effect. The inner recirculation zone was observed to exist further downstream in oxy-flames, compared to air-flames; the former thus have inferior stability. The 21% oxygen fraction of air was not attainable in oxy-flames; i.e., a greater minimum was needed, which was attributed to lower burning velocities and combustion temperatures in oxy flames due to the greater heat capacity of $CO_2$ compared to $N_2$. Multiple other studies supported these findings [44–49]; Song et al. [44] reported matching temperature profiles in $CH_4$/air and $CH_4/36\%O_2/64\%CO_2$ flames. Oh et al. [45] even examined the addition of carbon monoxide (CO) as a secondary fuel to accompany $CH_4$ in an oxy-fuel non-premixed burner and showed poorer stability and longer flame length as the CO mole fraction was increased. In the next section of this chapter (Sect. 4.3), the setup and testing of oxy-fuel combustion flames in a model gas-turbine combustor geometry are presented. The stability and shape of swirl-stabilized premixed $CH_4/O_2/CO_2$ flames are examined at constant inlet bulk velocity, which represents the operating conditions in actual air-fuel DACRS DLE combustors that are typically operated at constant pressure drop across the combustor headend. This constant pressure drop results in fixed inlet bulk velocity if the effects of flow compressibility are neglected, which is a widespread practice among gas-turbine manufacturers.

## 4.3 Combustor Operability of Premixed Oxy-methane Flames

In this section, an experimental and numerical investigation is presented on Stability map and shape of premixed $CH_4/O_2/CO_2$ flames in a model gas-turbine combustor [50]. Oxy-fuel combustion was investigated in a model combustor that mimics the geometry of operational air-fuel gas-turbine combustors implementing the LPM DLE DACRS technology (Lean Premixed, Dry Low Emissions, and Dual Annular Counter-Rotating Swirlers). Premixed swirl-stabilized $CH_4/O_2/CO_2$ flames were tested over ranges of equivalence ratio ($\varphi = 0.2$–$1.0$) and oxygen fraction ($OF =$ 25–70%). The non-preheated flow rates of $CH_4$, $O_2$, and $CO_2$ were adjusted for each flame to maintain a common bulk velocity of 5.2 m/s at burner throat throughout the entire study. The combustor stability map was identified by quantifying the flashback and blowout limits within the $\varphi$-$OF$ space. The adiabatic flame temperature ($T_{ad}$) was also mapped over the same ranges. Selected flames were imaged to analyze the effects of $\varphi$, $OF$, and $T_{ad}$ on flame shape.

### *4.3.1 Test Conditions*

The objective of this experimental study [50] is to examine oxy-fuel combustion in a model gas-turbine combustor headend that mimics DACRS air-fuel premixers [24, 25] in operational DLE gas turbines. The stability (blowout/flashback) as well as shape of confined swirl-stabilized premixed $CH_4/O_2/CO_2$ flames are studied at atmospheric pressure. To achieve this objective, the test rig shown in Fig. 4.1 was utilized. A 55° swirler imparts swirl to premixed reactant mixtures ($CH_4/O_2/CO_2$) of controlled compositions. Immediately downstream of the swirler, a converging-diverging throat insert accelerates the swirling flow to a throat diameter of 2 cm. The gradual convergence (30°) reduces the loss of swirl, while the 45° divergence downstream of the throat prevents sudden expansion into the quartz confinement. Flow acceleration is necessary for achieving flame stability by balancing the flow velocity and flame speed at the throat [24]. A solid conical center body suppresses vortex breakdown at the burner centerline, which, otherwise, could be a potential path for flame flashback [24].

The examined premixed reactant mixtures were prepared by injecting non-preheated (300 K) metered flows of $CH_4$, $O_2$, and $CO_2$ more than one meter upstream

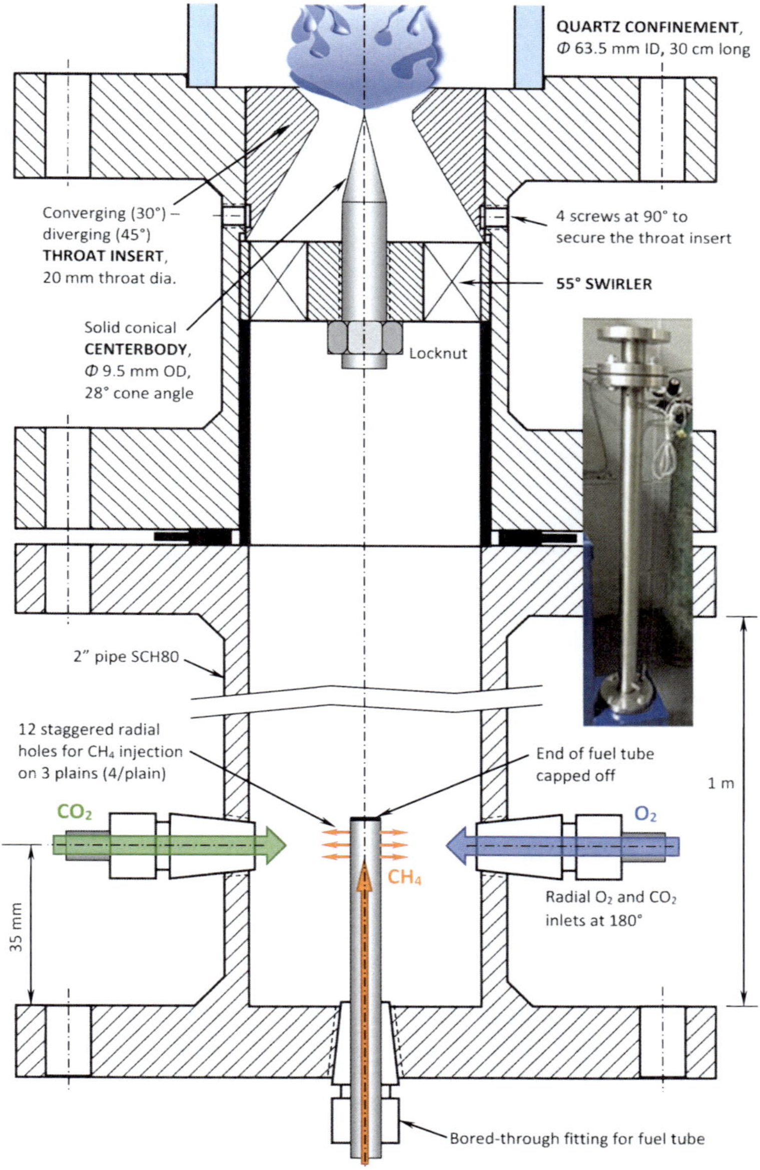


**Fig. 4.1** Schematic drawing of the experimental setup

of the swirler inlet. The mixing plenum is a 2-inch pipe, which creates an aspect ratio L/D of ~20 to maximize the degree of premixing. $O_2$ and $CO_2$ were injected radially from two opposite side inlets, while $CH_4$ was injected from a central fuel tube. The end of this tube is capped off, and fuel was injected from 12 radial holes arranged in a staggered fashion on three successive plains that align axially with the $O_2/CO_2$ inlets, see Fig. 4.1. Reactant mixedness was not directly measured at the combustor throat, but the following test was made on multiple stable flames of different reactant compositions to check on the degree of premixing: The axial location of fuel injection holes was varied over an axial distance of +100/−25 mm from the nominal position shown in Fig. 4.1. This corresponds to L/D of 18–20.5. No discernible changes in flame shape, size, color, or noise were observed, which shows that this mixing-plenum geometry does not create partially premixed flames. Full premixing was thus assumed.

All three gases were supplied from compressed cylinders. Thermal mass flow controllers were used to control/meter the gases through a computer interface, in order to maintain a common bulk throat velocity of 5.2 m/s for all flames examined in this study. An emergency shut-off valve on the fuel line was used to stop fuel flow rapidly after flame flashback. More information about the equipment can be found in previously published studies by the authors [50].

The effect of operating conditions on visual flame shape was analyzed by capturing photographic images of visual flame appearance using a 10-Mega-pixel camera with 1/60-s shutter speed, 5.6 f-stop, and 1600 ISO. Measurements of local temperature within the quartz confinement were also conducted using an R-type (PtRh13%-Pt) thermocouple of 1 mm junction diameter. The readings were corrected for radiation error to the surroundings using the model of Brohez et al. [51].

This experimental study investigates premixed oxy-methane flames [50]. While $CH_4$, $O_2$, and $CO_2$ were all premixed at one station for simplicity in this study, the practical implementation of exhaust-gas recirculation in oxy-combustion involves mixing $CO_2$ with $O_2$ only to create a diluted oxidizer upstream of combustor. The diluted oxidizer then enters the combustor headend, where $CH_4$ is injected into it and mixed with it to create a combustible $CH_4/O_2/CO_2$ flow. The inlet $CO_2$ is thus treated as a reactant in this study, i.e., the terms "bulk Reynolds number," "bulk flow rate," "bulk velocity," and "bulk viscosity" refer to these respective quantities of the $CH_4/O_2/CO_2$ reactant mixture as a whole and not of $CH_4$ and $O_2$ only.

All examined flames have a common bulk velocity $v = 5.2$ m/s at the throat. The following analysis explains how the flow rates of $CH_4$, $O_2$, and $CO_2$ were set to maintain constant bulk throat velocity. The bulk mass flow rate and velocity of reactant mixture at the burner throat are related according to:

$$\dot{m}_{mix} = \dot{m}_{CH_4} + \dot{m}_{O_2} + \dot{m}_{CO_2} = \rho_{mix} A v_a = \left(\frac{pM_{mix}}{R_u T}\right) A v_a = \left(\frac{pAv_a}{R_u T}\right) M_{mix} \tag{4.1}$$

Where the combustor pressure is atmospheric, and all gases are at 300 K (no preheating). Since $M_{mix} = 1/\sum(x_i/M_i)$, then substituting into Eq. 4.1 and rearranging:

$$\frac{\dot{m}_{CH_4}}{16} + \frac{\dot{m}_{O_2}}{32} + \frac{\dot{m}_{CO_2}}{44} = \left(\frac{pAv_a}{R_u T}\right) \tag{4.2}$$

For oxy-fuel combustion, the stoichiometric combustion equation is expressed as follows:

$$CH_4 + 2\left[O_2 + \left(\frac{1}{OF} - 1\right)CO_2\right] \rightarrow \left[1 + 2\left(\frac{1}{OF} - 1\right)\right]CO_2 + 2H_2O \tag{4.3}$$

The stoichiometric mass ratio of fuel to diluted oxidizer is calculated from:

$$(F/O)_{stoichiometric} = \frac{16}{2\left[32 + \left(\frac{1}{OF} - 1\right) * 44\right]} = \frac{2}{8 + 11\left(\frac{1}{OF} - 1\right)} \tag{4.4}$$

The classical definition of equivalence ratio is implemented as:

$$\varphi = \frac{(F/O)_{actual}}{(F/O)_{stoichiometric}} = \left(\frac{\dot{m}_{CH_4}}{\dot{m}_{O_2} + \dot{m}_{CO_2}}\right)\left[4 + 5.5\left(\frac{1}{OF} - 1\right)\right] \tag{4.5}$$

Based on the definition of oxygen fraction, the following relationship can be inferred for the $O_2/CO_2$ diluted oxidizer:

$$\frac{\dot{m}_{O_2}}{\dot{m}_{O_2} + \dot{m}_{CO_2}} = \frac{32OF}{32OF + 44(1 - OF)} \tag{4.6}$$

Equations 4.2, 4.5, and 4.6 can be solved simultaneously for any given combination of equivalence ratio and oxygen fraction to yield the set points needed to program the flow controllers of $\dot{m}_{CH_4}$, $\dot{m}_{O_2}$, and $\dot{m}_{CO_2}$. A test matrix was thus developed in Microsoft Excel to cover the ranges $\varphi = 0.2$–$1.0$ and $OF = 25$–$70\%$.

The bulk dynamic viscosity of reactant mixture was calculated using the formula [52]:

$$\mu_{mix} = \frac{\sum y_i \mu_i \sqrt{M_i}}{\sum y_i \sqrt{M_i}} \tag{4.7}$$

The bulk density of reactant mixture was determined from

$$\rho_{mix} = \frac{p \sum \dot{m}_i}{R_u T \sum (\dot{m}_i / M_i)} \tag{4.8}$$

The Reynolds number based on bulk throat velocity was thus calculated from

$$\mathrm{Re} = vD\rho_{mix}/\mu_{mix} \tag{4.9}$$

It is important to note here that changing the mixture composition (by varying either $\varphi$ or $OF$), changes the Reynolds number despite of keeping the bulk throat velocity constant throughout this entire study; Re ranges from 7382 at ($\varphi = 1.0$ and $OF = 70\%$) to 10,419 at ($\varphi = 0.2$ and $OF = 25\%$).

The adiabatic flame temperature ($T_{ad}$) was also estimated over the same examined ranges of $\varphi$ and $OF$. Since the dissociation of $CO_2$ was proven in past studies to reduce the stoichiometric $T_{ad}$ by 100–200 K only [53, 54], such error was tolerated and species dissociation was neglected, as it did not affect the main trends and findings to be reported here. For further simplicity, complete combustion was assumed, especially that lean conditions are examined here ($\varphi = 0.2$–$1.0$).

The simplified non-stoichiometric combustion equation can thus be expressed as:

$$\begin{aligned} CH_4 + \frac{2}{\varphi}\left[O_2 + \left(\frac{1}{OF} - 1\right)CO_2\right] &\rightarrow \left[1 + \frac{2}{\varphi}\left(\frac{1}{OF} - 1\right)\right]CO_2 \\ &+ 2H_2O + 2\left(\frac{1}{\varphi} - 1\right)O_2 \end{aligned} \tag{4.10}$$

The composition of exhaust gas, and consequently $T_{ad}$, depend on both $\varphi$ and $OF$. $T_{ad}$ was estimated from a thermodynamic analysis of Eq. 4.10 using tabulated values of temperature-dependent sensible enthalpies for $CO_2$, $H_2O$, and $O_2$ as well as tabulated enthalpies of formation for $CH_4$, $CO_2$, and $H_2O$ [55].

As previously explained, the flow rates of $CH_4$, $O_2$, and $CO_2$ were not varied independently; constant bulk throat velocity of 5.2 m/s was maintained for all flames, which linked the three flow rates for any desired combination of $\varphi$ and $OF$. Consequently, the transition from one data point to the next involved the adjustment of all three flows, even if both points shared the same $OF$ or $\varphi$.

The combustor stability map was scanned one oxygen fraction at a time, from 25% up to 70% in steps of 5%. For each $OF$ the flashback and blowout limits were quantified using the following approach:

1. Igniting the combustor at a stable point that is safely away from flashback or blowout—this point was obtained iteratively during the very first light-off attempts of the combustor.
2. Slowly adjusting the flow rates of $CH_4$, $O_2$, and $CO_2$ to target a combination of the desired $OF$ and an equivalence ratio that is presumed stable. If the combustor was observed to approach flashback during adjustment, a leaner $\varphi$ was sought for the same desired $OF$. The tendency for flashback was identified by higher audible noise, indicative of combustion dynamics prior to flashback. If, on the other hand, the combustor was observed to approach blowout during adjustment, a richer $\varphi$ was sought, also for the same target $OF$.
3. Once the desired $OF$ was reached safely, the stable $\varphi$ was noted. Transition was then made to a richer $\varphi$ at constant $OF$. As the flashback limit was approached, $\varphi$ was increased carefully in smaller increments between data points.
4. At the flashback limit, noise reached a maximum level before the flame was observed to flash rapidly upstream of burner throat, igniting the incoming fresh

reactants, which was accompanied by a small explosion. Fuel flow was stopped immediately by activating the emergency shut-off valve, then $O_2$ was stopped as well, leaving $CO_2$ flowing for five seconds to purge the mixing plenum before being stopped. The value of $\varphi$ that caused flashback was noted against that particular *OF*.

5. The burner was then reignited at the stable $\varphi$ obtained earlier, and the blowout limit was sought by reducing $\varphi$ for the same *OF*. As the limit was approached, $\varphi$ was decreased carefully in smaller increments between data points. The value of $\varphi$ that caused blowout was noted against that particular *OF*.
6. Steps 2–5 were repeated for different oxygen fractions, until the entire stability map was quantified at a fixed bulk throat velocity of 5.2 m/s.

### 4.3.2 Combustor Stability Maps

The stability map (flashback and blowout limits) of the examined model DACRS DLE gas-turbine combustor geometry is quantified for premixed $CH_4/O_2/CO_2$ flames over ranges of oxygen fraction (25–70%) and equivalence ratio (0.2–1.0). Figure 4.2 shows the obtained stability map. For each *OF*, the flashback point is indicated by a diamond symbol, while the blowout point is depicted using a circular one. Exponential fits were found to represent the flashback and blowout limits best, with $R^2$ values (Coefficient of determination) of 0.9978 and 0.9867, respectively. The

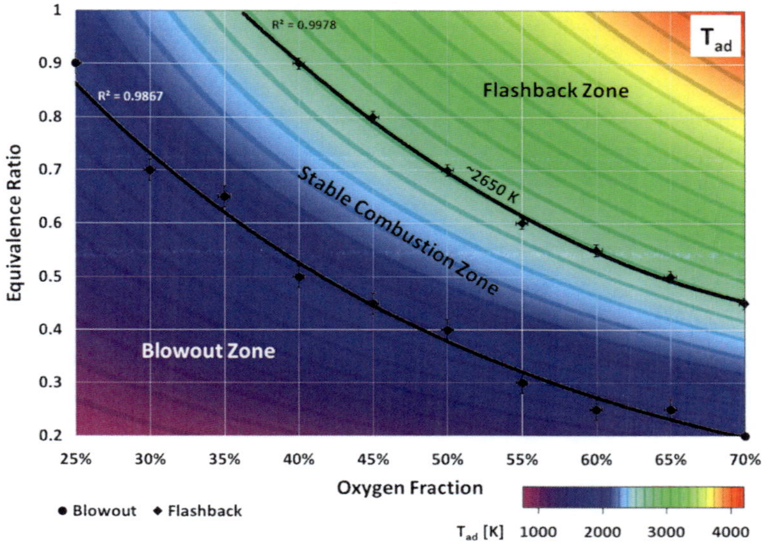


**Fig. 4.2** Combustor stability map plotted against the contours of constant adiabatic flame temperature. The bulk throat velocity is fixed at 5.2 m/s throughout

zone of stable combustion is the area between the fit lines, whereas the outside areas are the flashback and blowout zones on the top and bottom, respectively. The reason behind the lower fit quality of blowout limit is that intermittent flame lifting and reattachment occurred prior to blowout, which made it difficult to quantify this limit accurately. The flashback limit was thus more distinctly identifiable than the blowout one during experiments.

The background of Fig. 4.2 is colored by the contours of constant adiabatic flame temperature ($T_{ad}$). Figure 4.2 provides a novel finding for $CH_4/O_2/CO_2$ flames, namely that flame flashback occurs consistently at constant $T_{ad}$ if the bulk throat velocity is kept constant. This has direct application in designing and operating actual premixed gas-turbine combustors that are typically operated at constant pressure drop across the combustor headend, which results in fixed bulk throat velocity. Gas-turbine manufacturers commonly follow this practice to ensure a safe margin to flashback throughout different modes of combustor operation. It should be noted here, of course, that the depicted value of flashback $T_{ad}$ (~2650 K) is slightly higher than the true value for all data points, because of the aforementioned assumptions in $T_{ad}$ estimation. Nonetheless, the trend still holds, i.e., flashback occurring at constant $T_{ad}$. It can also be seen in Fig. 4.2 that the blowout limit roughly occurs at constant $T_{ad}$ as well (~1800 K for the examined geometry). This leads to the conclusion that $T_{ad}$ is a more relevant parameter choice than $\varphi$ or $OF$ to quantify the combustor stability map.

The consistent flashback $T_{ad}$ provides valuable insights into flame speed ($FS$) as well. It is known that $FS$ is function of $\varphi$ and $OF$, and so is $T_{ad}$, as Fig. 4.2 shows. At the flashback limit, $FS$ and bulk throat velocity ($v$) are delicately balanced. Since all examined flames have a common $v$ of 5.2 m/s, it can be deduced that $FS$ is constant at 5.2 m/s at the flashback limit. The contours of constant $T_{ad}$ can thus be used to infer the loci of constant $FS$; in other words, a single correlation exists between $FS$ and $T_{ad}$, so if $T_{ad}$ is known, $FS$ can be estimated directly, without having to investigate more complicated correlations that involve $\varphi$ and/or $OF$.

The value of $FS$ near blowout can be estimated from an analysis of flame shape in Fig. 4.3. Shown are flames at ($OF = 55\%$ & $\varphi = 0.35$) and ($OF = 60\%$ & $\varphi = 0.30$), which are close to blowout. These tall weak flames have conical shapes with ~18° and ~16° cone angles, respectively. $FS$ at flame base can thus be approximated as (5.2 * sin (0.5 * cone angle)) $\approx$ 0.8 and 0.7 m/s for the depicted flames. Consequently, it can be concluded that $FS$ decreases from 5.2 m/s at the flashback limit (~2650 K) to ~0.8 m/s at 1860 K to ~0.7 m/s at 1800 K. If $FS$ is plotted against $T_{ad}$ (Fig. 4.4), it can be seen that $\log(FS) \propto 1/T_{ad}$, which is in excellent agreement with the findings of Kaskan [33] and Van Maaren et al. [34]. The lower $FS$ of the 1800-K flame is also evidenced in its greater length, which indicates a longer reaction time scale and thus lower flame speed.

To check if similar correlations exist between $FS$ and other operability parameters, the stability map of Fig. 4.2 was replotted against the contours of combustor power density (Fig. 4.5), throat Reynolds number (Fig. 4.6), and bulk mass flow rate $\dot{m}_{mix}$ (Fig. 4.7). The power density ($PD$) in MW/m$^3$/bar is defined here as

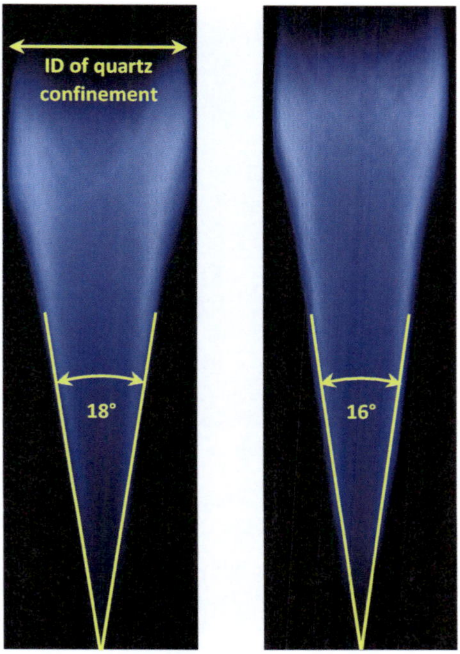


**Fig. 4.3** Visual flame images used for estimating the flame speed near blowout

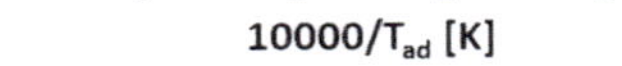


**Fig. 4.4** Flame speed as a function of adiabatic flame temperature, as derived from Figs. 4.2 and 4.3

$$PD = \frac{\dot{m}_{CH_4} \times H_C}{p \times Volume\, of\, quartz\, confinement} \tag{4.11}$$

Where $H_C$ is the standard tabulated enthalpy of combustion (55.5 MJ/kg) [55], and the combustor pressure is atmospheric. The power density is thus directly proportional to fuel flow rate. Figure 4.5 shows that *PD* ranges here from ~3.5 MW/m$^3$/bar at the blowout limit to ~5.5 MW/m$^3$/bar at the flashback one, which is comparable to most industrial gas turbines, as they are operated typically in the range 3.5–20 MW/m$^3$/bar [50]. It can be clearly seen in Fig. 4.5 that the flashback limit does not follow a contour of constant *PD*, which indicates that *FS* does not correlate to

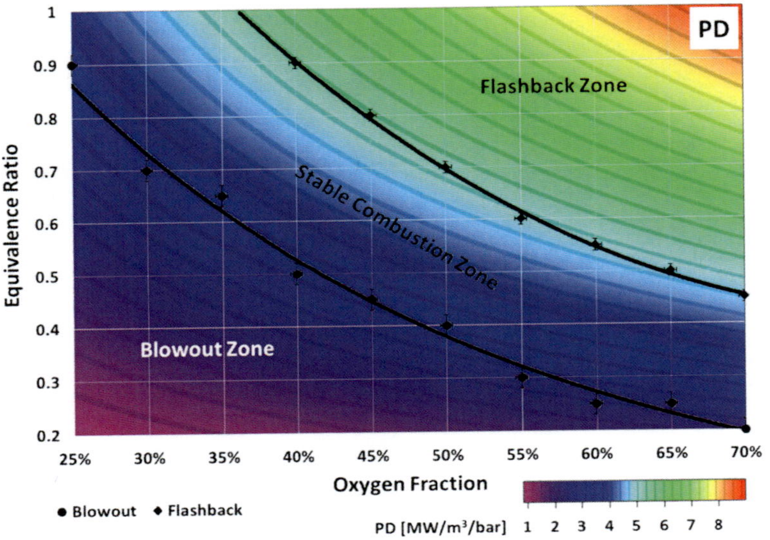


**Fig. 4.5** Combustor stability map plotted against the contours of constant combustor power density. The bulk throat velocity is fixed at 5.2 m/s throughout

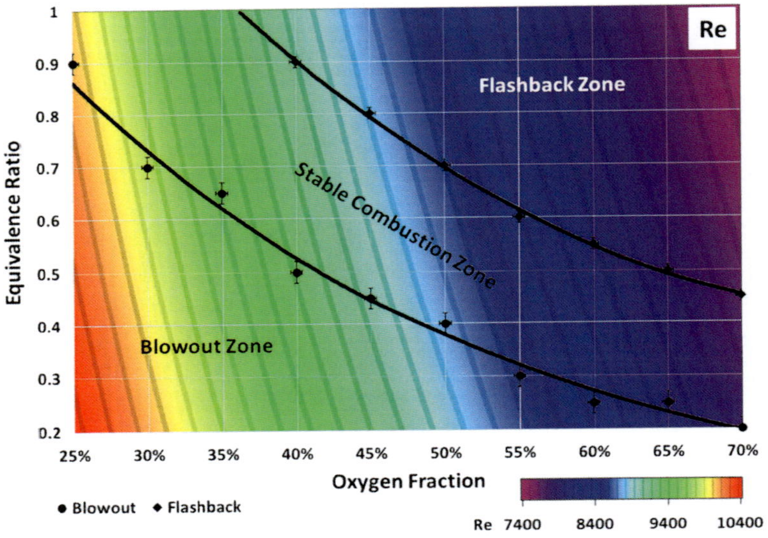


**Fig. 4.6** Combustor stability map plotted against the contours of constant throat Reynolds number. The bulk throat velocity is fixed at 5.2 m/s throughout

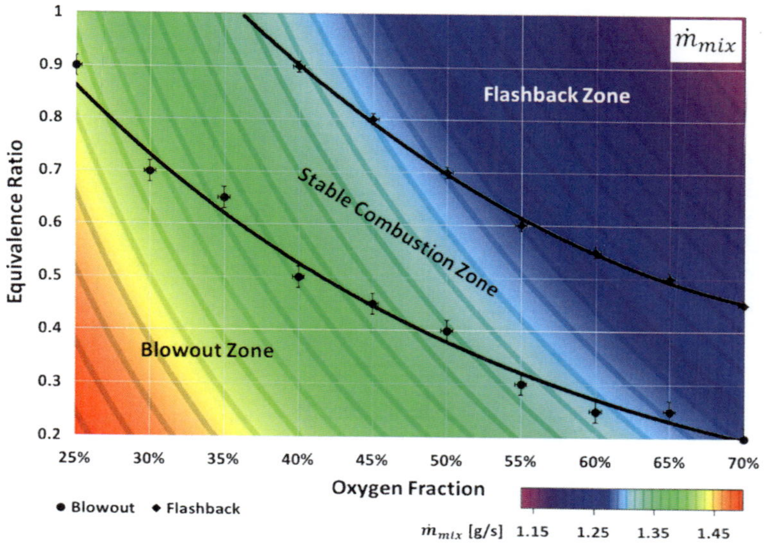


**Fig. 4.7** Combustor stability map plotted against the contours of constant bulk mass flow rate. The bulk throat velocity is fixed at 5.2 m/s throughout

*PD*. Thus, the assertion that combustor flashback always occurs at constant fuel flow rate is inaccurate.

Similar analyses of Figs. 4.6 and 4.7 show that flame speed does not correlate to throat Reynolds number or bulk flow rate either, which agrees with the findings of Taamallah et al. [9]. The Reynolds number was examined here because a previous study by the current authors revealed that the structure and stability of coaxial non-premixed $CH_4/O_2/CO_2$ flames were affected by the Reynolds number of $O_2/CO_2$ diluted oxidizer [56].

### 4.3.3 Flame Macrostructure

To explore the correlation between flame speed and $T_{ad}$ in more depth, a detailed analysis of flame shape was performed. Figure 4.8 and Table 4.1 show the test conditions of 13 distinct stable flames that were imaged to analyze their shapes. Flames 1–5 examine the effect of $T_{ad}$ by changing the equivalence ratio at constant $OF = 60\%$, while flames 5–8 also target the effect of $T_{ad}$ but through different oxygen fractions at constant $\varphi = 0.5$. Flames 9–13, on the other hand, were carefully chosen to have a common $T_{ad}$ (~2380 K) by varying both $OF$ and $\varphi$. The images of flames 1–5 (constant $OF$) are depicted in Fig. 4.9 along with their respective values of $T_{ad}$. Flames 5–8 (constant $\varphi$) are also shown but in reverse order to line up flame 8 with 2, 7 with 3, and 6 with 4, because each of these "couples" is roughly at the

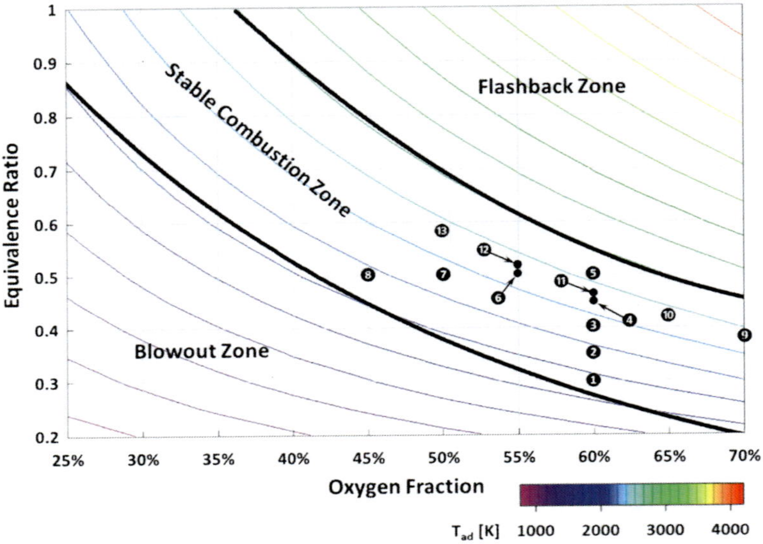


**Fig. 4.8** Combustor stability map highlighting the selected data points for which flame images were captured to analyze flame shape. Points 1–5 are at $OF = 60\%$, points 5–8 are at $\varphi = 0.5$, and points 9–13 are at $T_{ad} \approx 2380$ K. The contours of $T_{ad}$ are plotted in the background for a clearer visualization of points 9–13

**Table 4.1** Test conditions of the selected data points for which flame images were captured

| # | $OF$ (%) | $\varphi$ | $T_{ad}$ (K) |
|---|---|---|---|
| 1 | 60 | 0.30 | 1806 |
| 2 | | 0.35 | 1995 |
| 3 | | 0.40 | 2173 |
| 4 | | 0.45 | 2342 |
| 5 | | 0.50 | 2501 |
| 6 | 55 | | 2336 |
| 7 | 50 | | 2169 |
| 8 | 45 | | 2001 |
| 9 | 70 | 0.38 | 2380 |
| 10 | 65 | 0.42 | |
| 11 | 60 | 0.46 | |
| 12 | 55 | 0.52 | |
| 13 | 50 | 0.58 | |

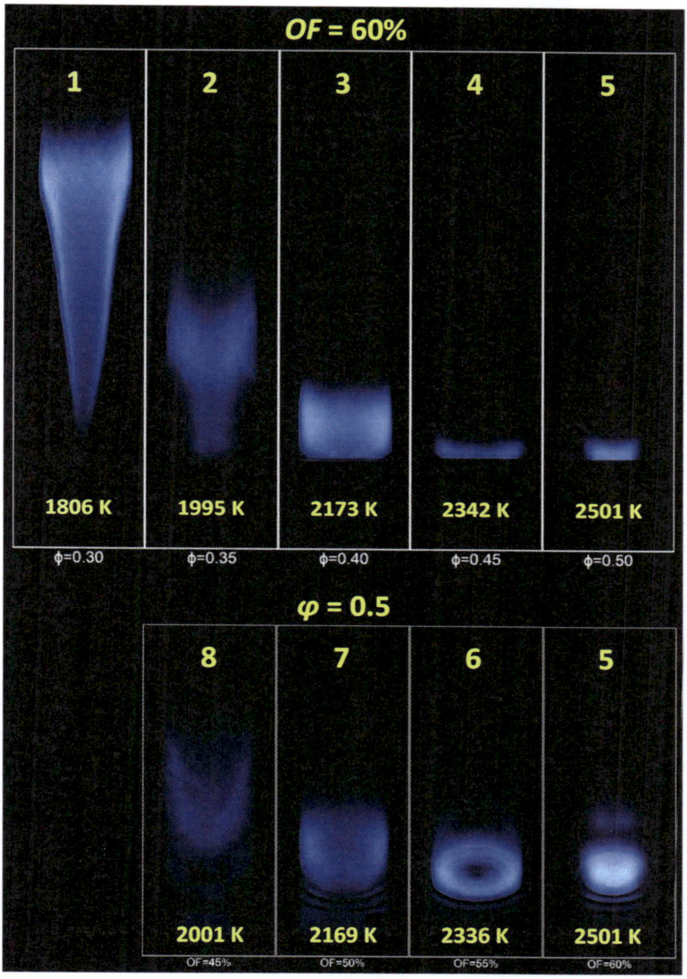


**Fig. 4.9** Visual images of flames 1–5 at $OF = 60\%$ (top) and 5–8 at $\varphi = 0.5$ (bottom). $T_{ad}$ is listed for each flame

same $T_{ad}$. Flames 2 and 8 are at ~2000 K, while 3 and 7 are at ~2170 K, and 4 and 6 are at ~2340 K. Side view is depicted for the top flames, while the bottom ones are shown in isometric view, to display the flame shapes from different angles.

It can be seen from Fig. 4.9 that increasing $T_{ad}$ from near blowout to near flashback drastically changes flame shape, which is in direct agreement with the findings of Shroll et al. [42]. Flame 1 in Fig. 4.9 is one of the cold weak tall flames analyzed previously in Fig. 4.3 for flame speed near blowout. As $T_{ad}$ is increased (flames 2 and 8), *FS* increases which shortens flame length. Further increase in $T_{ad}$ (flames 3 and 7) increases *FS* to the extent that the flame overwhelms the corner recirculation zone

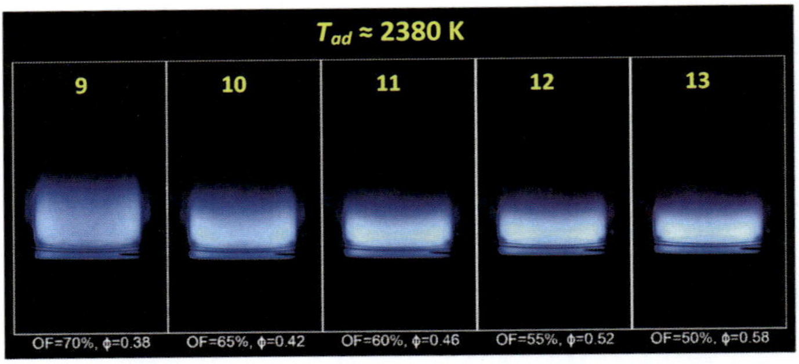


**Fig. 4.10** Visual images of flames 9–13 at $T_{ad} \approx 2380$ K. $OF$ and $\varphi$ are listed for each flame

and acquires a cup shape with shorter overall length. Higher $T_{ad}$ (flames 4 and 6) accelerates the reaction even more (higher $FS$), and the significantly shorter flame is now mostly contained within the corner recirculation zone, similar to the conditions of Taamallah et al. [9]. Finally, as the flashback limit is approached (flame 5), $T_{ad}$ and $FS$ become high enough to create a very compact flame that barely protrudes beyond the divergent section of throat insert. Beyond flame 5 the flame shape was noticed not to change much; the audible noise of combustion dynamics increased, however, while approaching flashback.

In addition to this distinct effect of flame speed and $T_{ad}$ on flame shape, it can also be concluded from Fig. 4.9 that the same variation in flame shape can be attained by changing either $OF$ or $\varphi$, i.e., both parameters have very similar individual effects on flame shape. This strongly affirms the aforementioned correlation between flame speed/shape and $T_{ad}$, because both $OF$ and $\varphi$ can be represented by $T_{ad}$. Changing either of the two parameters changes $T_{ad}$, which, consequently, affects flame speed and shape.

The logical question to ask here is: If $T_{ad}$ is kept constant, will the flame shape remain unchanged? Figure 4.10 answers this question, where it can be clearly seen that flames 9–13 have almost identical shapes despite of changing $OF$ and $\varphi$ over the ranges 50–70% and 0.38–0.58, respectively. Keeping $T_{ad}$ constant thus resulted in no changes in flame shape, which is another novel finding for oxy-methane flames.

A concluding remark to be made here is that the observed direct correlation between $T_{ad}$ and flame shape/size is of key importance in designing future oxy-fuel gas-turbine combustors operating at constant headend pressure drop (or inlet bulk velocity). $T_{ad}$ can be easily estimated for any given set of operating conditions. Once $T_{ad}$ is known, the flame shape/size as well as proximity of flame tip to combustor exit can be inferred during operation, even if the combustor is not optically accessible. Thus, combustor length can be easily designed to accommodate the entire flame under all conditions, in order to ensure complete combustion prior to expansion in the turbine.

### 4.3.4 Flame Temperature

To shed more light on the correlation between $T_{ad}$ and flame structure, measurements of local temperature were conducted inside the quartz confinement. Figure 4.11 shows the axial distribution of centerline temperature in the plumes of flames 5–7 at $\varphi = 0.5$ and $OF = 50$–$60\%$. It can be seen that the higher the $T_{ad}$, the greater are centerline temperatures. Interestingly, flame 5 has the largest $OF$ and thus the most compact reaction zone, so all measurements are significantly downstream the flame tip in the plume; nonetheless, this flame demonstrates the hottest temperature profile in Fig. 4.11. On the contrary, the tip of flame 7 is close to the location of its first data point, but the profile of this flame is coldest throughout. These observations confirm that $T_{ad}$ is the controlling parameter here.

Since these flames cannot be simplified as one-dimensional plug-flow reactors, radial temperature profiles were also measured in the plumes of flames 5–7 at $Z = 5$ cm, see Fig. 4.12. The previously observed dependence of local temperature on $T_{ad}$ can also be seen at most, but not all, radial locations, which indicates the two-dimensional nature of such flow fields. Figure 4.13 focuses on flame 7 ($OF = 50\%$, $\varphi = 0.5$, $T_{ad} \approx 2169$ K) by plotting radial temperature profiles at multiple axial locations. A detailed axial-radial map is thus provided for validation of future CFD models of this combustor. The same trends of Figs. 4.11 and 4.12 can be observed in Fig. 4.13; local temperature decreases as the gases flow downstream, and the maximum temperature at a given $Z$ is not at the centerline but rather at $r/R = 0.4$–$0.6$. The radial profiles at $Z = 5$ and 17.5 cm show exceptions from the systematic trend of other profiles, which indicates the possibility of local flow recirculation. The flowfield would thus be quantified in more detail using CFD tools to delve into the structure and locations of recirculation zones and provide additional confirmation to the findings of this study.

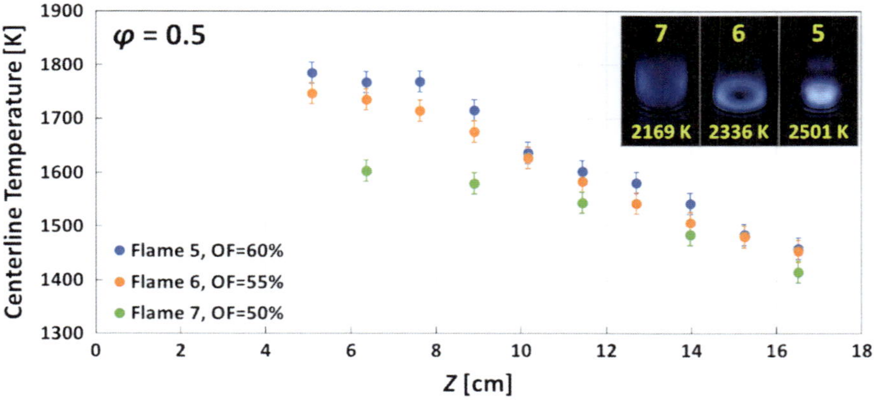


**Fig. 4.11** Axial distribution of centerline temperature in the plumes of flames 5–7 at $\varphi = 0.5$

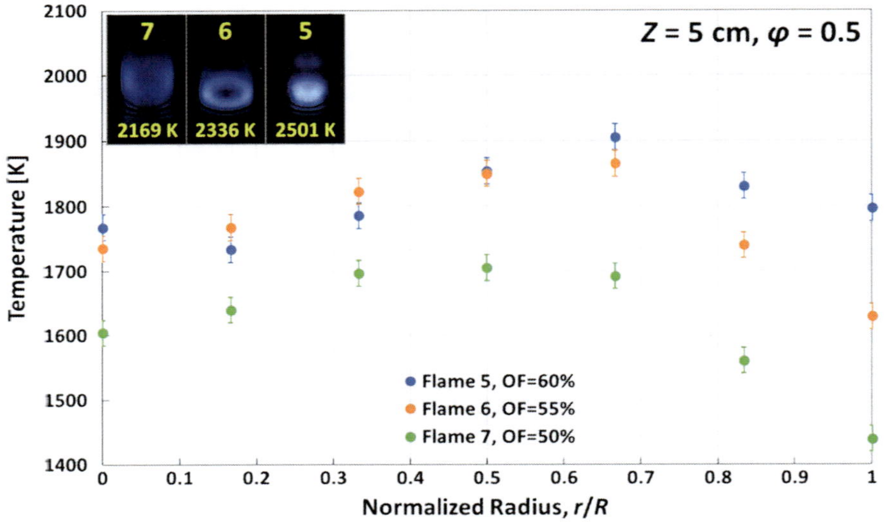


**Fig. 4.12** Radial profiles of local temperature at $Z = 5$ cm in the plumes of flames 5–7 ($\varphi = 0.5$)

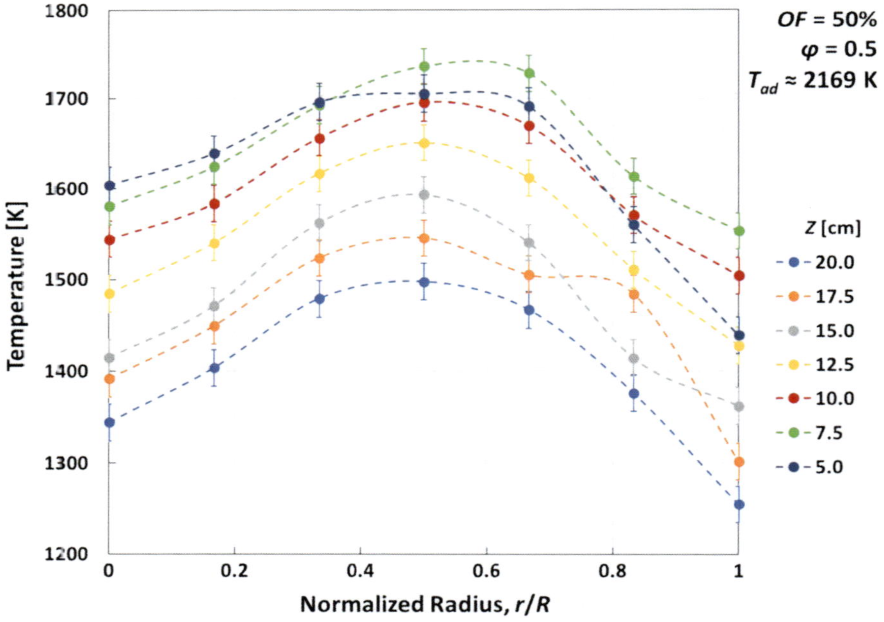


**Fig. 4.13** Radial profiles of local temperature at different $Z$ values in the plume of flame 7 ($OF = 50\%$, $\varphi = 0.5$, $T_{ad} \approx 2169$ K)

### 4.3.5 LES of Premixed Oxy-flames

This section presents the numerical geometry, grid independency study, governing equations, and models used to simulate the examined swirling oxy-methane flames. The numerical model used is a combination of the presumed probability density function (PDF) and the large eddy simulation (LES) viscous model, while the premixed combustion model was applied for simulating oxy-fuel flames stabilized over a 55° swirler. A 3-D cylindrical-based geometry of the gas turbine model combustor was established using Gambit to be used in the full 3-D numerical simulations performed using Ansys-Fluent 19.2. The swirl-stabilized premixed gas turbine combustor model is represented in Fig. 4.14 with all dimensions and the swirler location. The length of combustion chamber is 30 cm.

#### 4.3.5.1 Model Setup

Solving the conversation equation is challenging, particularly in the case of turbulent reactive flows that involve a wide spectrum of length and time scales. The multi-scale and multi-physics attributes of turbulent combustion necessitate the use of modeling approaches in order to simplify the pathetical description of the complex physical phenomena. Three different approaches may be identified: (i) Direct Numerical Simulation (DNS), (ii) Reynolds-Averaged Navier-Stokes (RANS), and (iii) Large-Eddy Simulation (LES); a detailed description of each approach was reviewed by Denis and Vervisch [57].

In case of LES, rather than averaging the effect of turbulence, the equations are filtered: The larger turbulent eddies are explicitly resolved and computed, as they have a more significant influence on the flow and are more dependent on geometry, whereas the effect of smallest eddies is modeled using sub-filter-scale models. Thus, in contrast to RANS, there is partial resolution of the turbulent fluctuations in LES, which decreases the uncertainty associated with modeling but increases the computational cost.

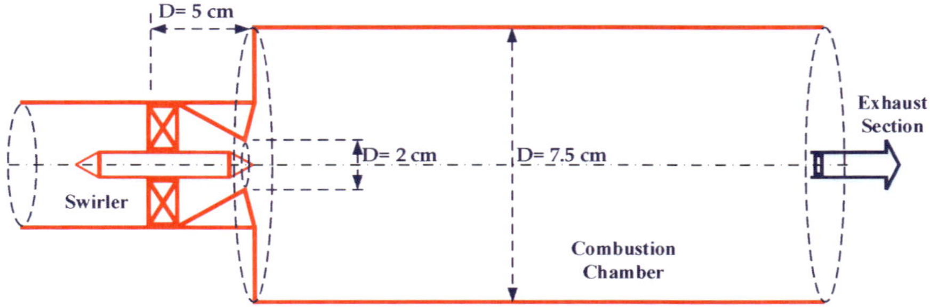


**Fig. 4.14** Dimensions of the combustor geometry

Turbulence transfers energy from large eddies to the small eddies where energy is dissipated according to the following spectrum [58].

$$E(k) = C_k \varepsilon^{2/3} k^{-5/3} \tag{4.12}$$

Where, $C_k$ is the Kolmogorov constant. In LES, the various flow quantities, Φ, are filtered in the spectral space (by suppressing the components greater than a given cut-off length), or in the physical space (weighted averaging in a given volume). The filtered quantity Φ is expressed as follows:

$$\overline{\Phi}(x) = \int \Phi(x') g(x - x') dx' \tag{4.13}$$

Where, g is a filter function. For variable density flows, Favre, or density-weighted, filtered quantity, Φ, can be expressed as follows:

$$\bar{\rho}\bar{\Phi}(\mathrm{x}) = \overline{\rho\Phi} \tag{4.14}$$

Often, the grid is used as the spatial filter. The governing equations for LES are obtained by Favre filtering operation to each term in the conservation equations, including, continuing equation, momentum equation, energy equation, and species transport equation. These equations are presented below as following:

Continuity Equation [59]:

$$\frac{\partial \bar{\rho}}{\partial t} + \frac{\partial(\bar{\rho}\tilde{u}_j)}{\partial x_j} = 0 \tag{4.15}$$

Where $\tilde{u}_j$ is the filtered velocity vector.

Conservation of momentum [59]:

$$\frac{\partial \bar{\rho}\tilde{u}_i}{\partial t} + \frac{\partial(\bar{\rho}\tilde{u}_j\tilde{u}_i)}{\partial x_j} = -\frac{\partial \bar{P}}{\partial x_i} + \frac{\partial \bar{\sigma}_{ji}}{\partial x_j} + \frac{\partial \tau_{ij}}{\partial x_j} \tag{4.16}$$

Where $\bar{\sigma}_{ji}$ and $\tau_{ij}$ refer to the filtered viscous stress tensor and the corresponding sub-grid scale (SGS) term, respectively.

Species Transport Equation [59]:

$$\frac{\partial \bar{\rho}\tilde{Y}_k}{\partial t} + \frac{\partial\left(\bar{\rho}\tilde{Y}_k\tilde{u}_j\right)}{\partial x_j} = \frac{\partial}{\partial x_j}\left(\bar{\rho}D_k\frac{\partial \tilde{Y}_k}{\partial x_j}\right) - \frac{\partial}{\partial x_j}\left[\bar{\rho}\left(Y_k u_j - \tilde{Y}_k\tilde{u}_j\right)\right] + \overline{\dot{\omega}_k} \tag{4.17}$$

Where D is the molecular diffusivity, and $Y_k$ and $\omega_k$ refer to the specie mass fraction and reaction rate, respectively.

Conservation of Energy [59]:

$$\frac{\partial \bar{\rho} \tilde{E}}{\partial t} + \frac{\partial \left( \bar{\rho} \tilde{u}_j \tilde{E} + \tilde{u}_j \bar{P} \right)}{\partial x_j} = -\frac{\partial \bar{q}_j}{\partial x_j} + \frac{\partial \overline{\sigma_{jk} u_k}}{\partial x_j} + \frac{\gamma R}{\gamma - 1} \frac{\partial \tau_j T}{\partial x_j} + \frac{\partial \tau_{jk} u_k}{x_j} \tag{4.18}$$

Where $\tilde{E}$ refers to the filtered total specific energy, and $\bar{q}$ is the filtered heat flux. More details on modeling the unresolved terms can be found in Ref. [60, 61]. All of the aforementioned transport equations were discretized by a second-order upwind scheme using the finite volume and under-relaxation method provided in Ansys fluent.

During turbulent combustion, chemical reactions are confined to thin reacting layers at small scales that cannot be resolved on typical LES grids. Therefore, most of the turbulence-chemistry interactions need to be modeled. This modeling of the reaction rates presents a major challenge in turbulent premixed combustion, because reaction rates are highly nonlinear functions of temperature and species mass fractions. An integral component of LES is thus the turbulent combustion sub-grid model, which is necessary to incorporate the effect of turbulence-chemistry interactions at the under-resolved scales on the reaction rate. Researchers to model the filtered reaction rates corresponding to specific flame regime have adopted various approaches.

The probability density function (PDF) model is used in this study. In this presumed-PDF framework, the reaction rate and the thermo-chemical properties are averaged over the presumed probability density function, which does not require solving a transport equation. Furthermore, the reaction chemistry can be taken into account using the flamelet generated manifold (FGM) approach. This technique focuses on including detailed chemistry effects within the combustion model by reducing the chemical subspace using a set of reference laminar flamelet computations and storing the chemical flame structure inside a lookup table. Hence, this approach can reduce the computational cost of preforming reacting flow with extensive chemical kinetic mechanisms. In this work the focus is on the presumed-PDF approach coupled with the FGM technique and a description of the method can be found in Ref. [59].

LES is very attractive for inherently unsteady simulations, especially those dominated by large-scale turbulent structure, including combustion instability and fire spread. LES is compelling for reacting flows because of several reasons. One reason is that the large turbulent eddies are anisotropic, long lasting, and depend on the geometry. The small eddies, on the other hand, are isotropic, more easily modeled, and can be determined by the large eddies as well. In addition to that, the LES is a more accurate turbulence model because LES has few or no adjustable constants as compared to the RANS model constants, which require adjustment to free jets, jets in cross flow, swirl, buoyancy, etc. In contrast to RANS, the LES modeling error decreases as grid resolution is increased. Furthermore, combustion is often controlled by the rate the large eddies mix reactants with products, and these large eddies are directly resolved in LES. Unlike other turbulence models, the LES model treats effectively the flow near walls.

In the current study, large-eddy simulations were conducted using the presumed probability density function (PDF) coupled with the flamelet generated manifold

**Table 4.2** Summery of the operating conditions and configuration parameters

| Oxy-fuel combustion ($CH_4/O_2/CO_2$) with oxygen fraction (O.F.) = 60% | |
|---|---|
| Equivalence ratio, φ | 0.35–0.70 |
| Combustor power (based on equivalence ratio) | 2.25–4.25 kW |
| Inlet mixture bulk velocity | 5.2 m/s |
| Operating pressure | 1 bar |
| Inlet mixture temperature | 300 K |
| Turbulence intensity | 5% |
| Viscosity ratio | 10% |

(FGM). The effect of equivalence ratio (φ) is examined. All simulations were performed at a common oxygen fraction (O.F.) of 60% and inlet bulk velocity of 5.2 m/s, similar to the experimental tests. Table 4.2 summarizes the considered operating conditions.

To consider convection and radiation heat transfer, a mixed thermal boundary condition was chosen for the combustor wall. A coupled scheme was selected for the pressure-velocity coupling, while the first-order upwind scheme was selected for the spatial discretization, such as momentum, turbulent kinetic energy, etc. The convergence criterion was $10^{-3}$ for the continuity, momentum, and species equations and $10^{-6}$ for the energy equation.

#### 4.3.5.2 Model Validation

For the sake of validating the numerical work, Fig. 4.15 presents a comparison between the predicted centerline temperature distribution and the corresponding experimental data at $\varphi = 0.5$ and O.F. = 60% with a theoretical adiabatic flame temperature (AFT) of ~2500 K. It has to be mentioned here that the theoretical adiabatic flame temperature does not account for $CO_2$ dissociation, which was proven in past studies to reduce that theoretical value by 100–200 K only [62]. Good agreement can be observed between the numerical and experimental results with only 5% deviation. A peak temperature of about 1900 K is predicted at an axial location $Z \approx 2.5$ cm, which indicates the presence of an inner recirculation zone (IRZ) where the local temperature is below AFT. A radial profile would thus reveal that the temperature does not peak at the centerline.

Figure 4.16 presents another comparison between predicted and experimental data in form of radial temperature profiles at $Z = 5$ cm, $\varphi = 0.5$, and O.F. = 60% and 50% with respect to the normalized combustor radius. The corresponding values of AFT are ~2500 and 2170 K, respectively. Good agreement can again be observed with <5% deviation. The analysis of Fig. 4.15 is confirmed by the fact that the radial profiles do not peak at the centerline, which infers the typical V-shape of swirl-stabilized flames with an inner recirculation zone (IRZ) about the centerline, where colder products are recirculated back upstream. Colder temperatures also exist near the combustor

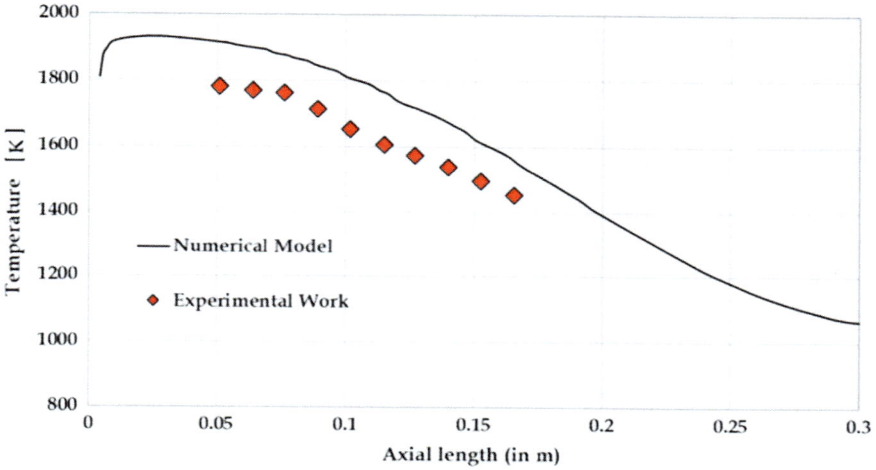


**Fig. 4.15** Predicted centerline temperature distribution compared to the corresponding experimental data

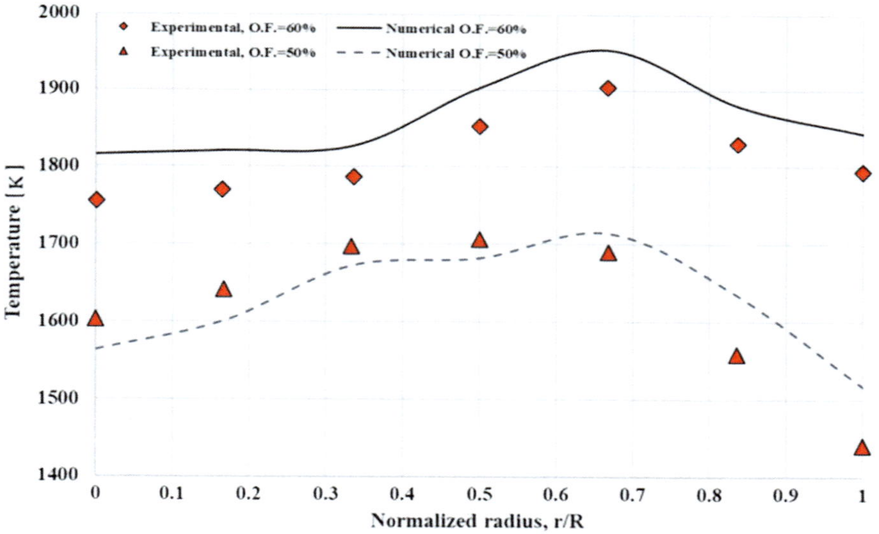


**Fig. 4.16** Predicted radial temperature profile at Z = 5 cm, compared to the corresponding experimental data

wall due to heat transfer by convection and radiation. A detailed analysis of flame structure will follow in the next sections.

An additional comparison can be made between the predicted map of the hydroxyl radical (OH*) and the corresponding experimental image of flame shape. Weigand et al. [63, 64] reported that the visual flame appearance is best represented by the OH* and CH* radicals. Figure 4.17 thus compares the predicted molar concentration map of OH with the corresponding experimental flame image at $\varphi = 0.35$ and O.F. = 60%. Excellent agreement is obtained, wherein two major reaction zones are observed experimentally and captured numerically: a small initial zone within the diverging section that follows the burner throat, and a larger conical downstream zone that anchors at the tip of burner center body. The two zones are separated by the corner recirculation zone (CRZ), where combustion products are circulated with no reactions and zero OH concentration. This flame structure has thus been described as "double conical" in past studies [65, 66].

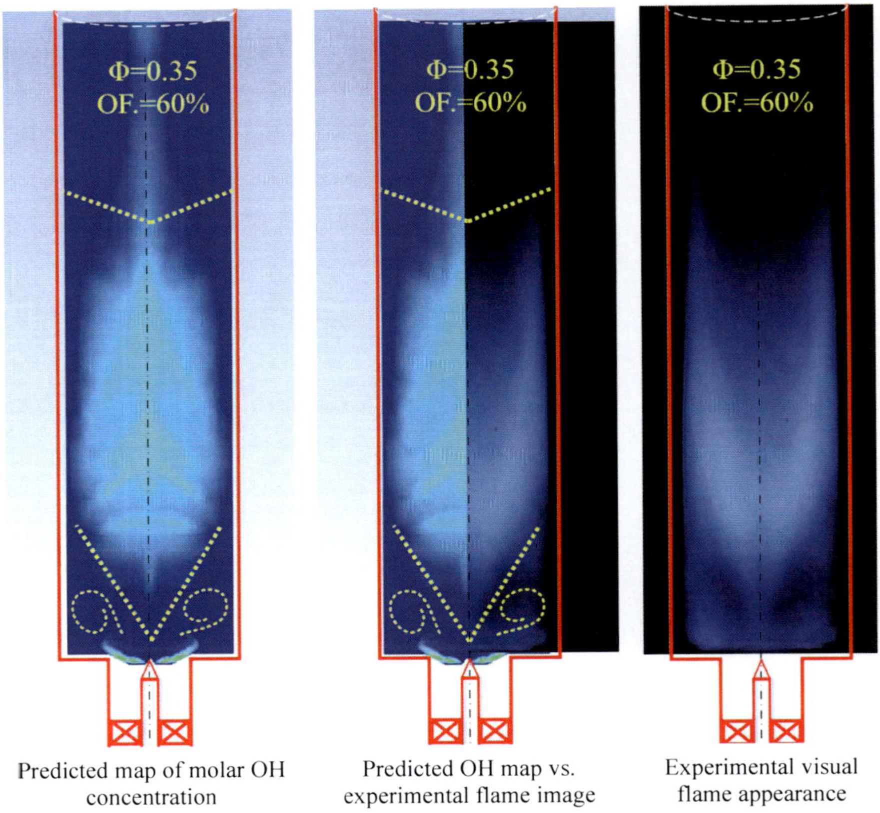


**Fig. 4.17** Comparison between the predicted map of molar OH concentration and the corresponding experimental visual flame appearance at O.F. = 60% and $\varphi = 0.35$

#### 4.3.5.3 Flow and Flame Characteristics

After successful validation of the numerical work, the predicted flow fields are analyzed in this section to quantify the effect of equivalence ratio at constant O.F. of 60%. Figure 4.18 compares the centerline temperature distributions at different equivalence ratios, namely 0.2, 0.3, 0.4, 0.5, and 0.6, which correspond to AFT of 1389, 1806, 2173, 2501, and 2796 K, respectively. It can be observed that increasing $\varphi$ generally increases the centerline temperature as expected, because the combustor power, heat release, and AFT increase. The temperature distribution at $\varphi = 0.2$, however, triggers an interesting observation, namely that the centerline temperature early on is as cold as 500 K and increases up to ~1100 K only towards the combustor exit. This is a clear representation of the blowout limit, where the flame cannot stabilize inside the combustor. The $\varphi = 0.3$ and 0.4 cases mimic the $\varphi = 0.35$ flame presented in Fig. 4.17, where the centerline temperature starts out cold then increases to approach AFT within the conical downstream reaction zone that anchors at the tip of burner center body. The observed temperature fluctuations within the early sections of the combustor indicate the three-dimensional nature of the swirling flow field; a detailed analysis of flame structure will thus follow in the next section. The $\varphi = 0.6$ flame is the only one where the centerline temperature peaks early on and approaches its AFT value. This infers that the flame is most compact, i.e., too small to house a large IRZ.

Figure 4.19 presents the centerline distribution of the axial velocity component normalized by the value of bulk inlet velocity (5.2 m/s). At all equivalence ratios (with the exception of 0.6), swirl induces an IRZ in the first ~10 cm within the combustor, where the axial velocity is negative indicative of reverse flow or recirculation. The fact that the $\varphi = 0.6$ case behaves differently is a clear indication that this case

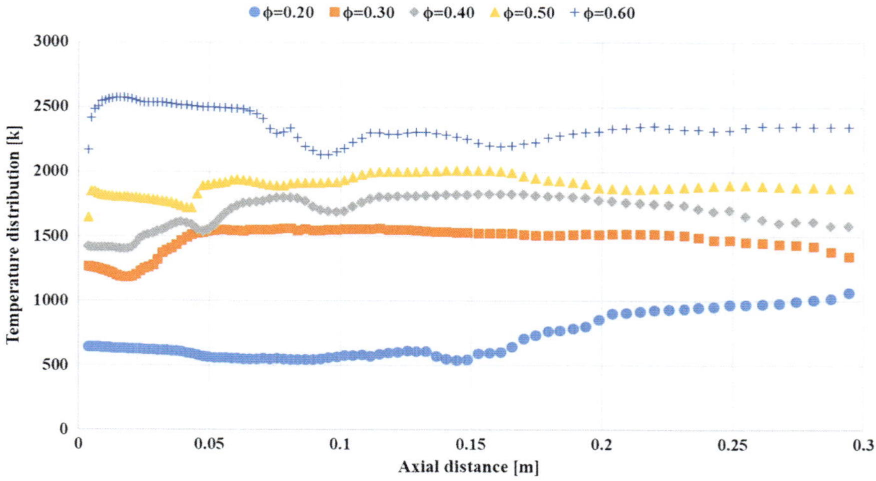


**Fig. 4.18** Centerline temperature distribution at different equivalence ratios

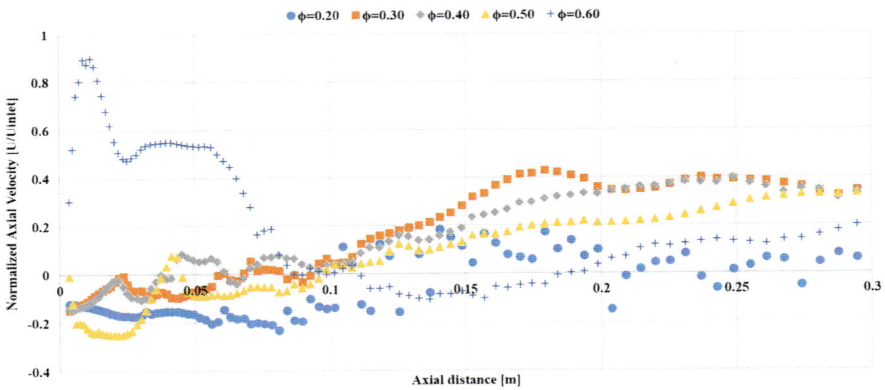


**Fig. 4.19** Centerline distribution of the axial velocity component normalized by the value of bulk inlet velocity (5.2 m/s) at different equivalence ratios

represents the flashback limit, where the flame is most compact, i.e., too small to house a large IRZ, as inferred from Fig. 4.18. Note that the axial velocity reaches +0.9 of the bulk inlet value at Z ≈ 1 cm, and that the centerline temperature peaks at this axial location (Fig. 4.18). Reacting flow thus exists at the centerline, and the bulk inlet velocity prevails within this flow in the absence of IRZ. This infers that a delicate balance exists at the burner throat between the bulk velocity and flame speed before the onset of flashback if φ were increased. This balance was discussed in an earlier study by the authors [67], where it was concluded that the combustor flashback limit occurred consistently at a constant flame speed of 5.2 m/s within the examined φ-O.F. space, because the bulk inlet velocity was fixed at 5.2 m/s. The upcoming analysis of flame structure will shed more light on the limiting flashback case.

Another interesting observation in Fig. 4.19 is that the $\varphi = 0.2$ case of the blowout limit has a significantly large swirl-induced IRZ, which is, nonetheless, not sufficient to stabilize this extremely lean case.

Figure 4.20 complements Fig. 4.18 by comparing the radial temperature profiles at different equivalence ratios (AFTs) and different axial locations (Z = 2.5, 5, and 10 cm) within the combustor. Increasing φ again increases the radial temperatures in general, because the combustor power, heat release, and AFT increase. The cold profiles at $\varphi = 0.2$ are further proof of the absence of a stable flame inside the combustor at this low φ. The $\varphi = 0.3$ and 0.4 cases show hotter radial profiles at greater Z, which agrees with the findings of Fig. 4.18 and confirms the presence of the conical downstream reaction zone that anchors at the tip of burner center body (see Fig. 4.17). The difference between centerline and wall temperatures is maximum at $\varphi = 0.6$, especially at Z = 2.5 and 5 cm, which supports the analysis of Fig. 4.19, namely that this flame is most compact.

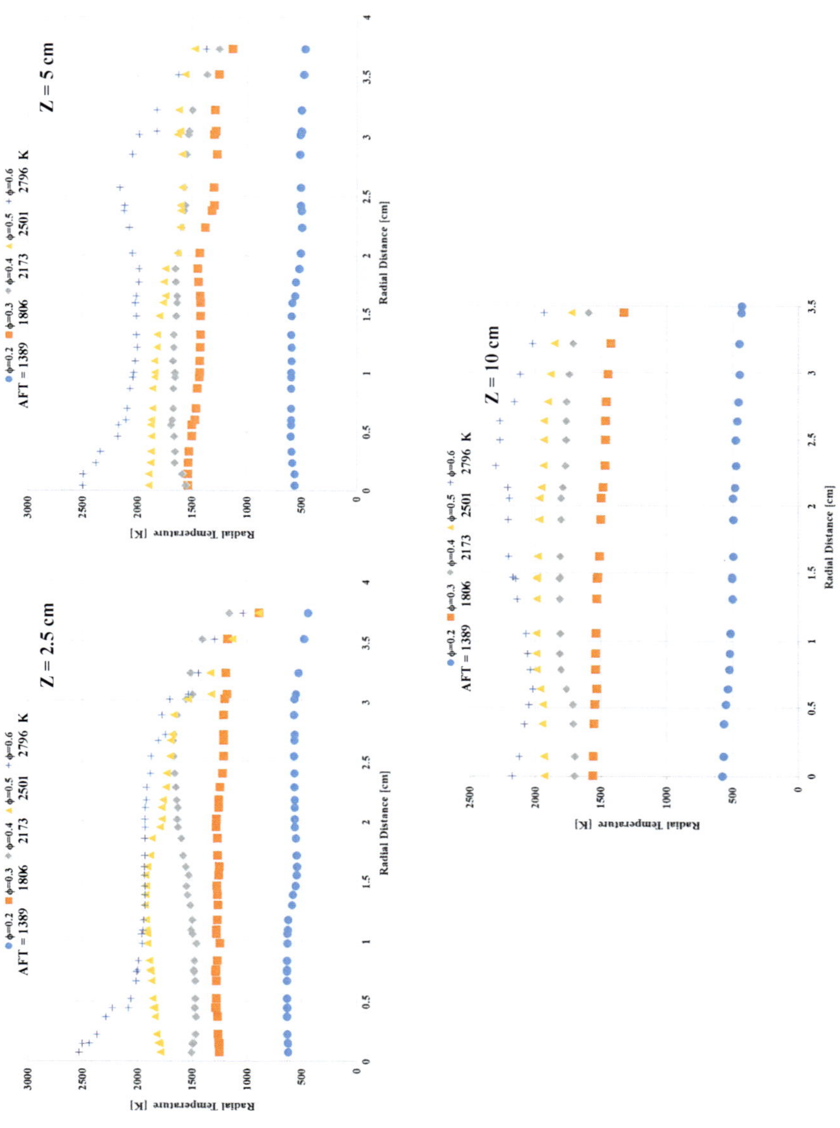


**Fig. 4.20** Radial temperature profiles at different axial locations (Z = 2.5, 5, and 10 cm) and different equivalence ratios

**Table 4.3** Summary of the flame shape, appearance description and stabilization method

| | Flame shape | Description of flame appearance and stabilization method |
|---|---|---|
| I | Double conical | Flame is stabilized within both inner and outer shear layers [36, 37] |
| II | Corner stabilized | Flame is stabilized along the outer shear layer near the combustor walls [19, 39] |
| III | Swirl stabilized (V-shape) | Typical V-shape of swirl-stabilized flames [40, 41] |

Flame structure will be analyzed in this section by presenting experimental flame images and numerical 2-D maps of different parameters, with comparisons to quantify the effect of equivalence ratio.

Previous studies on premixed swirl-stabilized flames [36–38] described three distinct flame shapes/structures, namely (I) double conical flame [68], (II) corner-stabilized flame, and (III) swirl-stabilized flame (V-shape). Table 4.3 summarizes these shapes and gives brief descriptions of the corresponding stabilization mechanisms.

These flame shapes were produced experimentally in this study by varying the equivalence ratio from 0.3 to 0.5 at fixed O.F. = 60%, see Fig. 4.21. At $\varphi = 0.3$ (AFT = 1806 K), the flame takes the double conical shape (I), which is characteristic of cold flames approaching the blowout limit. The "double conical" description comes from the fact that the flame is divided into two reaction zones by a reaction-free CRZ; the first reaction zone stabilizes within the diverging section of burner throat, while the second zone acquires a conical shape and stabilizes at the tip of burner center body. The flame thus stabilizes within both the inner and outer shear layers. The reaction-free zone gradually diminishes by increasing the equivalence ratio ($\varphi = 0.35$, AFT = 1995 K), and the two reaction zones merge into one flame that acquires a cup shape and stabilizes firmly within the CRZ ($\varphi = 0.4$, AFT = 2173 K), hence the description corner stabilized (II). The intense light-blue color of CRZ in the isometric view of the $\varphi = 0.4$ flame in Fig. 4.21 is proof that this flame stabilizes within CRZ. Increasing the equivalence ratio further ($\varphi = 0.45$, AFT = 2342 K) makes the flame transition from the corner-stabilized shape to the typical V-shape (III) of swirl-stabilized flames ($\varphi = 0.5$, AFT = 2501 K). This V-shape has high turbulence intensity, prevails up to the flashback limit, and is stabilized with help of a strong IRZ that becomes more and more compact as flashback is approached ($\varphi = 0.6$, AFT = 2796 K); recall the analysis of Fig. 4.19. Note how the intense light-blue color is confined to the burner throat in the isometric view of the $\varphi = 0.5$ flame as the flashback limit is approached.

Figure 4.22 shows the predicted velocity vectors and streamlines overlaying the maps of axial velocity for the $\varphi = 0.3$ (shape I), $\varphi = 0.4$ (shape II), and $\varphi = 0.5$ (shape III) flames, in addition to the blowout limit ($\varphi = 0.2$) and the flashback limit ($\varphi = 0.6$). Figure 4.23 shows the corresponding predicted maps of OH* mole fraction. At $\varphi = 0.3$ (double conical flame), two distinct IRZs can be identified by the "blue" negative axial velocities in Fig. 4.22: an IRZ within the first reaction zone at the

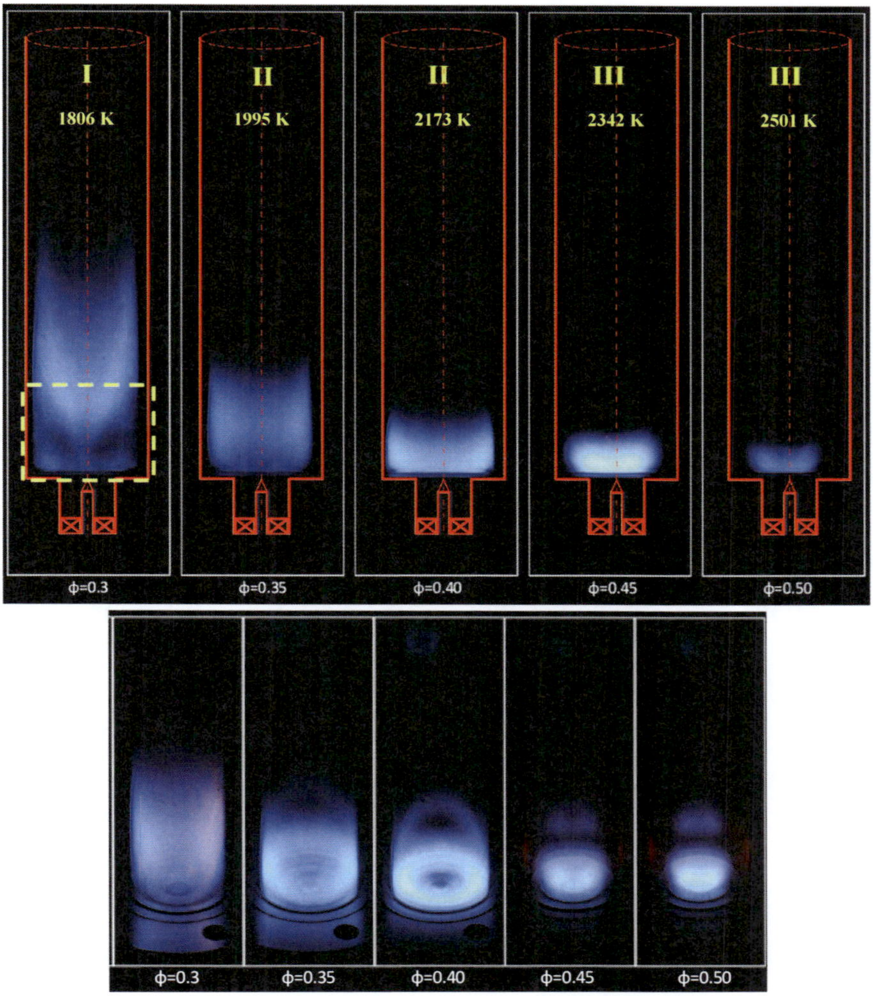


**Fig. 4.21** Visual images of $CH_4/O_2/CO_2$ flames at O.F. = 60% and different equivalence ratios. Top: elevation view, bottom: isometric view

diverging section of burner throat, and the second IRZ within the second reaction zone that takes a conical shape the tip of which touching the first IRZ. The two IRZs are separated by the "green" CRZ. Figure 4.23 highlights these reaction zones using the mole fraction of OH* radical. The first reaction zone is clearly located within the diverging section of burner throat, while the second reaction zone assumes a conical shape the tip of which extending upstream towards the first zone. An OH-free CRZ separates the two reaction zones, which confirms the findings of Fig. 4.22.

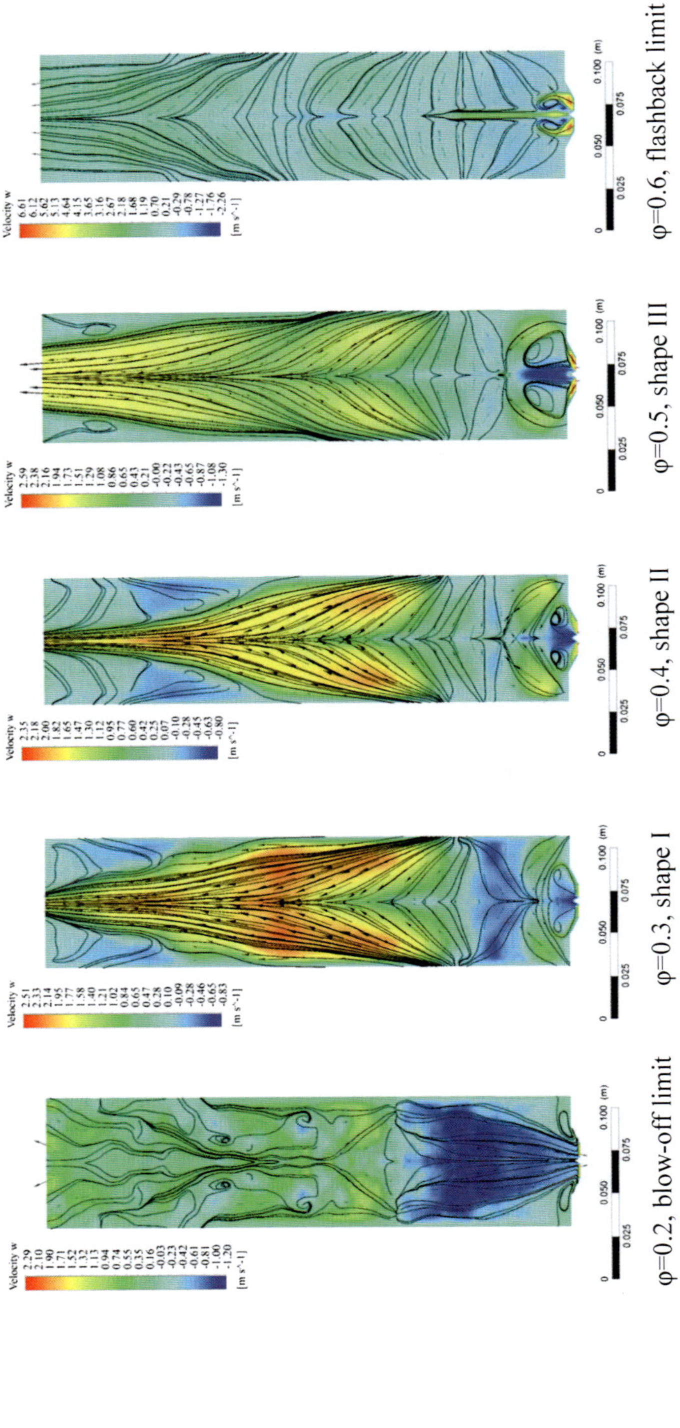


**Fig. 4.22** Predicted velocity vectors and streamlines overlaying the maps of axial velocity at different equivalence ratios

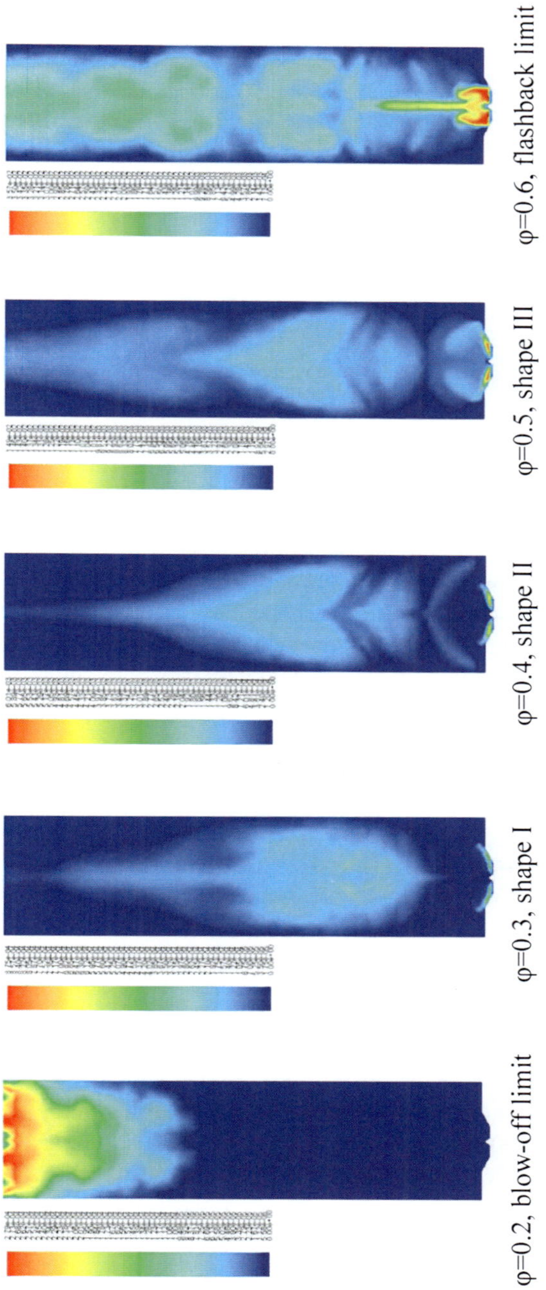


**Fig. 4.23** Predicted maps of OH* mole fraction at different equivalence ratios

At $\varphi = 0.4$ (corner-stabilized flame), the second IRZ diminishes (see Fig. 4.22), and the two reaction zones merge into one zone that moves closer to the burner throat and stabilizes within the "yellow" dominant CRZ. Figure 4.23 also confirms that this CRZ has significant OH* concentration, which indicates reactivity. Increasing $\varphi$ to 0.5 induces a V-shaped flame that moves even closer to the burner throat with significantly stronger IRZ and weaker CRZ (see Fig. 4.22), which explains the transition from corner-stabilization (shape II) to swirl-stabilization (shape III). The CRZ is free of OH* radicals now (Fig. 4.23), which confirms that this flame is stabilized by its IRZ as expected for swirling flames. At the flashback limit ($\varphi = 0.6$), the IRZ becomes strongest yet most compact with almost no CRZ (see Fig. 4.22), which is in direct agreement with the findings of Figs. 4.19 and 4.20. The OH* concentration is now significantly high throughout the entire IRZ (Fig. 4.23) because of its high turbulence intensity. The significantly large swirl-induced IRZ of the $\varphi = 0.2$ case can be clearly observed in Fig. 4.22, but it is not sufficient to stabilize this extremely lean case; Fig. 4.23 shows that the OH* concentration peaks near the combustor exit with reactivity starting very late inside the combustor, which provides solid proof that this case depicts the blowout limit.

Figure 4.24 depicts the predicted temperature maps for the same five equivalence ratios (0.2–0.6), while Fig. 4.25 overlays the velocity streamlines on the temperature maps to link the different flow features to the zones of heat release. At $\varphi = 0.3$ (double conical flame), two distinct "red" hot zones are identifiable, representing the two aforementioned reaction zones characteristic of this flame shape (I). A colder (reaction-free) CRZ separates the reaction zones as explained earlier. As $\varphi$ increases to 0.4, the second reaction zone becomes colder, despite the increase in combustor power, because this zone merges with the first one to form flame shape II, which stabilizes in a hotter CRZ, hence the description corner-stabilized. Increasing $\varphi$ to 0.5 induces a transition to the V-shape with significantly hotter IRZ and colder CRZ, which is additional proof that this flame is stabilized by its IRZ as expected for swirling flames. At the flashback limit ($\varphi = 0.6$), the IRZ is hottest and most compact with concentrated heat release that is aided by the high turbulence intensity. The CRZ, on the other hand, is significantly colder. The high temperature of IRZ increases the local flame speed, until it exceeds the bulk inlet velocity of fresh reactants at the burner throat, and flashback occurs. The blowout limit at $\varphi = 0.2$ is also observed, as the temperature peaks near the combustor exit, because no flame can stabilize within the combustor.

It is worth noting here that the theoretical values of AFT depicted at the top of Fig. 4.24 are ~160 K higher than the peak temperatures within their respective predicted flow fields. Previous studies have shown that the dissociation of $CO_2$ makes the actual AFT 100–200 K lower than its theoretical value [62]. This demonstrates the ability of the numerical code to accurately predict the examined flames.

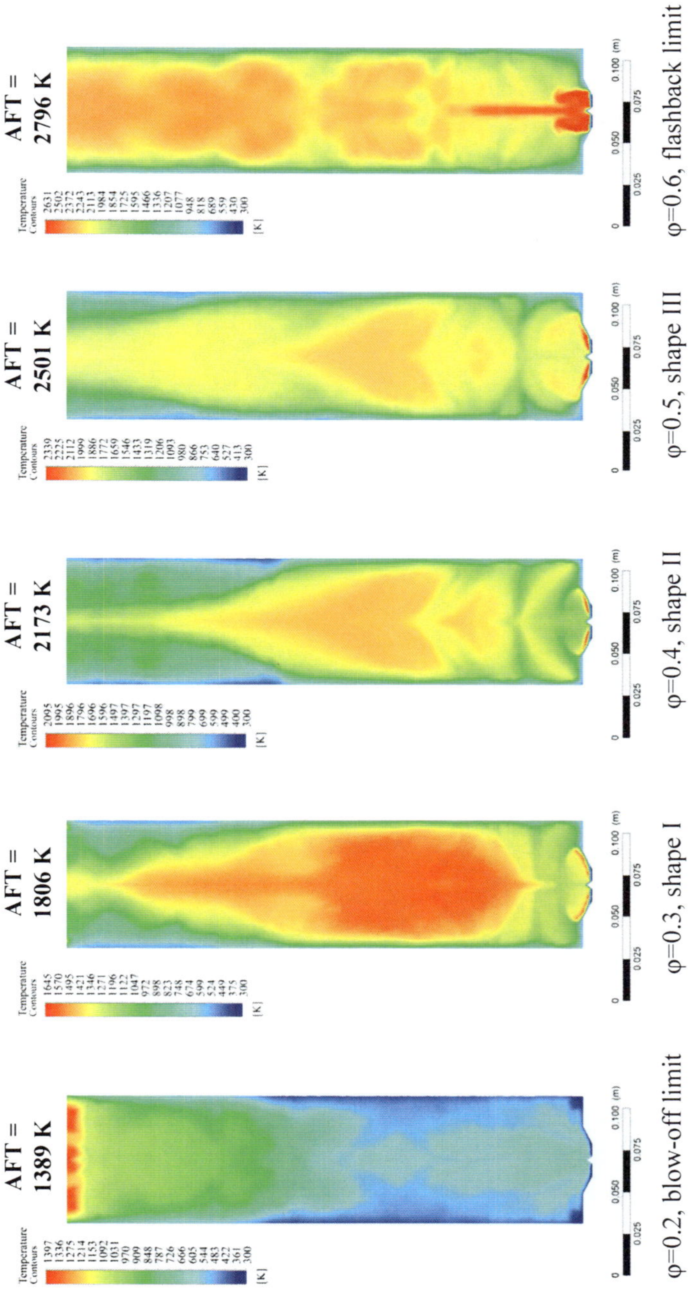


**Fig. 4.24** Predicted temperature maps at different equivalence ratios

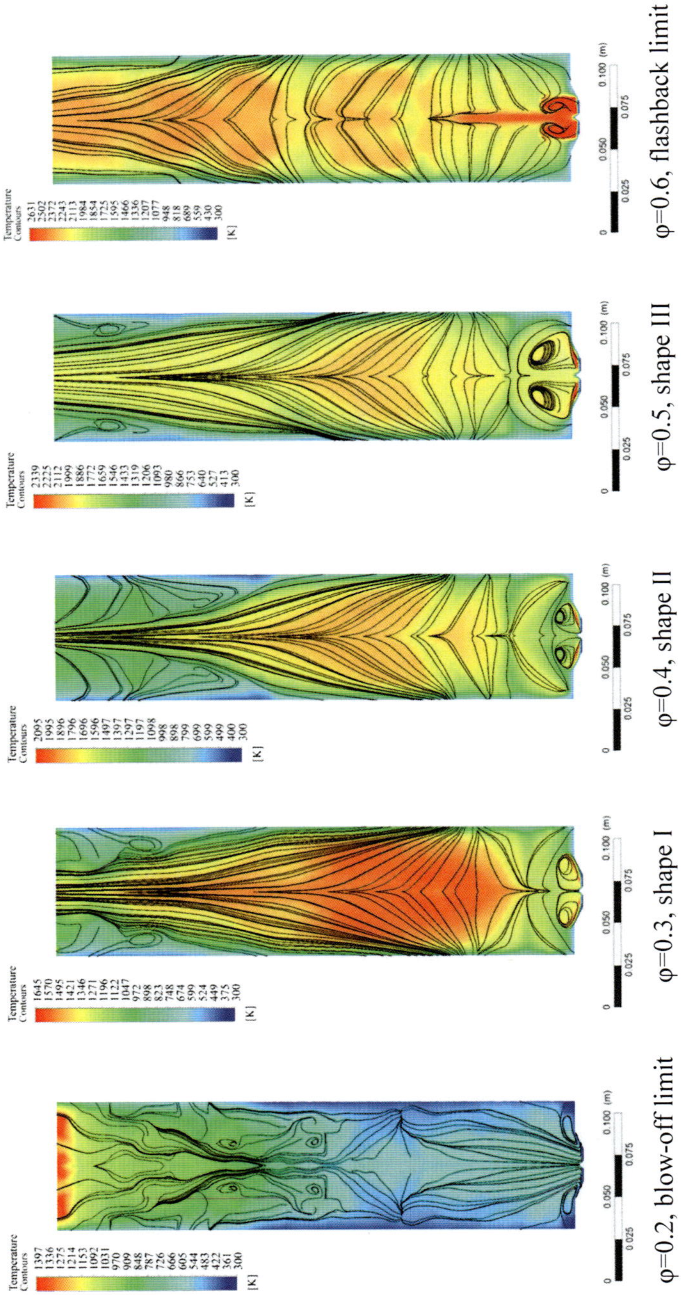


**Fig. 4.25** Predicted velocity streamlines overlaying the temperature maps at different equivalence ratios

## 4.4 Oxy-methane Versus Oxygen-Enriched-Air Flames for Gas Turbine Applications

Almost all of today's combustion applications use air as oxidizer. The resulting exhaust consists primarily of $CO_2$, $H_2O$, $O_2$, and $N_2$. $CO_2$ is a greenhouse gas, and its emissions are a global concern because of the escalating climate-change problem. Fossil fuels are the main source of energy in many countries and will remain so for several decades [69], but the resulting $CO_2$ emissions are inescapable. To prevent global temperature from rising more than 2 °C, international conventions, like the Paris Agreement, have strongly recommended carbon capture and sequestration as an irreplaceable solution [70]. The separation of $CO_2$ from the exhaust stream of air-combustion is, however, complicated and costly [71]. Oxy-fuel combustion, on the other hand, uses pure oxygen as oxidizer, so the resulting products are primarily $CO_2$ and $H_2O$. The former can thus be captured at significantly lower cost after condensing the latter. The inherent absence of air-based $N_2$ also offers two additional advantages: First, nitrogen oxides ($NO_x$) emissions are drastically reduced, and second, the exhaust volume is considerably smaller, which allows for using compact and more efficient treatment equipment.

### *4.4.1 Air Flames Versus Oxy-flames*

Switching from air to oxy-combustion, however, entails different degrees of freedom for combustor operability. Air is an ample low-cost oxidizer, so lean operation is typical in air-fuel gas turbines to control the combustor exit temperature. The separation of air to generate oxygen, on the other hand, poses an energy penalty [72], so oxy-fuel applications need to be operated near stoichiometry to utilize both fuel and oxygen efficiently. However, flames of fuel and pure $O_2$ have extreme temperatures beyond the metallurgical limits of most materials. Dilution is thus implemented to control flame temperature by recirculating part of the exhaust flow back into the combustor. The reactants are thus fuel plus an $O_2/CO_2$ diluted oxidizer.

The replacement of $N_2$ with $CO_2$ as diluent also affects the flame characteristics, primarily because $N_2$ and $CO_2$ have different properties, such as heat capacity, thermal conductivity, mass diffusivity, and dynamic viscosity [73–75]. $CO_2$ is known to have an adverse effect on the reaction rates, flame speed, and combustion efficiency [76, 77]. Consequently, oxy-fuel flames have narrower stability windows compared to air-fuel flames [48, 78]. Kutne et al. [15] attributed the reduced stability of oxy-fuel flames to the larger heat capacity of $CO_2$ compared to $N_2$, which reduces the flame temperature and burning velocity. This conclusion was substantiated by the observation that the inner recirculation zones of oxy-fuel flames exist further downstream compared to those of air-fuel ones. Amato et al. [79] examined the extinction mechanisms of premixed flames and showed that $CO_2$ adversely affects the reaction kinetics. This finding was also affirmed by Jerzak and Kuznia [80] who investigated the upper and lower flammability limits of natural-gas combustion in different oxidizer environments, i.e., $O_2/CO_2$ and $O_2/N_2$. The retarded kinetics and lower burning velocities of the $CO_2$ flames allowed them to resist flashback better than the $N_2$ ones by being stable at greater oxygen fractions before the onset of flashback. The minimum oxygen fraction of $O_2/CO_2$ diluted oxidizer needed to replicate the stability of air-fuel flames (21% $O_2$) was found in the literature to be in the range of 30% [81] to 40% [48]. Song et al. [44] reported that 36% is needed to match the temperature profiles.

Most of the literature studies on oxy-fuel combustion reported their findings of flame characteristics, stability, and reaction kinetics based on the effect of oxygen fraction in the $O_2/CO_2$ diluted oxidizer. The authors of this work, however, recently published a study of $CH_4/O_2/CO_2$ flames [67], where they took a different approach in analyzing flame stability and shape based on the adiabatic flame temperature ($T_{ad}$). It was shown that $T_{ad}$ is the parameter that best represents the stability map of the examined premixed swirl-stabilized model gas-turbine combustor at constant inlet bulk velocity. The flashback limit was observed to coincide with a contour of constant $T_{ad}$ within the investigated two-dimensional space of oxygen fraction ($OF$) and equivalence ratio ($\varphi$). Flames of the same $T_{ad}$ were found to have almost identical sizes and shapes, despite having different $OF$ and $\varphi$. This clearly demonstrated that both combustor operability and flame characteristics should be analyzed based on $T_{ad}$, instead of $\varphi$ or $OF$ separately, because $T_{ad}$ is an umbrella parameter that combines the individual effects of $\varphi$ and $OF$ and best describes the prevailing effect of reaction kinetics under similar cold flow conditions.

Similar analyses of oxy-fuel flames based on $T_{ad}$ were rarely performed in the literature. Shroll et al. [82] investigated stoichiometric swirl-stabilized premixed $CH_4/O_2/CO_2$ and $CH_4$/air flames with focus on their dynamic stability characteristics. $T_{ad}$ was controlled through $OF$. It was found that $T_{ad}$ is the parameter that governs the transition from one stability mode to another in both $CH_4/O_2/CO_2$ and $CH_4$/air flames, although they have different flame speeds at a given $T_{ad}$. High-speed imaging was used to confirm this finding by showing comparable vortex breakdown modes indicative of analogous turbulent flame geometries that change with $T_{ad}$. Jourdaine et al. [68] performed an experimental study of the stabilization mechanisms of premixed $CH_4/O_2/CO_2$ and $CH_4$/air flames. They revealed that, at the same swirl number, $CO_2$ and $N_2$ flames of the same $\varphi$ have similar shapes while maintaining common $T_{ad}$ and incoming jet velocity. They also reported that no combustor design modifications are needed to switch from air to oxy-fuel operation.

It is worth noting here that manufacturers of conventional LPM DLE air-fuel gas turbines commonly quantify combustor performance in terms of $T_{ad}$ (and not $\varphi$), whether on the scale of the entire combustor, of each ring/can separately, or even of each lit fuel-air premixer. Another common practice among gas-turbine manufacturers is to consider constant pressure drop across the combustor headend, which maintains constant bulk velocity of reactant mixture at the premixer throat and, consequently, creates a safe margin to flashback. This practice has been implemented by for $CH_4/O_2/CO_2$ flames [67] and is employed here for $CH_4/O_2/N_2$ flames as well. The same bulk throat velocity of 5.2 m/s is considered again here to have direct comparisons of combustor stability and flame macrostructure. Both studies were performed on the same model gas-turbine combustor, as described in the above Sect. 4.3. The combustor is based on the LPM DLE DACRS technology of operational air-fuel gas turbines [83, 84]. The choice of this bulk throat velocity is also based on the desired power density of the combustor, which is within the range of typical operation of industrial gas turbines, i.e., 3.5–20 MW/m$^3$/bar [50]. In addition to investigating how the combustor stability limits correlate with $T_{ad}$, a detailed analysis of flame macrostructure is performed here to highlight the effect of $T_{ad}$. The macrostructures of premixed $CH_4$/air flames were examined in recent literature studies [65, 66, 80, 85] using dump combustor geometries similar to the one considered here. The flame was found to assume one of four distinct shapes based on $\varphi$. The same classification is implemented here for both the $CH_4/O_2/N_2$ and $CH_4/O_2/CO_2$ flame sets; however, this analysis is based on $T_{ad}$ as an umbrella parameter that combines both effects of $\varphi$ and $OF$.

### 4.4.2 *Role of Adiabatic Flame Temperature for Controlling Flame Stabilization*

Nitrogen and oxygen have very similar properties, i.e., comparable densities and heat capacities in particular. Changing the oxygen fraction of $O_2/N_2$ diluted oxidizer is thus expected not to affect the apparent oxidizer properties much. Carbon dioxide, on the other hand, has significantly different properties; the density and heat capacity of $CO_2$ are both about 1.5 times their $O_2$ and $N_2$ counterparts at the same pressure and temperature. Oxygen fraction is thus expected to play a more significant role in oxy-fuel flames. These facts affect key combustion parameters like the adiabatic flame temperature ($T_{ad}$), which is of pivotal importance in the analyses of premixed flames as will be shown here.

Figure 4.26 shows the stability maps (flashback and blowout limits) of the examined model gas-turbine combustor geometry for both the $CH_4/O_2/N_2$ and $CH_4/O_2/CO_2$ flames at fixed flame-inlet bulk velocity (similar cold flow conditions). The color backgrounds are the respective $T_{ad}$ maps plotted against the same color scale for a one-to-one comparison. Flashback data points are indicated by triangular symbols in Fig. 4.26, while the blowout points are depicted using circular symbols. The stable combustion zone on each map is the area between the limits, whereas the outside areas are the flashback and blowout zones on the top and bottom, respectively. The reason behind the scattered nature of blowout data points is that intermittent flame lifting and reattachment occurred prior to blowout, which made it difficult to quantify this limit accurately. The flashback limit was thus more distinctly identifiable than the blowout one during experiments.

Two preliminary observations can be made from Fig. 4.26. Firstly, $CO_2$ flames always have lower $T_{ad}$ than their $N_2$ counterparts at the same $\varphi$ and $OF$, which is

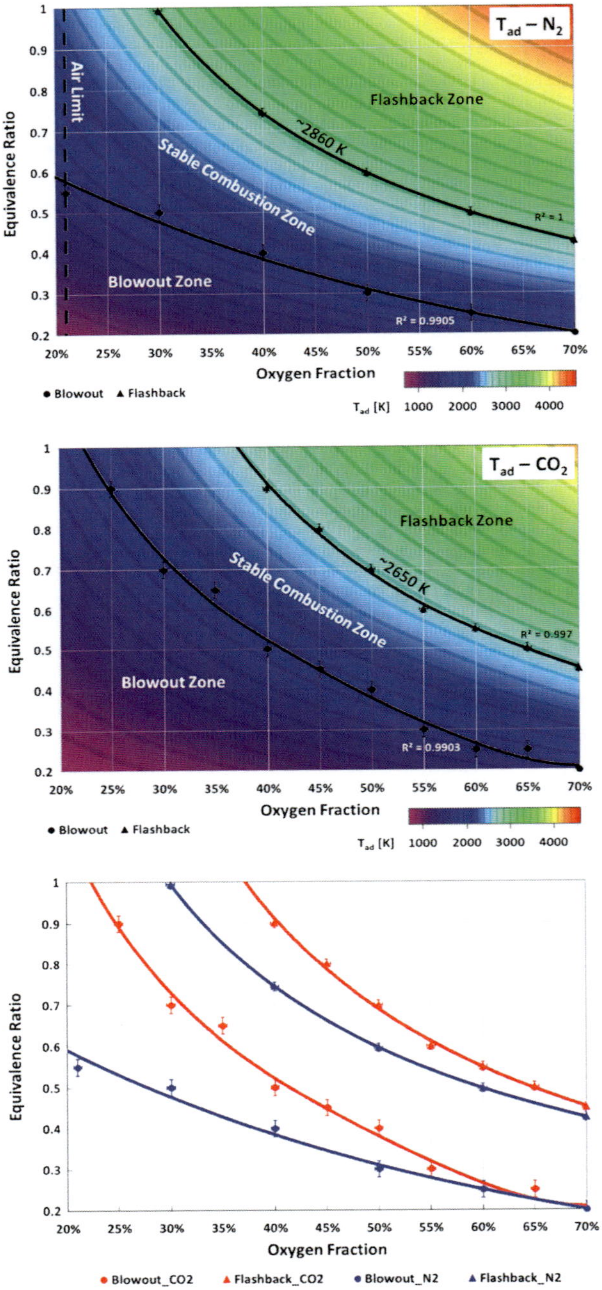


**Fig. 4.26** Combustor stability maps plotted against the respective contours of constant $T_{ad}$ for $CH_4/O_2/N_2$ flames (top) and $CH_4/O_2/CO_2$ flames (middle). Both background maps of $T_{ad}$ share a common color scale. The bottom plot compares the stability limits directly. The bulk throat velocity is fixed at 5.2 m/s throughout

attributed to the higher heat capacity of $CO_2$ compared to $N_2$. Secondly, the slopes of constant $T_{ad}$ contours $[\partial\varphi/\partial(OF)_{T_{ad}}]$ are steeper in $CO_2$ flames, indicating a greater influence of $OF$ on $T_{ad}$ in oxy-fuel flames as mentioned earlier, again because of the higher heat capacity of $CO_2$ compared to $N_2$.

Figure 4.26 also provides one of the primary findings of this study, namely that flame flashback occurs consistently at constant $T_{ad}$ for both the $CH_4/O_2/N_2$ and $CH_4/O_2/CO_2$ flame sets at constant flame-inlet bulk velocity. Although the contours of constant $T_{ad}$ have different slopes in the $N_2$ and $CO_2$ maps as discussed earlier, both $N_2$ and $CO_2$ flashback limits occur consistently at constant $T_{ad}$. This was reported recently by the authors for $CH_4/O_2/CO_2$ flames only [67], but the novelty of this work is proving that this finding is universal and applies to $CH_4/O_2/N_2$ flames as well. This behavior can be attributed to the fact that the flashback limits are controlled mainly by reaction kinetics under similar cold flow conditions (constant inlet bulk velocity). A delicate balance between flame speed and flow velocity exists at the flashback limit. The constant bulk velocity examined here allows for inferring that each flashback limit occurs at constant flame speed, which further proves the dominant role of reaction kinetics. The flame speed was shown to depend directly on $T_{ad}$ [67, 86–88], which completes the picture and explains why both $N_2$ and $CO_2$ flashback limits occur at constant $T_{ad}$. It can also be observed in Fig. 4.26 that $T_{ad}$ is generally the optimum parameter to describe the stability maps of premixed natural-gas combustors operating at constant inlet bulk velocity. Some deviation is observed near the blowout limit, which indicates that the blowout mechanism is not function of $T_{ad}$ only but depends on other flow parameters as well. The aforementioned intermittent lifting and reattachment of the flame prior to blowout is another proof of this.

Based on this analysis, it can be recommended to design and operate premixed oxy-fuel gas-turbine combustors based on $T_{ad}$, particularly at high and medium loads away from blowout, similar to the existing state-of-the-art LPM air-fuel gas-turbine combustors. These are commonly designed and operated based on $T_{ad}$, whether on the scale of the entire combustor, of each ring/can separately, or even of each lit fuel-air premixer. The flame temperature of an air-fuel combustor is controlled through the equivalence ratio, of course, but gas-turbine manufacturers design, operate, and evaluate their combustors based on flame temperature and not equivalence ratio. This study thus recommends the same practice for oxy-fuel gas turbines as well, especially that maintaining constant flame-inlet bulk velocity is another common practice in LPM air-fuel gas turbines. The combustor is typically operated at constant pressure drop across its headend, which maintains constant bulk velocity of reactant mixture at the premixer throat and, consequently, creates a safe margin to flashback.

Multiple other important observations can also be made from Fig. 4.26 regarding the reaction kinetics of $CH_4/O_2/N_2$ versus $CH_4/O_2/CO_2$ flames:

- The 2320-K $T_{ad}$ of a stoichiometric $CH_4$/air flame is matched at $OF = 30\%$ in a stoichiometric $CH_4/O_2/CO_2$ flame, which agrees with the findings of Ditaranto et al. [81], who reported this 30% $OF$ value but did not explain it in terms of $T_{ad}$.

- Stable $CH_4$/air flames ($OF = 21\%$) were attainable for $\varphi > 0.55$, whereas the same $OF$ was beyond the blowout limit of oxy-flames. This agrees directly with the findings of all past studies that investigated the minimum $OF$ needed to match the stability of air-fuel flames [44, 48, 81].
- $CO_2$ flames blowout at higher $\varphi$ than $N_2$ ones for any given $OF$, i.e., $CO_2$ flames generally exhibit inferior resistance to blowout compared to $N_2$ ones, which narrows the stable combustion zone of oxy-flames significantly and shows the negative impact of $CO_2$ on reaction kinetics and blowout. This agrees with the findings of Rashwan et al. [48], Kutne et al. [15], and Amato et al. [79].
- Flashback of $CO_2$ flames occurs at higher $OF$ compared to $N_2$ flames of the same $\varphi$. This is further evidence of the retarded chemical kinetics of $CO_2$ flames, as more oxygen is needed to trigger flashback, which is in direct agreement with the findings of Jerzak and Kuznia [80].
- It is only at high oxygen fractions (70%) where the $CO_2$ and $N_2$ flames exhibit almost identical flashback and blowout equivalence ratios, because the concentrations of $CO_2$ and $N_2$ in their diluted oxidizers have dropped enough to undermine the effect of their different properties.

To further prove that $T_{ad}$ is the parameter that best describes the stability maps of premixed natural-gas combustors ($CH_4/O_2/CO_2$ and $CH_4/O_2/N_2$) operating at constant flame-inlet bulk velocity, the stability maps of Fig. 4.26 were replotted against the contours of constant power density (*PD*, Fig. 4.27) and throat Reynolds number (Re, Fig. 4.28). Recall from the analysis of governing equations that the $N_2$ and $CO_2$ flame sets have identical *PD* maps, as seen in Fig. 4.27, because the combustor consumes the same amount of fuel in both sets for a given common combination of $\varphi$ and $OF$.

It can be seen in Fig. 4.27 that the flashback limit of $CH_4/O_2/N_2$ flame set follows a contour of constant *PD*, but the same does not apply to $CH_4/O_2/CO_2$ flames, which proves that, unlike $T_{ad}$, *PD* is not the optimum parameter for describing the stability maps. Interestingly, the observed trend of $N_2$ flashback limit is merely an artifact

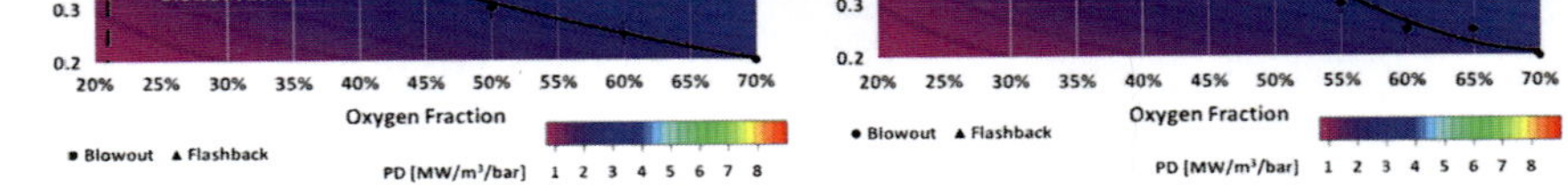


**Fig. 4.27** Combustor stability maps plotted against the contours of constant *PD* for $CH_4/O_2/N_2$ flames (left) and $CH_4/O_2/CO_2$ flames (right). The bulk throat velocity is fixed at 5.2 m/s throughout. Both flame sets have identical *PD* maps in the background

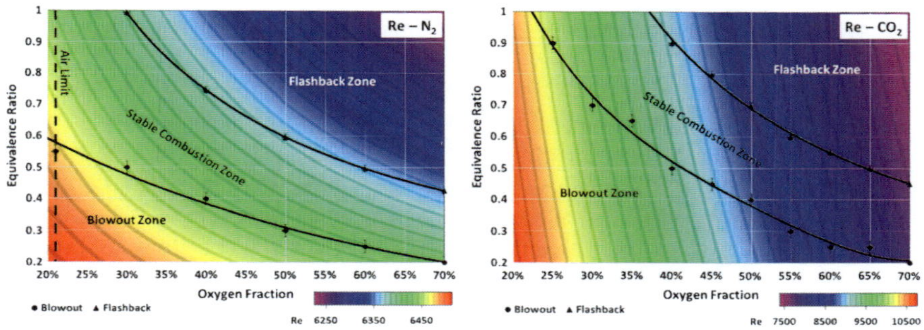


**Fig. 4.28** Combustor stability maps plotted against the respective contours of constant bulk throat Re for $CH_4/O_2/N_2$ flames (left) and $CH_4/O_2/CO_2$ flames (right). The bulk throat velocity is fixed at 5.2 m/s throughout. Each background map of Re has its own color scale

of the physical properties of $O_2$ and $N_2$. As discussed earlier, $O_2$ and $N_2$ have very similar densities and heat capacities. So, changing the *OF* in $O_2/N_2$ diluted oxidizer at constant $\varphi$ changes $T_{ad}$ because the concentrations of exhaust $O_2$ and $N_2$ change and not because of the minute difference in heat capacities of $O_2$ and $N_2$. The *PD* map of $CH_4/O_2/N_2$ set is also primarily affected by species concentrations because of the comparable densities of $O_2$ and $N_2$. This explains why the $T_{ad}$ and *PD* maps look so similar. To demonstrate this, Fig. 4.29 replots the stability map of $CH_4/O_2/N_2$ flames against the contours of the ratio of $T_{ad}$ to *PD*. It can be clearly seen that the

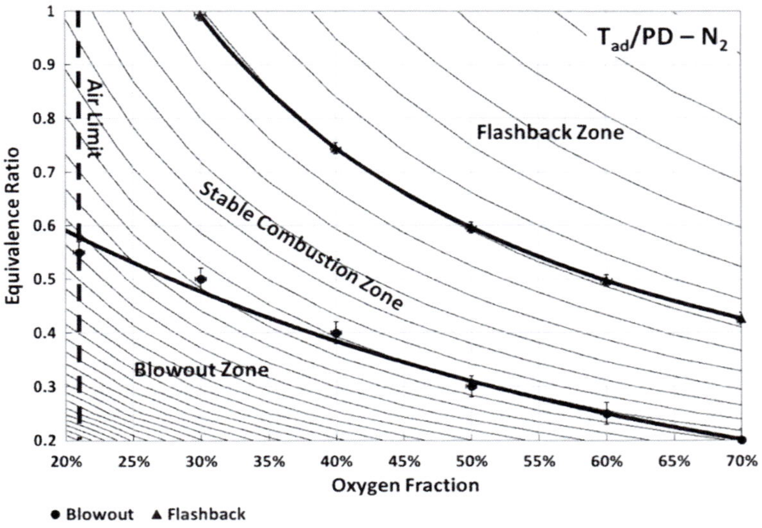


**Fig. 4.29** Combustor stability map of $CH_4/O_2/N_2$ flames plotted against the contours of constant ratio of $T_{ad}$ to *PD*

map of this ratio is almost identical to those of $T_{ad}$ and $PD$; consequently, the $N_2$ flashback limit follows a contour of constant $T_{ad}$-to-$PD$ ratio.

By comparison, $O_2$ and $CO_2$ have different properties, which results in dissimilar $T_{ad}$ and $PD$ maps for the $CH_4/O_2/CO_2$ set. Therefore, the observed dependence of flame flashback on $T_{ad}$ (Fig. 4.26) is not observed with $PD$ (Fig. 4.27) as well. This study thus agrees with the findings of Abdelhafez et al. [67], namely that combustor flashback does not always occur at constant fuel flow rate. The effect of $O_2$ versus $CO_2$ or $N_2$ densities is also evidenced in the Re maps of Fig. 4.28. The slopes of constant Re contours $[\partial\varphi/\partial OF)_{Re}]$ are significantly steeper in $CO_2$ flames, indicating that the Re map is dominated by $OF$ in oxy-fuel flames, similar to what was observed earlier with $T_{ad}$. This is attributed to the higher density of $CO_2$. It can also be seen in Fig. 4.28 that none of the stability limits in either map follows a contour of constant Re, which further proves that $T_{ad}$ is indeed the optimum parameter to describe the stability maps of premixed natural-gas combustors operating at constant flame-inlet bulk velocity. The effect of Re was examined here because a previous study by the authors [56] revealed that the stability and structure of coaxial non-premixed $CH_4/O_2/CO_2$ and $CH_4/O_2/N_2$ flames were affected by the Reynolds number of diluted oxidizer.

### 4.4.3 *Role of Adiabatic Flame Temperature for Controlling Flame Structure*

Past studies have identified four distinct stabilization modes of premixed $CH_4$/air flames, which are described in Table 4.4. Figure 4.30 shows the images of selected $CH_4/O_2/N_2$ flames at $OF = 40\%$, which were captured to visualize the mode classification of Table 4.4. Flame A ($T_{ad} = 2094$ K) is a tall unstable flame that assumes shape I (single cone) near the blowout limit with intermittent lifting and reattachment, because this limit is not function of $T_{ad}$ only, as discussed earlier. Increasing $T_{ad}$ induced a transition to shape II (double cone, flame B, $T_{ad} = 2241$ K). This metastable shape was found to exist within a narrow range of $T_{ad}$ with intermittent transition back to shape I. Further increase of $T_{ad}$ stabilized the flame within the corner recirculation zone and boundary layer at the quartz walls (shape III, flame C,

**Table 4.4** Shapes and stabilization modes of premixed $CH_4$/air flames from past studies

| | Flame shape | Description/stabilization mode |
|---|---|---|
| I | Single cone | At low $\varphi$. Flame stabilized within the inner shear layer [65, 66] |
| II | Double cone | Flame stabilized within both the inner and outer shear layers [65, 66] |
| III | Corner-stabilized | Flame stabilized within the corner recirculation zone and the boundary layer at the combustor walls [80] |
| IV | Swirl-stabilized V-shape | Typical V-shape of swirl-stabilized flames [85] |

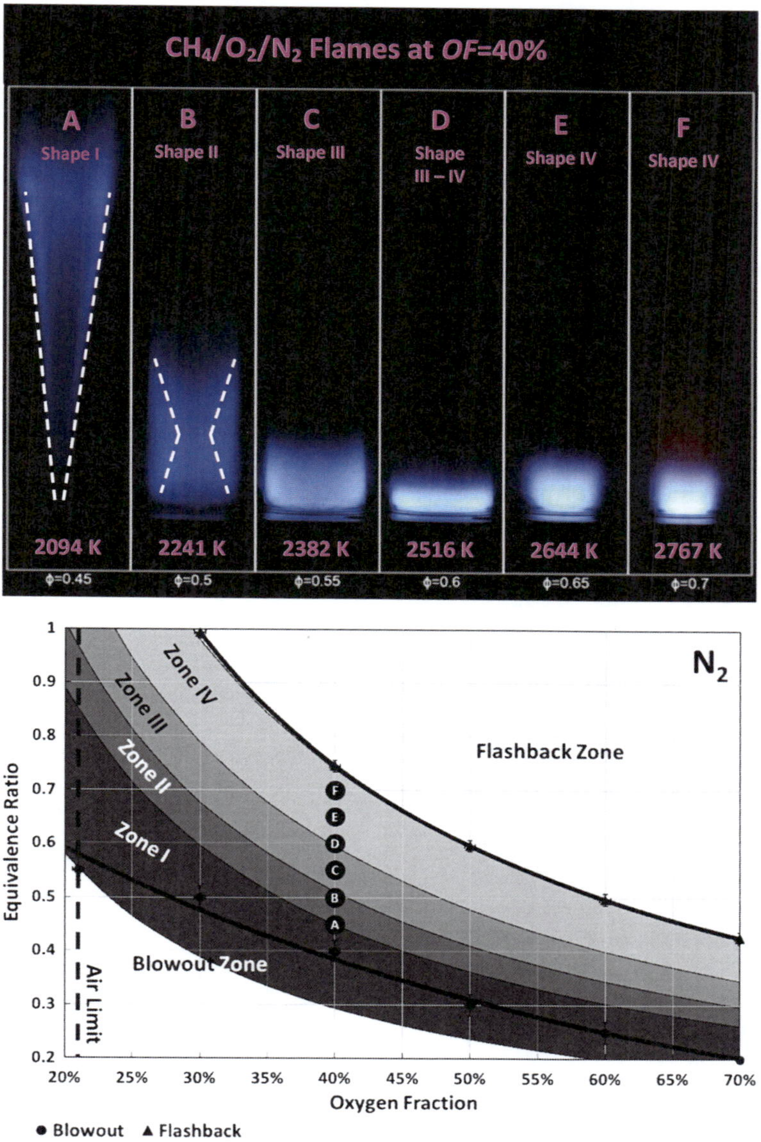


**Fig. 4.30** Top: Visual images of six $CH_4/O_2/N_2$ flames (A–F) at $OF = 40\%$; $T_{ad}$ is listed for each flame. Bottom: Combustor stability map highlighting data points A–F and the four sub-zones of the stable combustion zone. The bulk throat velocity is fixed at 5.2 m/s throughout

$T_{ad} = 2382$ K). This cup shape was observed to enjoy consistent stability and to span a wider range of $T_{ad}$. Flame D ($T_{ad} = 2516$ K) was carefully selected as it visualizes the upper limit of zone III, where the flame is considerably shorter, adopting a plate shape that is stabilized within the corner recirculation zone only. Intermittent transition to shape IV was observed here. Further increase of $T_{ad}$ made the flame acquire shape IV (flame E, $T_{ad} = 2644$ K) consistently, which is the typical V-shape expected for swirl-stabilized flames. Approaching the flashback limit (flame F, $T_{ad} = 2767$ K), the flame shape did not change much, but the audible noise increased, indicative of combustion dynamics prior to flashback.

This mode classification of flames based on their $T_{ad}$ will be used to subdivide the stable combustion zone into four sub-zones I–IV, bounded by contours of constant $T_{ad}$, as seen on the stability map in Fig. 4.30. Each zone is named after the flame shape that prevails throughout it. The upcoming analysis will then answer the question: Do all flames within each zone have a common shape irrespective of $\varphi$ and $OF$? In other words, is the knowledge of $T_{ad}$ only enough to accurately predict and characterize the macrostructures and stabilization modes of stable flames?

The effect of $T_{ad}$ on flame macrostructure and stabilization mode will be analyzed from three different aspects: changing $T_{ad}$ by varying $\varphi$ at constant $OF = 60\%$, changing $T_{ad}$ by varying $OF$ at constant $\varphi = 0.5$, and keeping $T_{ad}$ constant by varying both $OF$ and $\varphi$ simultaneously. All three approaches will be performed on both $CH_4/O_2/N_2$ and $CH_4/O_2/CO_2$ flames. The selected data points for which flame images were captured to execute this analysis are shown on the stability maps of Fig. 4.31. Figure 4.32 compares the $CO_2$ flame images side by side to their $N_2$ counterparts, and the following observations and conclusions can be made from Fig. 4.32:

- The same distinct 4 shapes observed with the $N_2$ flames are also observed with the $CO_2$ ones, namely, single cone (shape I), double cone (shape II), corner-stabilized (shape III), and the swirl-stabilized V-flame (shape IV). This agrees with the findings of Jourdaine et al. [68].
- Flames $(2\,\&\,8)_{CO_2}$ have a common $T_{ad}$ of ~2000 K, and both have shape II. Flames $(3\,\&\,7)_{CO_2}$ at $T_{ad} \approx 2170$ K have shape III. Even the transitional plate-shaped flames $(4\,\&\,6)_{CO_2}$ have a common $T_{ad}$ of ~2340 K. The same analogy is observed for flames $(4\,\&\,5)_{N_2}$ at $T_{ad} \approx 2690$ K, $(3\,\&\,6)_{N_2}$ at $T_{ad} \approx 2500$ K, and $(2\,\&\,7)_{N_2}$ at $T_{ad} \approx 2300$ K. The only exception is flame $8_{N_2}$ (shape II) versus flame $1_{N_2}$ (shape I), but even this discrepancy is justifiable because the blowout limit is not function of $T_{ad}$ only.
- It is very important to note here that the two flames within each of these "couples" have different $OF$ and $\varphi$ but the same $T_{ad}$. This singles out the direct dependence of flame macrostructure on reaction kinetics and $T_{ad}$ at constant inlet bulk velocity and agrees directly with the findings of Shroll et al. [82].
- Approaching the flashback limit, flames $5_{CO_2}$, $4_{N_2}$, and $5_{N_2}$ have shape IV, and approaching the blowout limit, flames $1_{CO_2}$ and $1_{N_2}$ have shape I.

All these observations can be combined to answer the aforementioned questions and provide two novel conclusions for stable flames (away from blowout) at constant

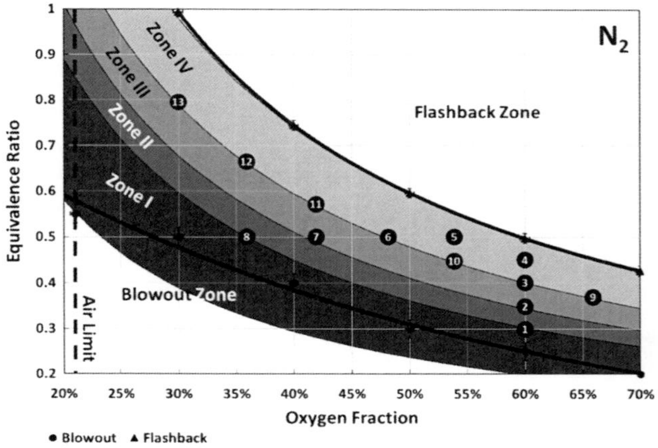


| # | *OF* | $\varphi$ | $T_{ad}$ [K] |
|---|---|---|---|
| 1 | 60% | 0.30 | 2080 |
| 2 | | 0.35 | 2297 |
| 3 | | 0.40 | 2499 |
| 4 | | 0.45 | 2688 |
| 5 | 54% | 0.50 | 2694 |
| 6 | 48% | | 2509 |
| 7 | 42% | | 2311 |
| 8 | 36% | | 2097 |
| 9 | 66% | 0.37 | 2500 |
| 10 | 54% | 0.45 | |
| 11 | 42% | 0.57 | |
| 12 | 36% | 0.66 | |
| 13 | 30% | 0.79 | |

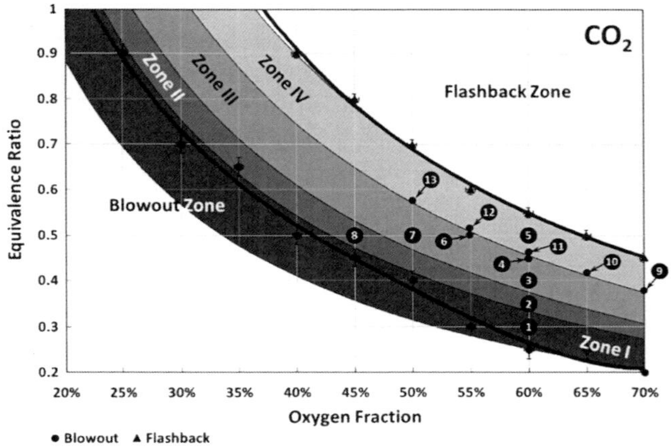


| # | *OF* | $\varphi$ | $T_{ad}$ [K] |
|---|---|---|---|
| 1 | 60% | 0.30 | 1806 |
| 2 | | 0.35 | 1995 |
| 3 | | 0.40 | 2173 |
| 4 | | 0.45 | 2342 |
| 5 | | 0.50 | 2501 |
| 6 | 55% | | 2336 |
| 7 | 50% | | 2169 |
| 8 | 45% | | 2001 |
| 9 | 70% | 0.38 | 2380 |
| 10 | 65% | 0.42 | |
| 11 | 60% | 0.46 | |
| 12 | 55% | 0.52 | |
| 13 | 50% | 0.58 | |

**Fig. 4.31** Combustor stability maps highlighting the selected data points for which flame images were captured to analyze flame shape. The bulk throat velocity is fixed at 5.2 m/s throughout

inlet bulk velocity. First, the knowledge of $T_{ad}$ alone is indeed enough to accurately predict and characterize flame macrostructure and stabilization mode. Second, the stable combustion zone of a premixed natural-gas combustor can be subdivided into $T_{ad}$-based sub-zones, where each zone has a single prevailing flame macrostructure irrespective of $\varphi$ and *OF*, because $T_{ad}$ is an umbrella parameter that combines the individual effects of $\varphi$ and *OF* and best describes the prevailing effect of reaction kinetics under similar cold flow conditions. Figure 4.31 illustrates these sub-zones for the $CH_4/O_2/N_2$ and $CH_4/O_2/CO_2$ flame sets.

The soundness of these conclusions can be further attested by comparing more than just two flames at the same $T_{ad}$. Flames $(3,\ 6,\ 9-13)_{N_2}$ are all zone-IV flames at $T_{ad} \approx 2500$ K, while flames $(9-13)_{CO_2}$ are all III–IV transition flames at $T_{ad} \approx 2380$ K, see Fig. 4.31. The images of these flames are portrayed in Fig. 4.33. It

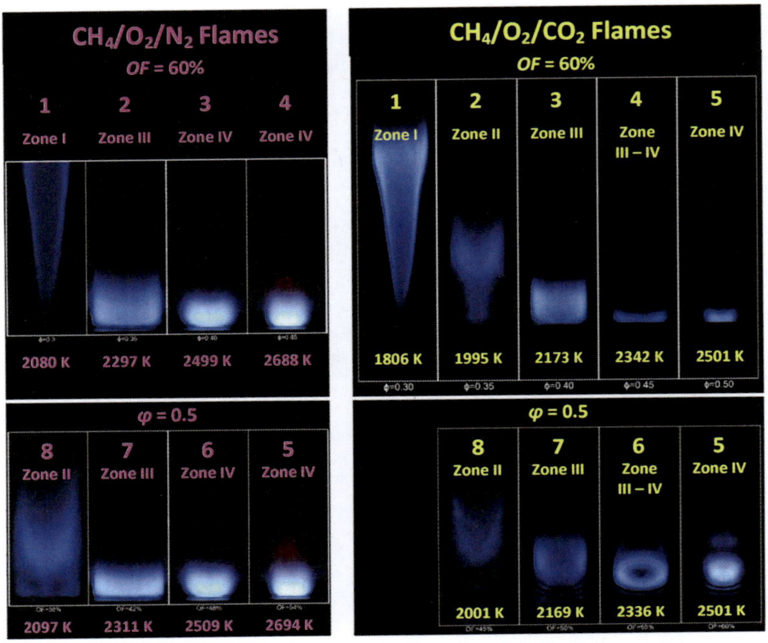


**Fig. 4.32** Visual images of $CH_4/O_2/N_2$ flames (left) and $CH_4/O_2/CO_2$ flames (right) at $OF = 60\%$ (top) and $\varphi = 0.5$ (bottom). The $CO_2$ images were adopted from Abdelhafez et al. [67]. $T_{ad}$ is listed for each flame

is undeniably clear that all seven $N_2$ flames have identical shapes at common $T_{ad}$, although the $OF$ ranges from 30 to 66% and $\varphi$ ranges from 0.37 to 0.79. The same applies to the five $CO_2$ flames with $OF = 50$–$70\%$ and $\varphi = 0.38$–$0.58$. This highlights the pivotal importance of $T_{ad}$ as the sole parameter needed to predict, characterize, and control the macrostructure and stabilization mode of stable premixed natural-gas flames ($CH_4/O_2/N_2$ and $CH_4/O_2/CO_2$) at constant inlet bulk velocity.

A concluding remark to be made here is that the findings of this study trigger the recommendation to design and operate future oxy-fuel gas-turbine combustors based on $T_{ad}$ and not dilution ratio or percent exhaust recirculation. $T_{ad}$ can be easily estimated for any given set of operating conditions. With the knowledge of $T_{ad}$ only, the shapes and stabilization modes of stable flames can be easily inferred. The heat shields at combustor headend can thus be designed accordingly, and the interactions of adjacent flames (and the resulting instabilities) in multi-premixer combustors can also be predicted.

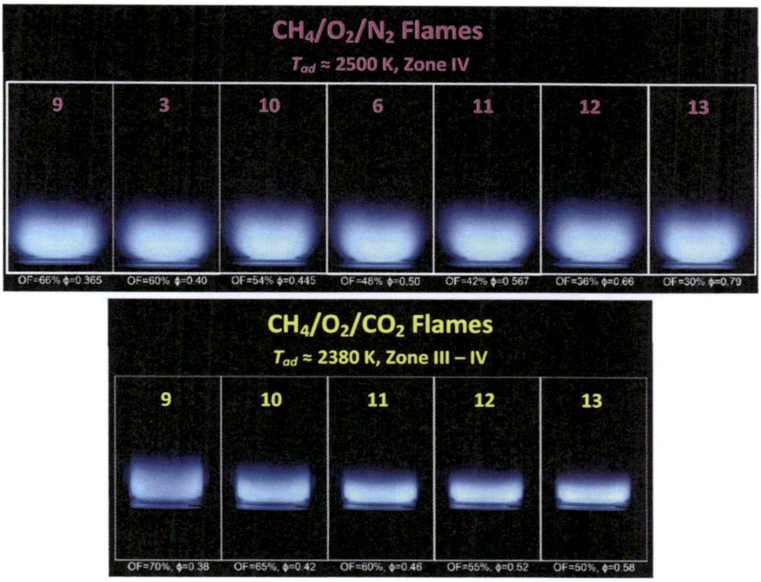


**Fig. 4.33** Visual images of $CH_4/O_2/N_2$ flames (top) and $CH_4/O_2/CO_2$ flames (bottom) at constant $T_{ad}$. The $CO_2$ images were adopted from Abdelhafez et al. [67]. *OF* and $\varphi$ are listed for each flame

## 4.5 Role of Flow Reynolds for Controlling Flame Structure and Stabilization

Few research studies considered the coupling between LPM and oxy-combustion technologies for dry low emissions (DLE) gas turbine applications with carbon capture, and fewer research studies considered the effect of flow Reynolds number on flame stability and shape. At constant Re operation (ranging from 6700 to 14,200), high pressure turbulent burning velocity measurements were carried out by Liu et al. [89]. The results showed that, at constant Re, turbulent burning velocity reduces while increasing the pressure in a minus exponential manner. Also, the results showed that as the Re increases, the turbulent burning velocities increases at any pressure value. Jourdaine et al. [90] tested both $N_2$-diluted and $CO_2$-diluted $CH_4$–$O_2$ flames under fixed swirling conditions and concluded that similar flame shapes can be obtained for both flames provided that both the adiabatic flame temperature and the ratio of laminar burning velocity to bulk flow velocity are kept the same. They reported also reduced operability range of the $CO_2$-diluted flames. Yoon et al. [91] studied the effect of inlet mixture velocity on combustion instability of a model gas turbine combustor. The results showed a key role of the fluid dynamical vortex frequency and structure on generating unstable flames at lower inlet velocities. Very recently, we investigated experimentally the stability limits of premixed oxy-flame on the same swirl stabilized combustor utilized in this study at fixed inlet bulk flow velocity [92]. The results revealed a leading role of Tad in quantifying the combustor stability map.

In this section, a detailed study is presented considering the same swirl-stabilized model gas turbine combustor as described in the above section. In this study, the changes in macrostructure and static stability limits of $CH_4/O_2/CO_2$ flames were recorded experimentally in a premixed swirl combustor over ranges of equivalence ratio ($\varphi$) and oxygen fraction (*OF*: volumetric percentage of $O_2$ in the $O_2$–$CO_2$ oxidizer mixture) and under fixed Reynolds (Re) operation to isolate its dynamic effect on flame stabilization. In this section, 2-D operability maps are presented over wide ranges of operating conditions, function of $\varphi$ and *OF*, on a gas turbine model combustor of similar power density (*PD*) as commercial gas turbines. This means that the obtained stability maps can be extrapolated for controlling the operability ranges of industrial-scale *DLE* gas turbine combustors adapting premixed oxy-combustion technology. In this study, the changes in flame macrostructure and static stability limits of premixed $CH_4/O_2/CO_2$ flames were recorded under fixed Re operation to isolate its dynamic effect on flame stabilization and understand the physics behind flame extinction. The results are presented in the form of two-dimensional stability maps, at fixed Re and variable inlet bulk flow velocity ($U_{in}$), presenting the limits as function of operating *OF* and $\varphi$. The maps are presented on the contours of $T_{ad}$, $U_{in}$ and *PD* to correlate these parameters to the blow out and flashback limits. In addition, the changes in flame macrostructure and flame transition toward unstable operation were recorded as function of $\varphi$, *OF*, Re, $T_{ad}$ and $U_{in}$.

Three sets of experiments were performed, each at fixed Re, in a swirl-stabilized DLE gas turbine model combustor under premixed oxy-combustion conditions. Three Re values were 7000, 9000 and 11,000. Methane ($CH_4$) was used as a fuel, and a mixture of oxygen ($O_2$) and carbon dioxide ($CO_2$) was used as oxidizer. At Re of 7000, the stability limits were obtained through running the combustor at fixed $U_{in}$ and variable *OF* (from 29% to 70% by volume) and $\varphi$ (from 0.2 up to 1.0) until the flashback and blow out limits occur. At a given $U_{in}$, Re, and *OF*, an excel sheet was used to iteratively obtain the value of $\varphi$ that keeps Re fixed at the desired value. Based on that, the flow rates of $CH_4$, $O_2$, and $CO_2$ were calculated for a given operating point and the mass flow controllers were adjusted to provide the required flow rates. Then, the value of *OF* was increased gradually, and the same procedure was repeated until the flashback limit is approached. The same procedure was repeated in the reverse direction until the blow out limit was approached. The values of $\varphi$ and *OF* at these limits were then recorded on the stability map. $U_{in}$ was then changed at the same Re, and the *OF*-$\varphi$ procedure was repeated. At Re of 7000, the full stability map was obtained considering the four values of $U_{in}$, namely 3.7, 4.0, 4.3 and 4.6 m/s. The same approach was followed to obtain the full stability maps at the other two Re values, 9000 and 11,000. The stability map at Re of 9000 was obtained at four values of $U_{in}$, namely 4.8, 5.2, 5.6 and 6.0 m/s. Similarly, the considered inlet velocities at Re of 11,000 were 5.8, 6.3, 6.8 and 7.3 m/s. Following a previous study by the authors of this work [92], $T_{ad}$ was calculated as function of the operating $\varphi$ and *OF*. More details of the experimental procedure and testing method are provided in [92], considering the same experimental set-up.

### 4.5.1 Effect of Inlet Flow Conditions on Flame Stability

In this section, the combustor stability mapping is presented under fixed Re operation considering three sets of stability maps. The three sets of stability maps present the stability limits as function of $\varphi$ and $OF$ on the contour plots of $U_{in}$, $T_{ad}$ and $PD$ of the combustor, respectively. For each set, three stability maps are presented, each at fixed Re. The considered Re values for the operation of the combustor were 7000, 9000 and 11,000. The combustor stability mapping was determined by recognizing the blow out and flashback limits of the flames at fixed Re, $U_{in}$ was fixed and $\varphi$ and $OF$ were varied until flashback and blow out limits were approached, and the procedure was repeated for different values of $U_{in}$.

Figure 4.34 presents the first set of stability maps on the contour plots of the corresponding $U_{in}$ for different inlet Re. three stability maps are presented, at Re = 7000 (top map), Re = 9000 (middle map), and Re = 11,000 (bottom map). The areas above the flashback limit and below the blow out line present unstable flame operation zones, while the area between the two limits presents the stable operability window of the combustor. For fixed Re operation, the flashback limits were best fit with second order curves consistently with negligible differences in slope. The blowout limits were best fit with exponential and second-order fits. For all fits, the $R^2$ (coefficient of determination) values are close to 1.0, indicating reasonable fit qualities.

The results in Fig. 4.34 show almost identical flashback conditions at all operating Re values indicating insignificant effect of the operating Re on the flashback limit of the generated flames. However, Re has a role in determining the blow out limit of the generated flames depending on the $U_{in}$. This indicates that the rate of reaction kinetics near flashback is a more relevant parameter than the flow dynamics (associated with change in Re) for determining the flame speed and, consequently, the flashback limit. On the other hand, near blow out limit, the stability of the flame depends mainly on the generated flow dynamics within the combustor (effect of Re), which control flame anchoring under lean operating conditions. The curves of flashback and blow out limits do not follow constant $U_{in}$ contours as shown in Fig. 4.34. However, at any operating Re, increasing $U_{in}$ reduces the operability window of the flame and shifts the flashback limit to the leaner side. For fixed $U_{in}$ operation, flashback and blow out phenomena are mainly function of the reaction time scale (flame speed) based on the definition of the Damkohler number [50, 93]. Increasing the $OF$ in the oxidizer mixture results in improved flame speed [52, 94]. Equivalence ratio must thus be leaner to reduce flame speed, and $U_{in}$ must be increased, to balance the flame speed at burner throat.

Figure 4.35 shows the same stability maps plotted on a background of combustor $PD$, in order to examine if the static stability limits correlate with the combustor $PD$ under fixed Re operation. Three maps are also presented in this Figure considering the same values of operating Re, namely 7000, 9000 and 11,000. The contours of $PD$ in Fig. 4.35, for all operating Re values, do not correlate with the line presenting the flashback limit. However, for all operating Re values, the curves presenting the blow

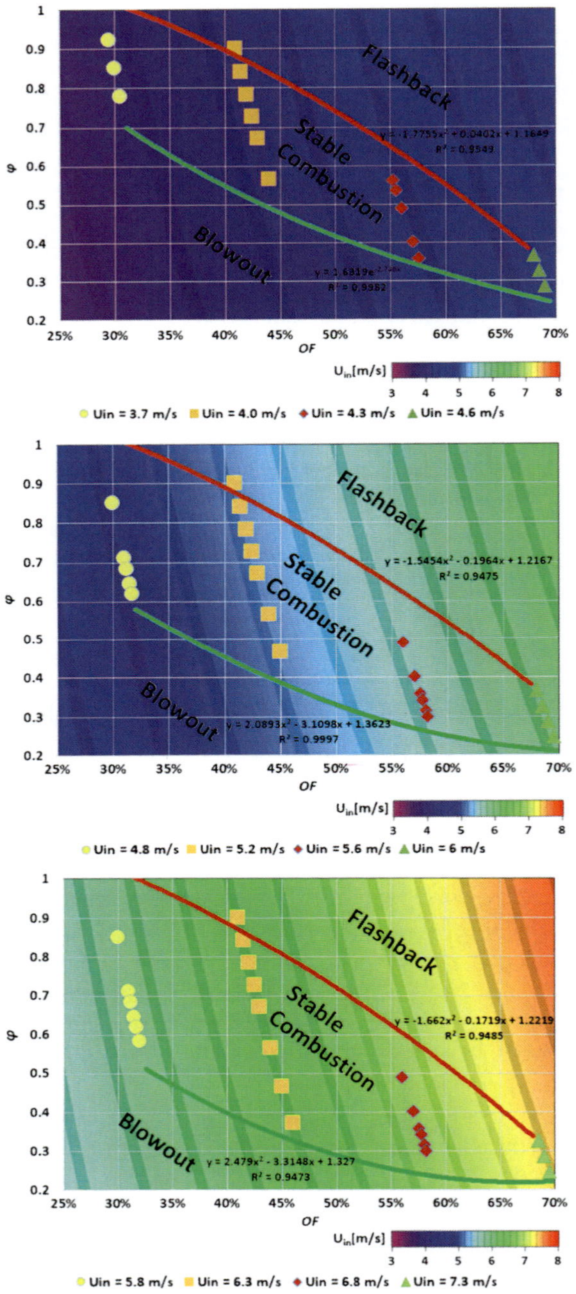


**Fig. 4.34** Stability maps on a background of corresponding $U_{in}$ for different inlet Re: (top) Re = 7000, (middle) Re = 9000, and (bottom) Re = 11,000

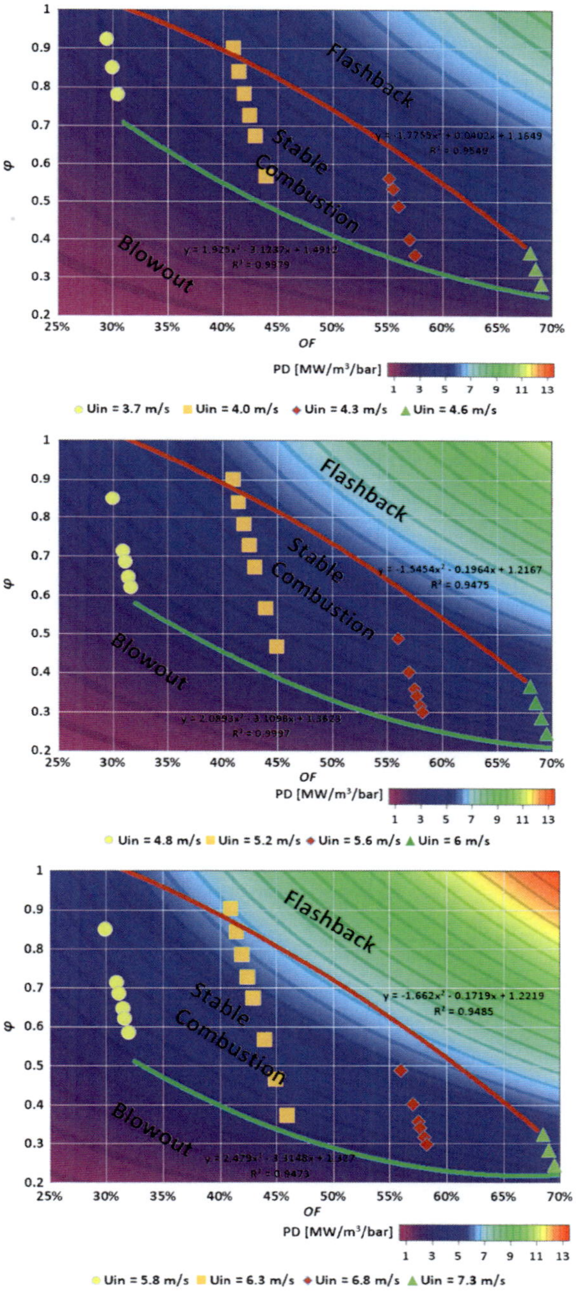


**Fig. 4.35** Stability maps on a background of corresponding combustor *PD* for different inlet Re: (top) Re = 7000, (middle) Re = 9000, and (bottom) Re = 11,000

out limits fit well with contours of constant *PD*. For the three operating Re values, the flame blows-out at the same operating combustor *PD*, about 3.0 MW/m$^3$/bar. On the other hand, the combustor *PD* at flashback increased from about 5.0 MW/m$^3$/bar at Re of 7000 to about 7.0 MW/m$^3$/bar at Re of 11,000. These results indicate that the fuel mass flow rate, based on a given size of the combustor, plays a key role in controlling flame stability near the lean blow out limit, and should not drop below a certain value to avoid flame blow out if the flow dynamics within the combustor are kept similar by keeping the same operating Re.

The stability limits were also plotted on a background of the contours of $T_{ad}$ for the three operating Re values, as presented in Fig. 4.36. As per Fig. 4.36, the plots do not show a direct link between $T_{ad}$ and the flashback and blow out phenomena under fixed Re operation. Interestingly, based on previous work by the authors of this work on the same combustor set-up [92], the results showed that both flashback and blow out limits followed contours of fixed $T_{ad}$ but under fixed $U_{in}$ operation, which means that $T_{ad}$ is the most relevant parameter controlling both flashback and blow out phenomena while operating under fixed $U_{in}$ [92] only. However, for operation under fixed Re, the fuel mass flow rate is the most relevant parameter controlling flame blow out limit, while the flash back limit under fixed Re operation is mainly controlled by the rates of reaction kinetics, based on operating $\varphi$ and *OF*. In addition, it was observed that it is not possible to sustain premixed oxy-methane flames in the ranges of *OF* below 29% and above 70%. These ranges match with results reported in another previous study by the authors on oxy-flames stabilized over a perforated plate burner [88].

### 4.5.2 Role of Flow Reynolds for Controlling Flame Stabilization

To have a clearer comparison of stability limits under different operating Re, the curves presenting the flashback limits and blow out limits under the different operating Re are all presented in Figs. 4.37 and 4.38 respectively. It should be noted here that the stability limits depicted on Figs. 4.34, 4.35 and 4.36 are curve fits, whereas the limits plotted in Figs. 4.37 and 4.38 are the actual data points without fitting to quantify the fine differences accurately. As per Fig. 4.37, the series presenting the flashback limits are almost identical under all operating Re. The results in Fig. 4.37 confirm our finding based on the above discussion that Re has an insignificant effect on the flashback limits of the tested premixed flames under oxy-combustion conditions. In fact, flashback occurs when the flame speed of the reacting mixture exceeds $U_{in}$. While varying the Re during the experiments, the value of $U_{in}$ is accordingly varied, which means that the flame speed at which flashback occurs is also varied. This means that the flashback phenomenon is mainly controlled by the rates of reactions kinetics, which control the flame speed based on operating $\varphi$ and *OF*. Amato et al.

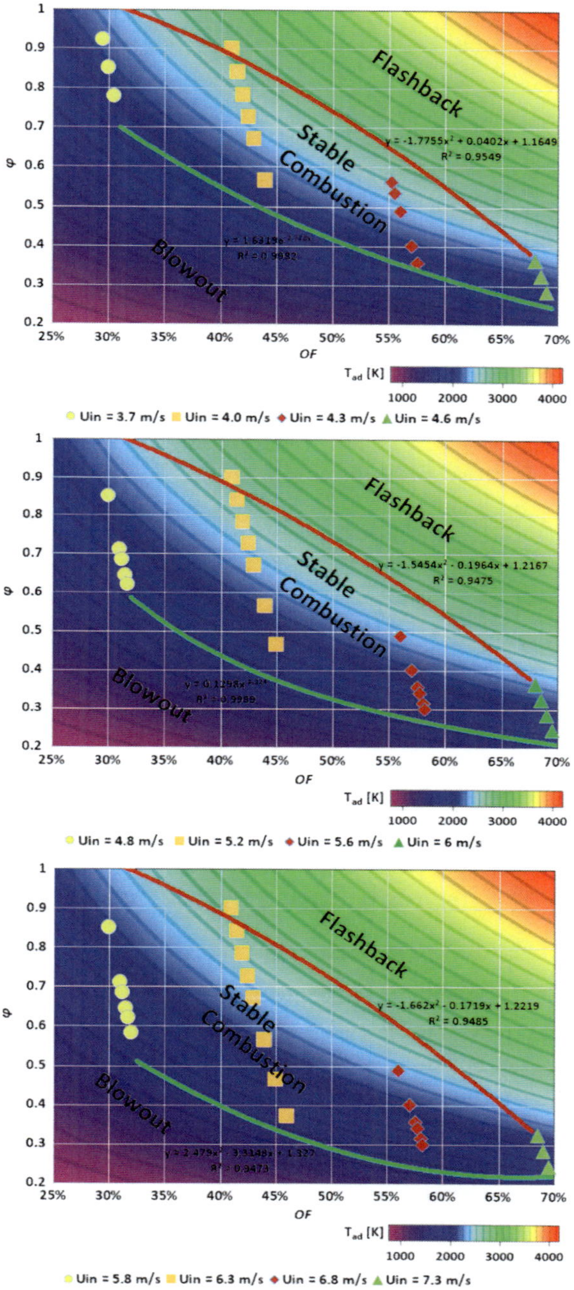


**Fig. 4.36** Stability maps on a background of corresponding $T_{ad}$ for different inlet Re: (top) Re = 7000, (middle) Re = 9000, and (bottom) Re = 11,000

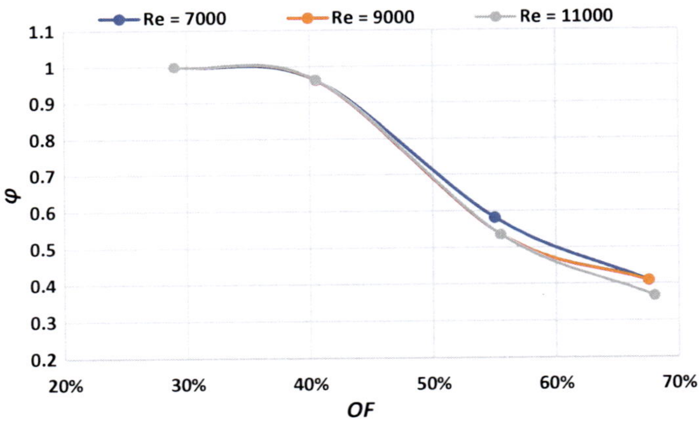


**Fig. 4.37** Obtained flashback limits at different inlet Re

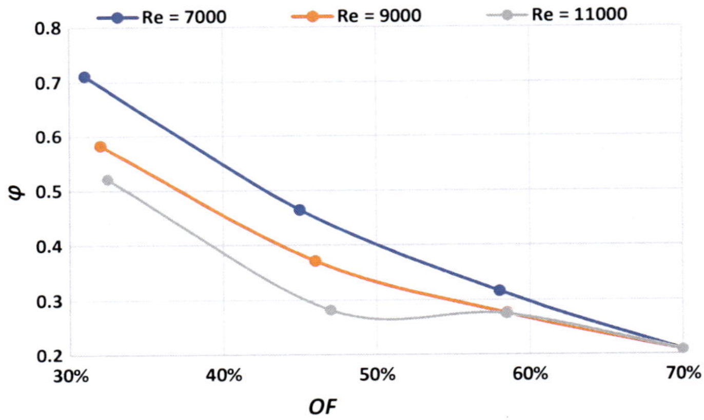


**Fig. 4.38** Obtained blow out limits at different inlet Re

[16] reached a similar conclusion on the stability limits of premixed oxy-methane flames.

Similarly, the series presenting the blow out limits under the same operating conditions are presented in Fig. 4.38. As per Fig. 4.38, increasing the Re widens the operability range of the combustor by shifting the blow out limit toward leaner conditions. This may be attributed to the improved flame anchoring under such limiting operating conditions due to the improved flow swirl in the ORZ because of the corresponding increase in $U_{in}$ at higher operating Re. These results indicate a key role of the flow dynamics (mainly function of $U_{in}$) within the combustor for controlling flame lifting and reattachment encountered before the blow out of the flame.

To have more insight on the mechanism of flame blow out, the changes in flame stabilization behavior from the flashback limit to the blow out one was traced visually through flame imaging. It was found at each operating $U_{in}$ that there is a value of $\varphi$, at which there is a transition in the flame stabilization mode from the ISL to the ORZ. This transition was found to be associated with the onset of thermos-acoustic instability and to occur prior to flame blow out. The values of $\varphi$ at which this transition occurs are presented in Fig. 4.39 as function of $U_{in}$ at different operating Re. As shown in the Figure, the flame transition occurs at higher velocities for higher operating Re $_{in}$but almost at the same $\varphi$ indicating insignificant effect of Re on flame stability. At fixed Re, the results in Fig. 4.39 show that the increase of $U_{in}$ retards the flame transition toward leaner conditions, closer to the lean blow out limit. This may be attributed to the enhanced flow recirculation in the ORZ at higher $U_{in}$. The slopes of the curves in Fig. 4.39 are high indicating a leading role of $U_{in}$ in controlling flame stabilization behavior. The transition can occur at different $T_{ad}$ and flame speed. This flame transition from the ISL to the ORZ was found in past studies to be mainly governed by the extinction strain rate if similar flow conditions (similar $U_{in}$) are imposed in the ORZ [9]. The main dimensionless parameter characterizing the flow is the ORZ in the Strouhal number ($St = f \cdot D_{in}/U_{in}$). This number is function of the azimuthal spinning frequency in the ORZ ($f$), $U_{in}$ and inlet diameter ($D_{in}$), so the Strouhal number is independent of the flow Re. These results can lead us to an important implication that $U_{in}$ is a more relevant parameter choice than Re for the control of the flow conditions in the ORZ and flame stability.

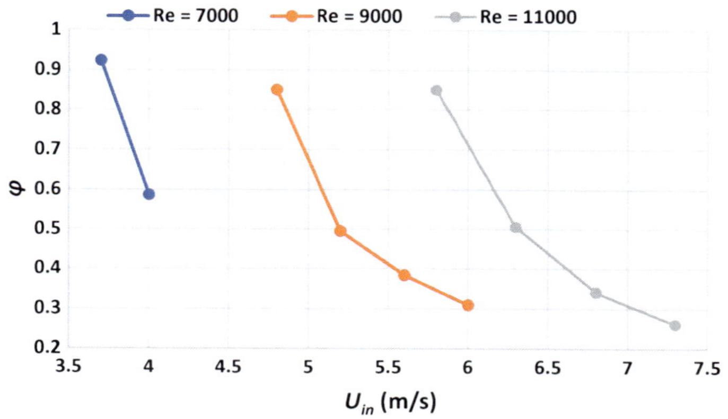


**Fig. 4.39** Equivalence ratio at the onset of flame transition from the ISL to the ORZ as function of $U_{in}$ for different operating Re

### 4.5.3 Role of Flow Reynolds for Controlling Flame Structure

In this section, changes in flame macrostructure are analyzed to understand the modes of flame stability and transition from one mode to another toward the lean blow out limit. In Fig. 4.40, three sets of flame images are compared, under different operating Re, captured at steady state conditions. Each set of flame images was captured under fixed Re and fixed $U_{in}$ while varying $OF$ and $\varphi$. The operating Re was increased from one set to another through increasing $U_{in}$ and keeping the same $OF$ and $\varphi$ for the corresponding flames. The results confirm that the operation under higher Re can extend the flame operability limit to leaner conditions before flame blow out. At Re values of 9000 and 11,000, the flame encountered a lifted unstable operation mode before being blown-out; however, at low operating Re of 7000, the flame encountered blow out directly without being lifted-off. The flames at higher Re can sustain better flame anchoring under extra lean conditions compared to those at lower Re. This may be attributed the improved circulation in the ORZ at higher $U_{in}$ corresponding to higher Re operation. Comparing flames of the same $OF$ and $\varphi$ (i.e., same $T_{ad}$)

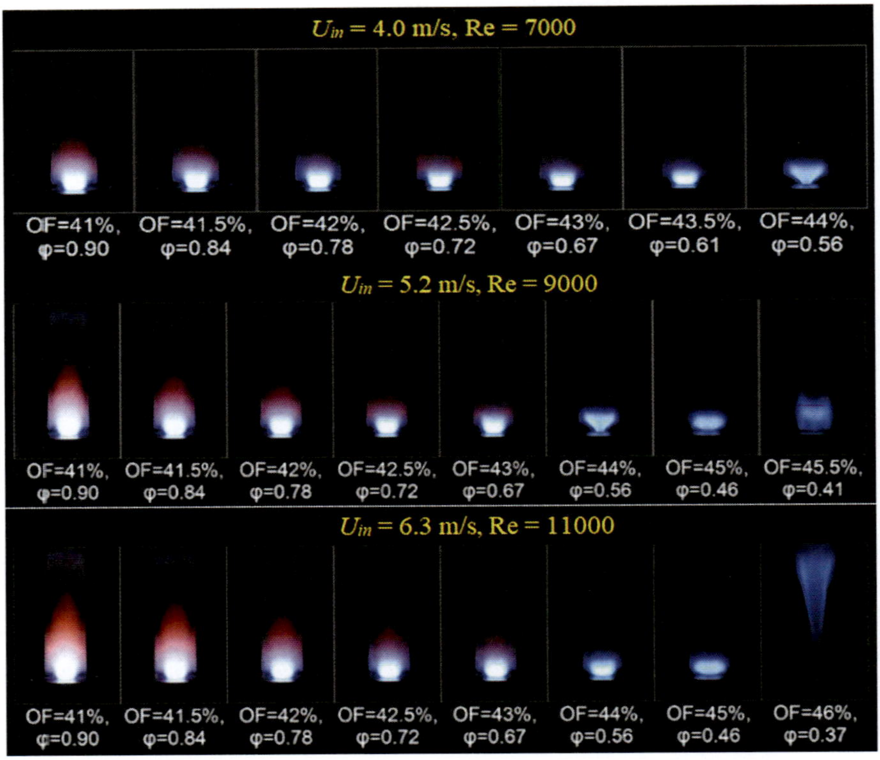


**Fig. 4.40** Flame macrostructure at fixed $U_{in}$ from near flashback (left) to near blow out (right) limits for different inlet Re: (top) Re = 7000, (middle) Re = 9000, and (bottom) Re = 11,000

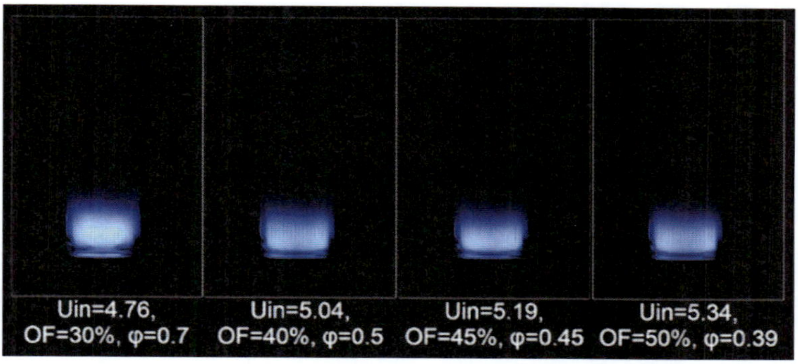


**Fig. 4.41** Comparison of flame macrostructures while fixing both, the flow Re at 9000 and $T_{ad}$ at 2000 K

in Fig. 4.40 indicates that flames of the same $T_{ad}$ do not necessarily have the same flame macrostructure. Flames at higher Re (higher $U_{in}$) have bigger flame size, which indicates that $T_{ad}$ does not have a sole role for controlling flame macrostructure and stability. To specify the controlling parameters of flame macrostructure, flame images are compared while fixing both flow Re at 9000 and $T_{ad}$ at 2000 K, as shown in Fig. 4.41. The results show that similar flame macrostructures can be obtained while fixing one flow characteristic parameter (Re) and one flame characteristic parameter ($T_{ad}$). This conclusion agrees with the findings of the authors, in our previous study on the same combustor set-up [92], similar flame macrostructures were obtained but while fixing both $U_{in}$ and $T_{ad}$.

To explore the effect of *OF* on flame macrostructure and stability, the flame images in the stable operation zone are compared at constant $\varphi = 0.5$, as presented in Fig. 4.42. The Figure presents three sets of flame images, each at fixed Re. The results show significant effect of *OF* on flame macrostructure and color. Increasing *OF* results in more compact and violent strong flames with brighter color, thus, indicating improved reaction kinetics. This is due to the associated increase in flame speed. As per the measurements of flame speed of oxy-methane flames by Kutne et al. [15], the flame speed increased by a factor of 2.4, when the *OF* was increased from 30 to 40% in the oxidizer mixture. Increasing the *OF* moves the flame upstream toward the burner throat, as per Fig. 4.42, indicating higher flame speed. Thus, the *OF* has a major controlling role on the flame macrostructure and stability. At the same operating $\varphi$, Re seems to have minor effect of flame macrostructure as per Fig. 4.42.

To quantify the effects of *OF* and Re on the rates of reaction kinetics within the combustor, local temperature was recorded at fixed $\varphi = 0.5$ along the combustor centerline in Fig. 4.43, as well as radially (at different axial locations, typically $h_1 = 6.35$ cm, $h_2 = 7.62$ cm and $h_3 = 8.89$ cm) in Fig. 4.44. While performing temperature measurements, the flame was approached as close as possible to avoid interference of the thermocouple probe with the generated turbulent reacting flow field and to prevent melting of the probe due to the elevated temperature within the flame core. The

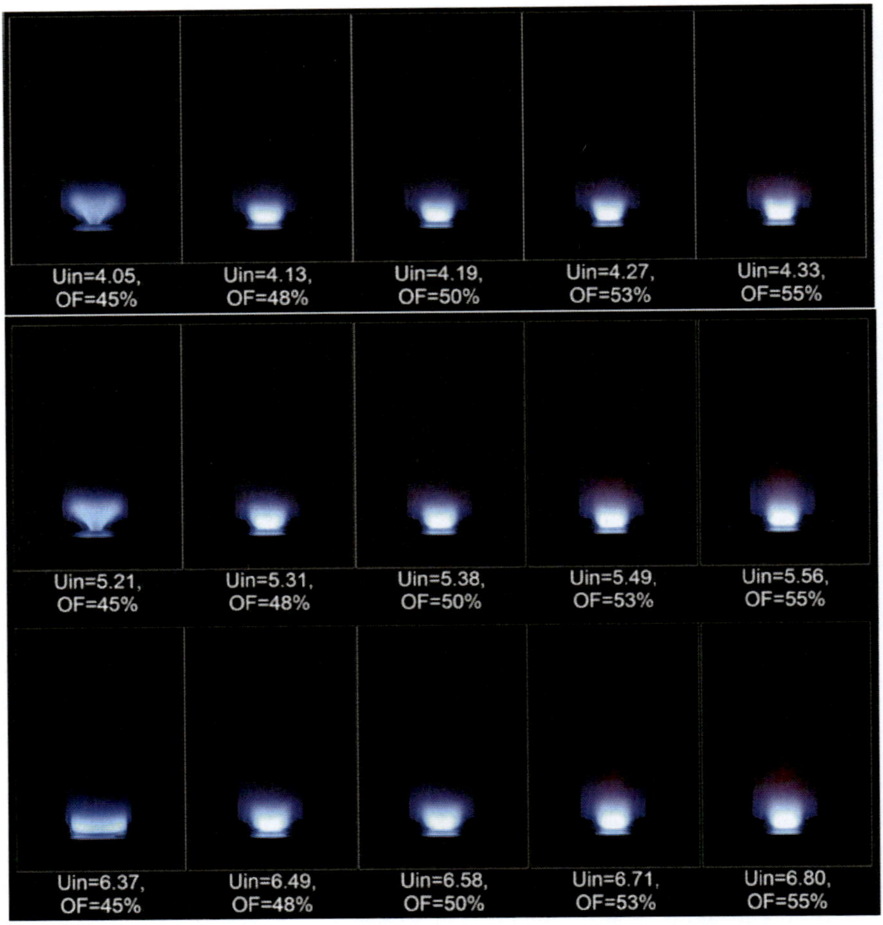


**Fig. 4.42** Comparison of flame macrostructures at fixed $\varphi$ of 0.5 as function of $OF$ for different inlet Re: (top) Re = 7000, (middle) Re = 9000, and (bottom) Re = 11,000

temperature plots in Figs. 4.43 and 4.44 serve as data base for validating numerical models of turbulent reacting flow under premixed oxy-combustion conditions, in addition to providing quantification of the effects of $OF$ and Re on flame temperature. Axial and radial temperature profiles, recorded at fixed Re of 7000, show hotter flame at higher $OF$, as per the top plots in Figs. 4.43 and 4.44. Very close to the flame, at an axial distance of 6.35 cm, increasing $OF$ from 50 to 55% resulted in significant temperature rise of about 200 degrees as per Fig. 4.43 (top plot). The flames are V-shaped (see Fig. 4.42) and stabilized in the ISL, where the radial temperature increases then drops towards the reactor wall, as per Fig. 4.44. An inner recirculation zone (IRZ) exists in the central zone of the combustor, mixing burned and unburned gases and keeping the centerline temperature below the maximum value within the

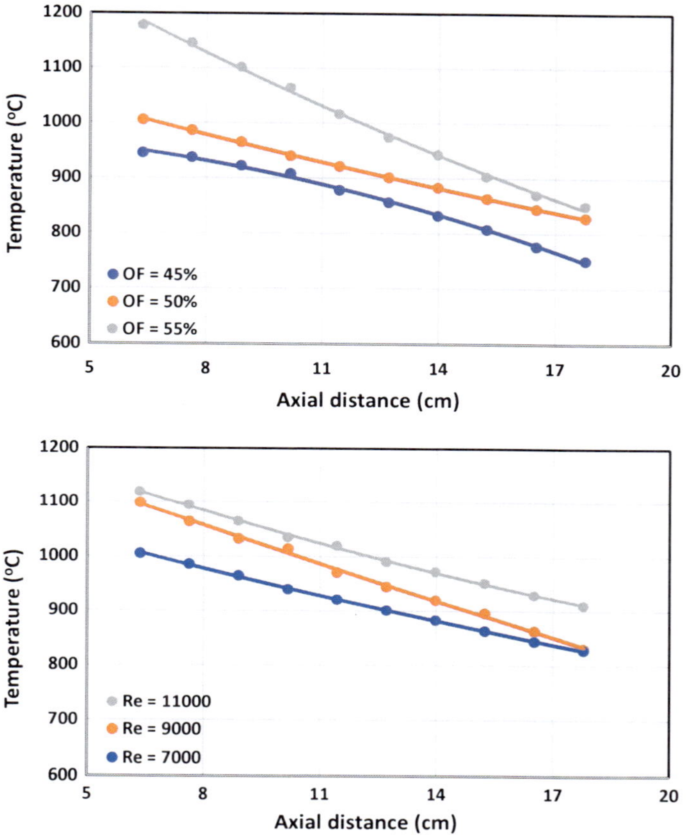


**Fig. 4.43** Axial temperature distributions (from the burner tip) at fixed $\varphi$ of 0.5; (top): effect of $OF$ at Re = 7000 and (bottom): effect of Re at $OF = 50\%$

inner shear layer (ISL), as per Fig. 4.44. These results confirm the key role of $OF$ in controlling reaction rates within the combustor under oxy-combustion conditions. As per Figs. 4.43 and 4.44 (bottom plots), the results recorded at fixed $OF$ of 50% show slightly hotter flames at higher Re. At axial distance of 6.35 cm, a ~100° increase in temperature is obtained when Re increases from 7000 to 9000 as shown in Fig. 4.43 (bottom plot). This may be attributed to the increase in $U_{in}$, which results in higher turbulence intensity and improved flame stability.

Flame macrostructures near the blow out and flash back limits are presented in Figs. 4.45 and 4.46, respectively, to explore whether the flames hold fixed macrostructure near the stability limits under fixed Re operation. Three sets of flame images are presented in each Figure, each set is recorded at fixed Re. The results in both Figures show that the flames do not hold fixed macrostructure near either stability limit. The flames at lower Re blow out without being lifted for the considered range of $U_{in}$. The

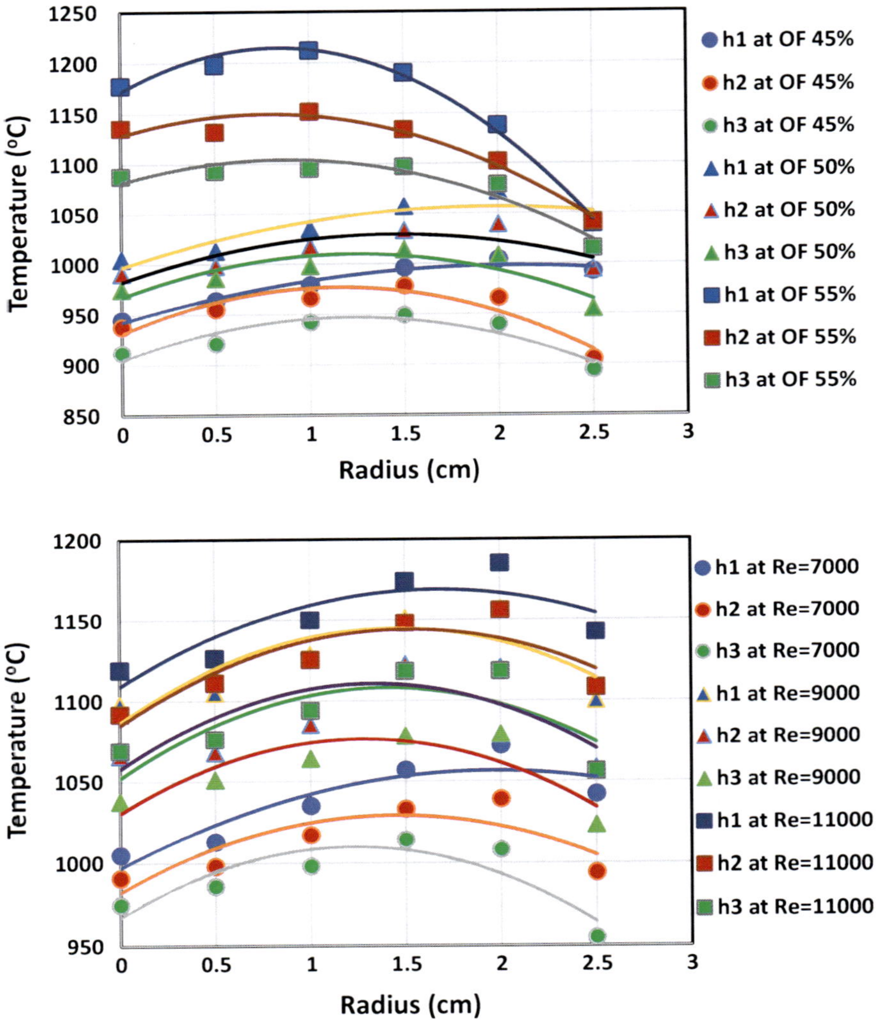


**Fig. 4.44** Radial temperature distributions at different heights ($h_1 = 6.35$ cm, $h_2 = 7.62$ cm and $h_3 = 8.89$ cm) and at fixed $\varphi$ of 0.5; (top): effect of $OF$ at Re = 7000 and (bottom): effect of Re at $OF = 50\%$

flames near blow out are V-shaped with varied size, while the flames at higher Re and lower $U_{in}$ encounter stabilization in the ORZ followed by lift-off before being blown-out. Increasing $U_{in}$ at higher Re operation, the flame lift-off mode is skipped, and the flame becomes stronger and shorter before it blows-out. Thus, increasing $U_{in}$ under higher Re operation helps better flame anchoring near blow out limit as presented in the bottom set of flame images in Fig. 4.45. Under any operating Re, the

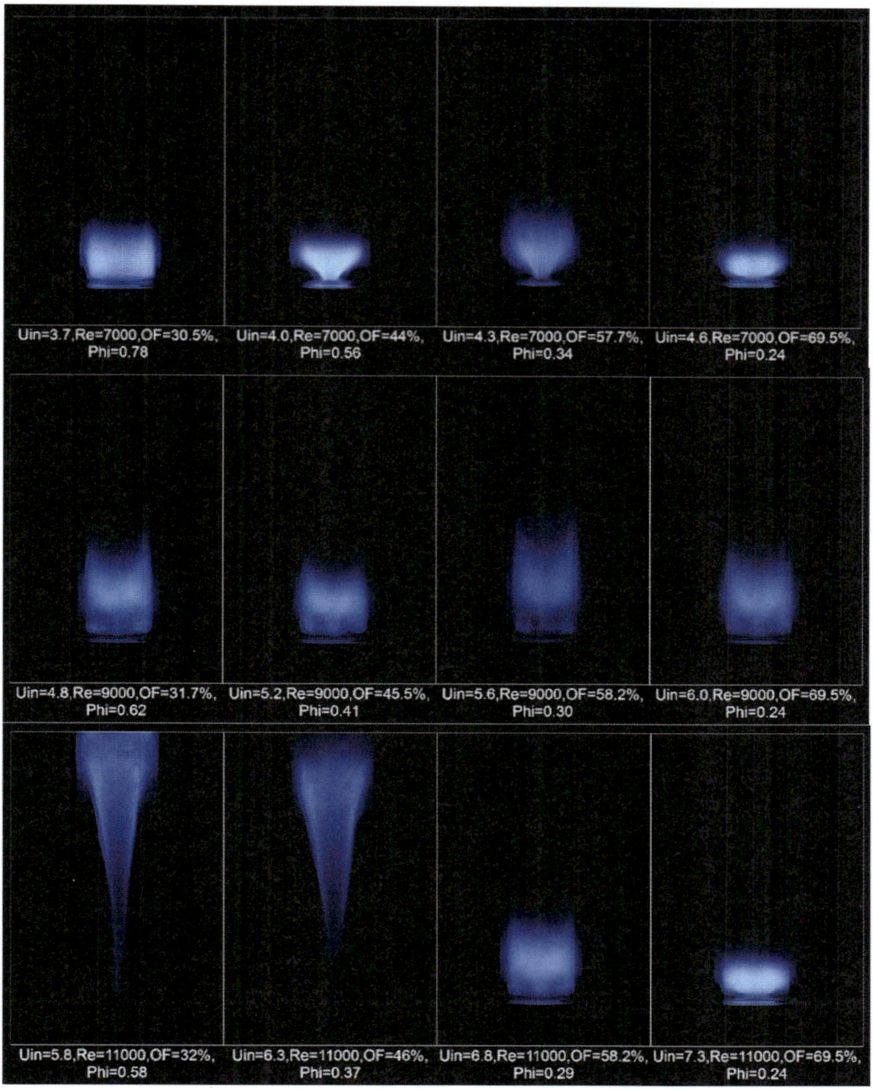


**Fig. 4.45** Comparison of flame shapes near blow out limit at different inlet Re: (top) Re = 7000, (middle) Re = 9000, and (bottom) Re = 11,000

flames at lower $U_{in}$ blow out at lower *OF* but higher $\varphi$; however, the flames at higher $U_{in}$ blow out at higher *OF* but lower $\varphi$. Increasing the operating Re enhances flame anchoring, especially at lower $U_{in,}$ and shifts the blow out limit to leaner conditions. Near the flashback limit, the flame macrostructure is almost of V-shape of varied size and color as shown in Fig. 4.46. The flame macrostructure depends on the reacting

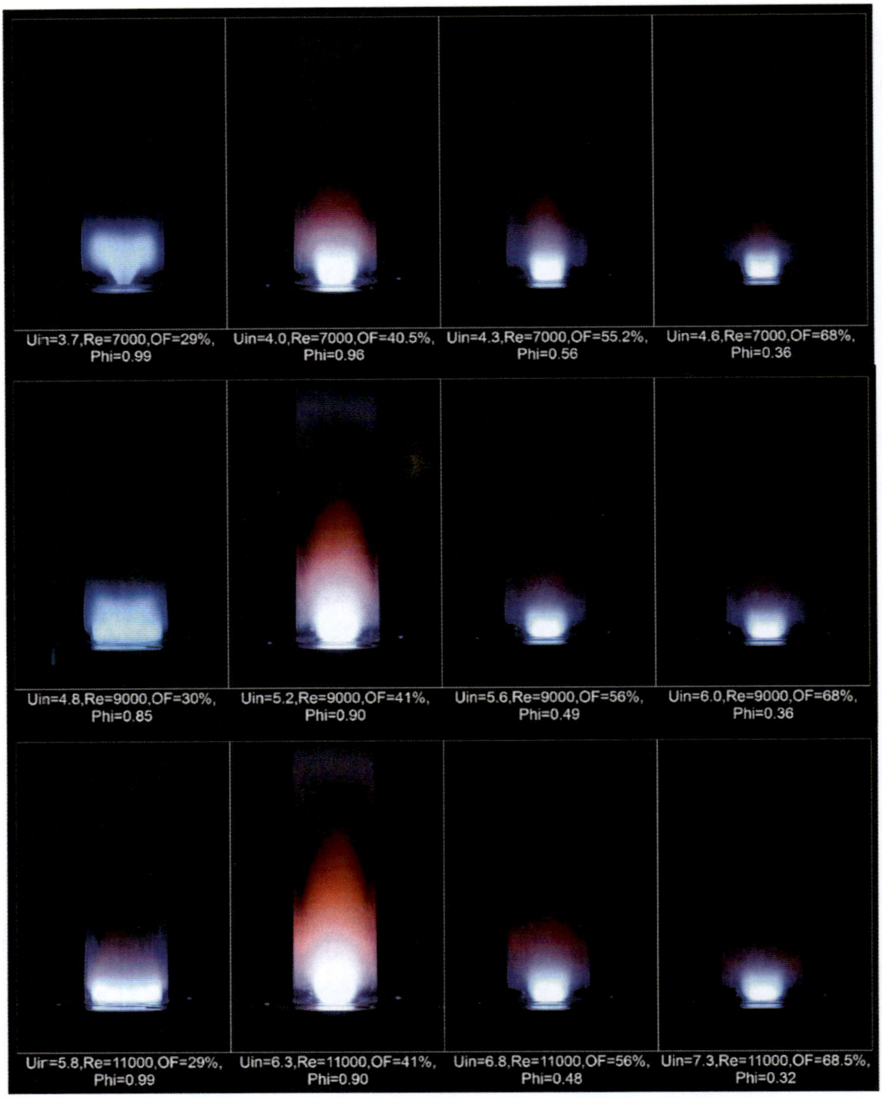


**Fig. 4.46** Comparison of flame shapes near flashback limit at different inlet Re: (top) Re = 7000, (middle) Re = 9000, and (bottom) Re = 11,000

mixture composition and reaction kinetic rates, which determine the flame speed near the flashback limit. Increasing $U_{in}$ at constant Re reduces the flame height and increases the noise level due to increased flame speed near the flashback limit. This increase in flame speed can be attributed mainly to the associated increase of $OF$ at higher operating velocity for a fixed Re operation.

## 4.6 Micromixer Burners for Gas Turbines

At the present stage, the world is largely dependent on fossil energy, yet combustion of fossil fuels emits the primary greenhouse-gas carbon dioxide ($CO_2$) causing global warming. Fossil fuels are expected to remain the primary source of energy in Saudi Arabia for a long time; more than 95% of the Kingdom's electricity has been produced with fossil energy in 2017 using gas turbines (GTs). The fact that KSA is on the world-top-10 list of countries producing electricity (311 TWh in 2017) and generating $CO_2$ emissions (589 $MtCO_2$ in 2017) makes KSA the largest user of GTs worldwide and highlights the critical need to implement efficient carbon capture and sequestration.

Traditionally, GT manufacturers used non-premixed combustion technology with air as oxidizer. Stoichiometric high-temperature zones exist within the non-premixed flame resulting in excessive $NO_x$ emissions. This motivated the development of lean-premixed (LPM) combustors, which premix fuel and oxidizer in fuel-lean proportions before admitting the mixture into the flame. This eliminates the creation of high-temperature zones within the flame, which reduces $NO_x$ emissions. However, unlike non-premixed flames, LPM ones require an efficient means of flames stabilization, as the combustor is operated close to its lean blowout limit. Over the past five decades, GT manufacturers switched from traditional non-premixed combustion to lean premixed (LPM) combustion technologies to reduce $NO_x$ emissions; General Electric successfully developed the micromixer (MM) LPM technology in their F-class utility GTs (TRL of 9). The use of air as oxidizer is, however, not beneficial from a carbon-capture perspective, as unreacted nitrogen from air forms the primary constituent of GT exhaust, which reduces the $CO_2$ concentration in it, thus complicating the $CO_2$ separation process and increasing the associated costs of $CO_2$ capture. Oxy-fuel combustion (i.e., burning fuel in oxygen instead of air), on the other hand, produces $CO_2$ and $H_2O$ as primary exhaust constituents, so once the latter is condensed out, the former can readily be captured at the lowest cost.

Micromixer stabilization is a successful emerging LPM combustion technology that is already implemented on the utility scale (e.g., General Electric's F-class GTs), and it has a technology-readiness level of 9 [95–97]. Micromixers (MMs) got their name from the fact that fuel is mixed with oxidizer on a small scale. An array of several straight tubes (each of a diameter of few millimeters) is supported between two faceplates (inlet and exit), which creates a gas plenum around the tubes between the plates. Compressor-discharge air flows through the tubes, while fuel is supplied from the outer gas plenum through several side holes (each <1 mm in diameter) in the tube walls. Fuel injection into the airflow, therefore, occurs in a jet-in-crossflow manner. The fuel holes are located axially near the inlet faceplate, in order to allow for a premixing length inside each tube; the reactant jets exiting the MM are thus fully premixed and developed. The faceplate geometry (jet diameter and interspacing) is optimized to minimize emissions and widen the operability window of the combustor. Flame flashback is prevented by carefully setting the operational ranges of flame temperature and airflow pressure drop across MM.

The use of air as oxidizer in combustion is, however, still the key source of $NO_x$ generation in both non-premixed and premixed flames because of the reaction of air-based nitrogen and oxygen. Moreover, unreacted nitrogen from air forms the primary constituent of GT exhaust, which increases the flow rate of exhaust and reduces the $CO_2$ concentration in it, thus complicating the $CO_2$ separation process and increasing the associated costs of carbon capture. If oxy-combustion is implemented instead of air-combustion, air is replaced with pure oxygen as oxidizer. Air-based nitrogen is absent, so $NO_x$ emissions are inherently eliminated. Moreover, GT exhaust will consist mainly of $CO_2$ and water vapor, so once the latter is condensed out (latent heat recovery), the former can readily be captured at the lowest cost.

## 4.7 Operability of Micromixer Combustor Holding Premixed Oxy-methane Flames

### *4.7.1 Combustor Design*

In this section, the results of an experimental study on micromizer combustor holding premixed oxy-methane flames are presented. The static stability (blowout) and macrostructure of confined premixed $CH_4/O_2/CO_2$ flames were examined at atmospheric pressure and temperature (300 K). For the set objectives to be achieved, the test rig used by Abdelhafez et al. [67] was considered with modification of the burner headend, as shown in Fig. 4.47. The 316-L stainless-steel multi-hole headend shown in Fig. 4.48 was considered. Sixty-one holes (3.18 mm diameter) are distributed on a hexagonal matrix to maintain consistent spacing of 5.5 mm between each core jet and the six surrounding jets. The inlet of each hole is of smaller diameter (1.6 mm) to act as a flame arrestor that prevents flashback into the mixing plenum upstream of the burner headend; the flashback limit was thus not examined in this study for safety purposes. The height of burner headend is 50 mm, while the main holes (3.18 mm) are 40 mm deep, so the arrestor section is ~10 mm in height. The aspect ratio (L/D) of the main holes is thus 40/3.18 = 12.6, which is large enough to facilitate fully developed flow at the flame base. The main holes were machined with a drill bit of 120° tip angle, which created gradual expansion from the arrestor diameter to the main diameter. Such gradual expansion aids in achieving fully developed flow within the height of main holes. All the flames were examined at a common jet velocity of 5.2 m/s to maintain similar flow conditions and turbulence intensities for all isothermal flow fields.

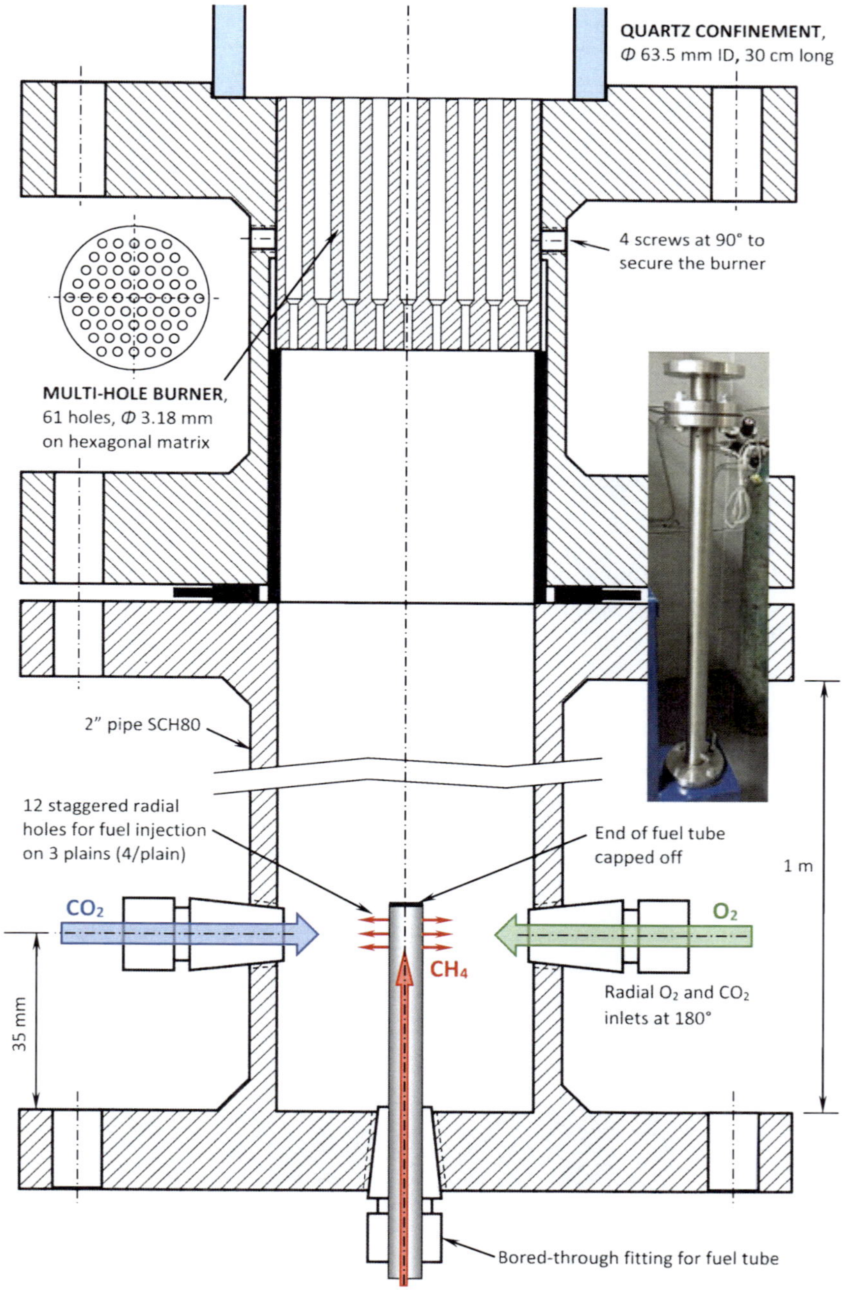


**Fig. 4.47** Micromixer combustor test rig [67]

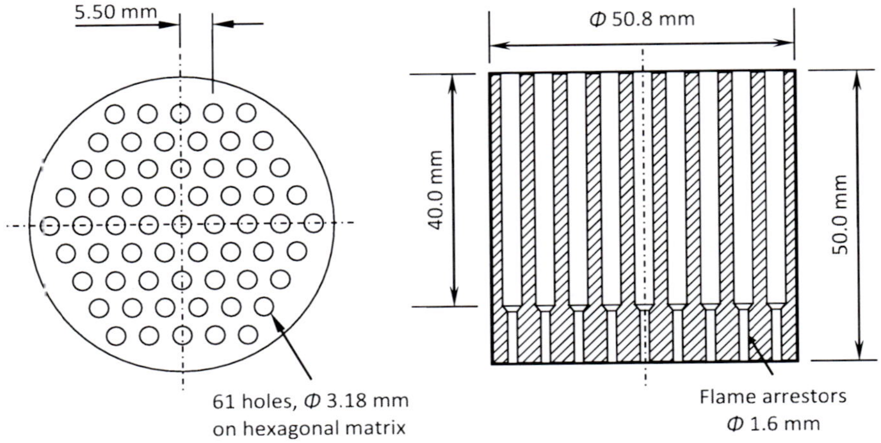


**Fig. 4.48** Multi-hole burner headend [67]

## 4.7.2 *Combustor Stability Maps*

The stability map of premixed $CH_4/O_2/CO_2$ flames in the examined micromixer-like multi-hole burner is presented within the *OF*-$\varphi$ space in Fig. 4.49. The black spherical data points indicate the blowout limit, which is represented by the black solid fit curve

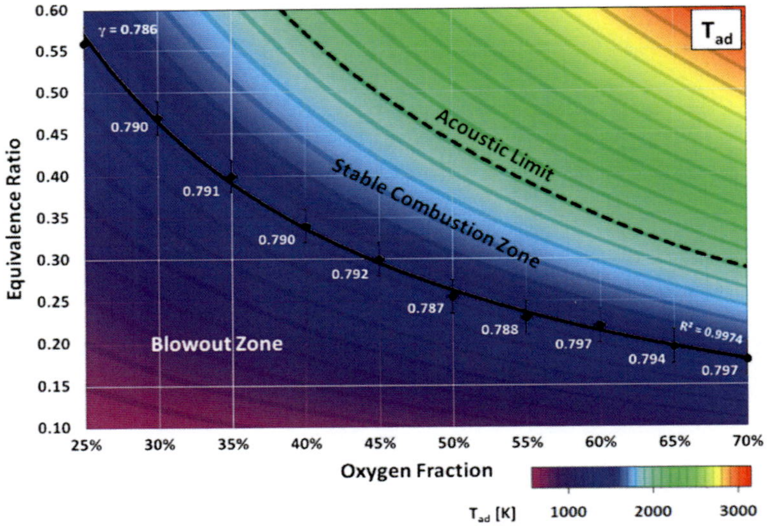


**Fig. 4.49** Combustor stability map within the OF-$\varphi$ space plotted against the contours of constant $T_{ad}$ at constant inlet jet velocity

with an $R^2$ value of 0.9974. The region below the blowout limit is an unattainable blowout zone. Also shown in Fig. 4.49 is a dashed curve labeled "Acoustic Limit" and bounding another unachievable region of high *OF* and high $\varphi$. This region is characterized by loud screech noise indicative of combustion dynamics, hence the name acoustic limit. This limit is presented as a dashed line because it is not based on a measured threshold of noise amplitude, i.e., this limit was identified by the onset of noise as arbitrarily captured by the human ear. The zone between the blowout and acoustic limits indicates the window of stable operation. The stability map is plotted against background color contours of constant adiabatic flame temperature ($T_{ad}$). Figure 4.49 provides the primary novel finding of this study, namely that, under similar cold-flow conditions, the stability limits of premixed oxy-methane flames in a micromixer-like burner follow contours of constant $T_{ad}$.

This conclusion has been proven previously for premixed swirl-stabilized $CH_4/O_2/CO_2$ [67] and $CH_4/O_2/N_2$ [67] flames, but the novel finding here is that the same conclusion applies to another burner technology as well, namely multi-hole burners. This has key implication in burner design. Manufacturers of LPM air-based gas turbines already use $T_{ad}$ and the pressure drop across combustor head-end ($\Delta p$) to control load and flame stability. $\Delta p$ is typically kept constant within a given load module, which results in constant inlet bulk air velocity. This is the reason why this study targeted fixed inlet jet velocity. The novel recommendation here, however, is that $T_{ad}$ should be the primary parameter in the design and operation of future oxy-fuel gas-turbine combustors, and not OF or $\varphi$, especially that the current findings confirm the previously reported ones [67] in proving that the static stability of premixed oxy-methane and air-methane flames is primarily controlled by $T_{ad}$ in different combustor technologies.

### *4.7.3 Flame Temperature*

In an effort to shed more light on the relationships between the flame structure, $T_{ad}$, $\varphi$, and *OF*, local temperature measurements were conducted along the axial and radial directions of the quartz tube. Figure 4.50 shows the domain of temperature measurements, in which point *0* corresponds to the burner surface at the centerline, while *Z* and *R* are the overall length and radius of the quartz tube, respectively. Local axial and radial positions were denoted with *z* and *r*, respectively. The local temperature (*T*), axial (*z*), and radial (*r*) locations were normalized (i.e. non-dimensionalized) with $T_{ad}$, *Z*, and *R*, respectively. The details of the flames considered for the temperature measurements are given in Table 4.5.

Figures 4.51 and 4.52 show the absolute and normalized local temperature distributions along the axis (i.e. centerline) of the combustor. Looking closely at Fig. 4.51, it could be realized that the stronger flames (i.e. flames 5, 7, and 9) exhibit higher temperature profiles while flames 2 and 12 show lower profiles. This is simply a reflection of $T_{ad}$, 1900 versus 1600 K.

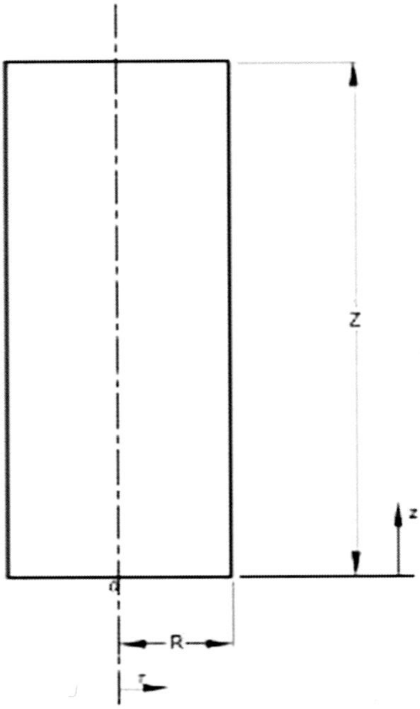


**Fig. 4.50** Temperature measurements domain [67]

**Table 4.5** Considered flames for the temperature measurements [67]

| Flame | Oxygen fraction (OF) (%) | Equivalence ratio (ϕ) | Adiabatic flame temperature ($T_{ad}$) (K) |
|---|---|---|---|
| 1 | 33 | 0.500 | 1589 |
| 2 | 60 | 0.250 | 1605 |
| 3 | 42 | 0.500 | 1899 |
| 4 | 50 | 0.407 | 1901 |
| 5 | 60 | 0.325 | 1902 |

Interestingly, when $T_{ad}$ is almost constant (flames 5, 7, 9), there is some difference in temperature profile. Although the images of these three flames have the same physical appearance, there is another parameter responsible for the differences in the axial temperature distribution. These flames have almost the same value of $T_{ad}$ but with different values of $\varphi$ and *OF*, where flame 9 has the highest value of $\varphi$. This implies that $\varphi$ has minor effect on the temperature profile, especially that the observed differences are comparable to the scale of the error bars.

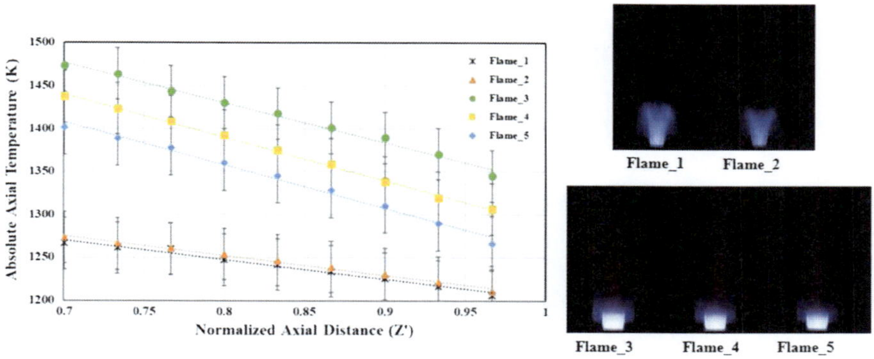


**Fig. 4.51** Absolute local temperature distributions along the axis of the combustor and images of the considered flames

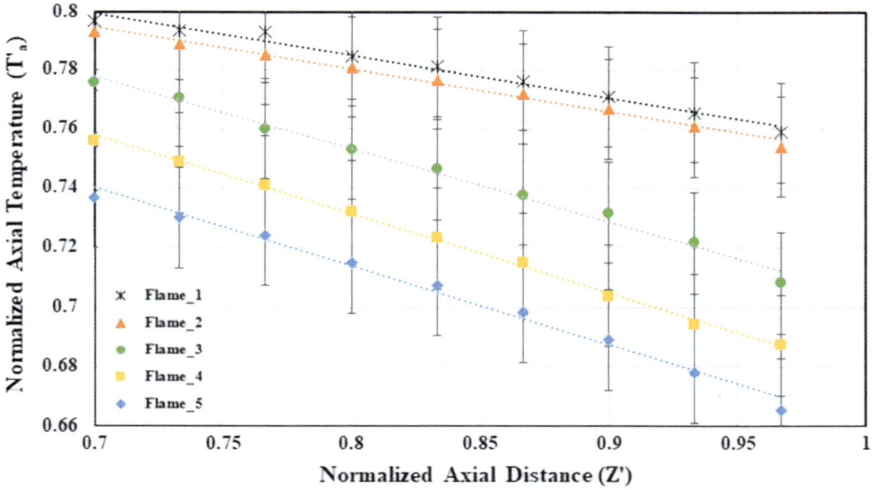


**Fig. 4.52** Normalized local temperature distributions along the axis of the combustor for the five flames considered

## 4.8 Performance of Solar-Integrated Oxy-combustion Cycles Adopting Membrane Reactors

### *4.8.1 Oxygen Separation Techniques for Oxy-combustion Cycles*

Various experimental and numerical studies have been conducted on the oxy-combustion of methane, and deeper insights on the flame characteristics and emissions were presented [98–105]. It has been found that combustion in $CO_2$ rich environment can drastically inhibit the flame speed and reaction rates. Several studies show that hydrogen-enriched methane has better flame characteristics and can effectively be applied to an oxy-fuel combustor [106–113]. Thermodynamic analysis of oxy-fuel combustion based power generation system can be found in the literature [114–117]. Due to low second law efficiency of commercial cryogenic air separation technology, the thermal efficiency of a carbon capture and sequestration (CCS) based system may suffer an efficiency loss of up to 8.5% as compared to the natural gas combined cycle (NGCC). Hence, cost-effective air separation technology has remained the focus of research for many of the scientists working on CCS integrated technologies.

The ability of ion transport membranes (ITMs) to selectively allow for oxygen separation with high purity at low energy cost, has made this technology for oxygen separation as one of the most promising alternatives of conventional cryogenic separation [118, 119]. These perovskite membranes require oxygen chemical gradient across the surfaces to drive the oxygen permeation through the membranes. Moreover, the operating temperature should be maintained in the range from 800 to 950 °C to ensure proper oxygen ion mobility in the lattice. Many attempts have been made to operate these membranes for oxygen separation using sweep inert (non-reacting) gases with reported oxygen permeation flux in the range from 0.01 to 1.0 $\mu mol/cm^2$ s [120–122]. However, for the ITM technology to be economically viable for the existing power generation systems, the required oxygen flux should be around 7 $\mu mol/cm^2 s$, which is significantly higher. The oxygen permeation rate can, however, be enhanced significantly by creating high oxygen partial pressure gradient across the membrane. This can be achieved through combing both the oxygen separation and the combustion processes in one unit called oxygen transport reactor (OTR), where in situ oxy-combustion of the sweeping fuel occurs at the permeate side of the membrane using the permeated oxygen. Oxygen consumption in the combustion process creates low oxygen partial pressure at the permeate side resulting in significant improvement of the oxygen permeation flux [121].

Thermodynamic analysis of a system is crucial to assess its feasibility and success as a power generation technology. Several attempts have been made to evaluate the practicality of oxy-combustion based plants utilizing air separation units (ASUs) via conventional cryogenic separation [123–126] as well as membrane technologies [127–129]. Xiong et al. [123] found an exergy efficiency loss of 4.08% in a pulverized-coal fired power plant, using cryogenic air separation, as compared to a conventional system. They further concluded that the exergy cost of the components is 10% higher in the oxy-combustion system than the conventional system due to additional power consumption for the air separation [124]. Moreover, the unit thermo-economic cost of the product for a component in the oxy-combustion system was found to be 1.22 times higher than that for the same component in the conventional system due to the additional monetary cost incurred due to the implementation of the oxy-combustion system. Effect of implementation of membrane technology for air separation has also been reported in some studies with varying opinions. Pfaff and Kather [127] reported slightly better exergetic efficiency of the oxy-combustion system with cryogenic air separation unit as compared to with the membrane ASU. However, thermo-economic analysis conducted by Osikowska et al. [129] shows 1.1% higher efficiency of a hybrid membrane-cryogenic separation based system as compared to only cryogenic separation based system. Castillo [128] also reported that with proper thermal integration, around 4% higher efficiency can be achieved using high-temperature membrane separation technology than through conventional cryogenic separation.

Hence, the integration of membrane technology with the oxy-combustion carbon capture system seems to have a very promising future. The literature survey shows numerous studies on the flame characteristics of oxy-combustion and several works on the thermodynamic analysis of the power-generation cycles. However, the thermodynamic investigations of an oxy-combustion based power generation system integrated with ITM separation technique demand further explorations to make its implementation technologically and economically feasible. The study in this section proposes two novel power systems incorporated with carbon capture via oxy-combustion for clean energy production applications. The thermodynamic analysis of the power generation system and air separation unit is presented. The exergy destruction in various components of the power generation system is presented, and different cycle designs are studied accordingly trying to improve the overall efficiency of the proposed plants.

### 4.8.2 *Proposed Power Generation Cycles*

The study in this section considers different power generation systems and compares the results in terms of exergy destruction of all the components and first and second law efficiencies of each system. Two oxy-combustion based power generation systems, premixed and non-premixed, are evaluated. Both the systems employ ITM based oxygen separation units to aid the combustion in a nitrogen-free environment in a gas turbine cycle. The two oxy-combustion cases are compared with a conventional base case system based on air combustion of methane to obtain useful work through a gas turbine cycle. The overall thermal efficiencies of the systems are assessed and compared to the base case. The second law efficiency of the systems, as well as exergy destructions in the components, are calculated and the sources of maximum exergy losses are identified. Based on the obtained thermal and exergy losses, an attempt is made to minimize the losses and increase the overall efficiencies of the systems through the modification of the system design. Based on that, the additional two modified cycles, considering premixed and non-premixed conditions, are also proposed.

A simple power generation system based on conventional combustion in a gas turbine combustor is selected as the base case for this study. The fuel used is methane which is the primary fuel used in all of the cases presented in this study. The system considered is a combined cycle gas turbine comprising of basic components to eliminate the intricacies of the thermodynamic calculations of a real gas turbine cycle. The power generation system consists of a compressor, a combustor, and a turbine with heat recuperation from the exhaust gases through a high-pressure and a low-pressure steam turbine. The heat recuperation system also consists of three heat exchangers, a condenser and a pump, configured as presented in the schematic diagram of the combined cycle power system in Fig. 4.53. The first step is to suck the atmospheric air and compress it to high pressure to be fed to the combustor. The oxidizer and fuel are mixed and burned in the high-pressure combustion chamber of the gas turbine which delivers the running power to the rotor during expansion, and the exhaust gas from the gas turbine is fed to the steam turbine cycle for heat recuperation. The conventional air combustion base case is selected such that the proposed carbon capture systems can be compared in terms of thermodynamic efficiencies and exergy losses.

Two oxy-combustion cycles are investigated under premixed and non-premixed combustion conditions. In both systems, a separation unit is used for oxygen separation either from the air in case of the premixed combustion cycle or from $CO_2/H_2O$ in case of the non-premixed combustion cycle. The black box modeling assumption was applied to model the separation of oxygen from the air in the premixed oxy-combustion cycle, and from $CO_2/H_2O$ stream in the non-premixed integrated

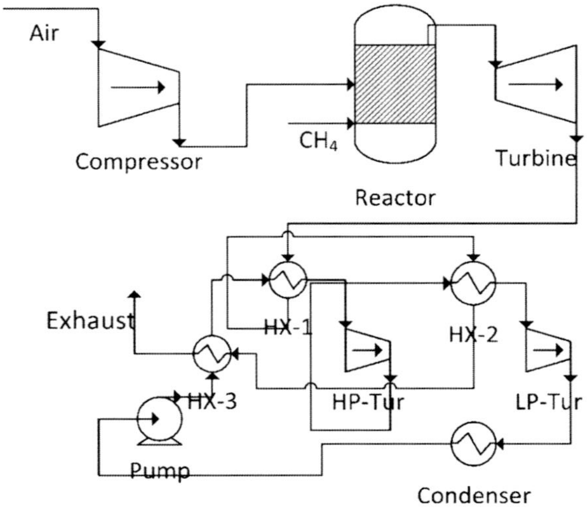


**Fig. 4.53** Schematic diagram of the base case power system

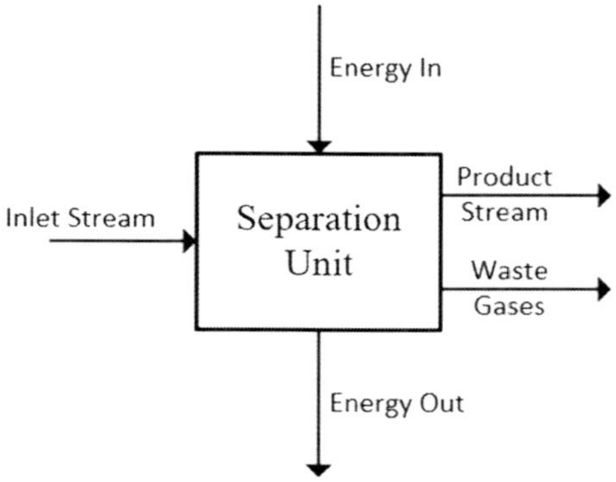


**Fig. 4.54** Black box model of the separation unit

oxy-combustion cycle. The model ignores the actual mechanism of the separation process but rather deals with the exergy balance to the control volume of the separation unit with the unit's second law efficiency as the dominant variable of the process. The separator considered in this study is an ITM-based unit, which works on the oxygen partial pressure gradient across the membrane maintained at an elevated temperature in the range from 1100 to 1200 K. The separation units produce two product streams of oxygen and waste gas from a single inlet stream. The energy stream is supplied to the unit to achieve the desired temperature and pressure suitable for efficient oxygen permeation through the membrane. A diagram of the separation unit is shown in Fig. 4.54. The product stream is high purity oxygen as the ITM allows highly selective permeation. The waste stream consists of residual gases in the stream. The inlet condition of the stream is calculated from the upstream and by defining the outlet temperature, pressure and composition of the outlet stream and the thermodynamic state of the system are well defined. First law analysis requires energy balance across the control volume and can be given as:

$$m_i h_i + W + Q = m_p h_p + m_w h_w \quad (4.19)$$

The exergy balance equation for the separation unit is given by:

$$\begin{aligned} m_i[(h_i - h_o) - T_o(s_i - s_o)] + W + Q\left(1 - {}^{T_o}/_{T}\right) &= Ex_{des} \\ + m_p\left[(h_p - h_o) - T_o(s_p - s_o)\right] + m_w[(h_w - h_o) - T_o(s_w - s_o)] \end{aligned} \quad (4.20)$$

The proposed cycle is based on premixed oxy-combustion of methane in a gas turbine cycle, as shown in Fig. 4.55. An air separation unit (ASU) based on the oxygen permeation through an ITM is implemented to provide the oxygen for the oxy-combustion process. The unit comprises a compressor to supply high-pressure air to the separation chamber, where oxygen is separated through the membrane at an elevated temperature. The model assumes a heat source for the ASU which can be extracted from a renewable energy source, most probably solar energy. The separated

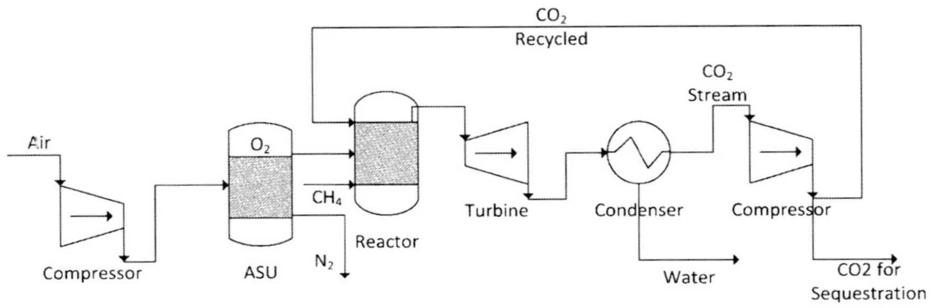


**Fig. 4.55** Schematic diagram of a single stage premixed oxy-combustion based cycle

pure oxygen is used directly in the combustor for the premixed oxy-combustion of methane. A portion of the exhaust $CO_2$ is recycled into the combustion chamber to maintain the desired temperature for the gas turbine material. The flue gas is expanded in the gas turbine to extract the useful work and then directed to a condenser. The flue gas, which is a mainly carbon dioxide and water vapor, is cooled down to condense the water vapor out, and a highly-concentrated stream of $CO_2$ is obtained. The captured $CO_2$ is further compressed and a portion of it is recirculated into the combustion chamber, and the rest is stored for sequestration. The first and second law efficiencies of the system can be given respectively as:

$$\eta_{th} = \frac{W_{useful}}{\left(LHV_{CH_4} + Q_{in,ASU}\right)} \tag{4.21}$$

$$\eta_{II} = \frac{W_{useful}}{\sum W_{rev,i}} \tag{4.22}$$

Where $W_{useful}$ is the net useful work obtained from the cycle, $LHV_{CH_4}$ is the lower heating value of the fuel stream, $Q_{in,ASU}$ is the heat required for the air separation unit, and $W_{rev,i}$ is the theoretical reversible work that can be obtained from the component '*i*' by bringing the system to the environmental dead state.

Another oxy-combustion cycle integrated with a steam reformer and an ITM reactor is also proposed [130]. The cycle diagram is presented in Fig. 4.56. The cycle implements pre-treatment of methane in a solar reformer to obtain syngas to be utilized in the combustion chamber. As discussed earlier, the reactions kinetics is significantly diminished in an oxy-combustion flame, and several studies suggested the introduction of hydrogen in the fuel stream to overcome this issue. Syngas, which basically comprises carbon monoxide and hydrogen, has the potential to deal with this issue and, hence, has been implemented in the cycle. The syngas produced in the reformer is introduced into the combustion chamber which is integrated with an ITM separation unit. The idea is to create a high oxygen partial pressure gradient for higher oxygen flux due to rapid consumption of oxygen on the permeate side through combustion [121]. The exhaust gases are utilized to run the turbine to extract the useful work and, then, water is separated out via condensation to obtain $CO_2$

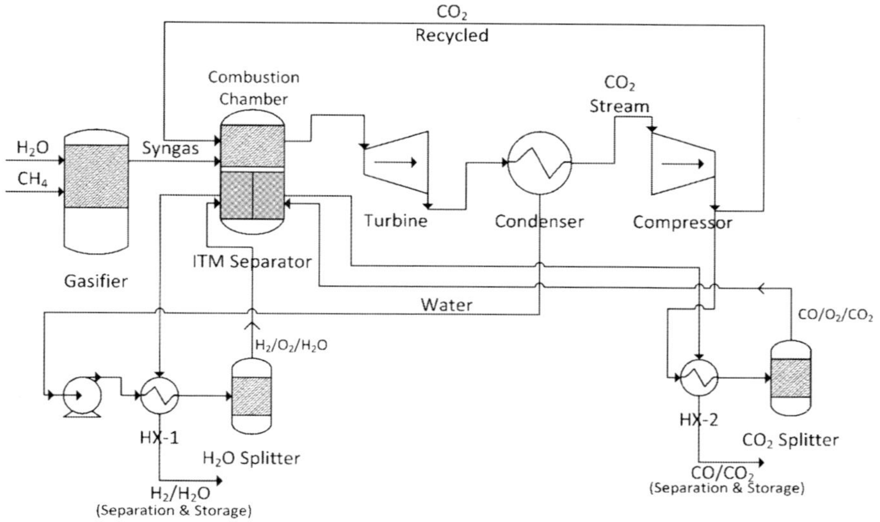


**Fig. 4.56** Schematic diagram of a single stage non-premixed oxy-combustion based cycle

rich stream. A portion of $CO_2$ is recycled into the combustion chamber after being compressed and the rest is transferred to the $CO_2$ splitter. Similarly, condensate water is fed to the $H_2O$ splitter. The purpose of both splitters is to provide oxygen to the ITM reactor for combustion while the residual gases from the splitters, which is basically concentrated $CO/H_2$, can be stored as an energy source when solar energy is not available for the steam reformer during the night mode of operation. Both splitters are equipped with heat exchangers for heat recuperation from the residual gases. The first law efficiency of the system can be given as:

$$\eta_{th} = \frac{W_{useful}}{\left(LHV_{CH_4} + \sum Q_{in,i} - LHV_{H_2} - LHV_{CO}\right)} \tag{4.23}$$

Where, $W_{useful}$ is the net-work output of the system, $LHV_{CH_4}$ is the lower heating value of methane fed to the gasifier, $\sum Q_{in,i}$ is the heat required in the components namely gasifier, $H_2O$ splitter, and $CO_2$ splitter, and $LHV_{H_2}$ and $LHV_{CO}$ are the lower heating values of hydrogen and carbon monoxide obtained from the $H_2O$ and $CO_2$ splitters, respectively. Second law efficiency of the system can be given as:

$$\eta_{II} = \frac{W_{useful}}{\sum W_{rev,i}} \tag{4.24}$$

Where, $W_{useful}$ is the net-work output of the system, and $W_{rev,i}$ is the theoretical reversible work that can be extracted from the individual components by bringing the system to the environmental dead state.

The operating parameters of each system were chosen based on the operational range of the different units and were kept the same for all the cases for a fair comparison. The inlet temperature and pressure were set to the environmental dead state conditions for all the cases. For a combustor working under oxy-combustion conditions, the equivalence ratio must be maintained close to stoichiometry, as lean combustion will significantly affect the purpose of carbon capture as well as the energy cost associated with oxygen separation. Hence, an equivalence ratio of one (stoichiometric) was set for both the oxy-combustion cases. Heat loss from the gas turbine system for all the cases has been assumed to be 100 kW. The operating parameters of the base case air-combustion cycle, the premixed oxy-combustion cycle, and the non-premixed oxy-combustion cycle are presented in Tables 4.6, 4.7 and 4.8, respectively. The exhaust temperature of the reactor was set to 1200 K, which is within the operating temperature range of a gas turbine combustor. Moreover, the non-premixed case utilizes the integrated ITM reactor, which needs to be operated within this temperature range too for efficient permeation. The temperature of the reactor is controlled by adjusting the recycled $CO_2$ flow rate into the combustion chamber. For the conventional air combustion base case, the temperature is controlled by adjusting the lean ratio of the reacting mixture.

The isentropic efficiency of all turbines and compressors is assumed to be 90%, and the effectiveness of all heat exchangers is assumed to be 0.95. Fuel consumption is normalized by the net useful power obtained from the system, and other flow rates are adjusted accordingly. All the calculations and results are for per megawatt (MW) of useful work. Commercial process simulator Aspen Hysys V7.3 was used to model and simulate the different proposed systems. Peng Robinson equation of state was

**Table 4.6** Operating parameters for base case air-combustion cycle

| Components | Temperature (K) | | Pressure (bar) | | Isentropic efficiency |
|---|---|---|---|---|---|
| | Inlet | Outlet | Inlet | Outlet | |
| Compressor | 298 | 664.2 | 1 | 15 | 90% |
| Reactor | 664.2 | 1200 | 15 | 15 | – |
| Turbine | 1200 | 672.4 | 15 | 1 | 90% |

**Table 4.7** Operating parameters for the premixed oxy-combustion cycle

| Components | Temperature (K) | | Pressure (bar) | | Isentropic efficiency |
|---|---|---|---|---|---|
| | Inlet | Outlet | Inlet | Outlet | |
| ASU-compressor | 298 | 664.2 | 1 | 15 | 90% |
| ASU | 664.2 | 1200 | 15 | 15 | – |
| Reactor | 573.9 | 1200 | 15 | 15 | – |
| Turbine | 1200 | 826.5 | 15 | 1 | 90% |
| Condenser | 826.5 | 298 | 1 | 1 | – |
| $CO_2$-compressor | 300 | 543 | 1 | 15 | 90% |

**Table 4.8** Operating parameters for the non-premixed oxy-combustion cycle

| Components | Temperature (K) | | Pressure (bar) | | Isentropic efficiency |
|---|---|---|---|---|---|
| | Inlet | Outlet | Inlet | Outlet | |
| Gasifier | 298 | 873 | 15 | 15 | – |
| Reactor | 596.4 | 1200 | 15 | 15 | – |
| Turbine | 1200 | 824.6 | 15 | 1 | 90% |
| Condenser | 824.6 | 298 | 1 | 1 | – |
| $CO_2$-compressor | 300 | 544.4 | 1 | 15 | 90% |
| $CO_2$ splitter | 1000 | 1200 | 15 | 15 | – |
| $H_2O$ splitter | 610 | 1200 | 15 | 15 | – |

selected for the calculation of the properties of gases and mixtures for all proposed systems.

### 4.8.3 *Performance of the Proposed Cycles*

The base case assumed in this study works on the conventional methane combustion gas turbine cycle for power generation. All the results are normalized for 1.0 MW power output of the turbine. The thermal efficiency of the single stage cycle has been found to be 42.7% based on the lower heating value of the fuel. Methane consumption for this cycle is 0.047 kg/s per unit MW of power generation. Figure 4.57 presents the exergy destructions in the various components of the system. It is obvious to find the highest exergy losses in the reactor due to highly irreversible reactions taking place inside the combustion chamber. Exergy destruction in the reactor is found to be 870 kW per unit MW of useful power, which constitutes 82.7% of the total exergy

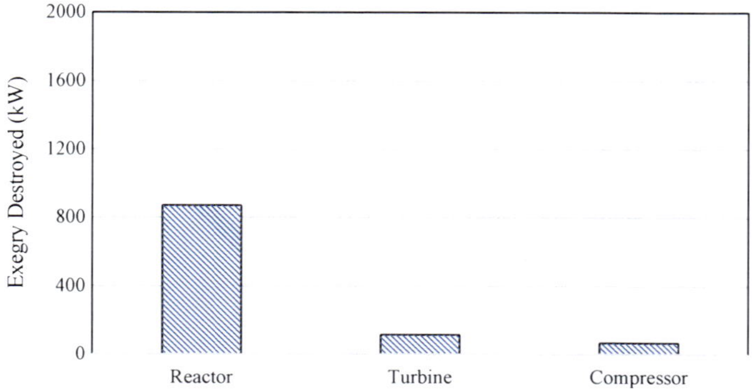


**Fig. 4.57** Exergy destruction per MW of useful power output in various components of the base case system

destruction in the system. The exergy destruction in the turbine is calculated to be 10.9% and that value in the compressor is calculated to be 6.4%, which has a second law efficiency of 0.95 each. The second law efficiency of the whole system has been calculated to be 48.7%.

The proposed premixed oxy-combustion cycle integrated with an ASU based on ITM technology, as shown in Fig. 4.54, is presented in this section. The first law efficiency of the system is calculated to be 15.9%, which is quite low when compared to the conventional cycle. The cycle consumes fuel in the rate of 0.1 kg/s per unit MW of power production, which is around 75% higher than the base case requirement. The lower thermodynamic efficiency can be attributed to the energy cost associated with the process of air separation. It has been found that a total energy requirement for the ASU, including the air compressor, is 1770 kW per unit MW of power generation, which is basically 25.4% of the total energy requirement for the cycle. The energy cost associated with the oxygen production using the ITM separator is found to be 4.3 MJ/kg of oxygen. Although exergy loss in the ASU contributes only a minor value, as can be seen from Fig. 4.58, it is to be noted that most of the energy is wasted to the environment through nitrogen waste at high temperature. Highest exergy losses can be observed in the reactor and in the condenser, constituting 48.5% and 44.6% of the total exergy destruction of the system, respectively. Exergy destruction in the reactor is 920 kW higher than that value of the base case for the same 1.0 MW of power generation. High exergy loss in the condenser can be attributed to the large temperature gradient in the condenser as compensation to achieve high $CO_2$ purity of about 94.5%. The second law efficiency of the whole cycle has been found to be 21.3%, which is again much lower than that of the base case. Significant reduction in the second law efficiency of the system is mainly due to the large exergy loss in the condenser, which is basically bypassed in the base case, where there is no need to separate flue gases for sequestration. Hence, the second law efficiency of the

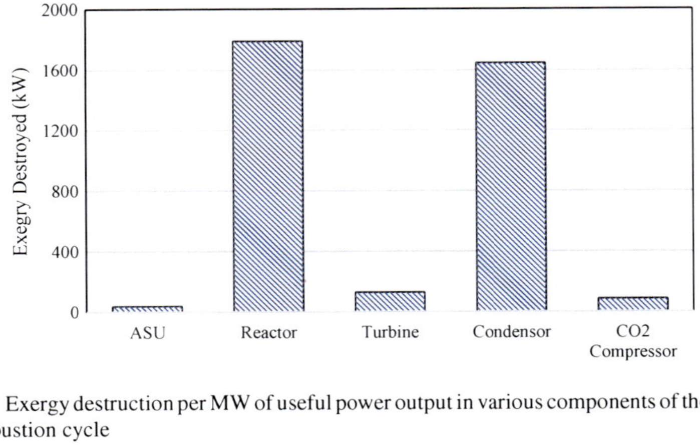


**Fig. 4.58** Exergy destruction per MW of useful power output in various components of the premixed oxy-combustion cycle

system can be achieved by reducing the temperature gradient across the condenser by integrating it with a recuperation cycle.

The ITM integrated reactor system working on the non-premixed oxy-combustion gas turbine cycle is presented in this section. The system is also equipped with a gasifier as shown in Fig. 4.55. Thermodynamic efficiency of the system was found to be 26.4%, which shows much improved efficiency than the premixed case; however, still lagging behind the base case by 16.3%. The most energy intensive unit in the system is the $H_2O$ splitter, followed by the gasifier, and the $CO_2$ splitter. However, these units are generally equipped with solar power and, hence, add to the $CO_2$ mitigation efforts. Methane consumption is calculated to be 0.046 kg/s per unit MW of power produced, which is much lower than that value of the premixed case. This can be attributed to the addition of the gasifier, where the energy content of the fuel is increased by converting it into syngas. The exergy destruction in the various components is shown in Fig. 4.59. Highest exergy loss was encountered in the $H_2O$ splitter followed by the condenser, the reactor, the gasifier, and the $CO_2$ splitter. Large exergy losses in the $H_2O$ splitter, the gasifier, and the $CO_2$ splitter are due to highly endothermic reactions taking place within the units. Condenser contributes to 939 kW of exergy loss per unit MW of power generation, constituting 21.6% of the total exergy loss. Exergy destruction in the reactor was found to be 803 kW as compared to 870 kW per unit MW of power generation for the base case. The reduction in exergy loss in the reactor can be attributed to the increased energy content of the fuel due to gasification. The overall second law efficiency of the cycle was found to be 36.0%, which resulted in a significant improvement from the premixed cycle; however, still 12.7% lower than that of the base case. As can be seen from Fig. 4.59, the exergy losses in the reactor and condenser are reduced by around 55% and 43%, respectively, as compared to the non-premixed case, thereby increasing the second law efficiency of the non-premixed case by 14.7%.

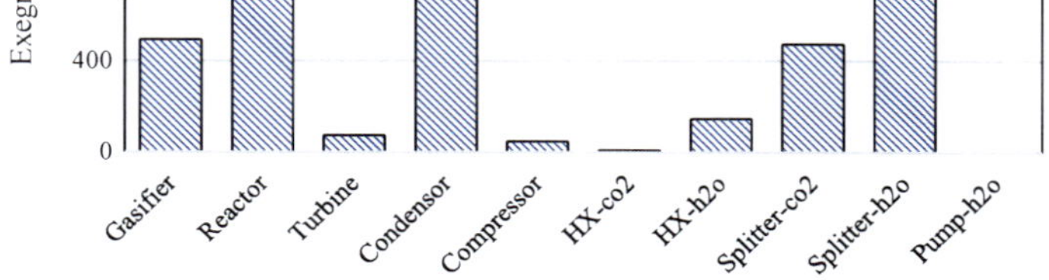


**Fig. 4.59** Exergy destruction per MW of useful power output in various components of the non-premixed oxy-combustion cycle

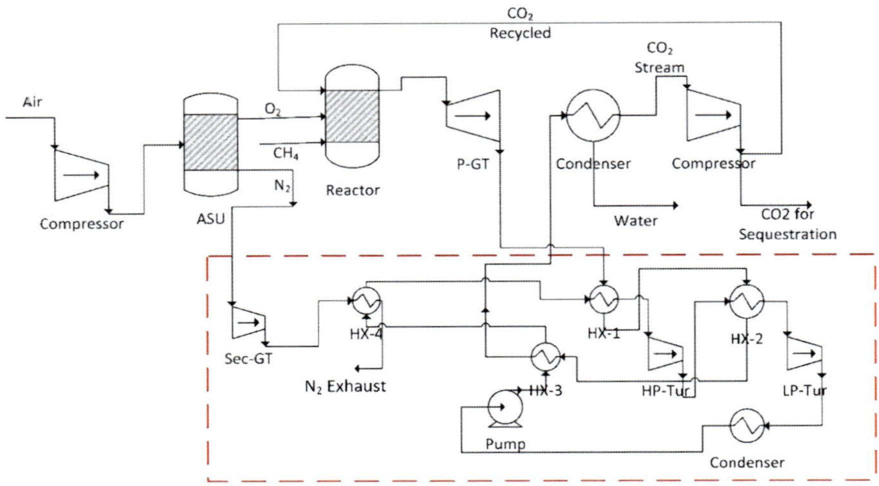


**Fig. 4.60** Schematic diagram of the premixed oxy-combustion system with heat recuperation

### *4.8.4 Modified Power Generation Cycles*

The premixed oxy-combustion cycle discussed earlier, was found to have very low first law and second law efficiencies. The results presented above show that the energy can be extracted from the ASU and the condenser using heat recuperation cycles to increase the efficiency of the system. The waste gas from the ASU, which is mainly the separated nitrogen, is exited at a temperature of 1200 K and has high energy content that can be used to run a secondary gas turbine. Similarly, energy from flue gases exhausted from the gas turbine can be utilized to generate some power using steam cycles. The schematic of the modified cycle is presented in Fig. 4.60. The exhausts from the ASU and the gas turbine are directed to a set of heat recuperation units consisting of four heat exchangers and two steam turbines (a high-pressure turbine and a low-pressure turbine). After heat extraction, nitrogen is dumped into the environment and flue gases are redirected to the condenser for the separation of $CO_2$. The system has been normalized such that the power output of the combined system is 1.0 MW.

The first law efficiency of the modified system was found to be 42.3%, which is more than twice of the previous cycle and very close to the base case cycle efficiency. Fuel consumption was calculated to be 0.049 kg/s/MW as compared to 0.047 kg/s/MW for the base case cycle. Total energy requirement by the ASU has been reduced to 408 kW per unit MW of useful work obtained from the system, which constitutes 17.3% of the total energy input. This share has been reduced by more than half as compared to the previous case, thanks to energy recovery from the waste gases. The oxygen separation cost was calculated to be 2.2 MJ/kg of oxygen, which is still much higher than that of the conventional separation cost of 0.725 MJ/kg of

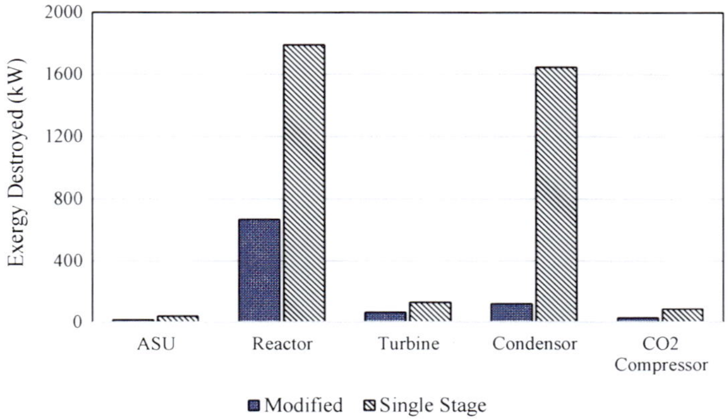


**Fig. 4.61** Exergy destruction per MW of useful power output in various components of the modified pre-mixed combustion cycle

oxygen [131]. It can be seen from Fig. 4.61 that the exergy loss in the condenser has been reduced significantly by implementing the heat recuperation. The exergy loss in the condenser constitutes on 13.2% of the system total, as compared to 44.6% for the previous case. The second law efficiency of the cycle, thus, improved by 31.5% with an overall efficiency of 52.8%, which is even higher than the base case cycle efficiency.

The non-premixed oxy-combustion cycle has been modified to implement the heat recuperation from the exhaust gas out of the gas turbine, as shown in Fig. 4.62. The flue gas is fed to a series of heat exchangers to produce steam to obtain work in two stages via high-pressure and low-pressure turbines. The whole system is normalized for 1.0 MW of useful work generated from the system.

The thermal efficiency of the modified system was calculated to be 37.8%, which is 4.9% lower than that of the assumed base case cycle. The energy requirement of all the components has been decreased by around 30% per unit MW of the power produced, due to the improved overall efficiency of the plant. It can be seen that the first law efficiency of the modified non-premixed cycle is, however, lower than that of the modified pre-mixed cycle. Figure 4.63 presents the exergy destruction in all of the cycle components. The cycle is able to extract much work from the flue gases, thus, reducing the exergy loss in the condenser. The exergy destruction in the condenser has been reduced to only 4.9% of the total value as compared to 21.6% for the previous case. The second law efficiency of the system, thus, approached 50.4%, which is better than all the cases considered in this study [130].

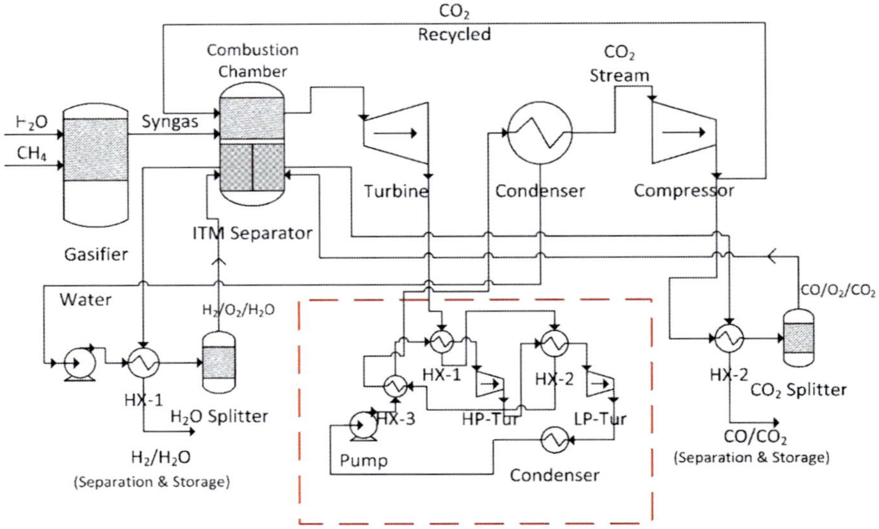


Fig. 4.62 Schematic diagram of the non-premixed oxy-combustion system with heat recuperation

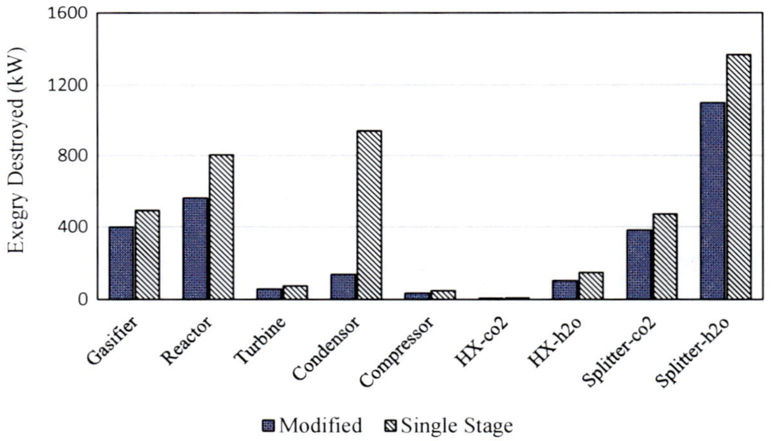


Fig. 4.63 Exergy destruction per MW of useful power output in various components of the modified non-pre-mixed combustion cycle

## 4.9 Concluding Remarks

In this chapter, performance and emissions of various gas turbine combustors adapting different burner designs are evaluated under both air- and oxy-fuel combustion conditions. Oxy-fuel combustion was investigated based on experimental and numerical studies on a model combustor that mimics the geometry of operational air-fuel gas-turbine combustors implementing the LPM DLE DACRS technology (Lean Premixed, Dry Low Emissions, Dual Annular Counter-Rotating Swirlers). The data indicated that $T_{ad}$ is a more relevant parameter choice than $\varphi$ or $OF$ to quantify the combustor stability map. Flames of the same $T_{ad}$ demonstrated almost identical shapes, which is another novel finding for oxy-methane flames. Three different flame structures were observed while varying the equivalence ratio, namely (I) double conical flame, (II) corner-stabilized flame, and (III) swirl-stabilized (V-shaped) flame. The double conical flame is observed at low equivalence ratios near the blowout limit. The corner-stabilized flame is observed at higher equivalence ratios. Increasing the equivalence ratio further induces a transition to the typical V-shape of swirl-stabilized flames with significantly stronger IRZ and weaker CRZ. Premixed oxygen-enriched air-methane flames ($CH_4/O_2/N_2$) are compared to their oxy-methane counterparts ($CH_4/O_2/CO_2$) on the same model gas-turbine combustor and under identical conditions of oxygen fraction and equivalence ratio. The following novel conclusions were made for both $N_2$ and $CO_2$ flames at common inlet bulk velocity. (a) Their stable combustion zones can both be characterized by $T_{ad}$ only, although they have different $T_{ad}$ maps. The stable combustion zones are thus governed mainly by the reaction kinetics (especially near the flashback limits) under similar cold flow conditions. (b) Both $N_2$ and $CO_2$ flames undergo the same changes in macrostructure and stabilization mode as $T_{ad}$ is increased from the blowout limits to the flashback ones. (c) Stable flames of different $\varphi$ and $OF$ but the same $T_{ad}$ have identical shapes, which proves that the knowledge of $T_{ad}$ alone is enough to accurately predict and characterize flame macrostructure and stabilization mode under similar cold flow conditions. (d) Both $N_2$ and $CO_2$ stability maps can be subdivided into sub-zones based on $T_{ad}$ only, where each zone has a single prevailing flame macrostructure irrespective of $\varphi$ and $OF$. This is yet another proof that $T_{ad}$ is an excellent tool for predicting flame macrostructure at constant inlet bulk velocity. Based on these findings, it is recommended to design and operate future oxy-fuel gas-turbine combustors based on $T_{ad}$ (and not $OF$ or $\varphi$), particularly at high and medium loads away from blowout, following the existing common practice among manufacturers of conventional LPM DLE air-fuel gas turbines to quantify combustor performance in terms of $T_{ad}$. Considering the same swirl-stabilized combustor, the static stability limits of oxy-methane flames were recorded experimentally in a premixed swirl combustor under fixed Reynolds (Re) operation. The aim was to isolate the dynamic effect of Re on flame stabilization and understand the physics behind flame extinction. The results showed that the operation of premixed swirl-stabilized oxy-methane flame is not possible for $OF$ ranges below 29% and above 70%. The flashback limit is mainly controlled by the reaction kinetics rates, which control the flame speed at flashback. All flames under all operating Re values blow out at the same combustor power density (*PD*) indicating its leading role in controlling the lean blow out stability limit of the oxy-flames. Increasing Re shifts the blow out limit to the leaner side resulting in wider stable flame operability. Flame transition from the ISL to the ORZ is observed at certain value of $\varphi$ controlled mainly by the value of $U_{in}$.

Micromixer burner for gas turbine applications was also discussed in this chapter. Oxy-methane combustion in a premixed multi-hole model gas-turbine combustor that mimics air-fuel pre-mixers in operational DLE gas turbines was investigated. The results show that: (i) the acoustic and blowout limits follow a path of constant $T_{ad}$, which implies that $T_{ad}$ is a key control parameter of combustor stability. This agrees with what was reported in past studies for swirl-stabilized combustors, which highlights the universal nature of this conclusion. Such conformity was not observed with any of the other parameters. (ii) the examined micromixer-like geometry exhibits superior performance compared to swirl-based technologies, especially at low combustor loads, which is of critical importance to the operation of future oxy-fuel gas turbines.

The feasibility of oxy-combustion cycles with various configurations was demonstrated in this chapter to investigate the effect of carbon capture on the cycle efficiency. The proposed cycles are the pre-mixed oxy-combustion with a separate ASU based on ITM technology, and the non-premixed oxy-combustion cycle with ITM integrated reactor and a gasification unit. The performance of both the cycles has been compared to a basic air-combustion cycle to demonstrate the effect of additional energy cost for oxygen separation on the cycle performance. The thermal efficiency of the assumed base case system was calculated to be 42.7%. The highest exergy loss was found to be in the reactor with an overall second law efficiency of 48.7%. The non-premixed oxy-combustion system, based on the integrated ITM reactor, has lower efficiencies than that of the assumed base case air-combustion cycle. The thermal efficiency of the cycle was found to be 37.8% as compared to 42.7% for the base case cycle. The second law efficiency was around 50%, with the largest exergy destructions in the $H_2O$ splitter and the combustion chamber, where highly irreversible reactions taking place. The proposed integrated ITM reactor system is capable of producing syngas via $CO_2/H_2O$ splitting, which can be used as an auxiliary power source when renewable energy for gasification and splitting is not available. The premixed oxy-combustion system was evaluated based on the thermodynamic parameters and was compared to the base case. The cost of oxygen separation via ASU dominates as the parameter to affect the cycle efficiency of the proposed premixed oxy-combustion system. It has been found that a significant amount of energy is lost through the waste gases from the ASU and, hence, heat recuperation was employed to the unit along with the heat recovery from the flue gases. The modification was able to improve the first law (42.3%) and the second law (52.8%) efficiencies of the system significantly to be much closer to those of the base case air-combustion cycle. Effect of hydrogen blending on the efficiencies of the premixed system has also been evaluated. It has been found that the hydrogen addition increases the overall first and second law efficiencies of the system; however, the change is only marginal.

**Acknowledgements** The authors wish to acknowledge the support received from King Fahd University of Petroleum & Minerals under Grant number BW191002 for the preparation of this book chapter.

# References


1. Hoeven MV (2011) $CO_2$ emission from fuel combustion. IEA, Paris, France, pp 12–27
2. Hansen J, Sato M, Ruedy R, Lo K, Lea DW, Elizade M (2006) Global temperature change. Proc Natl Acad Sci 103(39):14288–14293
3. http://cornerstonemag.net/overview-of-oxy-fuel-combustion-technology-for-CO2-capture/
4. Oh J, Noh D (2012) Laminar burning velocity of oxy-methane flames in atmospheric condition. Energy 45:669–675
5. Ghoniem AF (2011) Needs, resources and climate change: clean and efficient conversion technologies. Prog Energy Combust Sci 37:15–51
6. Habib MA, Nemitallah M, Ben-Mansour R (2013) Recent development in oxy-combustion technology and its applications to gas turbine combustors and ITM reactors. Energy Fuels 27(1):2–19
7. http://web.mit.edu/16.unified/www/FALL/thermodynamics/notes/node111.html
8. Warnatz J, Mass U, Dibble RW (1996) Combustion. Springer, New York, Chap. 17
9. Taamallah S, Chakroun W, Watanabe H, Shanbhogue J, Ghoniem AF (2017) On the characteristic flow and flame times for scaling oxy and air flame stabilization modes in premixed swirl combustion. Proc Combust Inst 36(3):3799–3807
10. Kim K (2016) Combustion instability feedback mechanisms in a lean-premixed swirl-stabilized combustor. Combust Flame 171:137–151
11. Runyon J, Marsh R, Bowen P, Pugh D, Giles A, Morris S (2018) Lean methane flame stability in a premixed generic swirl burner: isothermal flow and atmospheric combustion characterization. Exp Thermal Fluid Sci 92:125–140
12. Li B, Shi B, Zhao X, Ma K, Xie D, Zhao D, Li J (2018) Oxy-fuel combustion of methane in a swirl tubular flame burner under various oxygen contents: operation limits and combustion instability. Exp Thermal Fluid Sci 90:115–124
13. Nemitallah MA, Rashwan SS, Mansir B, Abdelhafez A, Habib MA (2018) Review of novel combustion techniques for clean power production in gas turbines. Energy Fuels 32(2):979–1004
14. Rashwan SS, Nemitallah MA, Habib MA (2016) Review on premixed combustion technology: stability, emission control, applications, and numerical case study. Energy Fuels 30(12):9981–10014
15. Kutne P, Kapadia K, Meier W, Aigner M (2011) Experimental analysis of the combustion behavior of oxyfuel flames in a gas turbine model combustor. Proc Combust Inst 33:3383–3390
16. Amato A, Hudak B, D'Carlo P, Noble D, Scarborough D, Seitzman J, Lieuwen T (2011) Methane oxy-combustion for low $CO_2$ cycles: blow-off measurements and analysis. J Eng Gas Turbines Power 33:061503–1
17. Peng Q, Jiaqiang E, Zhang Z, Hu W, Zhao X (2018) Investigation on the effects of front-cavity on flame location and thermal performance of a cylindrical micro combustor. Appl Therm Eng 130:541–551
18. Peng Q, Jiaqiang E, Chen J, Zuo W, Zhao X, Zhang Z (2018) Investigations on effects of wall thickness and porous media on the thermal performance of a non-premixed hydrogen fueled cylindrical micro combustor. Energy Convers Manag 155:276–286
19. Zuo W, Jiaqiang E, Peng Q, Zhao X, Zhang Z (2017) Numerical investigations on a comparison between counterflow and coflow double-channel micro combustors for micro-thermophotovoltaic system. Energy 122:408–419
20. Zuo W, Jiaqiang E, Peng Q, Zhao X, Zhang Z (2017) Numerical investigations on thermal performance of a micro-cylindrical combustor with gradually reduced wall thickness. Appl Therm Eng 113:1011–1020
21. IEA (2016) 20 Years of carbon capture and storage. Available: https://www.iea.org/publications/freepublications/publication/20YearsofCarbonCaptureandStorage_WEB.pdf
22. UN (2015) Paris agreement. Available: https://unfccc.int/files/essential_background/convention/application/pdf/english_paris_agreement.pdf

23. Li H, Yan J, Yan J, Anheden M (2009) Impurity impacts on the purification process in oxy-fuel combustion based $CO_2$ capture and storage system. Appl Energy 86(2):202–213
24. Joshi ND, Epstein MJ, Durlak S, Marakovits S, Sabla PE (1994) Development of a fuel air premixer for aero-derivative dry low emissions combustors. In: International gas turbine and aeroengine congress and exposition, vol 94-NaN-253, pp 1–9
25. Huang Y, Yang V (2009) Dynamics and stability of lean-premixed swirl-stabilized combustion. Prog Energy Combust Sci 35(4):293–364
26. Xie Y, Wang J, Zhang M, Gong J, Jin W, Huang Z (2013) Experimental and numerical study on laminar flame characteristics of methane oxy-fuel mixtures highly diluted with $CO_2$. Energy Fuels 27(10):6231–6237
27. Oh J, Noh D (2012) Laminar burning velocity of oxy-methane flames in atmospheric condition. Energy 45(1):669–675
28. Konnov AA, Dyakov IV (2004) Measurement of propagation speeds in adiabatic flat and cellular premixed flames of $C_2H_6 + O_2 + CO_2$. Combust Flame 136:371–376
29. Konnov AA, Dyakov IV (2007) Experimental study of adiabatic cellular premixed flames of methane (ethane, propane) + oxygen + carbon dioxide mixtures. Combust Sci Technol 179(4):747–765
30. Konnov AA, Meuwissen RJ, De Goey LPH (2011) The temperature dependence of the laminar burning velocity of ethanol flames. Proc Combust Inst 33(1):1011–1019
31. Gu X, Huang Z, Wu S, Li Q (2010) Laminar burning velocities and flame instabilities of butanol isomers-air mixtures. Combust Flame 157(12):2318–2325
32. Van Lipzig JPJ, Nilsson EJK, De Goey LPH, Konnov AA (2011) Laminar burning velocities of n-heptane, iso-octane, ethanol and their binary and tertiary mixtures. Fuel 90(8):2773–2781
33. Kaskan WE (1957) The dependence of flame temperature on mass burning velocity. Symp Combust 6(1):134–143
34. Van Maaren A, Thung DS, De Goey LPH (1994) Measurement of flame temperature and adiabatic burning velocity of methane/air mixtures. Combust Sci Technol 96(4–6):327–344
35. Metghalchi M, Keck JC (1980) Laminar burning velocity of propane-air mixtures at high temperature and pressure. Combust Flame 38:143–154
36. Liao SY, Jiang DM, Huang ZH, Zeng K, Cheng Q (2007) Determination of the laminar burning velocities for mixtures of ethanol and air at elevated temperatures. Appl Therm Eng 27(2–3):374–380
37. Galmiche B, Halter F, Foucher F, Dagaut P (2011) Effects of dilution on laminar burning velocity of premixed methane/air flames. Energy Fuels 25(3):948–954
38. Yi B, Zhang L, Huang F, Mao Z, Zheng C (2014) Effect of $H_2O$ on the combustion characteristics of pulverized coal in $O_2/CO_2$ atmosphere. Appl Energy 132:349–357
39. Toporov D, Bocian P, Heil P, Kellermann A, Stadler H, Tschunko S, Förster M, Kneer R (2008) Detailed investigation of a pulverized fuel swirl flame in $CO_2/O_2$ atmosphere. Combust Flame 155(4):605–618
40. Khare SP, Wall TF, Farida AZ, Liu Y, Moghtaderi B, Gupta RP (2008) Factors influencing the ignition of flames from air-fired swirl pf burners retrofitted to oxy-fuel. Fuel 87(7):1042–1049
41. Khalil AEE, Gupta AK (2017) Flame fluctuations in Oxy-$CO_2$-methane mixtures in swirl assisted distributed combustion. Appl Energy 204:303–317
42. Shroll AP, Shanbhogue SJ, Ghoniem AF (2012) Dynamic-stability characteristics of premixed methane oxy-combustion. J Eng Gas Turbines Power 134(5):51504
43. Kutne P, Kapadia BK, Meier W, Aigner M (2011) Experimental analysis of the combustion behaviour of oxyfuel flames in a gas turbine model combustor. Proc Combust Inst 33(2):3383–3390
44. Song Y, Zou C, He Y, Zheng C (2015) The chemical mechanism of the effect of $CO_2$ on the temperature in methane oxy-fuel combustion. Int J Heat Mass Transf 86:622–628
45. Oh J, Noh D, Lee E (2013) The effect of CO addition on the flame behavior of a non-premixed oxy-methane jet in a lab-scale furnace. Appl Energy 112:350–357
46. Liu CY, Chen G, Sipöcz N, Assadi M, Bai XS (2012) Characteristics of oxy-fuel combustion in gas turbines. Appl Energy 89(1):387–394

47. Yin C, Yan J (2016) Oxy-fuel combustion of pulverized fuels: combustion fundamentals and modeling. Appl Energy 162:742–762
48. Rashwan SS, Ibrahim AH, Abou-Arab TW, Nemitallah MA, Habib MA (2016) Experimental investigation of partially premixed methane-air and methane-oxygen flames stabilized over a perforated-plate burner. Appl Energy 169:126–137
49. Ramadan IA, Ibrahim AH, Abou-Arab TW, Rashwan SS, Nemitallah MA, Habib MA (2016) Effects of oxidizer flexibility and bluff-body blockage ratio on flammability limits of diffusion flames. Appl Energy 178:19–28
50. Nemitallah MA, Habib MA (2013) Experimental and numerical investigations of an atmospheric diffusion oxy-combustion flame in a gas turbine model combustor. Appl Energy 111:401–415
51. Brohez S, Delvosalle C, Marlair G (2004) A two-thermocouples probe for radiation corrections of measured temperatures in compartment fires. Fire Saf J 39(5):399–411
52. Carr NL, Kobayashi R, Burrows DB (1954) Viscosity of hydrocarbon gases under pressure. J Pet Technol 6(10):47–55
53. Marsh R, Runyon J, Giles A, Morris S, Pugh D, Valera-Medina A, Bowen P (2017) Premixed methane oxycombustion in nitrogen and carbon dioxide atmospheres: measurement of operating limits, flame location and emissions. Proceedings of the Combustion Institute. Proc Combust Inst 36(3):3949–3958
54. Li M, Tong Y, Thern M, Klingmann J (2017) Investigation of methane oxy-fuel combustion in a swirl-stabilised gas turbine model combustor. Energies 10(5)
55. Cengel YA, Boles MA (2015) Thermodynamics: an engineering approach 8th edn
56. Habib MA, Rashwan SS, Nemitallah MA, Abdelhafez A (2017) Stability maps of non-premixed methane flames in different oxidizing environments of a gas turbine model combustor. Appl Energy 189:177–186
57. Veynante D, Vervisch L (2002) Turbulent combustion modeling. Prog Energy Combust Sci 28:193–266
58. Peters N (2001) Meas Sci Technol. https://doi.org/10.1088/0957-0233/12/11/708
59. Kewlani G, Supervisor T (2014) Large eddy simulations of premixed turbulent flame dynamics: combustion modeling, validation and analysis. http://hdl.handle.net/1721.1/93863
60. Emanuel F, Pérez H (2011) Subfilter scale modelling for large eddy simulation of lean hydrogen-enriched turbulent premixed combustion. Thesis, pp 1–186
61. Herná FE, Groth CPT, Gü L (2014) Large-eddy simulation of lean hydrogen–methane turbulent premixed flames in the methane-dominated regime. Int J Hydrogen Energy 39:7147–7157
62. Marsh R, Runyon J, Giles A, Morris S, Pugh D, Valera-Medina A et al (2017) Premixed methane oxycombustion in nitrogen and carbon dioxide atmospheres: measurement of operating limits, flame location and emissions. Proc Combust Inst 36:3949–3958
63. Weigand P, Meier W, Duan XR, Stricker W, Aigner M (2006) Investigations of swirl flames in a gas turbine model combustor: I. Flow field, structures, temperature, and species distributions. Combust Flame. https://doi.org/10.1016/j.combustflame.2005.07.010
64. Meier W, Duan XR, Weigand P (2006) Investigations of swirl flames in a gas turbine model combustor: II. Turbulence–chemistry interactions. Combust Flame. https://doi.org/10.1016/j.combustflame.2005.07.009
65. Taamallah S, LaBry ZA, Shanbhogue SJ, Habib MA, Ghoniem AF (2015) Correspondence between "Stable" flame macrostructure and thermo-acoustic instability in premixed swirl-stabilized turbulent combustion. J Eng Gas Turbines Power 137:071505. https://doi.org/10.1115/1.4029173
66. Taamallah S, Labry ZA, Shanbhogue SJ, Ghoniem AF (2015) Thermo-acoustic instabilities in lean premixed swirl-stabilized combustion and their link to acoustically coupled and decoupled flame macrostructures. Proc Combust Inst 35:3273–3282
67. Abdelhafez A, Rashwan SS, Nemitallah MA, Habib MA (2018) Stability map and shape of premixed CH4/O2/CO2 flames in a model gas-turbine combustor. Appl Energy 215:63–74. https://doi.org/10.1016/j.apenergy.2018.01.097

68. Jourdaine P, Mirat C, Caudal J, Lo A, Schuller T (2017) A comparison between the stabilization of premixed swirling $CO_2$-diluted methane oxy-flames and methane/air flames. Fuel 201:156–164
69. IEA (2016) 20 years of carbon capture and storage. https://www.iea.org/publications/freepublications/publication/20YearsofCarbonCaptureandStorage_WEB.pdf
70. UN (2015) Paris agreement. https://unfccc.int/files/essential_background/convention/application/pdf/english_paris_agreement.pdf
71. Li H, Yan J, Yan J, Anheden M (2009) Impurity impacts on the purification process in oxy-fuel combustion based CO2 capture and storage system. Appl Energy 86:202–213
72. Cao Y, He B, Ding G, Su L, Duan Z (2016) Performance modeling of integrated chemical looping air separation and IGCC with $CO_2$ capture. Energy Fuels 30:9953–9961
73. Nemitallah MA (2016) A study of methane oxy-combustion characteristics inside a modified design button-cell membrane reactor utilizing a modified oxygen permeation model for reacting flows. J Nat Gas Sci Eng 28:61–73
74. Nemitallah MA, Habib MA, Mezghani K (2015) Experimental and numerical study of oxygen separation and oxy-combustion characteristics inside a button-cell LNO-ITM reactor. Energy 84:600–611
75. Habib MA, Salaudeen SA, Nemitallah MA, Ben-Mansour R, Mokheimer EMA (2016) Numerical investigation of syngas oxy-combustion inside a LSCF-6428 oxygen transport membrane reactor. Energy 96:654–665
76. Williams TC, Shaddix CR, Schefer RW (2007) Effect of syngas composition and $CO_2$-diluted oxygen on performance of a premixed swirl-stabilized combustor. Combust Sci Technol 180:64–88
77. Zhang J, Mi J, Li P, Wang F, Dally BB (2015) Moderate or intense low-oxygen dilution combustion of methane diluted by $CO_2$ and $N_2$. Energy Fuels 29:4576–4585
78. Nemitallah MA, Rashwan SS, Mansir IB, Abdelhafez A, Habib MA (2018) Review of novel combustion techniques for clean power production in gas turbines. Energy & Fuels 32(2):979–1004
79. Amato A, Hudak B, D'Carlo P, Noble D, Scarborough D, Seitzman J, Lieuwen T (2011) J Eng Gas Turbines Power 133:061503
80. Jerzak W, Kuźnia M (2016) Experimental study of impact of swirl number as well as oxygen and carbon dioxide content in natural gas combustion air on flame flashback and blow-off. J Nat Gas Sci Eng 29:46–54
81. Ditaranto M, Hals J (2006) Combustion instabilities in sudden expansion oxy-fuel flames. Combust Flame 146:493–512
82. Shroll AP, Shanbhogue SJ, Ghoniem AF (2012) Dynamic-stability characteristics of premixed methane oxy-combustion. J Eng Gas Turbines Power 134:051504
83. Joshi ND, Epstein MJ, Durlak S, Marakovits S, Sabla PE (1994) Development of a fuel air premixer for aero-derivative dry low emissions combustors. International Gas Turbine Aeroengine Congr Expo 94-GT-253:1–9
84. Huang Y, Yang V (2009) Dynamics and stability of lean-premixed swirl-stabilized combustion. Prog Energy Combust Sci 35:293–364
85. Taamallah S, Shanbhogue SJ, Ghoniem AF (2016) Turbulent flame stabilization modes in premixed swirl combustion: physical mechanism and Karlovitz number-based criterion. Combust Flame 166:19–33
86. Kaskan WE (1957) The dependence of flame temperature on mass burning velocity. Symp Combust 6:134–143
87. Van Maaren A, Thung DS, De Goey LPH (1994) Measurement of flame temperature and adiabatic burning velocity of methane/air mixtures. Combust Sci Technol 96:327–344
88. Galmiche B, Halter F, Foucher F, Dagaut P (2011) Effects of dilution on laminar burning velocity of premixed methane/air flames. Energy Fuels 25:948–954
89. Peng Q, Chen J, Zuo W, Zhao X, Zhang Z (2018) Investigations on effects of wall thickness and porous media on the thermal performance of a non-premixed hydrogen fueled cylindrical micro combustor. Energy Convers Manag 155:276–286

90. Zuo W, Peng Q, Zhao X, Zhang Z (2017) Numerical investigations on a comparison between counterflow and coflow double-channel micro combustors for micro-thermophotovoltaic system. Energy 122:408–419
91. Zuo W, Peng Q, Zhao X, Zhang Z (2017) Numerical investigations on thermal performance of a micro-cylindrical combustor with gradually reduced wall thickness. Appl Therm Eng 113:1011–1020
92. Bollinger M, Williams T (1948) Effect of Reynolds umber in the turbulent-flow range on flame speeds of bunsen-burner flames. Flight Propulsion Research Laboratory Cleveland, Ohio, September 1948
93. Joshi ND, Epstein MJ, Durlak S, Marakovits S (1994) Development of a fuel air pre-mixer for aero-derivative dry low emissions combustors. ASME Pap 94-GT 4:253
94. Brohez S, Delvosalle C, Marlair G (2004) A two-thermocouples probe for radiation corrections of measured temperatures in compartment fires. Fire Saf J 39:399–411
95. Funke HH-W, Beckmann N, Keinz J, Abanteriba S (2017) Numerical and experimental evaluation of a dual-fuel dry-low-nox micromix combustor for industrial gas turbine applications. In: Proceedings of the ASME turbo Expo
96. Haj Ayed A, Kusterer K, Funke HHW, Keinz J, Striegan C, Bohn D (2015) Experimental and numerical investigations of the dry-low-NOx hydrogen micromix combustion chamber of an industrial gas turbine. Propuls Power Res
97. Asai T, Dodo S, Karishuku M, Yagi N, Akiyama Y, Hayashi A (2015) Performance of multiple-injection dry low-NOx combustors on hydrogen-rich syngas fuel in an IGCC pilot plant. J Eng Gas Turbines Power
98. Liu H, Zailani R, Gibbs B (2005) Comparisons of pulverized coal combustion in air and in mixtures of $O_2/CO_2$. Fuel 84:833–840
99. Lupion M, Diego R, Loubeau L, Navarrete B (2011) CIUDEN CCS project: status of the $CO_2$ capture technology development plant in power generation. Energy Procedia 4:5639–5646. https://doi.org/10.1016/j.egypro.2011.02.555
100. Wall TF (2007) Combustion processes for carbon capture. Proc Combust Inst 31:31–47. https://doi.org/10.1016/j.proci.2006.08.123
101. Croiset E, Thambimuthu KV (2001) $NO_x$ and $SO_2$ emissions from $O_2/CO_2$ recycle coal combustion. Fuel 80:2117–2121. https://doi.org/10.1016/S0016-2361(00)00197-6
102. Hong J, Kirchen P, Ghoniem AF (2013) Interactions between oxygen permeation and homogeneous-phase fuel conversion on the sweep side of an ion transport membrane. J Memb Sci 428:309–322. https://doi.org/10.1016/j.memsci.2012.10.055
103. Van Blarigan A, Kozarac D, Seiser R, Cattolica R, Chen J-Y, Dibble R (2013) Experimental study of methane fuel oxycombustion in a spark-ignited engine. J Energy Resour Technol 136:012203. https://doi.org/10.1115/1.4024974
104. Becher V, Bohn J-P, Dias P, Spliethoff H (2011) Validation of spectral gas radiation models under oxyfuel conditions—Part B: natural gas flame experiments. Int J Greenh Gas Control 5:S66–S75. https://doi.org/10.1016/j.ijggc.2011.05.006
105. Blasiak W, Yang WH, Narayanan K, von Schéele J (2007) Flameless oxyfuel combustion for fuel consumption and nitrogen oxides emissions reductions and productivity increase. J Energy Inst 80:3–11. https://doi.org/10.1179/174602207X174379
106. Gülder ÖL, Snelling DR, Sawchuk RA (1996) Influence of hydrogen addition to fuel on temperature field and soot formation in diffusion flames. Symp Combust 26:2351–2358. https://doi.org/10.1016/S0082-0784(96)80064-6
107. İlbaş M, Yılmaz İ (2012) Experimental analysis of the effects of hydrogen addition on methane combustion. Int J Energy Res 36:643–647. https://doi.org/10.1002/er.1822
108. Hermanns RTE, Konnov AA, Bastiaans RJM, De Goey LPH (n.d.) Laminar burning velocities of diluted hydrogen-oxygen-nitrogen mixtures. https://doi.org/10.1021/ef060553g
109. Nakahara M, Kido H (2008) Study on the turbulent burning velocity of hydrogen mixtures including hydrocarbons. AIAA J 46. https://doi.org/10.2514/1.23560
110. Coppens FHV, De Ruyck J, Konnov AA (2007) The effects of composition on burning velocity and nitric oxide formation in laminar premixed flames of $CH_4 + H_2 + O_2 + N_2$. Combust Flame 149:409–417. https://doi.org/10.1016/j.combustflame.2007.02.004

111. Liu F, Guo H, Smallwood GJ (2003) The chemical effect of $CO_2$ replacement of $N_2$ in air on the burning velocity of $CH_4$ and $H_2$ premixed flames. Combust Flame 133:495–497. https://doi.org/10.1016/S0010-2180(03)00019-1
112. Imteyaz B, medhat A Nemitallah, Ahmed A Abdelhafez, Mohamed A Habib (2018) Combustion behavior and stability map of hydrogen-enriched oxy-methane premixed flames in a model gas turbine combustor. Appl Energy 63–74. https://doi.org/10.1016/j.ijhydene.2018.07.087
113. Abdelwahid S, Nemitallah MA, Imteyaz B, Abdelhafez AA, Habib MAM (2018) On the effects of $H_2$-enrichment and inlet velocity on stability limits and shape of $CH_4/H_2$-$O_2/CO_2$ flames in a premixed swirl combustor. Energy & Fuels 32:acs.energyfuels.8b01958. https://doi.org/10.1021/acs.energyfuels.8b01958
114. Park S, Kim T, Sohn J, Lee Y (2011) An integrated power generation system combining solid oxide fuel cell and oxy-fuel combustion for high performance and $CO_2$ capture. Appl Energy 88:1187–1196. https://doi.org/10.1016/j.apenergy.2010.10.037
115. Zhang N, Lior N (2008) Two novel oxy-fuel power cycles integrated with natural gas reforming and $CO_2$ capture. Energy 33:340–351. https://doi.org/10.1016/j.energy.2007.09.006
116. Simpson AP, Simon AJ (2007) Second law comparison of oxy-fuel combustion and post-combustion carbon dioxide separation. Energy Convers Manag 48:3034–3045
117. Kakaras E, Koumanakos A, Doukelis A, Giannakopoulos D, Vorrias I (2007) Oxyfuel boiler design in a lignite-fired power plant. Fuel 86:2144–50. https://doi.org/10.1016/j.fuel.2007.03.037
118. Hunt A, Dimitrakopoulos G, Ghoniem AF (2015) Surface oxygen vacancy and oxygen permeation flux limits of perovskite ion transport membranes. J Memb Sci 489:248–257. https://doi.org/10.1016/j.memsci.2015.03.095
119. Hunt A, Dimitrakopoulos G, Kirchen P, Ghoniem AF (2014) Measuring the oxygen profile and permeation flux across an ion transport membrane and the development and validation of a multistep surface exchange model. J Memb Sci 468:62–72. https://doi.org/10.1016/j.memsci.2014.05.043
120. Sunarso J, Baumann S, Serra JM, Meulenberg WA, Liu S, Lin YS et al (2008) Mixed ionic-electronic conducting (MIEC) ceramic-based membranes for oxygen separation. J Memb Sci 320:13–41. https://doi.org/10.1016/j.memsci.2008.03.074
121. Mancini ND, Mitsos A (2011) Ion transport membrane reactors for oxy-combustion—Part I: intermediate-fidelity modeling. Energy 36:4701–4720. https://doi.org/10.1016/j.energy.2011.05.023
122. Kirchen P, Apo DJ, Hunt A, Ghoniem AF (2013) A novel ion transport membrane reactor for fundamental investigations of oxygen permeation and oxy-combustion under reactive flow conditions. Proc Combust Inst 34:3463–3470. https://doi.org/10.1016/j.proci.2012.07.076
123. Xiong J, Zhao H, Zheng C (2011) Exergy analysis of a 600 MW oxy-combustion pulverized-coal- fired power plant. Energy Fuels 25:3854–3864. https://doi.org/10.1021/ef200702k
124. Xiong J, Zhao H, Zheng C (2012) Thermoeconomic cost analysis of a 600 MWeoxy-combustion pulverized-coal-fired power plant. Int J Greenh Gas Control 9:469–483. https://doi.org/10.1016/j.ijggc.2012.05.012
125. Scaccabarozzi R, Gatti M, Martelli E (2016) Thermodynamic analysis and numerical optimization of the NET power oxy-combustion cycle. Appl Energy 178:505–526. https://doi.org/10.1016/j.apenergy.2016.06.060
126. Ziółkowski P, Zakrzewski W, Kaczmarczyk O, Badur J (2013) Thermodynamic analysis of the double Brayton cycle with the use of oxy combustion and capture of $CO_2$. Arch Thermodyn 34:23–38. https://doi.org/10.2478/aoter-2013-0008
127. Pfaff I, Kather A (2009) Comparative thermodynamic analysis and integration issues of CCS steam power plants based on oxy-combustion with cryogenic or membrane based air separation. Energy Procedia 1:495–502. https://doi.org/10.1016/j.egypro.2009.01.066
128. Castillo R (2011) Thermodynamic analysis of a hard coal oxyfuel power plant with high temperature three-end membrane for air separation. Appl Energy 88:1480–1493. https://doi.org/10.1016/j.apenergy.2010.10.044

129. Skorek-Osikowska A, Bartela Ł, Kotowicz J (2015) A comparative thermodynamic, economic and risk analysis concerning implementation of oxy-combustion power plants integrated with cryogenic and hybrid air separation units. Energy Convers Manag 92:421–430. https://doi.org/10.1016/j.enconman.2014.12.079
130. Habib MA, Imteyaz B, Nemitallah MA (2020) Second law analysis of premixed and non-premixed oxy-fuel combustion cycles utilizing oxygen separation membranes. Appl Energy 259:114213
131. Dillon DJ, Panesar RS, Wall RA, Allam RJ, White V, Gibbins J et al (2005) Oxy-combustion processes for $CO_2$ capture from advanced supercritical PF and NGCC power plant. Greenh. Gas Control Technol 211–20

# Chapter 5
# Operability of Fuel/Oxidizer-Flexible Gas Turbine Combustors

## 5.1 Introduction

In this chapter, the operability of gas turbine combustors under fuel-flexible and oxidizer-flexible combustion conditions is evaluated based on experimental and numerical investigations, and the results are compared with those of non-flexible gas turbine combustors. Stability and combustion characteristics of the different flames in premixed combustors considering different fuels with/without hydrogen enrichment and under air or oxygen combustion conditions are discussed. Several combustion techniques have been proposed, reviewed, and evaluated aiming at decreasing the greenhouse gas emissions, primarily $CO_2$. Among these techniques, oxy-fuel combustion stands out as a promising carbon capture and storage (CCS) technology that can be implemented in new power plants and with little hardware retrofitting in the existing systems. However, fuel combustion in pure oxygen environment results in high combustion temperature, which may not be suitable for safe operation of gas turbine blades. Premixing the reactants in an environment of recirculated $CO_2$ under premixed oxy-combustion condition can control the combustion temperature for safe operation of the turbine blades while controlling the emissions. The main target of gas turbine manufacturers is to design versatile environmental-friendly gas turbine combustors that can handle different fuel and oxidizer compositions. This chapter evaluates the operability of gas turbine combustors of novel burner designs capable of handling different fuels and oxidizer compositions for the sake of better performance at the minimum emissions levels. This chapter investigates the stability and combustion features of $C_3H_8/O_2/CO_2$ flames in a model dry low emissions (DLE) gas turbine burner adopting lean premixed (LPM) oxy-fuel combustion technology. Relationship between the flame speed (FS) and the adiabatic flame temperature ($T_{ad}$) is premeditated for better analyzing and predicting the flame behavior. The stability of the flame is characterized by its blowout and flashback limits over ranges of equivalence ($\phi$) ratio and oxygen fraction (OF: volumetric concentration of $O_2$ in the $O_2/CO_2$



M. A. Nemitallah et al., *Approaches for Clean Combustion in Gas Turbines*,
Fluid Mechanics and Its Applications 122,
https://doi.org/10.1007/978-3-030-44077-0_5

oxidizer mixture). Flame shapes at different sets of conditions are investigated. The effects of fuel type and hydrogen-enrichment are investigated based on experimental and numerical investigations considering $C_3H_8/O_2/CO_2$ and $CH_4/H_2/O_2/CO_2$ flames operated on the same gas turbine model combustor. In addition, results out of three-dimensional (3-D) transient large eddy simulations (LES) are presented considering premixed $C_3H_8/O_2/CO_2$ and $CH_4/H_2/O_2/CO_2$ flames under various combinations of $\phi$ and OF to have more insight on flow, combustion, and emission characteristics. The temperature and species concentration distributions, flow field, and flame structure are also presented and discussed.

## 5.2 Oxidizer Flexibility in Gas Turbines

Global climatic changes are one of the greatest challenges of twenty-first century and $CO_2$ emissions are the top among the greenhouse gases contributing to this shift in climate. The major source of $CO_2$ emissions are the power plants, burning different fossil fuels. It has been forecasted by the International Energy Agency that about 38% of the global electricity generation would come from coal based power plants by 2020 [1]. There are several techniques being analyzed and reviewed for carbon capture and sequestration including pre-combustion capture, oxy-fuel combustion and post-combustion capture. For efficient utilization of fossil fuels and reduction in environmental damaging emissions, attention has recently been shifted towards oxy-fuel combustion. This technology shows promising prospects of carbon capture and sequestration by changing the current air-based power plants to oxygen-based ones with smaller hardware retrofitting.

Air is utilized for the combustion of fossil fuels in most of the existing combustion applications including gas turbines. Exhaust of fuel oxidized with air comprises mainly of $CO_2$, $H_2O$, $O_2$ and $N_2$ and $NO_x$. Among these exhaust gases; $CO_2$ is a greenhouse gas and hence the reason of ever rising global warming problems. Currently the global energy mix is heavily dependent on the fossil fuels for energy harnessing and is expected to remain as such for several years to come. While $CO_2$ is a certain product of fossil fuel combustion, with ever increasing stricter regulations on emissions, it becomes necessary to explore ways for its separation and capture. When $CO_2$ is along with other exhaust products, it is both expensive and complicated process to separate and capture it [2]. Oxy-fuel combustion stands out as one of the promising solutions to this problem. When pure $O_2$ is used as an oxidizer instead of air, the main products of fossil fuel combustion are $CO_2$ and $H_2O$ only. Simple condensation is a cost economic process, which removes $H_2O$ from exhaust, thereby leaving $CO_2$ ready for capture. Removal of $N_2$, which makes about 78% of air by volume, from the products, causes improvement in the efficiency of treatment equipment and it also, reduces the size of the hardware. Zeldovich mechanism is a prominent source of thermal $NO_x$ formation [3], which is drastically reduced in oxy-fuel combustion, as absence of $N_2$ in reactants reduces the probability of interaction of $N_2$ and $O_2$ molecules at elevated temperatures at first, which in turn nearly eliminates the possibility of $NO_x$ formation.

### 5.2.1 *Oxy-combustion Flames*

Air and oxygen-based combustion, on the other hand, have quite different degrees of freedom, in terms of their stable operations. Air is an abundantly available resource, so it is usually used in leaner proportions with the fuel in combustion applications. Lean operations also result in reducing the combustion temperature until safe hardware limits. While, pure oxygen is an expensive product in itself, therefore most of the oxy-fuel combustion applications are usually operated at or near stoichiometric conditions. But, in the absence of $N_2$ gas, which acts as a transporter of energy out of the combustor, this stoichiometric operation of oxy-fuel combustion consequently approaches to above hardware safety temperature limits. Due to this reason, pure $O_2$ is diluted with recycled $CO_2$ for reduction of combustion temperature, and this recirculation of exhaust $CO_2$ is generally known as exhaust gas recirculation (EGR). Oxy-fuel combustion thus, have fuel and $O_2/CO_2$ oxidizer as its reactants. With stricter environmental regulations and scarce nature of fossil fuels, it will be one of the significant motivations for the existing lean premixed (LPM) air-based combustion customers to switch towards oxy-fuel combustion without any substantial hardware changes. Understanding the behavior of stable flames is essential for switching in between oxidizers.

Different hydrocarbon fuels reacting with $O_2/CO_2$ oxidizer mixture have been studied so far including methane [4, 5], ethane [6], propane [7–9], ethanol [10], butanol [11] and heavier hydrocarbons like heptane and octane [12]. For leaner mixtures, all of these oxy-fuel studies agreed consistently that both key characteristics of flame, $T_{ad}$ and FS, are increased with $\phi$ of the reacting mixture. For fundamental combustion characteristics of propane air flames, reader is referred further to the references [13–15]. Kaskan et al. [16] observed that FS of the propane-air mixture reaches its maximum at $\phi = 1.1$. It peaks then drops off for both leaner and richer mixtures. Studies conducted on various fuels by Van Maaren [17] and Kaskan et al. [16] directly presented a correlation between FS and $T_{ad}$ as log (FS) $\alpha$ $1/T_{ad}$. For the case of increasing $T_{ad}$ by preheating the reactants of fuel-air mixture, FS was increased. Both the liquid [18] and gaseous [19] fuels have shown this similar relationship between $T_{ad}$ and FS. On the other hand, dilution of fuel oxidizer mixture with $N_2$, $CO_2$, $H_2O$, argon or helium resulted in lower $T_{ad}$ and therefore lowers FS, as indicated in the referred studies [20, 21]. Dilution represents a decrease of the OF in the oxidizer mixture of $O_2/CO_2$ for oxy-fuel combustion. It is consequently reported that any change in $T_{ad}$ by either preheating or dilution results in a similar changing trend of FS. Thus, it can be concluded that FS is a sole function of $T_{ad}$.

### 5.2.2 *Oxidizer Dilution*

High combustion efficiency, high melting capacity (as it reaches higher flame temperatures) and lower volume of exhaust gases are the prominent features of oxy-fuel combustion [22]. The physical and chemical properties of $N_2$ and $CO_2$ differ substantially and this causes undesirable flame instability [23]. It has been observed that the premixed oxidizer stream i.e. mixture of $O_2$ and $CO_2$, also affects the flame stability [24] and chemical kinetics of the reaction. Since the volumetric proportion of $O_2$ in air is 21% therefore, Heil et al. [25] investigated the behavior of $O_2/CO_2$ oxidizer mixture with this proportion. It was reported that this proportion resulted in poorer burnout and also, darker flames were observed. As oxygen fraction (OF) in the oxidizer is reduced, the amount of $CO_2$ in the oxidizer mixture is increased, resulting in reduced laminar flame speed and combustion efficiency [14–16]. The behavior of oxy-fuel combustion with increased $CO_2$ dilution was experimentally and numerically investigated by Glarborg and Bentzen [26]. In the results, higher concentration of CO was observed near-burner region and it was due to the fact that both $CO_2$ and $O_2$ compete for the atomic hydrogen. $CO_2$ also reacts with hydrocarbon molecules to form CO. The higher concentration of CO may cause corrosion and slagging problems.

## 5.3 Fuel Flexibility in Gas Turbines

In this section, focus is made on hydrogen-enriched combustion as one of the main fuel-flexibility approach for the control of flame stability in gas turbines. Inclusion of $H_2$ in the fuel not only results in extended lean flammability limits but also favorable flame temperatures and reduced $NO_x$ emissions [27]. Since, $H_2$ molecule shows higher power density on burning, so smaller fraction of $H_2$ can replace bigger proportions of other fuels (reduced fuel consumption) in gas turbine and internal combustion engine industry. Due to its far better and advantageous effects over other fuels, $H_2$ is considered as the future energy carrier. Currently it is an expensive gas, so it can be blended with natural gas mixture in small fractions. Several studies reported that the introduction of $H_2$ gas with different hydrocarbon fuels increases the flame speed and extends the lean flammability limits [28–30]. The effect of $H_2$ enrichment on methane fuel in a 50 kW laboratory scale swirl stabilized model combustor was experimentally investigated by Tuncer [31]. He reported that combustion of this mixture results in extended lean flammability limit but as a penalty of its chemical kinetics and higher propagation speed, flashback is reached earlier. Kim et al. [32] also studied the effect of $H_2$ enrichment on $CH_4$-air flames. They reported that the $NO_x$ emissions are reduced as $H_2$ enrichment of the fuel in mixture was increased. In addition, the temperature of the reaction zone was increased and due to higher combustibility of $H_2$ recirculation flow is reduced resulting in quicker reaction. But in upper region of flame the effect of swirl is more than the effect of $H_2$

enrichment of $CH_4$. Similarly, the stability limits of premixed $H_2$ enriched flames under various operating conditions were studied by Schefer [33]. It was reported that the stability map of blowout limit is shifted towards leaner equivalence ratios as HF is increased from 70 to 100%. This result of extension of blowout limit with addition of $H_2$ in small proportions was also consistent with the above discussed results of several studies [28–30]. Investigation of the laminar flame speed of $H_2$ enriched methane fuel was carried out by Yu et al. [34] and it was reported that as HF increases, the laminar flame speed of the mixture increases linearly. Gauducheau et al. [35] explained that $H_2$ addition results in higher resistance of the flame towards its stretching and this leads towards extended flammability limit of the blended mixture. It was also reported that $H_2$ enrichment increases the concentration of flame radicals (O, H and OH) and thereby the rate of reaction. Mansur et al. [36] studied $H_2$ enriched oxy-methane flames and reported that the flame stability increased with increasing hydrogen fraction. Binash et al. [37] found that inclusion of $H_2$ in the mixture of $CO_2$ diluted oxy-methane flames shifts both the static stability limits of blowout and flashback to learner side with increasing hydrogen fraction, as $H_2$ improves the chemical kinetics oxy-methane flames.

## 5.4 Stability of Oxy-flames

Combustion stabilization is one of the bottlenecks for the oxy-combustion technology in making its way to the replacement of air-based combustion technology. There are a number of affecting factors for flame stability including fuel composition, manner of fuel-oxidizer mixture introduction in the combustor, turbulence level and burner design (material, geometry etc.). Combustion stability is divided into two categories i.e. static and dynamic stabilities. The static stability of flames is defined by their blowout and flashback extinction limits, as these flames remain stable in between these two extremes. When the burning velocity of flame is less than the flow velocity of the combustible mixture, flame gets detached and propagates away, from the throat of the burner, resulting in blowout extinction. On the other hand, when the burning velocity of flame is more than the combustible mixture's flow velocity, flame propagates upstream of the burner, resulting in flashback flame extinction. The composition of the reactant mixture affects the flame speed drastically, and thus the blowout and flashback limits. The knowledge of both of these extinction limits is important for safe operation of the combustor. The present chapter investigates the static stability of $CO_2$-diluted oxy-fuel flames. It is therefore, mentioned here that wherever the term "stability of flame" is used in this document, it refers to the static stability of the flame.

High sound pressure waves are generated in combustion systems when the heat of combustion is coupled with the acoustic fluctuations. These pressure waves disturb the stability of flame, even when the flame is away from its static extinction limits. The flame stability in terms of these thermoacoustic waves is known as dynamic stability of the flame. Information of this dynamic stability is also important as it can cause wear to the parts of combustor, which on moving with flue gases can damage the gas turbine components downstream [38–40].

Previous studies of oxy-fuel combustion were majorly inclined towards coal combustion, as it is a carbon rich fuel and burning it alone with oxygen results in increased concentration of waste $CO_2$, which consequences in increased carbon capture efficiency [15, 41]. Majority of these oxy-fuel studies dealt with partially premixed and non-premixed conditions, while the effect of swirler has also been included. Premixed swirl-stabilized oxy-propane flames diluted with $CO_2$ have been rarely discussed, although research on the stability of non-premixed oxy-propane [8] flame is reported. Tamallah et al. [42] compared lean premixed swirl stabilized air and oxygen-based methane flames. Their study was conducted under flame stabilization encouraging conditions in the outer recirculation zone. The adiabatic flame temperature of oxy-methane and air-methane combustion was matched by adjusting the dilution ratio i.e. the oxygen fraction OF, of the oxy-fuel mixture. The flow was characterized by introducing Strouhal number, which is based upon the flow inlet velocity and the azimuthal spinning frequency of the outer recirculation zone. The value of Strouhal number was reported to be constant, irrespective of the inlet Reynolds number (Re), at about 0.12 for both the oxygen and air-based flames. It was therefore concluded that Re is not the governing parameter for maintaining similar flow conditions in the outer recirculation zone; instead the inlet flow velocity of the mixture controls it. This finding persuaded to conduct the experiments of the current study at fixed inlet velocity. In the survey of Tamallah et al. [42], temperature measurement of the flow in its outer recirculation zone was carried out, and then the local extinction rates were calculated and compared. It was observed in both air and oxygen-based flames that a single relationship exists between characteristic flame time (the reciprocal of extinction strain rate) and characteristic flow time (reciprocal of spinning frequency). Based on these outcomes, a similar analysis of premixed oxy-propane flame and flow time scales is presented in this chapter.

Shroll et al. [43] examined the dynamic stability characteristics of oxy-methane flames. They used OF for controlling $T_{ad}$. Although the burning velocities of both air and oxygen-based flames are quite different for a specific $T_{ad}$ value, but still both of these flames were similar in the fact that $T_{ad}$ of the reactant mixture controls the movement from one stability point to the other. Since turbulent flame geometry is a strong function of $T_{ad}$, they reinforced this finding by showing high speed images of the comparable vortex breakdown modes, which was representing similar turbulent flame geometries. They used high speed images for reinforcing this finding. These images represented analogous vortex breakdown modes indicating similar turbulent flame geometries, which are strong function of $T_{ad}$. Several studies have been conducted on propane-air flame stability [44, 45], but very few investigations have yet looked for the stable oxy-propane flames [9]. Abubakar et al. [8] investigated the non-premixed $CO_2$ diluted oxy-propane flames. They reported that as $CO_2$ exceeded 40% in the $O_2/CO_2$ mixture, the flame transitioned from jet like structure to the V-shaped one. Kutne et al. [46] used swirl-stabilized model gas turbine combustor for comparing the partially premixed oxygen and air-based methane flames. They considered range of 0.5–1 for $\phi$ and 20–40% for OF. Raman measurements and

OH*-chemiluminescence imaging techniques were used for quantifying the flame structure. It was concluded that the flame characteristics are strong function of OF, while $\phi$ affects them only slightly. They also reported that the inner recirculation zone of the oxy-flames existed further downstream as compared to the air ones, representing the inferior stability of oxy-flames. $CO_2$ has greater heat capacity and lower thermal diffusivity and chemical dissociation [42–46] as compared to $N_2$. The replacement of $N_2$ with $CO_2$ results in lower flame speeds and combustion temperatures in the oxy-fuel combustion when compared to the air-fuel one. Due to this reason, 21% $O_2$ in $O_2/CO_2$ oxidizer mixture is not achievable and therefore, $O_2$ composition should be more than this minimum threshold in the oxidizer mixture for achieving stable flames. These findings were supported by multiple other studies [4, 5, 24, 47].

## 5.5 Combustor Operability of $H_2$-Enriched Premixed Oxy-methane Flames

The insatiable demand for power is on rise worldwide to cope up with the emerging economies and technologies. In 2011, it has been reported that around 80% of the world's electricity production was based on the fossil fuels [48], indicating it as a major contributor to the greenhouse gases emissions. The Paris Agreement has reiterated the concerns of global warming and the implementation of carbon capture technologies to check the global average temperature rise [49]. In attempts to reduce the greenhouse gases emissions to the environment to curb the threats of global warming, several techniques to capture carbon dioxide resulting from burning hydrocarbon fossil fuels have been evolved. Oxy-combustion is a promising technology which can be easily retrofitted to an existing plant with slight modifications [50]. In contrast to the conventional air combustion, fuel is burnt with oxygen separated from air resulting in combustion products consisting mainly of carbon dioxide and water vapor. Carbon dioxide is, then, easily captured by condensing water vapor out of the flue gas stream. Combustion in pure oxygen environment is undesirable due to the excessive temperature rise in the flame zone, which can damage the combustor and the blades of the gas turbine. Hence, a part of carbon dioxide from the exhaust is recirculated back to the combustion chamber to control the flame temperature. Non-premixed flames have been used in gas turbines for power generation thanks to their strong stability behavior over wide ranges of loading conditions. However, non-premixed flames results in stoichiometric combustion zones within the combustor and, consequently, elevated temperature spots are created within the combustor, which can exceed the limit for gas turbine blades. The case may even get worse when oxy-combustion is adapted in non-premixed flames, as the temperature is expected to be excessively raised. Converting the combustion mode from non-premixed to premixed prevents the creation of stoichiometric combustion zones within the combustor as the reactants are premixed upstream of the combustor resulting in effective control of combustion temperature. Pre-mixed combustion is a promising approach for many industrial applications (such as large-scale gas turbine systems, automotive and aero-engines) primarily because of its benefits such as lower pollutant emissions and more efficient combustion when compared to non-premixed systems and configurations.

Due to higher specific heat of carbon dioxide, as compared to nitrogen, lower volumetric flow rates are required for similar flame characteristics [51–56], resulting in reduced reactor size. Moreover, oxy-combustion technology has an innate advantage of elimination of NOx formation as the combustion takes place in absence of nitrogen. However, the application of oxy-combustion technology for gas turbine is associated with a set of constraints. The main constraint for wider application of oxy-combustion technology is the separation of combustion oxygen from air, which adds additional cost to the process. However, the continuous advances in the air separation and hydrogen production technologies make the future look promising for this integration. Since the technology requires pure oxygen, which is obtained at the expense of extra energy, it is preferred based on economics to keep the operating equivalence ratio near unity i.e. stoichiometric combustion. Also, replacement of $N_2$ by $CO_2$ as a diluent within the combustor affects significantly the behavior and operability ranges of the generated flames because of the differences in chemical kinetics and thermo-physical and radiative properties [54, 56–59]. This may result in tight operability limits of the generated premixed oxy-flames when compared to air flames due to the slower kinetics and generated flame instabilities [57]. Fuel flexibility approach, mainly through hydrogen enrichment, is considered as an effective technique to control combustion instabilities within gas turbine combustors to avoid flashback, auto ignition and combustion dynamics [60–62]. In addition, hydrogen is becoming a paramount topic for storage of energy resources. Hydrogen-enriched premixed oxy-flame stability and its characteristics are vital for proper application of this technology in new or existing power plants, which is the subject of this section.

There have been in the literature a set of studies on determining the flame characteristics under oxy-combustion conditions [4, 5, 63–66]. In addition, a number of studies on hydrogen-enriched flames have been carried out; however, they are mainly focusing on the formation of NOx in the presence of hydrogen under air combustion conditions [67–69]. The characteristics of premixed oxy-combustion flames have also been examined extensively as per the open literature. A brief analysis of the most relevant ones is discussed in the study presented in this section. Abdelhafez et al. [70] conducted stability mapping of premixed oxy-methane combustion at various oxygen fraction and equivalence ratios. Their main finding was the dependence of stability limits on the adiabatic flame temperature. Mazas et al. [71] investigated the characteristics of oxygen-enriched methane flames under atmospheric pressure. The results showed that the laminar flame velocity decreases linearly with the water vapor fraction and is independent of the equivalence ratio for a given enrichment ratio. In the study by Li et al. [72], the adiabatic flame temperature calculation showed that at lower diluent/oxygen ratio the flame temperature difference between various inert diluents ($N_2$/$H_2O$/flue gas/$CO_2$) is small and the difference increases with increasing dilution ratio. Jourdaine et al. [73] investigated the stability limits of both $CH_4$/air and

$CH_4/CO_2/O_2$ flames and compared the results. The results showed that the transformation from both $CH_4$/air flame to $CH_4/CO_2/O_2$ flame in a burner can be realized by matching the adiabatic flame temperature and the swirl number without modifying the geometry. The results showed similar topology of both $CH_4$/air and $CH_4/CO_2/O_2$ flames at fixed injector diameter and fixed flame temperature. However, the operability limits are reduced in case of $CH_4/CO_2/O_2$ flame [74]. Similar findings were reported by Amato et al. [57], and they attributed this reduction in the operability limit of $CH_4/CO_2/O_2$ flames to the reduced kinetics. The blow-off in the $CO_2$-diluted flame was encountered at an adiabatic flame temperature of 300 K higher than that of air flame at a given nozzle exit velocity, proving that the adiabatic flame temperature does not govern flame blow-out [57]. Marsh et al. [75] reported that effect of dilution ratio on the flame location, heat release, and the operational response is far more significant in the $CO_2$ diluent system than in $N_2$ under lean operating conditions. Further, the overall rate of formation of CO is reduced and a higher rate of formation of H-radicals is obtained while increasing oxygen fraction in the oxidizing mixture. Taamallah et al. [76] studied the transition mechanism of the flame to stabilize in the outer recirculation zone (ORZ). The results showed that this transition depends on the extinction strain rate and does not depend on the adiabatic flame temperature and unstretched laminar burning velocity at similar bulk flow conditions in outer recirculation zone (ORZ).

Several studies on hydrogen enrichment in conventional methane combustion have been reported [27, 32, 77–79]. Hydrogenation of the fuel was found to affect combustion stability, soot formation, NO concentration and laminar burning velocities. Effect of hydrogen enrichment on the laminar burning velocity was studied over a wide range of equivalence ratio of methane-air flames by Hu et al. [80]. Three combustion regimes were identified namely methane dominated flames, transition flames, and hydrogen dominated flames, where unstretched laminar flame speed was found to be linearly increasing for methane dominated and hydrogen dominated flame, while exponentially increasing for the transition flames with the increasing hydrogen fraction. Schefer [81] found higher OH mole fraction with increasing hydrogen fraction in the fuel stream and increased lean blow-out limits. A linear correlation of laminar flame speed to the hydrogen molar fraction in the fuel blend was predicted by Sarli and Benedetto [82] in their numerical study on hydrogen-methane air premixed combustion. Effects of hydrogen addition to propane air flames were investigated by Tang et al. [44]. Though the laminar flame speed was found to be increasing with the hydrogen fraction in the fuel blends, the flame stability behavior was similar to propane air flames when the hydrogen fraction was less than 60%, while for higher hydrogen fractions the stability behavior was similar to that of the hydrogen air flames. Aliyu et al. [83] studied experimentally and numerically the effects of hydrogen enrichment on stability of non-premixed $CH_4$–$O_2$ flames in a swirl combustor. Their results showed that oxy-combustion of $H_2$–$CH_4$ are not achievable above equivalence ratio of 0.95 due to stability issues.

The results reported in the literature shows restricted operability window of the $CH_4/CO_2/O_2$ premixed flames as compared to air premixed flames. The difference in the physical and chemical properties of $CO_2$ as compared to $N_2$ makes oxy-combustion flame dynamics much different from the conventional air combustion. As the flame speed is reduced in $CO_2$ environment, hydrogen enrichment of the fuel stream may provide a counter effect to the chemical kinetics, balancing the effect of oxidizer dilution. Addition of hydrogen has been found to decrease the ignition delay time [84], increase the flame stability limits and reduce the soot formation [85–89]. Oxy-fuel combustion has emerged as a promising and feasible technology [55] for carbon capture, and hydrogen has the reputation of alternative clean source of energy. Hence, hydrogen-enriched oxy-combustion has the potential to be technically and economically viable. However, this new combustion configuration needs to be investigated to understand the flame dynamics and stability limits.

As per the open literature, the operability window and flame stabilization modes for industrial-scale gas turbines adapting premixed oxy-combustion with hydrogen-enrichment have rarely been investigated. In the study presented in this section, the effect of hydrogen enrichment level on stability limits as compared to the operation without hydrogen enrichment is investigated in a swirl-stabilized gas turbine model combustor of similar power density to the industrial-scale gas turbines. Detailed stability maps along with changes in flame macrostructure toward stability limits are well reported. The oxygen fraction in the oxidizer stream was kept at 30% by volume, resembling closest to the air combustion environment [54, 90–93]. The boundary conditions have been selected based on the desired power density of the combustor. Based on the provided flow conditions, this model gas turbine combustor can provide power density within the range of operation of industrial gas turbines, as they are operated typically in the range 3.5–20 MW/m$^3$/bar [90]. The stability maps are recorded at different inlet velocities to explore the effect of flow dynamics on operability limits of the hydrogen-enriched flames. The investigated velocities are in the range of 2–5 MW/m$^3$/bar of power density which fits with the current combustor design to generate similar flow conditions of industrial-scale gas turbines. The obtained stability map at fixed inlet velocity, function of HF and equivalence ratio at fixed OF, is compared with the obtained map at the same velocity, but function of (OF) and equivalence ratio for pure methane, to explore the effects of oxygen and hydrogen enrichments on shifting the operability range of the generated flame. The flame shapes near flashback limit, as well as near blow-out limit, are compared to explore the mechanisms of flame extinctions. Flame shapes are compared at fixed adiabatic flame temperature, fixed inlet velocity and fixed flow swirl to isolate their effects and investigate the effect of kinetic rates. The effect of inlet flow velocity on the flame shapes at fixed adiabatic flame temperatures and fixed inlet swirl is also investigated.

### 5.5.1 Operating Conditions

This study investigates fully premixed hydrogen-enriched methane oxy-combustion flames in a swirl-stabilized model gas-turbine combustor. The combustion takes place at atmospheric pressure, where the flame is enclosed by a quartz tube opened to the atmosphere. Schematic diagram of the combustor rig is shown in Fig. 5.1. A mixture of hydrogen and methane with different compositions enters the bottom of the rig

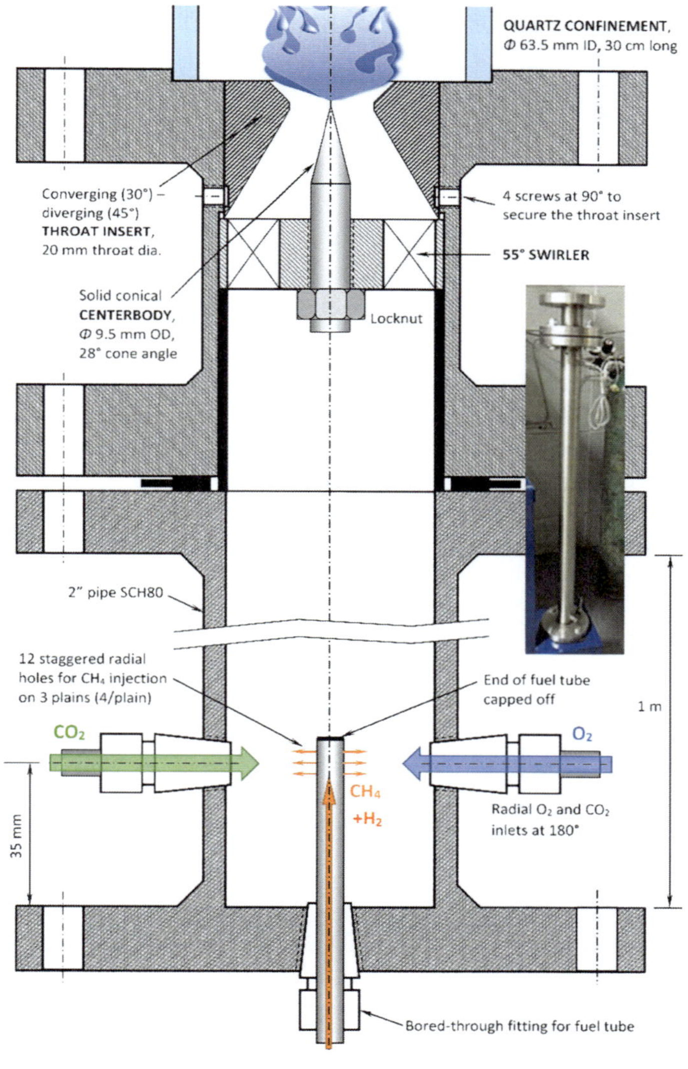


**Fig. 5.1** Schematic of the gas-turbine model combustor rig

through a central fuel tube. The end of the tube is closed with 12 radial staggered perforations in three successive planes ensuring cross flow mixing with the oxidizer stream. Oxygen and carbon dioxide enter the mixing plenum through two radial tubes at 35 mm above the base. The mixing plenum is 1 m in length with L/D ratio of ~20, ensuring well premixing of all the gases [70]. The gases are supplied through compressed gas cylinders and the flow rates are regulated using mass flow controllers by Bronkhorst-High-Tech with an uncertainty of ±0.5% of full scale. The mass flow controllers are operated through a computer interface to deliver the flows such that a particular throat velocity of the mixture is achieved. The fuel delivery tube-line is pre-attached to an emergency shut-off valve to cut the fuel supply in any case flashback explosion.

At the downstream of the mixing chamber and just below the inlet section to the combustor, a 55° swirler is attached to provide the necessary vortex for the swirl stabilization of the flame. The swirl number is calculated using following correlation [42]:

$$Sw = \frac{2}{3}\left[\frac{1-(D_{cb}/D_{in})^3}{1-(D_{cb}/D_{in})^2}\right]\tan(\alpha_{sw}) \sim 0.98 \quad (5.1)$$

The swirler is followed by a convergent-divergent nozzle which accelerates the gases through the throat with the diameter of 2 cm. The 30° convergent part ensures the gradual acceleration of the flow with minimal swirl reduction. The 45° divergent part in the inlet section inhibits the sudden expansion of the gases into the combustion chamber. A solid conical projection is placed at the center of the throat which has been found to eliminate the Combustion induced vortex breakdown (CIVB), limiting the potential path for flame flashback, especially at ~1.0 swirl number [94]. A high-definition camera is used to capture the images of the selected flames for flame shape processing. The camera is operated at night-vision mode with ISO sensitivity set to 1600, shutter speed of 1/60 s and f-stop setting at f/5.6.

The combustion of hydrogen-enriched methane is carried out in a vertical premixed burner at the atmospheric pressure conditions enclosed by a quartz tube opened to the atmosphere. For all generated flames, the oxidizer stream constitutes a mixture of 30% oxygen and 70% carbon dioxide by volume. This composition has been found to have similar flame characteristics as that of conventional air combustion [54, 76, 80–82] and hence, it has been fixed in this study. Dilution of the oxidizer stream with $CO_2$ ensures lower flame temperature, which otherwise is impractically high to be retro-fitted in an existing power plant.

In this study, three sets of experiments were performed, each at fixed inlet bulk velocity of the mixture ($O_2/CO_2/CH_4/H_2$) at the throat section, typically 4.4, 5.2 and 6.0 m/s. The non-preheated flow rates of $CH_4$, $H_2$, $O_2$, and $CO_2$ were adjusted for each flame to maintain particular inlet bulk velocity. The hydrogen fraction (HF) is varied from 0 to 75% against the equivalence ratios ($\phi$) ranging from 0.3 to 1.0 to achieve a 2-D stability mapping. The gas turbine model combustor can provide power density in the range of operation of industrial gas turbines. However, the power output of the model combustor (2–5 MW/m$^3$/bar) is still in the low range of operation of industrial combustors (3.5–20 MW/m$^3$/bar). This power output can be improved through considering a compact combustor design. This can be done either by sizing down the combustor confinement or enlarging the throat area to handle more flow rate at fixed inlet bulk velocity. The stability map for the operability of the gas-turbine combustor, in terms of flashback and blow-out limits, is recorded at each inlet velocity. The obtained stability map at fixed inlet velocity of 5.2 m/s, function of HF and equivalence ratio at fixed OF, is compared with the obtained map at the same velocity but function of oxygen fraction and equivalence ratio for pure methane. The flame shapes near flashback limit, as well as near blow-out limit, are recorded over a range of hydrogen enrichment level and compared to explore the physics behind flame extinctions. Flame shapes are also recorded and compared at fixed adiabatic flame temperature, fixed inlet velocity and fixed flow swirl to isolate their effects and investigate the effect of kinetic rates on flame shape. Also, flame shapes are recorded at a fixed adiabatic flame temperature and fixed inlet swirl while varying the inlet velocity is also investigated.

The non-stoichiometric combustion equation for the calculation of adiabatic flame temperature is defined as:

$$\mathrm{CH_4} + \left(\frac{HF}{1-HF}\right)\mathrm{H_2} + \frac{1}{2\phi}\left(\frac{4-3HF}{1-HF}\right)\left\{\mathrm{O_2} + \frac{7}{3}\mathrm{CO_2}\right\} \rightarrow \left[1 + \frac{1}{6\phi}\left(\frac{28-21HF}{1-HF}\right)\right]\mathrm{CO_2} + \left(\frac{2-HF}{1-HF}\right)\mathrm{H_2O} + \left(\frac{1}{\phi} - 1\right)\mathrm{O_2} \tag{5.2}$$

where, $HF$ is the volumetric hydrogen fraction in the fuel stream. The power density of the reactor is defined as:

$$Power\ density = \frac{\dot{m}_{\mathrm{CH_4}} \times CV_{\mathrm{CH_4}} + \dot{m}_{\mathrm{H_2}} \times CV_{\mathrm{H_2}}}{Volume\ of\ quartz\ confinement \times combustor\ pressure} \tag{5.3}$$

where, $CV$ is the standard calorific value and $\dot{m}$ represents the mass flow rates of the fuel.

The bulk dynamic viscosity of the reactant mixture can be given as:

$$\mu_{mix} = \frac{\sum y_i \mu_i \sqrt{M_i}}{\sum y_i \sqrt{M_i}} \tag{5.4}$$

Mixture bulk density can be determined from:

$$\rho_{mix} = \frac{p \sum \dot{m}_i}{R_u T \sum (\dot{m}_i / M_i)} \tag{5.5}$$

Reynolds number of the reactant mixture is thus calculated based on the bulk throat velocity ($v$) and characteristic diameter of the throat ($D$) as:

$$\mathrm{Re} = vD\rho_{mix}/\mu_{mix} \tag{5.6}$$

Three sets of experiments are conducted, each at fixed inlet velocity to determine the stability maps and variations of flame shape over a wide range of hydrogen enrichment level. As explained in the above section, the flow rates of all the reactant gases ($O_2/CO_2/CH_4/H_2$) were adjusted in a way to get the desired combined velocity for a given HF and equivalence ratio. Hence, a change in any parameter requires the re-adjustment of the flow rates of all the reactant gases simultaneously. The HF in the fuel stream was varied from 0 to 75% with the increment of 10% from 0 to 50% and with 5% increment from 50 to 75%. The stability mapping of the flames was scanned for the equivalence ratio ranging from 0.3 to 1.0 in a step of 0.05. The procedures adopted for the experimental measurements are as follows:

(1) Determining a safe ignition point of the burner which is far from the blow-out region and flashback point.
(2) Adjusting the flow rates of the gases gradually to avoid any sudden jump in the flow meters. The flow rates were adjusted for a particular starting set point at a certain HF which is presumed to be stable.
(3) After reaching the stable point for a particular HF, the equivalence ratio was increased gradually in steps until the flashback was observed.
(4) As the equivalence ratio increases, the flame gets shorter and more violent in nature with the increased noise level. At the flashback point, the noise reaches to the maximum and the flame reaches down to the throat causing a mild explosion. The fuel supply was immediately cut-off at this condition using the emergency valve to stop flame propagation upstream into the plenum. The value of $\phi$ at which flashback was observed is noted against the selected HF.
(5) The burner was reignited at the stable point achieved earlier and $\phi$ was reduced gradually for the same hydrogen fraction until the blow-out was observed. At the blow-out limit, the flame gets very lean and extinguishes. The value of $\phi$ is recorded against the particular HF where blow-out limit is observed.
(6) The procedures from step (2) up to step (5) are repeated for all the hydrogen fractions until the two-dimensional stability map for a particular throat velocity was achieved.
(7) The whole procedure was repeated to quantify the stability mapping for the other two throat velocities to achieve the mappings for the three sets at 4.4, 5.2 and 6.0 m/s.

## 5.5.2 *Combustor Stability Maps*

The stability mapping is quantified by identifying the flashback and blow-out limits of the flames at three sets of combined velocities of the reactant gases at the throat. Stability map at inlet bulk velocity of 5.2 m/s is first presented on the contours of adiabatic flame temperature (AFT), power density (PD), inlet Reynolds number (Re),

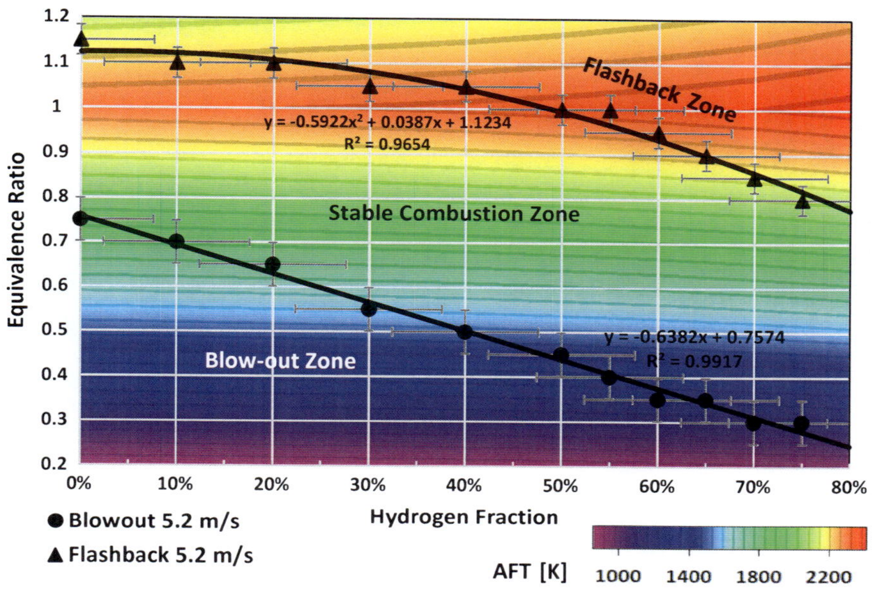


**Fig. 5.2** Flame stability mapping against the contours of adiabatic flame temperature at the throat velocity of 5.2 m/s

and mixture mass flow rate $\dot{m}_{mix}$. The obtained stability map at fixed inlet velocity of 5.2 m/s, function of HF and equivalence ratio at fixed OF, is compared with the obtained map at the same velocity, but function of oxygen fraction and equivalence ratio for pure methane. This is followed by the analysis of the effect of inlet flow velocity on flashback and stability limits of the generated flame.

Figure 5.2 presents the blow-out and flashback limits with varying the hydrogen fraction and equivalence ratios at oxygen fraction (OF) of 30% and throat velocity of 5.2 m/s against the contours of adiabatic flame temperature. It has been observed that the addition of hydrogen shifted both the blow-out and flashback limits towards the leaner side. This wider operability window is mainly due to the significant shift of the blow-out limit toward leaner conditions, as the flashback limit is tightening while increasing the hydrogen fraction due to the associated increase of flame speed. The blow-out limit increases from the equivalence ratio of 0.75 with pure methane combustion to 0.3 with 75% HF. The blow-out limit follows almost linear trend with increasing HF which can be attributed to increased chemical kinetics with the addition of hydrogen resulting in more stable flames in the leaner region. However, due to vigorous reaction with the percentage increase of hydrogen towards the richer region, the flashback occurs sooner than the methane-rich flames. The flashback limit reduces from the equivalence ratio of 1.0 at 50% HF to 0.8 at 75% HF. It is obvious that the adiabatic flame temperature has little effect of hydrogen addition and hence, the flame limits show no apparent correlation with it. Similarly, the stability mapping does not exhibit a relation with the power density (MW/m$^3$/bar) of the combustor

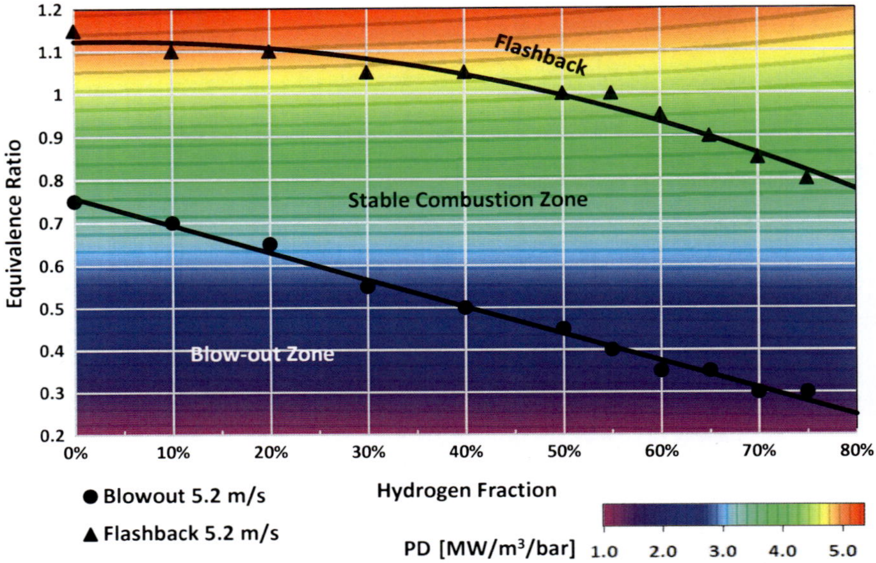


**Fig. 5.3** Flame stability mapping against the contours of power density at the throat velocity of 5.2 m/s

as depicted from Fig. 5.3. The flashback and blow-out limits are mainly functions of Damkohler number [57, 95]. Provided that the stability mapping was carried out at the constant throat velocity of 5.2 m/s, the stability limits are largely governed by the reaction time scale or the flame speed. It is known that the hydrogen fraction has a tremendous effect on the reaction kinetics although it hardy affects the flame temperature or power density. In fact, flashback is a very complicated phenomenon and is highly affected by flow regime and the flow-flame interaction. For low-swirl flow, the resistance to boundary layer flashback can be improved under low operating equivalence ratio. However, at higher equivalence ratio, flashback is mainly initiated by the combustion induced vortex breakdown (CIVB) mechanism. Bulk flow flame propagation mode can happen under low flow speeds, which we assume that it is not the case in the study presented in this section.

The hydrogen fraction has an inverse relation with both Reynolds number and reactant mixture mass flow rates. It is expected that a decrease in Reynolds number would bring both the flashback and blow-out limits to the richer side by increasing the turbulence level in the flow. However, it can be seen from Fig. 5.4 that the stability limits are following the Reynolds number and mass flow rates, although the relation is not direct. Based on previous work by the authors of this book on the same combustor setup [70] considering pure hydrogen operation, the results showed that similar flame macrostructures are obtained while fixing both inlet velocity and AFT. In addition, both flashback and blow-out limits followed contours of fixed AFT but under pure hydrogen operation. Interestingly, in the study presented in this section

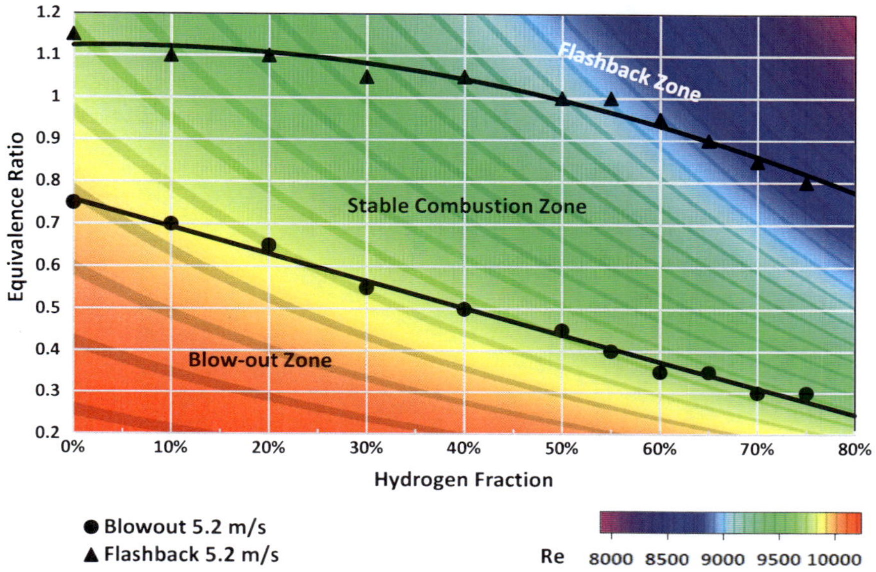


**Fig. 5.4** Flame stability mapping against the contours of Reynolds number at the throat velocity of 5.2 m/s

considering hydrogen-enriched combustion, the stability limits are observed to follow the Reynolds number based on reactant gases velocity at the throat and the throat diameter. This implies that hydrogen addition plays a leading role in controlling both flame shape and flashback mechanism due to associated changes in the reactants concentrations and flow inlet Re in order to keep the same operating inlet bulk flow velocity. The phenomenon can be understood by the fact that the addition of hydrogen does not only affect the flow pattern but also the chemical kinetics inside the combustor. In this study, two conflicting parameters are considered, one is the hydrogen enrichment which enhances the chemical kinetic rates and the second is the oxy-combustion mode in the presence of $CO_2$ which slow the chemical kinetic rates. Figure 5.5 explores the effects of HF and OF on shifting the stability map limits through comparing the stability limits of methane oxy-combustion (HF = 0.0) with those of hydrogen-enriched methane oxy-combustion (OF = 30%) at the same inlet velocity of 5.2 m/s. It is apparent from the Figure that both hydrogen enrichment at fixed OF and oxygen enrichment at fixed HF have a similar effect on shifting the blow-out limit to the very lean operation zone. This may be attributed to the increased reaction kinetic rates for both cases which enhances the burner capability of holding the flame at very lean burning conditions. A similar conclusion on the effect of $CO_2$-diluted mixture on limiting flame stability is reported by Amato et al. [57]. The flashback limits are shifted from the near stoichiometric combustion zone to the very lean combustion zone while increasing the OF from 30 to 70% for pure methane oxy-combustion. However, increasing HF from 30 to 70% at fixed oxygen fraction

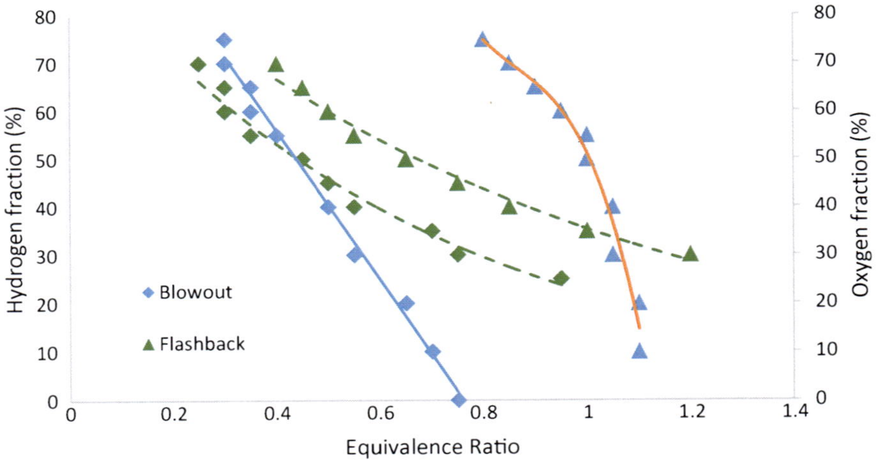


**Fig. 5.5** Comparison of flame stability limits of methane oxy-combustion (HF = 0.0), presented by dotted lines, with those of hydrogen-enriched methane oxy-combustion (OF = 30%), presented by solid lines, at the same inlet velocity of 5.2 m/s

of 30% shifts the flashback limit toward the lean condition but remains near the stoichiometric conditions. i.e. increasing HF results in slight changes in the flashback limits as compared to the significant changes in the flashback limits when increasing the OF, as shown from the slope of the flashback lines for both cases in Fig. 5.5. On the other hand, for the same oxygen fraction of 30%, hydrogen enrichment of 30% can significantly widen the blow-out limit in terms of equivalence ratio from 0.8 down to 0.55, and the flashback limit is slightly shifted from equivalence ratio of 1.2 down to 1.1. Increasing HF above 50% does not result in significant changes on static stability limits, blow-out and flashback, at fixed inlet velocity. This may be attributed to the reduction in the total inlet fuel mass flow rate when substituting methane with hydrogen at fixed inlet flow velocity. Hydrogen enrichment is effective in the zone from HF = 20% up to HF = 50%, and oxygen enrichment is also effective in a similar zone from OF = 20% up to 50%, with wider stability boundaries for hydrogen enrichment as compared to oxygen enrichment.

### *5.5.3 Effect of Inlet Velocity on Flame Stability*

In this section, the effects of changing inlet bulk velocity (4.4, 5.2 and 6 m/s) on shifting the stability limits, flashback and blow-out, as function of HF and equivalence ratio are explored. In fact, the performance of swirl combustors is mainly affected by combustor geometry which governs the flow pattern and flow residence time within the combustor. This should make some differences in flow and combustion characteristics from one combustor to another. Increasing velocity has a two-way effect on the flashback and lean blowout limits. The increased turbulence tends to increase the flame speed, resulting in the larger numerator of the Damkohler Number, while increased flow rate decreases the flow time scale, which tends to reduce the Damkohler Number. The asymptotic region found for the blow-out limits suggests the complex correlation of the flame velocity with the stability limits.

Figures 5.6 and 5.7 and presents the flame flashback and blow-out limits at various throat velocities. At a particular fuel composition, increase in throat velocity tightens

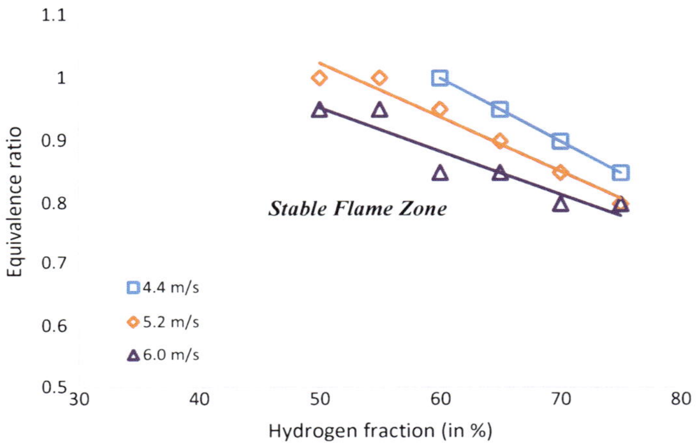


**Fig. 5.6** Flashback limits at the throat velocities of 4.4, 5.2 and 6.0 m/s

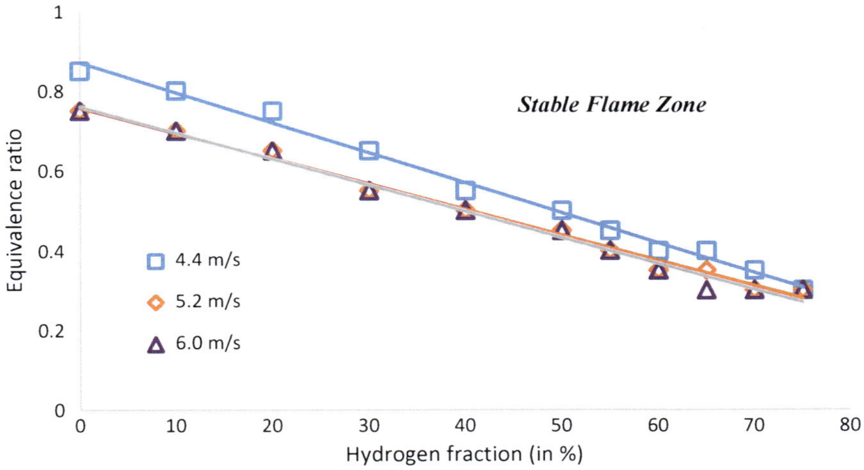


**Fig. 5.7** Blow-out limits at the throat velocities of 4.4, 5.2 and 6.0 m/s

the flashback limit and widen the blow-out limit, as per Figs. 5.6 and 5.7. The case may be different from combustor design to another depending on the generated flow characteristics within the combustor. Actually, flame propagation does not only depend on the fuel type but also on the flow regime. It has been established that the ratio of turbulent to laminar flame speed is a function of flame fronts in the two regimes as [96]:

$$S_T/S_L = A_T/A_L \tag{5.7}$$

where, $S_T$ and $S_L$ are the turbulent and laminar flame speeds, whereas $A_T$ and $A_L$ are the turbulent and laminar flame front area, respectively. The turbulent flame front structure gets distorted but intact and hence, the increase in flame front area is a function of turbulent intensity. It has been observed that the turbulent flame speed varies linearly with the turbulent fluctuation magnitude as [97, 98]:

$$S_T = S_L + K \cdot u' \tag{5.8}$$

where, $K$ is an empirical constant whereas $u'$ is the root mean square turbulent fluctuation magnitude. The inference can be taken from the above discussion to explain the observations on the effect of inlet velocity on shifting the stability limits. Increasing throat velocity affects the swirl number causing a highly turbulent flow, which evidently increases the flame speed. An increased flame speed would result in early flashback and elongated blow-out limits. For 60% HF, the flashback occurs as early as at an equivalence ratio of 0.85 for inlet velocity of 6.0 m/s compared to equivalence ratio of 1.0 for inlet velocity of 4.4 m/s. However, the blow-out limits for the throat velocities of 5.2 and 6.0 m/s seems to coincide, whereas, at 4.4 m/s, blow-out occurs at richer flames. It is suggested that this anomaly comes from the fact that the blow-out is observed at much higher Reynolds number as compared to flashback, indicating that the reaction rate dependence on turbulence is reaching the asymptotic value at the throat velocity of 5.2 m/s at lower equivalence ratios. Thus, the effect of turbulence is no longer a vital player in affecting the flame speed.

Temperature profiles are recorded in this set of experiments to give an indication about the progress and intensity of reactions within the combustor, and to serve as basis in the literature for validation and improvement of numerical models. The effect of hydrogen enrichment level on the distributions of temperature within the combustor is investigated. The temperatures are recorded as close as possible to the flame core due to the limitations of the thermocouple range of operation and the interference of the thermocouple probe with the generated turbulent reacting flow field. Figure 5.8 presents the axial temperature profiles at stoichiometric combustion conditions and inlet velocity of 5.2 m/s for a range of hydrogen enrichment level, from HF = 0% up to HF = 30%. Similarly, and under the same operating conditions, radial temperature profiles at a height of 5.87 cm form the base of the quartz tube, combustion chamber inlet, are presented in Fig. 5.9. It is obvious and expected that hydrogen enrichment should enhance the rates of reactions and improve the combustion temperature, and the temperature should reduce toward the combustor exit in the

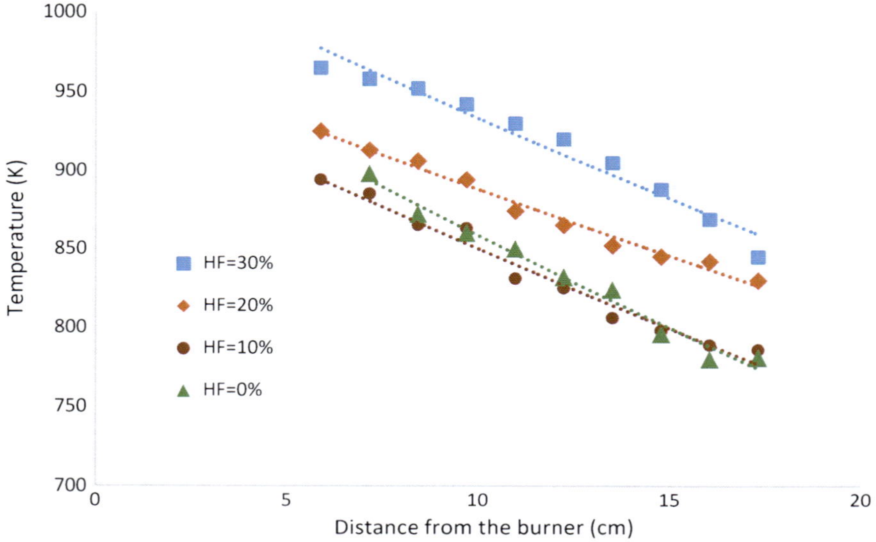


**Fig. 5.8** Axial temperature profiles at stoichiometric combustion conditions and inlet velocity of 5.2 m/s for a range of hydrogen enrichment level

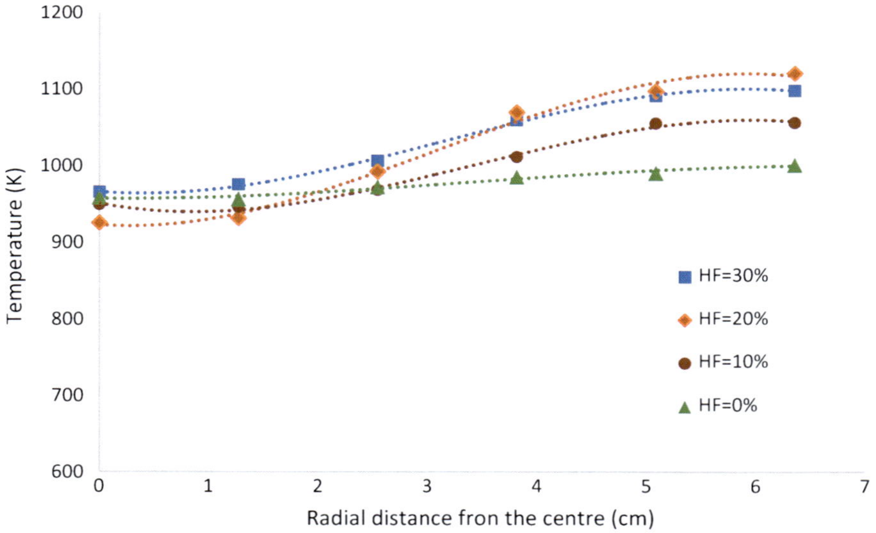


**Fig. 5.9** Radial temperature profiles at stoichiometric combustion conditions and inlet velocity of 5.2 m/s for a range of hydrogen enrichment level, and at 5.87 cm from the inlet of the combustion chamber

axial direction. However, the results do not show any significant effect on combustion temperature at hydrogen enrichment level of 10%. This may be attributed to the low inlet hydrogen mass flow rate at this level of enrichment to balance $CH_4$ in order to keep the same inlet velocity of 5.2 m/s. In a previous study by the authors, similar conclusion was reported for hydrogen enrichment level below 10% in our previous work on the non-premixed combustion mode [98]. As per Fig. 5.9, temperature levels at the center of the combustor is higher within the axial main combusting stream and flame core and decreases in the radial direction where flow recirculation and heat loss through the combustor wall are intense. At lower hydrogen fractions, the slope is almost linear while as the hydrogen fraction increases, the temperature gradient increases and the flame takes the shape of hill and valley. This can be explained by increased chemical kinetics with increasing hydrogen fraction resulting in more confined flames. The reduction in temperature at HF = 30% as compared to HF = 20% for radial distance above 3.5 cm can be attributed to the reduction of flame size at HF = 30%, at which the outer radius of the flame core is less than 3.5 cm as per Fig. 5.11.

### 5.5.4 Flame Shape Analysis

For a deeper understanding of the flame characteristics, visualization of flame shapes is presented in this section. Effect of hydrogen addition at the equivalence ratio of 1.0 for all the investigated throat velocities is displayed in Fig. 5.10. The effect of hydrogen fraction on AFT is marginal and a rise of around 200 K is observed from 0 to 75% HF as can be seen from Fig. 5.2. The effect of AFT is evident through increasing flame brightness as we go from the left to the right. Moreover, hydrogen addition has significant effect on the flame shape. With pure methane, the flames can be seen to spread over the reactor to take the shape of the confinement. However, increasing hydrogen percentage is characterized by more violent and compact flames, implicating higher chemical kinetics and faster flame speeds. With increasing hydrogen fractions, the audible noise of the combustion dynamics increases until the flashback is observed. For the throat velocity of 4.4 m/s the flashback occurs after 50% HF, while for 5.2 and 6.0 m/s the flashback was observed after 40% HF. This is due to increased turbulent flame speed, which is also characterized by higher audible noise, as explained in the earlier section. Other than that, no significant effect of throat velocity was observed on the flame shapes.

Figure 5.11 shows the effect of equivalence ratio ($\phi$) on the flame structure at 50% HF. Three prominent flame shapes can be observed including cup shape at lower value of $\phi$, v-shaped at intermediate $\phi$ and vase shape at higher $\phi$. Lower equivalence ratios are characterized by lower AFT and lower flame speed. The flames have lower brightness and indistinct shape near blow-out region. With increasing $\phi$, the flames go through a transition phase having v-shaped flames and then vase shaped high-noise flames with shorter flame height. The throat velocity does not seem to have much impact on the flame structure at higher values of $\phi$; however, for $\phi < 0.8$, the

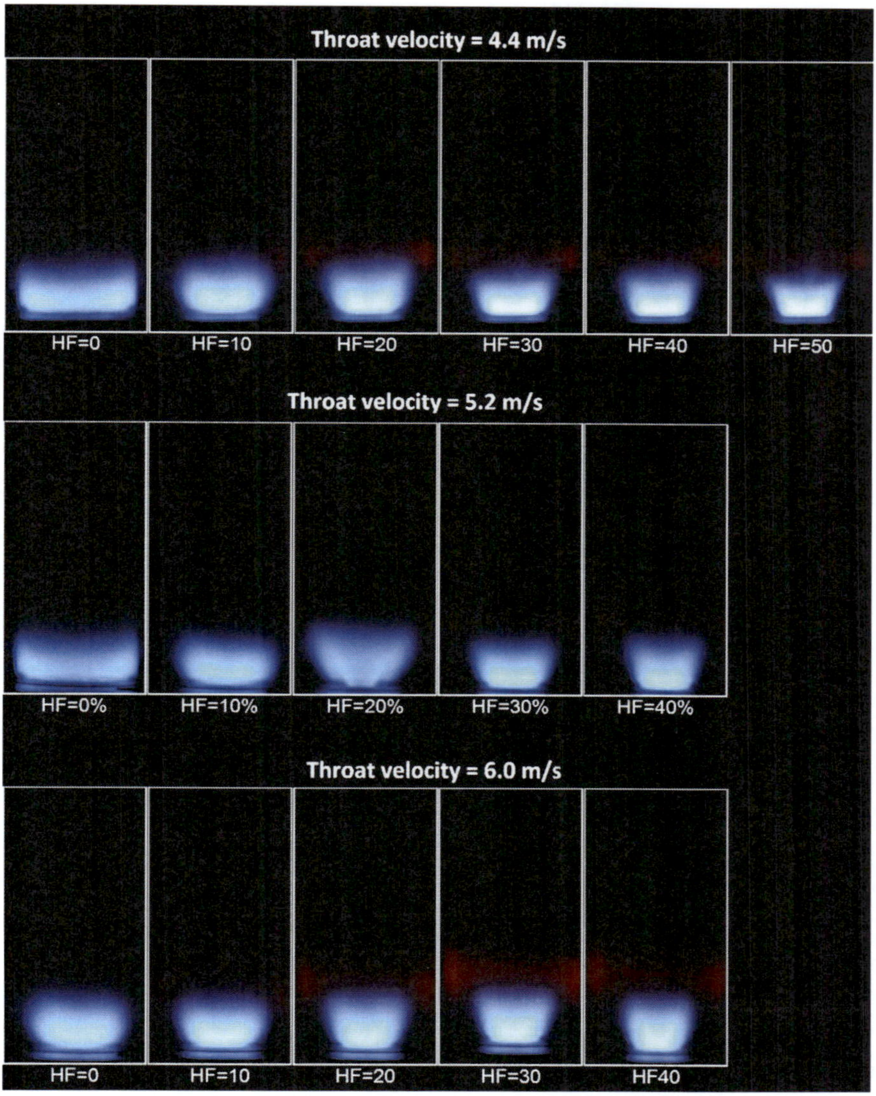


**Fig. 5.10** Flame images at an equivalence ratio of 1.0 over a range of hydrogen enrichment for different throat velocities

effect is more prominent. The transition from cup shape to vase shape occurs between $\phi = 0.75-0.80$ at the throat velocity of 4.4 m/s; however, at higher velocities, the transition is observed between $\phi = 0.65-0.70$ and $\phi = 0.60-0.65$ for the throat velocities of 5.2 and 6.0 m/s, respectively. Moreover, at lower equivalence ratios, the flame brightness is also found to be increasing with increasing throat velocity,

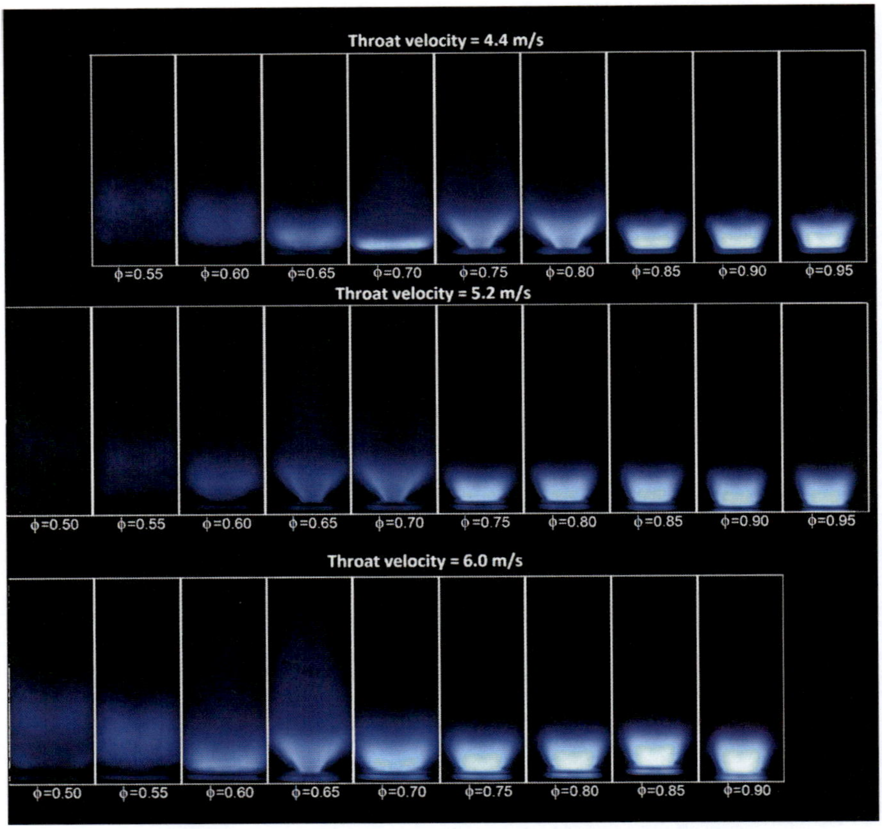


**Fig. 5.11** Flame images at the hydrogen fraction of 50% over a range of equivalence ratio for different throat velocities

albeit characterized by same AFT which is a function of $\phi$ and HF only and does not depend on the flow rates. The observations suggest that the effect of throat velocity is more prominent at a lower value of $\phi$, implicating that the turbulent flame speed comes into a major role at lower equivalence ratios.

It is observed from the presented flame shapes in Fig. 5.11 that there is an equivalence ratio at which a transition of flame stabilization from the inner shear layer (ISL) to the outer recirculation zone (ORZ) occurs. The value of this equivalence ratio is different from operating velocity to another. This transition was recorded based on the analysis of the visual flame appearance when the flame extends on the quartz tube walls and stabilizes above the hot recirculating gases beside the wall of the quartz tube. As per Fig. 5.11, this transition occurred at equivalence ratios of 0.7, 0.6 and 0.6 for the inlet velocities of 4.4, 5.2 and 6.0 m/s, respectively. Similar transition in the flame shape was also observed by Taamallah et al. [42] for both air and oxy-combustion modes. They reported that the flame transition is governed by the extinction strain rate and may occur at different laminar flame speed and adiabatic flame temperature. They defined a non-dimensional number characterizing the flow in the ORZ that is the Strouhal number. This number is independent of the flow Reynolds number and is mainly function of the inlet flow velocity. They showed that the inlet flow velocity is a more relevant parameter than the inlet flow Reynolds number in order to maintain similar flow conditions in the ORZ. These findings are confirmed in the study presented in this section showing the significant effect of inlet bulk velocity on flame stabilization behavior. However, it is observed here that this effect is significant when the inlet velocity is raised from 4.4 to 5.2 m/s and insignificant when the inlet bulk velocity is raised from 5.2 m/s up to 6 m/s. The value of the transition equivalence ratio is reduced while increasing the inlet bulk velocity. This may be attributed to the increased level of the azimuthal ORZ spinning frequency at higher velocities.

To explore whether or not the adiabatic flame temperature has an effect on the transition of flame shape and, accordingly, on the operability boundaries of the flame, flame shapes are captured at the fixed AFT of 2000 K and variable hydrogen fraction and compared as presented in Fig. 5.12. The results showed that the flame shapes do not depend on the AFT This implies that the adiabatic flame temperature does not govern the flame stability. Same conclusion was reported by Amato et al. [57]. Addition of hydrogen shortens the flame height due to increased chemical kinetics. Interestingly, based on previous work by the authors of this book on the same combustor set-up [70] considering pure hydrogen operation, the results showed that similar flame macrostructures are obtained while fixing both inlet velocity and AFT. In addition, both flashback and blow-out limits followed contours of fixed AFT but under pure hydrogen operation. This implies that hydrogen addition significantly affects both flame shape and flashback mechanism. Flame shapes near the blow-out and flashback limits are presented in Fig. 5.13. Both the limits are observed at lower equivalence ratios with increasing hydrogen fraction. Hence, the flame brightness

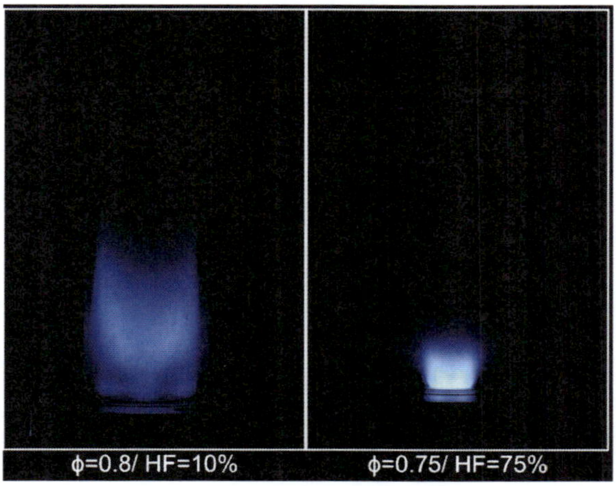


**Fig. 5.12** Flame shapes captured at the fixed AFT of 2000 K for inlet bulk velocity of 5.2 m/s

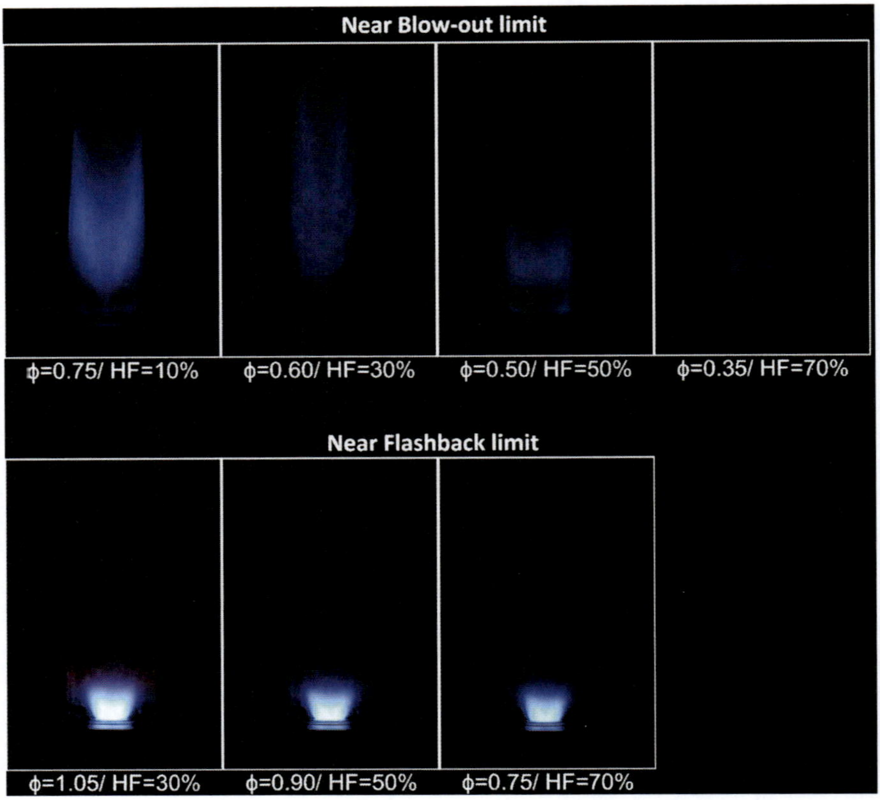


**Fig. 5.13** Flame shapes captured near blow-out and flashback limits for inlet bulk velocity of 5.2 m/s

decreases as we go towards higher hydrogen fraction due to both lower equivalence ratios and flame radiation characteristics at higher hydrogen fractions. Near the blow-out region, with increasing hydrogen fraction, flames appear to take the shape of the quartz confinement with the reduction in the flame height. However, flame shapes near the flashback region are observed to be invariant with the increasing hydrogen fraction, except that the brightness decreases.

## 5.5.5 *Modeling $H_2$-Enriched Oxy-methane Flames*

Using hydrogen blends and syngas in combined cycle power generation cycles is attracting interests of the research community as an alternative to natural gas. Hydrogen-enriched fuels are regarded as the prominent alternative fuels for the future energy mix [99–101]. Recycling of industrial hydrogen by-product and pre-combustion carbon removal from the fuel as a carbon capturing technique can be integrated with gas turbine combined cycle plants. Significant reduction in CO, $NO_x$ and soot formation with hydrogen addition to the hydrocarbons has been reported [102, 103]. Addition of hydrogen suppresses the soot inception and lowers the surface growth rate in the formation of soot [88, 104]. It has also been observed that hydrogen-enrichment enhances the local temperature levels [85, 86], can increase the laminar flame speed [28] and widen the lean flammability limits of the enriched flames [105, 106]. Moreover, the addition of even a small amount of hydrogen in the fuel mixture can have an impact on both the instantaneous and average characteristics of a turbulent premixed flame [77], necessitating a rigorous research on the combustion dynamics of such flames with varied fuel compositions.

Gersen et al. [84] investigated the ignition delay time of hydrogen/methane mixture at various compositions in rapid compression machine. They found that the ignition delay time decreases with the addition of hydrogen in the mixture, and can be correlated quantitatively to the mole fraction of hydrogen in the fuel in accordance with the mixing law proposed by Cheng and Oppenheim [107]. The laminar burning velocity tends to increase linearly with the addition of hydrogen in the fuel blends [80, 82]. The unstretched flame propagation speed and the unstretched laminar burning velocity were reported to increase with the addition of hydrogen in the fuel blend in a study conducted by Tang et al. [44]. They also found that this dependence gets even more remarkable at higher hydrogen fractions. Sankaran and Im [108] concluded that the high laminar burning velocity of $H_2$ is essentially because of higher molecular diffusivity of $H_2$ and the enhancement of formation of H, O, and OH radicals in hydrogen rich flames. Further, in lean mixtures, addition of hydrogen to a hydrocarbon has been found to increase the turbulent burning velocity monotonically [109]. In rich mixtures, hydrogen addition had no apparent effect on the burning velocity but was reported to be dependent on the type of hydrocarbon. Similar dependence of turbulent burning velocity on hydrogen fraction was reported by Halter et al. [77] in their investigation on hydrogen addition in premixed methane/air flames. They further concluded that the combustion intensity $S_T/S_L$ increases with hydrogen addition, implying that the turbulent burning velocity ($S_r$) increases faster than the laminar burning velocity ($S_L$) with hydrogen addition. This was attributed to the fact that with the hydrogen addition, flame front curvature distribution tends to enlarge slightly implying higher probability of larger curvatures. This enhances the smaller scale flame wrinkling resulting in higher turbulent burning velocity. Similar findings were reported by Daniele et al. [110, 111].

Addition of hydrogen to the fuel mixture enhances the reaction rate, which is due to the increased H, O, and OH radical mole fractions in the flame as reported by Hu et al. [80]. They concluded that the higher burning velocities are characterized by high radical formation in the reaction zone. Similar observations were reported in the study by Wang et al. [27] on the effect of hydrogen addition on methane/air combustion using GRI-Mech 3.0. Moreover, they suggested that the addition of hydrogen reduces the formation of $CH_2O$ and $CH_3CHO$, consequently, reducing the aldehyde emissions in methane combustion. An increase of around 20% of OH mole fraction with the blending of 20% hydrogen with methane was reported by Schefer [81]. Due to high combustibility of hydrogen, flames with higher hydrogen fractions are characterized by higher reaction zone temperature, although, the adiabatic flame temperature of the fuel blend decreases with the hydrogen addition. Due to the higher temperature, the reaction zone expands faster, consequently, inhibiting the cooler recirculation flow to the reaction zone [32].

Syred et al. [112] reported that various blends of hydrogen in methane can be configured to minimize the flashback behavior. Gas turbine operation with hydrogen enrichment has been shown to increase the lean stability and reduction in NOx emission [113]. Schefer [81] conducted an experimental study on flame stability with hydrogen addition. He found that the lean blowout limit (LBO) is maximum at stoichiometric condition and reduces for both the lean and rich conditions. Moreover, he reported that by the addition of up to 20% hydrogen in the fuel mixture, the lean blowout limit could be reduced by around 15%. Cheng and Littlejohn [97] conducted a laboratory scale investigation on low swirl injector (LSI) to study the combustion characteristics of pure hydrogen and $H_2$–$N_2$ dilution with air with the aim to implement it in the gas turbines of the integrated gasification combined cycle (IGCC). They found an optimum swirl number 0.51 for $H_2$ in the investigated various LSI configurations as compared to 0.54 for the hydrocarbon LSI. The LBO limit was found to occur at almost a constant equivalence ratio of 0.17 at varying bulk velocities, indicating that the LBO of LSI hardly depends on the Reynolds number. Another parameter which has been found to be fairly successful in predicting the LBO limit is Damkohler number. Zhang et al. [114] reported an interesting finding that the LBO was observed to occur at a Damkohler number of 0.4 with various compositions of $H_2/CO/CH_4$ fuel blends.

The addition of hydrogen to the fuel is advantageous in terms of flame stability and emissions; however, the chemical kinetics, reactor's local temperature, and flame dynamics change significantly compared to hydrocarbon combustion. The effect of hydrogen enrichment in $O_2/CO_2$ flames is even more interesting due to the distinct nature of oxy-fuel combustion. An in-depth literature survey indicates that the operability window and flame behavior of premixed oxy-flames with hydrogen enrichment has rarely been investigated. The static stability limits of hydrogen-enriched air flames, in terms of flashback and LBO limits, are reported in the literature; however, few number of studies have been performed under oxy-combustion conditions. The study presented in this section investigates experimentally and numerically the operability limits and flame characteristics of hydrogen-enriched methane combustion in $O_2/CO_2$ oxidizer in a model gas turbine combustor. The combustor can provide power density in the operation range of industrial gas turbines, typically 3.5–20 $MW/m^3$/bar. The combustor adopts the LPM-DLE combustion technology for controlled emissions gas turbine applications. The study contributes to the understanding of static stability and flame behavior over wide ranges of hydrogen blends in the fuel stream and equivalence ratio, for the purpose of the development of versatile environmental-friendly gas turbines. A detailed numerical study was performed considering LES under fixed *OF* at stoichiometric condition and over a range of *HF* to investigate the effect of hydrogen-enrichment on stability and characteristics of $CH_4$–$H_2$–$O_2$–$CO_2$ flames.

The experiments were carried out considering the same combustor and test procedure described in the previous section at atmospheric pressure with hydrogen-enriched methane as the fuel and $O_2 + CO_2$ as the oxidizer. The oxygen fraction ($OF$) of the oxidizer was fixed to 30% vol. for all cases under the current investigation. Previous studies showed that this oxidizer composition has similar flame characteristics to those of conventional air combustion [54, 81, 82, 93] and, hence, is important to be studied for retrofitting existing burners to adapt with the oxy-fuel combustion technology. The examined fuel flow rates correspond to power densities that fall within the range of operation of industrial gas turbines, typically 3.5–20 MW/m$^3$/bar. Experiments were performed at different hydrogen fractions ($HF$) of the fuel blend in the range 0–75% and different equivalence ratios ($\varphi$) in the range 0.3–1.0. The flow rates of the individual gases were adjusted for each flame to attain a common bulk inlet throat velocity of 5.2 m/s throughout this study. This procedure was reported in detail for premixed $CH_4/O_2/CO_2$ flames [70] and has been adopted here for $CH_4/H_2$–$O_2/CO_2$ flames as well.

#### 5.5.5.1 LES Model Setup

In this mathematical model, the conservation equations for mass, momentum, energy, and transport equation for scalar variables were solved numerically in the full three-dimensional (3-D) domain of the combustor using large eddy simulation (LES) model. ANSYS-Fluent 19.2 commercial software was used to perform all simulations. The three-dimensional and elliptic equations were solved simultaneously to predict the flow regime, heat transfer, and emission characteristics of the hydrogen-enriched oxy-methane flames in the model gas turbine reactor. The governing equations for the conservation of mass, momentum, energy, and species transport, can be found from the work by Chen and Ghoniem [115] based on the general conservation equation:

$$\frac{\partial}{\partial x_j}\left(\overline{\rho U}_j \Phi + \overline{\rho u_j \phi}\right) = \frac{\partial}{\partial x_j}\left[\Gamma_\phi \frac{\partial \Phi}{\partial x_j}\right] + \rho \overline{S}_\Phi \qquad (5.9)$$

where $\Phi$ and $\phi$ are the Reynolds averaged and fluctuating components of the dependent variable, $u_j$ is the velocity component in $j$ direction, $\Gamma_\phi$ is the diffusion coefficient and $S_\Phi$ is the source term.

The radiative heat transfer equation was solved using the discrete ordinate (DO) model with one radiative iteration per five energy iterations. DO model is valid for a wide range of optical thickness and is suggested for oxy-fuel combustion modeling [70]. The absorption coefficient of the gaseous mixture was calculated using weighted sum of gray gas model (WSGGM) which has been verified over a wide range of operating conditions [116, 117].

The turbulence in the flow can be categorized into large eddies, typically of the scale of the characteristic length of the mean flow, and small eddies, which are responsible for the dissipation of turbulence kinetic energy. Large eddy simulation

(LES) model was employed to solve the turbulence by filtering out the small eddies. Low-pass filtering operation for a quantity $\varphi$ can be obtained using equation:

$$\overline{\varphi}(x,t) = \int G(r,x)\varphi(x-r,t)dr \tag{5.10}$$

where the integration is done over the entire domain and $G$ is the filter function that determines the scale of the resolved eddies. Using filtering function, continuity and Navier-Stokes equations yield the following forms:

$$\frac{\partial \rho}{\partial t} + \frac{\partial}{\partial x_i}(\rho \bar{u}_i) = 0 \tag{5.11}$$

$$\frac{\partial}{\partial t}(\rho \bar{u}_i) + \frac{\partial}{\partial x_j}(\rho \bar{u}_i \bar{u}_j) = \frac{\partial}{\partial x_j}\left(\mu \frac{\partial \sigma_{ij}}{\partial x_j}\right) - \frac{\partial \bar{p}}{\partial x_i} - \frac{\partial \tau_{ij}}{\partial x_j} \tag{5.12}$$

where $\sigma_{ij}$ is the stress tensor due to molecular viscosity, and is given by:

$$\sigma_{ij} \equiv \left[\mu\left(\frac{\partial \bar{u}_i}{\partial x_j} + \frac{\partial \bar{u}_j}{\partial x_i}\right)\right] - \frac{2}{3}\mu \frac{\partial \bar{u}_l}{\partial x_l}\delta_{ij} \tag{5.13}$$

where $\tau_{ij}$ is the sub-grid-scale stress given by:

$$\tau_{ij} \equiv \rho \overline{u_i u_j} - \rho \bar{u}_i \bar{u}_j \tag{5.14}$$

The sub-grid-scale stresses resulting from the filtering operation have to be modeled. Isotropic part of the sub-grid-scale stress is merged with the filtered static pressure term and the deviator part is modeled using turbulent viscosity as:

$$\tau_{ij} = -2\mu_t \overline{S}_{ij} \tag{5.15}$$

where $\mu_t$ is the turbulent viscosity and $\overline{S}_{ij}$ is the rate of strain tensor. The turbulent viscosity was solved using Wall-Adapting Local Eddy-viscosity (WALE) model given by:

$$\mu_t = \rho L_s^2 \frac{\left(S_{ij}^d S_{ij}^d\right)^{3/2}}{\left(\overline{S}_{ij}\overline{S}_{ij}\right)^{5/2} + \left(S_{ij}^d S_{ij}^d\right)^{5/4}} \tag{5.16}$$

where $L_s$ (mixing length for sub-grid-scale) and $S_{ij}^d$ are given by:

$$L_s = \min\left(\kappa d, C_w V^{1/3}\right) \tag{5.17}$$

$$S_{ij}^{d} = \frac{1}{2}\left(\bar{g}_{ij}^{2}\bar{g}_{ji}^{2}\right) - \frac{1}{3}\delta_{ij}\bar{g}_{kk}^{2}, \quad \bar{g}_{ij} = \frac{\partial \bar{u}_i}{\partial x_j} \tag{5.18}$$

where $\kappa$ is the von Karman constant, $d$ is the distant to the closest wall, $C_w$ is the WALE constant, and $V$ is the volume of the computational cell.

The oxy-combustion of hydrogen-enriched methane was solved using the partially premixed combustion model that was modified to address the fully premixed combustion case. The transport equations of the mean reaction progress variable $\bar{c}$, the mean mixture fraction $\bar{f}$, and the mixture fraction variance $\overline{f'^2}$ were solved to obtain the other scalar variables. The reaction progress variable can be defined in terms of mass fraction of a suitable species in the gaseous mixture such that it assumes a value of zero in the unburnt mixture and a value of 1.0 in the burnt gas mixture. Mixture fraction is defined as:

$$f = \frac{sY_{fu} - Y_{ox} + Y_{ox,0}}{sY_{fu,1} + Y_{ox,0}} \tag{5.19}$$

where $s$ is the oxygen to fuel ratio and $Y$ is the mass fraction. $Y_{ox}$ and $Y_{fu}$ represent the mass fractions of oxidizer and fuel, while, the subscripts 0 and 1 refers to the inlet streams of oxidizer and fuel, respectively. The density weighted average scalars such as temperature and species fractions can be calculated from the probability density function of mixture fraction and progress variable from the following equation:

$$\bar{\phi} = \bar{c}\int_0^1 \phi_b(f)p(f)df + (1-\bar{c})\int_0^1 \phi_u(f)p(f)\,df \tag{5.20}$$

where $u$ and $b$ represent the burnt and unburnt mixtures, respectively.

Progress variable is defined on the basis of the mass fractions of the species as:

$$c = \left(\frac{\sum_{i=1}^{n} Y_i}{\sum_{i=1}^{n} Y_{i,ad}}\right) \tag{5.21}$$

where $Y_i$ represents the mass fraction of product species $i$, and the sub-script $ad$ represents the species after complete combustion. Hence, the value of $c = 0$ and $c = 1$ means the unburnt and burnt mixture, respectively.

Heat release factor is a non-dimensional parameter used to quantify the heat release in a combustion process and is given by:

$$\gamma = \left(\frac{T_{ad} - T_0}{T_{ad}}\right) \tag{5.22}$$

where $T_{ad}$ is the adiabatic flame temperature and $T_0$ is the unburnt temperature of the reactant gases.

To study the effect of flame stretching, the critical rate of strain is calculated using the equation:

$$g_{cr} = \frac{BS_L}{\alpha} \tag{5.23}$$

where $B$ is constant, $S_L$ is the laminar flame speed, and $\alpha$ is the unburnt thermal diffusivity of the reactant mixture.

Flame thickness has been estimated using the following correlation:

$$\delta = \frac{\lambda_u}{\rho_u c_p S_L} \tag{5.24}$$

where $\lambda_u$ and $\rho_u$ are the unburnt thermal conductivity and density, respectively.

A non-dimensional ratio of reaction rate to diffusion rate, known as Damkohler number, is given as:

$$Da = \frac{(l_t/u')}{(\delta/S_L)} \tag{5.25}$$

where $l_t$ is the integral length scale and $u'$ is the RMS of velocity fluctuation.

Karlovitz number is defined for the smallest eddies and represents the ratio of the chemical time scale to the Kolmogorov time given as:

$$Ka = \frac{\sqrt{\varepsilon/\nu}}{S_L/\delta} \tag{5.26}$$

where $\varepsilon$ is the rate of the turbulent dissipation and $\nu$ is the kinematic viscosity.

An axi-symmetric quadrant of the reactor geometry was modeled as presented in Fig. 5.14. After mesh independency check, a mesh of around 40,000 finite volume cells, having finer meshing in the near burner region and coarser meshing in the near outlet zone of the combustor, was selected to perform the numerical study presented in this section. A particular composition of the $CH_4/H_2/O_2/CO_2$ mixture was injected through the inlet of the combustor with a constant throat velocity. Axial and tangential components of the flow were set to induce a swirl flow with the swirl number of 0.98. Fluctuating velocity was solved using vortex method algorithm. Mean mixture fraction of the inlet stream was calculated according to the composition and equivalence ratio and was set under the inlet stream of the partially premixed model.

The combustor wall was assigned mixed thermal boundary condition with a convective heat transfer coefficient of 20 $W/m^2$ K, and free stream temperature and external radiation temperature of 300 K each. The wall thickness (6 mm) was set to semi-transparent boundary type. Non-gray band model with 4 wavelength bands was used to calculate the radiation heat transfer through the quartz tube. Both the symmetry faces of the quadrant were assigned periodic boundary conditions to take

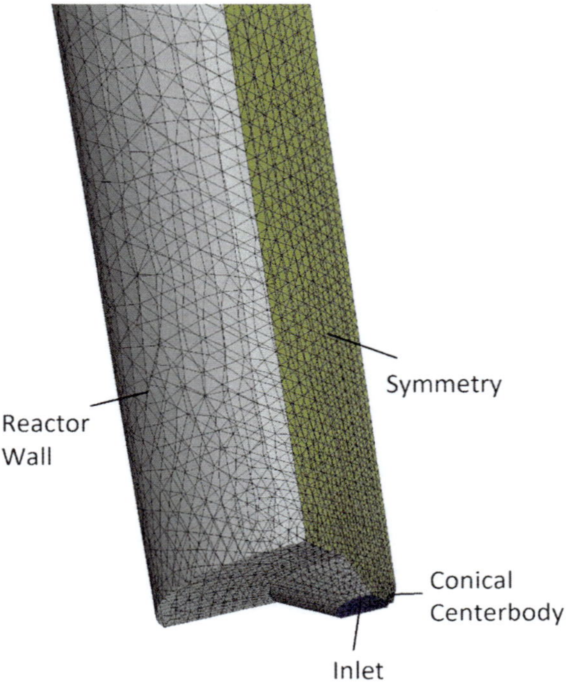


**Fig. 5.14** Schematic of the combustor quadrant mesh used for the numerical simulations

care of the tangential components of the swirling flow in the numerical calculations. All simulations were performed under a fixed oxygen fraction of 30% and at stoichiometric condition (equivalence ratio of 1.0).

#### 5.5.5.2 Model Validation

The behavior of premixed hydrogen-enriched oxy-methane flames was studied experimentally over a range of fuel blends in the model gas turbine combustor. The flame stability was mapped based on experimental measurements by identifying the flame lean blowout (*LBO*) and flashback (*FBL*) limits over ranges of hydrogen fraction and equivalence ratio at a fixed throat inlet velocity of 5.2 m/s to provide a constant volumetric flow rate of the reactant mixture. Stable flame region lies below the flashback curve and above the blowout curve as presented in Fig. 5.15. It can be seen that the operating window of the heat release factor ($\gamma$) increases significantly with hydrogen addition to the fuel stream. *LBO* limit decreases linearly with hydrogen addition, thus ensuring stable flames at lower flame temperatures. Moreover, flashback was observed at higher values of $\gamma$ with increasing the hydrogen fraction, attaining a maximum value at 55% *HF*. Then, any further increase in *HF* results in decrease of the value of $\gamma$ at which flashback occurs. This may be attributed to the fact that

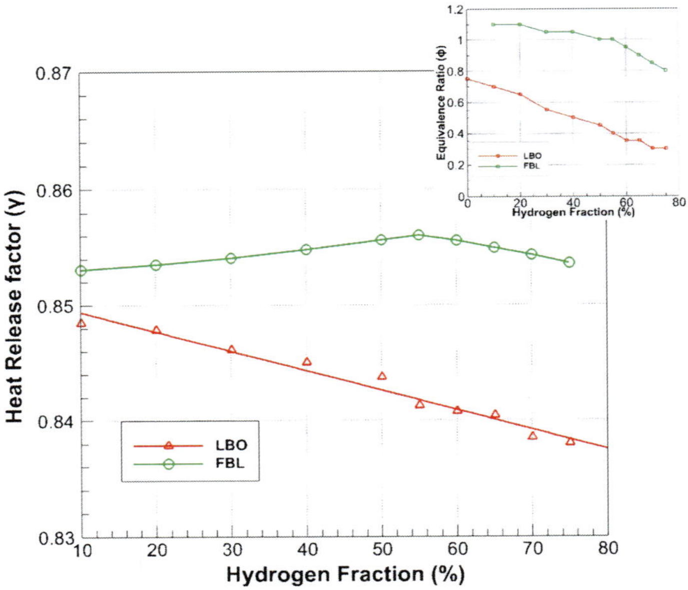


**Fig. 5.15** Experimental flame stability mapping as a function of heat release factor ($\gamma$) and hydrogen fraction, at a fixed throat inlet velocity of 5.2 m/s and fixed OF of 30%

flashback occurred at $\varphi = 1$ at 55% *HF* but at slightly rich $\varphi$ for $HF < 55\%$ and at lean $\varphi$ for $HF > 55\%$, as shown in Fig. 5.15. At 55% hydrogen enrichment of methane, the flame enjoyed the widest operability window. Any attempt to increase the hydrogen fraction beyond 55% under stoichiometric condition (equivalence ratio of 1.0) failed, and the flame extinguished because of flashback. This is unlike the pure methane oxy-flame, which was found to be stable at stoichiometric condition. The increase in hydrogen fraction in the fuel blend increases the laminar flame speed, and the flame attempts to propagate upstream of the reactant gases resulting in earlier flashback when compared to pure methane oxy-flame. Interestingly, the results showed that the $H_2/CH_4$–$O_2/CO_2$ flames do not follow constant Damkohler number at either the flashback or the blowout limits. This is unlike the $CH_4$–$O_2$–$CO_2$ flames, which were found to flashback and blow out at fixed Damkohler number [70]. This indicates that hydrogen enrichment results in different combustion and flame stability characteristics.

Figure 5.16 presents a comparison of the experimentally recorded and numerically calculated values of temperature along the central axis of the reactor at an equivalence ratio of 1.0 over a range of *HF*. The temperature levels were found to increase slightly with increasing hydrogen fraction in the fuel blend. This may be attributed to (a) the increased fuel diffusion and reactivity and (b) the slightly higher adiabatic flame temperature with the increased *HF*. The axial temperature levels decrease gradually downstream of the burner after a height of around $Z/D = 1.0$, i.e. downstream of the flame core. Due to the limited operating range of the thermocouple, temperature could

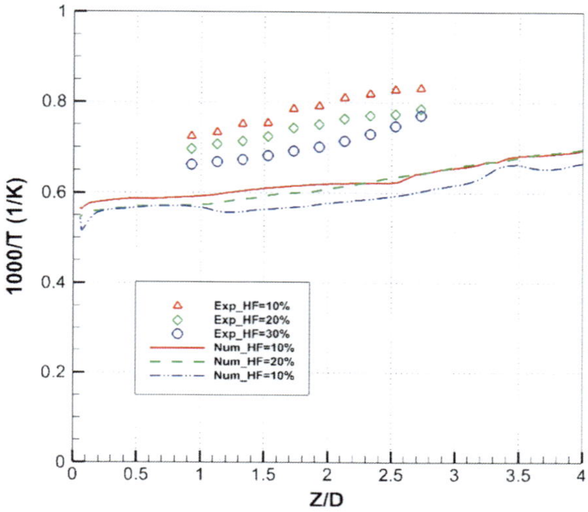


**Fig. 5.16** Experimental versus numerical temperature profiles along the axis of the combustor at equivalence ratio of 1.0 and at fixed OF of 30%

not be measured within the flame core, i.e. in the zone of $Z/D < 1.0$. However, the values of temperature within the flame core were calculated numerically as shown in Fig. 5.16. Although the flames are stabilized at both the inner and outer shear layers, the temperature profile reveals nearly constant temperature level within $0.2 < Z/D < 1.0$. This can be attributed to the creation of an inner recirculation zone (*IRZ*), which transfers the convective heat to the region. The radial temperature profiles, at an axial height of $Z/D = 0.92$ from the burner, are presented in Fig. 5.17. The differences in flame temperature were only marginal for the different hydrogen fractions. For all *HFs*, the temperature increases radially from the center of the combustor, where the

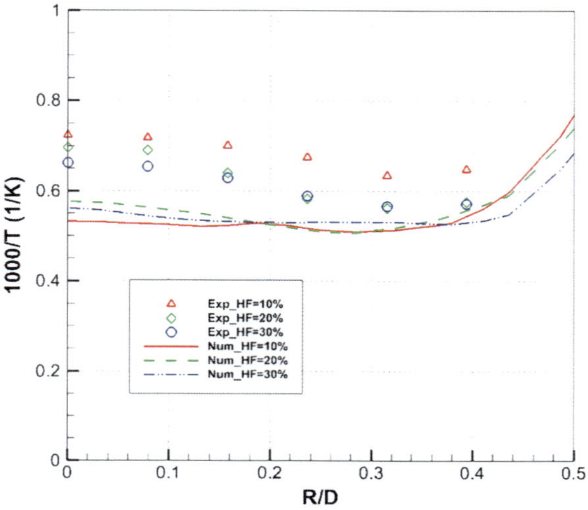


**Fig. 5.17** Experimental versus numerical temperature profiles along the radial direction at an axial height of Z/D = 0.92 at equivalence ratio of 1.0 and at fixed (*OF*) value of 30%

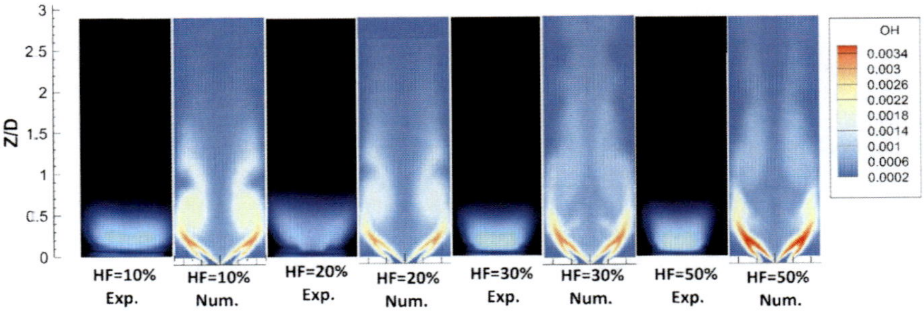


**Fig. 5.18** Comparison between experimental (photos taken by high-speed camera) and numerical (based on *OH* contour plots) flame shapes at different hydrogen fractions under stoichiometric conditions

*IRZ* is located, towards the zone near the combustor wall where the *ORZ* is located, passing across the main flame core. The temperature deceases near the wall due to the effect of *ORZ* and heat loss through the wall.

When comparing the measured and calculated temperature values as presented in Figs. 5.16 and 5.17, one can see that the numerical model slightly over-predicts the temperature within the combustor. This may be attributed to the cooling effect of the burner port, which quenches the reaction in its close vicinity, causing slight reduction of the combustion temperature [118]. This quenching effect of the burner metal wall was not addressed in the computational model, causing such slight differences between the experimentally recorded and the numerically predicted temperature values. The quenching effect of the burner wall is clearly identifiable in the experimental flame images (Fig. 5.18) by a dark zone beside the burner wall and below the flame base. This Figure compares the experimentally captured flame images with the numerically calculated flame shape presented by the center-plane contours of OH radicals. No reactions occur within the *ORZ* and the narrow zone beside the burner wall due to the wall cooling effect, which causes local quenching. At low hydrogen fractions ($HF = 10\%$), the flame is spread out, taking the shape of the confinement, as seen in Fig. 5.18. However, the flames with higher hydrogen fractions are characterized by more violent and compact shapes, indicating better reactivity of the fuel mixture.

#### 5.5.5.3 Flow Field Characteristics

The center-plane contours of velocity streamlines are plotted over the temperature contours in Fig. 5.19 for different hydrogen fractions. The flow is characterized by the creation of both *IRZ* and *ORZ*. At lower *HFs*, below 20%, the *IRZ* is characterized with one main large eddy spinning in the counter-clockwise direction. Increasing *HF* to 30% resulted in breakdown of the *IRZ* into two eddies: the larger primary eddy (*PE*) moved a little downstream, and a smaller secondary eddy (*SE*) was created upstream and spins in an opposite direction to the *PE*. Increasing *HF* to 50% resulted in further breakdown of the *IRZ*, with two *SE* spinning opposite to each other, in addition to the *PE*. The size and number of *SEs* are mainly controlled by the hydrogen fraction. This indicates better turbulence-reactions interaction within the flame with hydrogen enrichment at fixed equivalence ratio. Actually, at higher *HF* operation, the flame becomes more compact and well anchored to the burner inlet (see Fig. 5.18) due to the increased burning velocity. This compactness of the flame allows for a portion of the fresh reactant mixture to escape around the flame, flow along the combustor walls, and reflect back to interact with the main central stream, creating enhanced flow recirculation within the *IRZ*. In a previous study by Nemitallah et al. [119] on a backward-facing step combustor considering premixed propane-air combustion, the *SE* appeared only while operating under lower values of lean equivalence ratios, and the effect of equivalence ratio on the structure of recirculation zone was insignificant at near-stoichiometric operation. This highlights the key effect of hydrogen enrichment in controlling flame structure and stability in premixed stoichiometric oxy-flames. Moreover, the experimental observations at higher hydrogen fractions were characterized by high audible sound, conforming to the numerical results that predict increase in the vorticity, causing more intense and violent flames. The creation of an *ORZ* was also observed at the bottom end of the combustor beside the wall. The size of the *ORZ* was not affected significantly by hydrogen enrichment. This may be attributed to the high aspect ratio between the diameter of the combustor and the burner throat diameter.

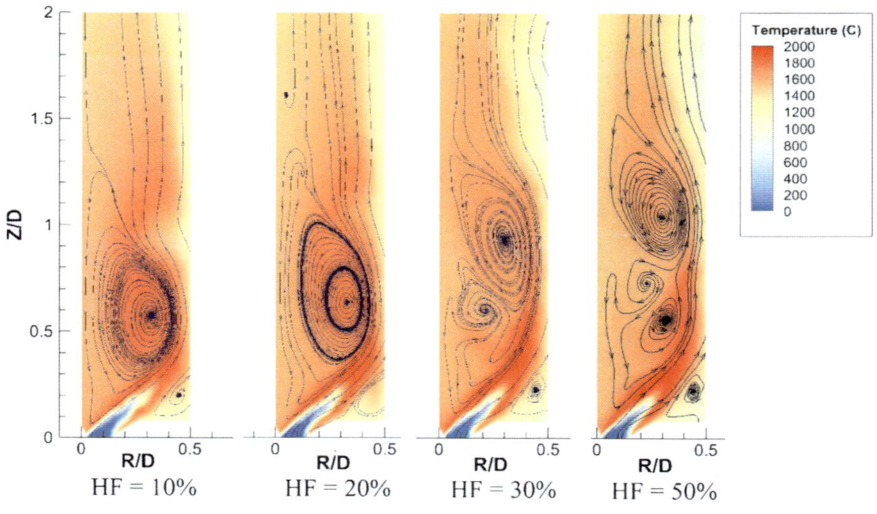


**Fig. 5.19** Velocity streamlines mapped over temperature contour for different hydrogen fractions at equivalence ratio of 1.0 and at fixed *OF* of 30%

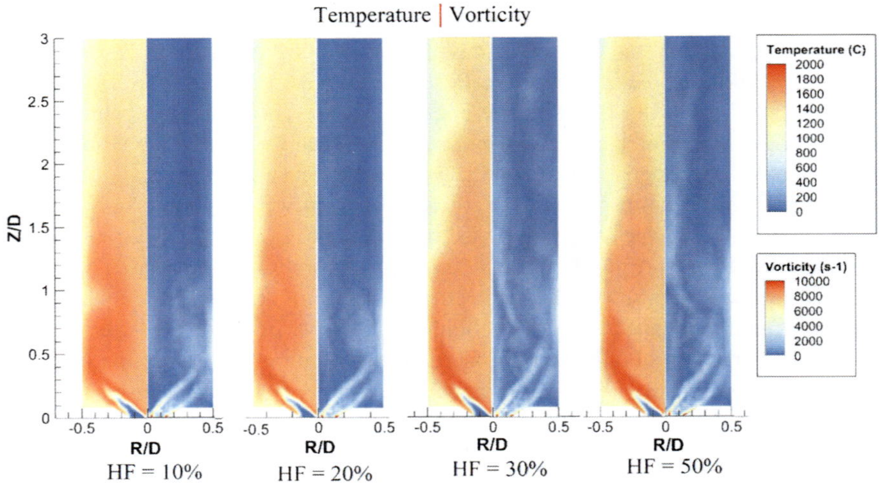


**Fig. 5.20** Contours of temperature (left) and vorticity (right) for various hydrogen fractions at equivalence ratio of 1.0 and at fixed *OF* of 30%

The contours of vorticity magnitude and static temperature are presented in Fig. 5.20. As shown in the figure, the flame is stabilized where reactions take place between the inner and outer shear layers. The length of the inner shear layer increases with hydrogen addition due to enhanced vorticity levels near the central axis. The temperature within the flame core approaches 2000 °C, then reaches moderate values at the combustor exit that are suitable for turbine blade operation; see Fig. 5.16 as well. The obtained temperature distributions indicate better combustion and emission characteristics of such kind of premixed oxy-methane flames with hydrogen enrichment. Figure 5.21 presents the variations of the predicted and critical strain rates with the hydrogen fraction. The rate of strain presented in this figure is the average value calculated on the iso-surfaces of $c = 0.9$ and $c = 0.99$. The critical rate of stain increases significantly from ~4300 $s^{-1}$ for pure methane combustion to ~9700 $s^{-1}$ at 50% *HF*. However, the rate of strain increases only slightly with hydrogen addition, indicating a wider stability range against LBO, as is evident from the experimental findings discussed earlier. Thus, hydrogen blending stabilizes the flames for higher strain rates before they quench and extinguish.

#### 5.5.5.4 Flame Characteristics

Figure 5.22 shows the contours of product formation rate (*PFR*) and progress variable near the burner region for different hydrogen fractions. It is observed that the reaction zone exists between the inner and outer shear layers, and the flame is stabilized on both layers. At higher hydrogen fractions, the *PFR* increases in the thinner inner shear layer, indicating more intense and compact flames. At lower hydrogen fractions, the

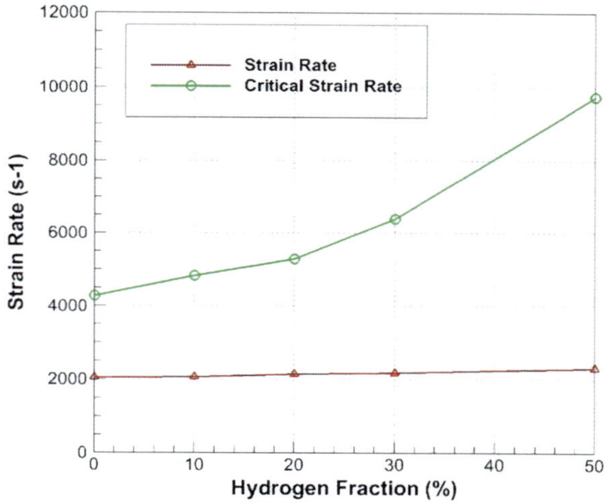


**Fig. 5.21** Variation of predicted and critical strain rates for various hydrogen fractions at equivalence ratio of 1.0 and at fixed *OF* of 30%

*PFR* decreases in the inner shear layer, and the flame thus tends to stabilize towards the corner of the combustor. Although the numerical model is incapable of capturing the switch from *ORZ* to *IRZ* stabilization, the switch is inferred from the trend of *PFR*. Recall that corner stabilization was observed experimentally at 10% *HF*, as depicted in Fig. 5.18. This indicates that the role of *ORZ* in flame stabilization is significant at lower *HFs* and diminishes at higher *HFs*. The model also predicts shorter and more compact flames at higher hydrogen fractions, as per the contours of the progress variable, which indicates the transition from unburnt mixture to completely burnt products, as presented in Fig. 5.22.

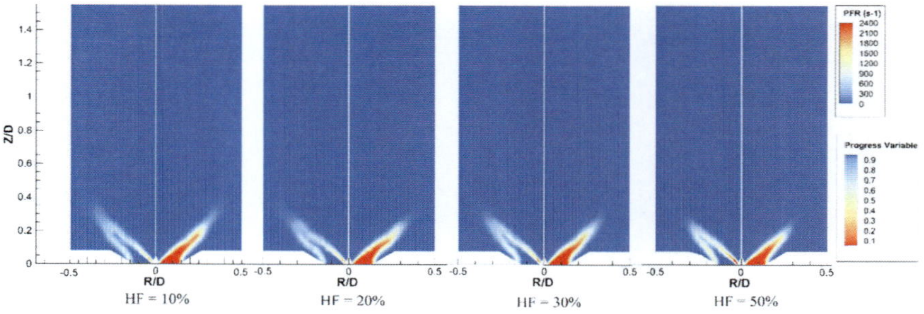


**Fig. 5.22** Contours of product formation rate (*PFR*) and progress variable for various hydrogen fractions at equivalence ratio of 1.0 and at fixed *OF* of 30%

For a deeper insight into the flame to see how the reactions are proceeding, the mole fraction distributions of $H_2$, $CH_4$, OH and CO are presented in Fig. 5.23 for 10% and 50% *HF*. The contours of fuel concentration show longer tails for both $H_2$ and $CH_4$ at 10% *HF* than those at 50% *HF*, indicating faster reactions and shorter flame lengths with hydrogen addition. This conclusion is also confirmed by enhanced OH radical formation at higher *HFs* as presented in the figure. It can also been observed that the oxidation of CO is retarded with hydrogen addition, which is attributed to the fact that at higher hydrogen concentrations, carbon has to compete with hydrogen for oxygen, resulting in retarded CO oxidation. The CO emission at the combustor exit increases from 46 ppm for pure methane to 378 ppm at 50% *HF*, as depicted in Fig. 5.24.

The effect of hydrogen enrichment on flame thickness is presented in Fig. 5.25. The calculated flame thickness ($\delta$) is 47.6 $\mu$m for pure methane, as compared to 35 $\mu$m at 50% *HF*. The flame thickness decreases almost linearly with hydrogen fraction as shown in the figure. The Damkohler (*Da*) and Karlovitz (*Ka*) numbers are plotted in Fig. 5.26 at different hydrogen fractions. The Damkohler number increases from 2600 for pure methane to 3900 at 50% *HF*, indicating significantly faster reactions with hydrogen addition. The Karlovitz number was found to decrease from 0.33

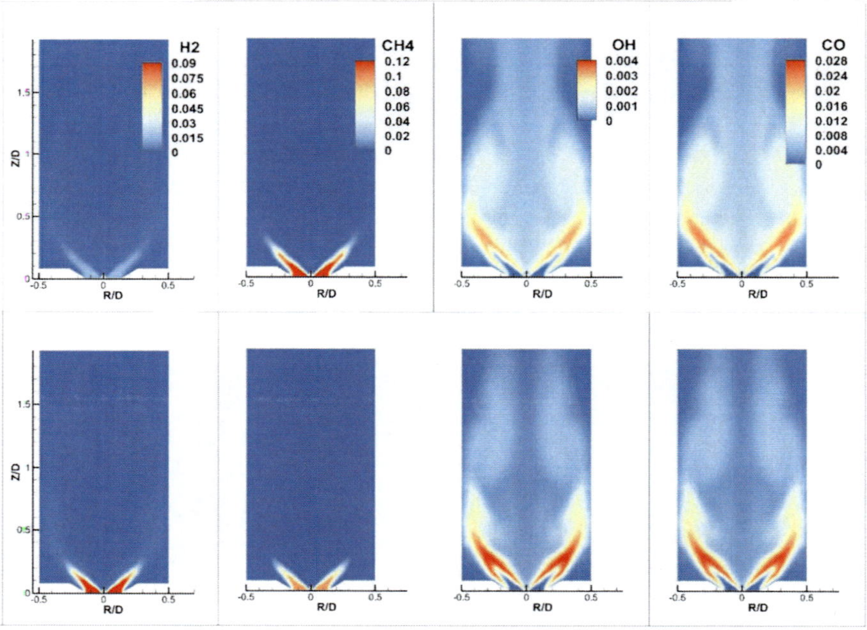


**Fig. 5.23** Mole fraction distributions of $CH_4$, $H_2$, CO, and OH within the combustor at equivalence ratio of 1.0 and at fixed OF of 30% for hydrogen fractions of 10% (top) and 50% (bottom)

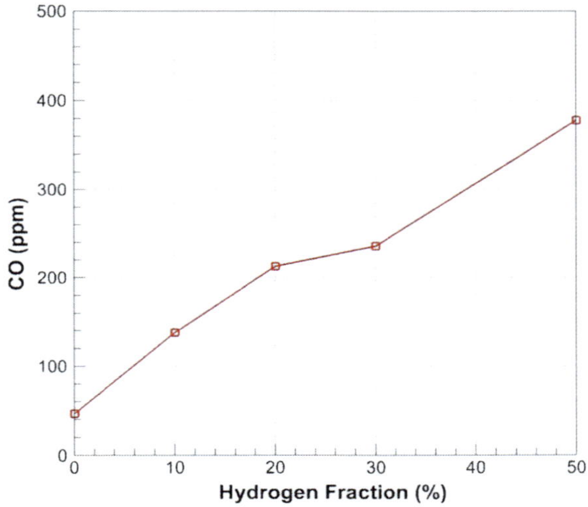


**Fig. 5.24** CO emission at the combustor exit for various hydrogen fractions at equivalence ratio of 1.0 and at fixed *OF* of 30%

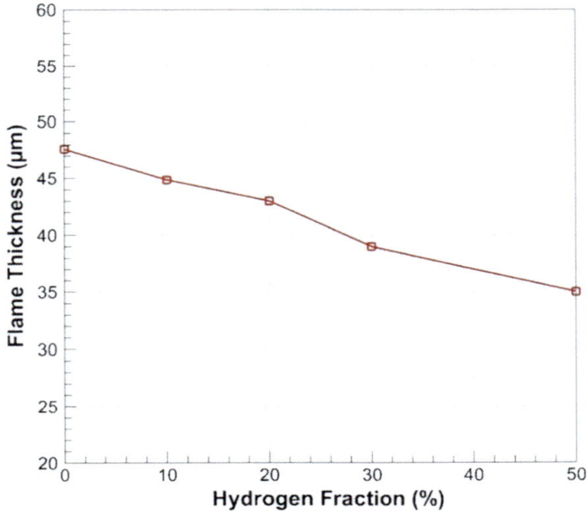


**Fig. 5.25** Effect of hydrogen fraction on the flame thickness at equivalence ratio of 1.0 and at fixed *OF* value of 30%

for pure methane to 0.25 at 50% *HF*; however, the correlation is not linear. Lower *Ka* value suggests that the chemical time scale is much less than the Kolmogorov time scale, and hence, the flame thickness of $CH_4/H_2$ oxy-flames is smaller than the smallest turbulent scale. In such flames, the flame front is thin, and the turbulence incapable of altering the inner structure of the flame, but it gets wrinkled because of the turbulence motion. Thus, the reactions are governed by the laminar burning rate.

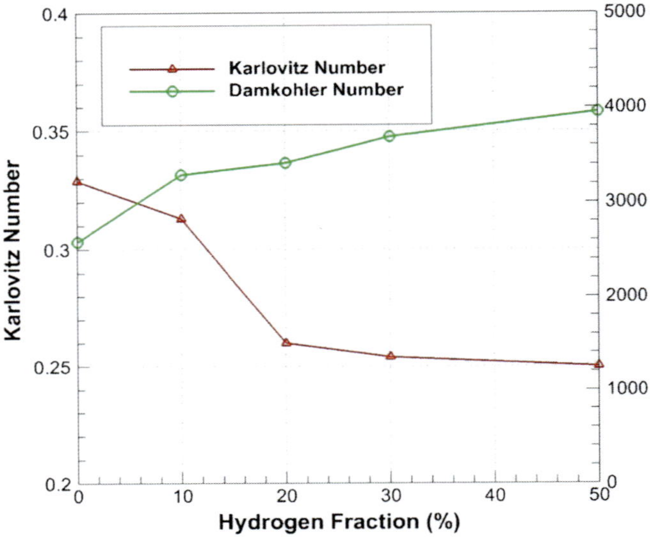


**Fig. 5.26** Variations of Karlovitz and Damkohler numbers with hydrogen fraction at equivalence ratio of 1.0 and at fixed *OF* value of 30%

## 5.6 Combustor Operability Under Stoichiometric $H_2$-Enriched Conditions

Considering the same swirl premixed combustor described in the preceding section of this book, the stability characteristics of swirl-stabilized hydrogen-enriched $CH_4$–$O_2$–$CO_2$ premixed flames were experimentally investigated to determine the effects of hydrogen addition and inlet velocity on flame stability under stoichiometric conditions ($\varphi = 1.0$) [120]. The stability limits in terms of blowout and flashback were identified over wide ranges of hydrogen fraction (*HF*: %$H_2$ in $H_2$–$CH_4$ mixture) and oxygen fraction (*OF*: %$O_2$ in $O_2$–$CO_2$ mixture) at fixed inlet velocity of 6 m/s. The lines of stability limits were plotted against the contours of adiabatic flame temperature (*AFT*), power density (*PD*) and Reynolds number (Re) to understand the physics behind the flame extinction mechanism. The effects of inlet velocity on flame stability limits under hydrogen-enrichment were investigated by comparing the stability maps of the combustor for different inlet velocities, namely 4.4, 5.2 and 6.0 m/s.

### 5.6.1 Flame Stability Mapping

Figure 5.27 presents the flashback and blowout limits with the variation of hydrogen and oxygen fractions at inlet bulk velocity of 6.0 m/s and under stoichiometric equivalence ratio against the contours of constant adiabatic flames temperature (*AFT*). It

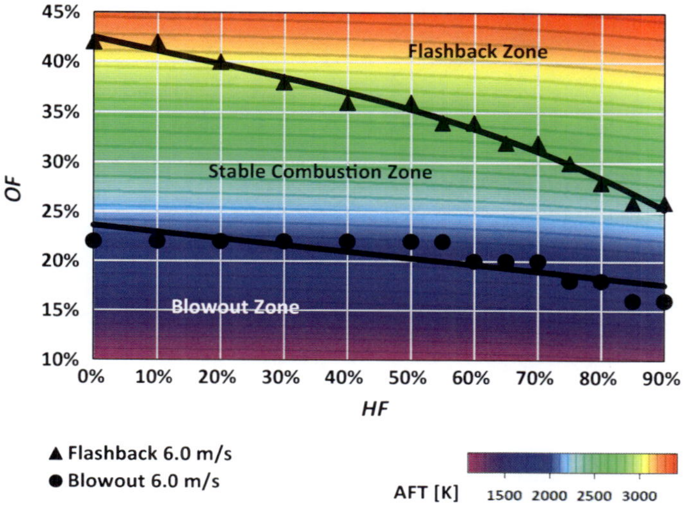


**Fig. 5.27** Flame stability mapping against the contours of adiabatic flame temperature at the throat velocity of 6.0 m/s

is observed that the hydrogen enrichment has an insignificant effect on *AFT*, which is dominated by the *OF* instead. Thus, the stability limits show no obvious correlation associated with *HF*. The flashback reduces from 45% *OF* with pure methane combustion to 25% *OF* at 90% *HF*. The observation can be explained through the fact that hydrogen has a high flame speed, which eventually increases the Damkohler number of the reactive mixture. Thus, the hydrogen-enriched flames have higher tendency to propagate upstream of the nozzle, resulting in earlier flashback. Higher levels of audible noise were also observed before flashback with hydrogen addition. This can be compensated by reducing the *OF* in the reaction mixture, which slows down the flame speed by reducing the reaction rate, to maintain the balance of flame speed and fixed inlet bulk velocity. However, the blowout limit is slightly and linearly extended with the addition of hydrogen, which can be attributed to the increased chemical reaction and turbulence levels of the reacting mixture, resulting in enhanced flame stability. The variation of blowout limit with hydrogen addition is not as significant as that of the flashback limit, which appears to be pertaining to the high Re range of the reactive mixture in the former case. The effect of higher flame speed, due to the addition of hydrogen, is partly neutralized by the high convective mass transport rate, resulting in only inappreciable variations in the blowout limit. Hence, the blowout limit shows a minor increase from 22% *OF* for pure methane combustion to 16% *OF* at 90% *HF*. Like the case of *AFT*, the *PD* of the combustor exhibits no apparent relation with the stability map as shown in Fig. 5.28. The stability map depends greatly on the Damkohler number and the flame speed. The addition of hydrogen to the reaction zone increases the reaction rate; however, it has almost no effect on the combustor *PD*.

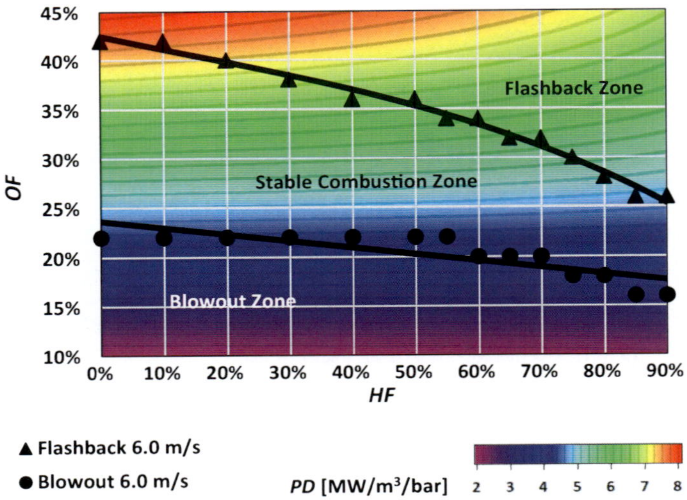


**Fig. 5.28** Flame stability mapping against the contours of *PD* at the throat velocity of 6.0 m/s

Flame shapes near the blowout and flashback limits are illustrated in Fig. 5.29. Near the blowout limit, the flame becomes shorter and more compact with the increase of *HF*. It is also observed that the flame spread within the reactor tends to take the

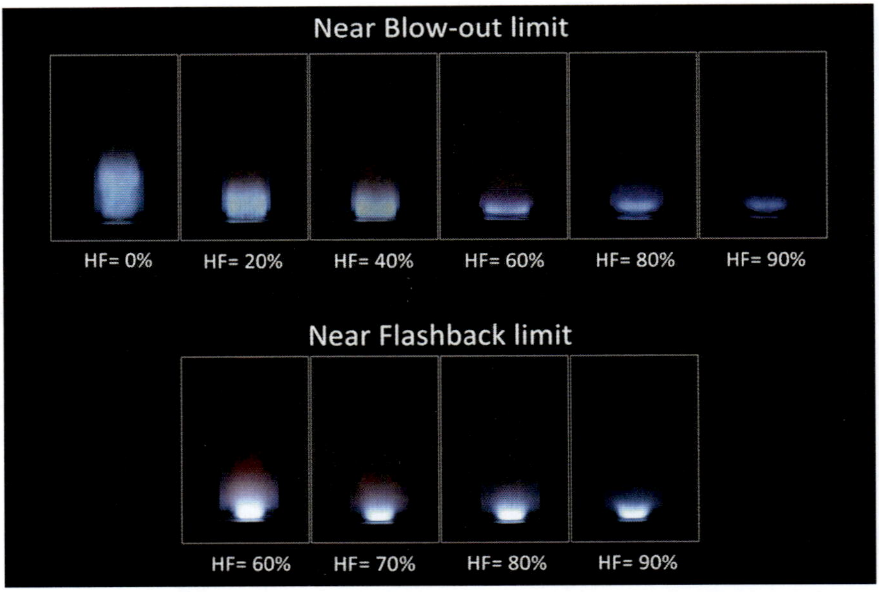


**Fig. 5.29** Flame shapes captured near blowout and flashback limits at inlet velocity of 6.0 m/s

shape of the confinement. No significant changes in flame shape are observed near the flashback limit. The effect of kinetic rates on flame shape is investigated by comparing flames shapes at fixed *AFT* of 2000 K in the stable flame operation zone over ranges of *HF* and *OF* and at inlet velocity of 6 m/s as presented in Fig. 5.30. The *OF* had to be slightly decreased while increasing *HF*, in order to maintain constant *AFT*. It is observed that flame becomes shorter and more intense near the bottom part of the combustion zone with increasing *HF*. This can be attributed to the enhanced chemical reactions and higher turbulent flame speed. Although all flames in Fig. 5.30 share the same bulk *AFT*, the local temperature within the flame varies, as it is obvious from the changes in flame brightness and compactness. Increased reaction rates with hydrogen enrichment increases the brightness and temperature of the flame.

Figure 5.31 depicts the variation of the stability limits with hydrogen addition with respect to the Re of the reactive mixture. Near-flashback region is characterized by

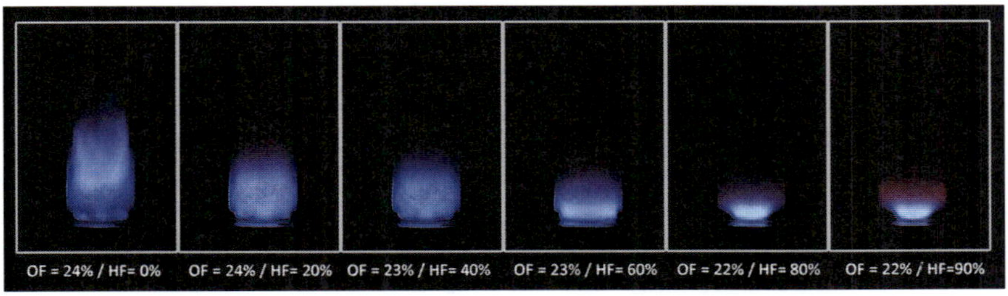


**Fig. 5.30** Flame shapes captured at fixed *AFT* of 2000 K at inlet velocity of 6.0 m/s

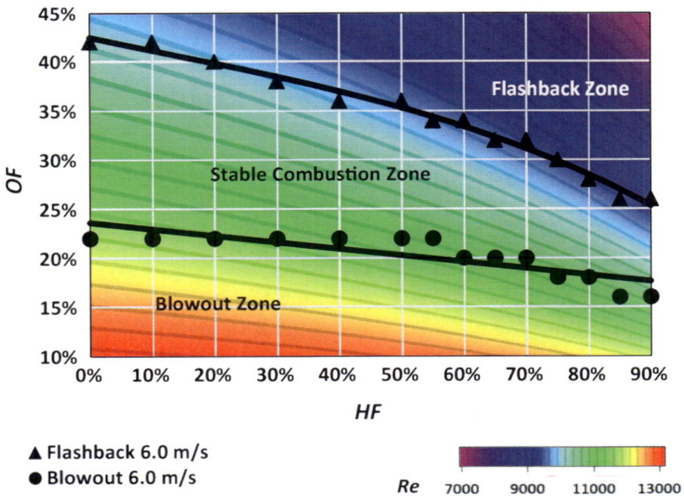


**Fig. 5.31** Flame stability mapping against the contours of Re at the throat velocity of 6.0 m/s

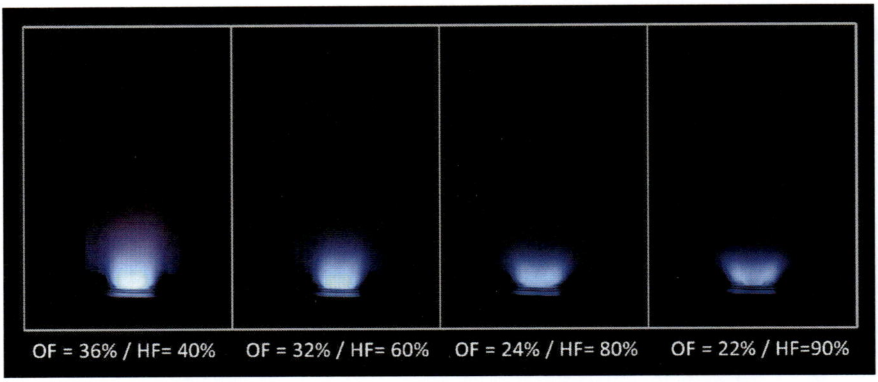


**Fig. 5.32** Flame shapes captured at fixed Re of 10,000 at inlet bulk velocity of 6.0 m/s

high Damkohler number, and the reaction is affected by the diffusive mass transport rate, suggesting the flashback limits to follow the Re trendline. Figure 5.31 confirms this proposition, where the flashback limit is observed to approach the constant Re contour. The near-blowout region, on the other hand, is characterized by lower Damkohler number, and the reaction rate is limited by the chemical reactions. Thus, the addition of hydrogen to the reactive mixture enhances the blowout limit to lower *OF*, without necessarily following the Re trendline. Finally, the effect of Re on flame shape is investigated by comparing flames shapes at fixed Re of 10,000 in the stable flame operation zone over ranges of *HF* and *OF* and at inlet velocity of 6 m/s as depicted in Fig. 5.32. It is observed that increased *HF* shortens the flame length at the same operating Re since the Damkohler number increases with hydrogen enrichment.

## 5.6.2 *Effect of Inlet Velocity on Flame Stability*

Three sets of experiments were conducted at three different inlet bulk throat velocities, namely 4.4, 5.2, and 6.0 m/s, to investigate the effect of inlet velocity on flame stability limits under stoichiometric premixed conditions and over wide ranges of *HF* and *OF*. The experiments were repeated several times to ensure good reproducibility of the results. Figures 5.33 and 5.34 show the flashback and blowout limits, of the swirl-stabilized gas turbine model combustor over ranges of *HF* (from 0 to 90%) and *OF* (from 16 to 44%). Figure 5.33 shows that the increase of *HF* decreases the values of stable *OF* compared to pure methane oxy-fuel combustion. This can be attributed to increased chemical reactions and turbulent flame speed which is characterized by high levels of audible noise before flashback occurrence. Also, this is expected since hydrogen has a high burning velocity, and at higher *HF*, the reaction zone becomes more sensitive to the addition of $O_2$ which causes flashback to occur at lower *OF*. Figure 5.34 illustrates clearly that the addition of hydrogen in the combustion reaction

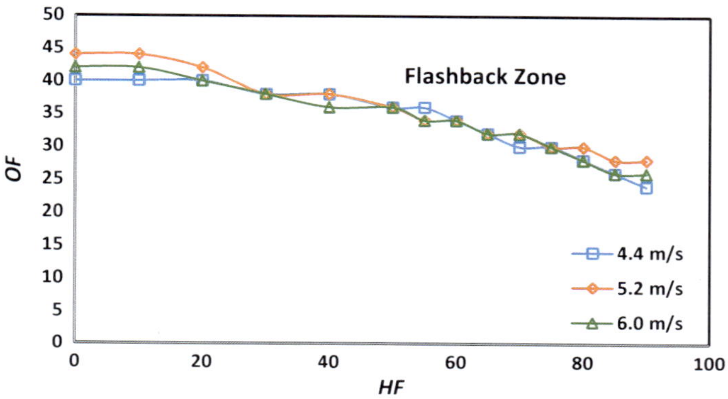


**Fig. 5.33** Flashback limits in the*OF-HF* space for different throat bulk velocities of 4.4, 5.2 and 6.0 m/s

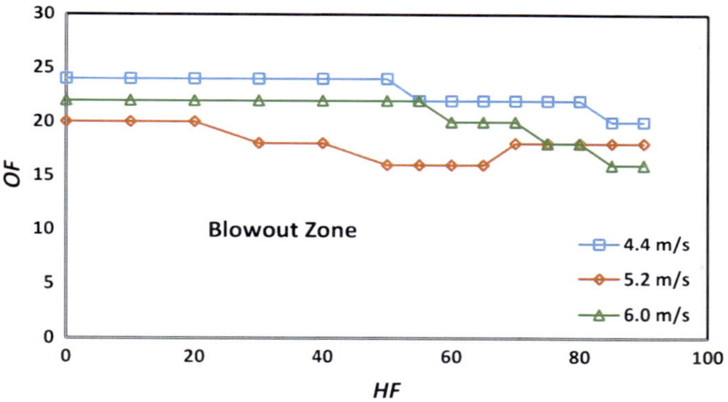


**Fig. 5.34** Blowout limits in the *OF-HF* space for different throat bulk velocities of 4.4, 5.2 and 6.0 m/s

zone extends the blowout limit compared to the case of pure methane combustion. The effect of increasing the inlet throat velocity is almost minimal on flashback; however, it has significant effect near the blowout limit. Figures 5.33 and 5.34 suggest that the inlet bulk flow velocity is not the optimum parameter to characterize the combustor stability map within the *HF-OF* space.

The Re of the mixture can be defined as:

$$\mathrm{Re} = \frac{Dv}{\upsilon_{mix}} = Const. \times \left(\frac{v}{\upsilon_{mix}}\right) \tag{5.27}$$

where $D$ and $\upsilon_{mix}$ are the throat diameter and the kinematic viscosity of the mixture, respectively. At fixed inlet velocity, Re and $\upsilon_{mix}$ are inversely proportional. Increasing the *OF* at fixed *HF* results in an increase in the value of $\upsilon_{mix}$, and consequently lower Re. Thus, flashback always occur at lower Re values than those of blowout. The effect of changing the throat velocity (4.4, 5.2, and 6.0 m/s) should thus be studied in terms of Re to incorporate the effect of mixture viscosity as well. Figures 5.35 and 5.36 plot the stability limits in terms of *HF* and Re. The effect of changing the throat velocity is now distinctly identifiable. Results show that increasing the mixture *HF* results in lower mixture Re values at stability limits. This is attributed to the increase of the dynamic viscosity of the mixture with the increase of *HF*. Equation (5.27) shows that the inlet mixture velocity is directly related to Re. Therefore, increasing the throat inlet velocity results in higher mixture Re values for both flashback and blowout limits. Figures 5.35 and 5.36 indicate that increasing Re leads to improved flashback limit while decreasing Re extends the blowout limit. Thus, the results suggest that

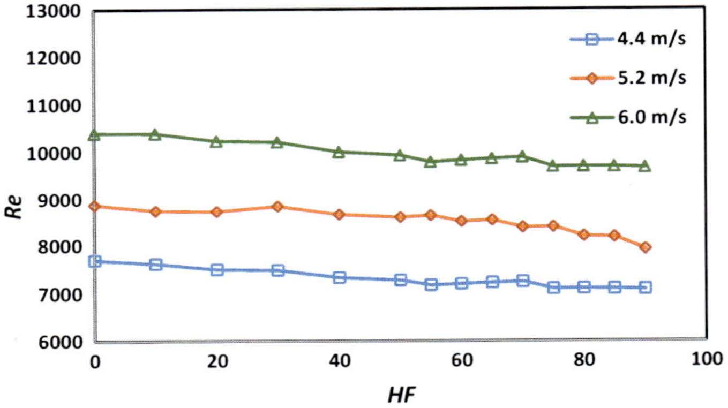


**Fig. 5.35** Flashback limits in the *Re-HF* space for different throat bulk velocities of 4.4, 5.2 and 6.0 m/s

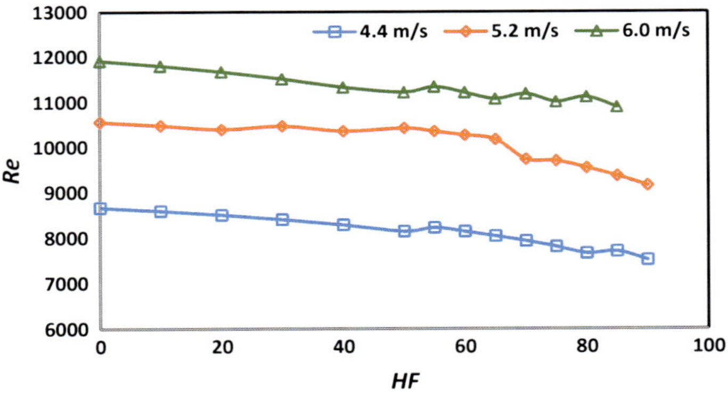


**Fig. 5.36** Blowout limits in the *Re-HF* space for different throat bulk velocities of 4.4, 5.2 and 6.0 m/s

mixture Re is a more relevant parameter than the inlet bulk velocity in controlling flame stability limits.

Flame visualization is an important approach for a better understanding of flame characteristics and extinction mechanism. Figure 5.37 illustrates the effect of hydrogen addition on flame macrostructure at fixed *OF* of 30% for different throat velocities. With pure methane operation, the flame is V-shaped and relatively homogenous

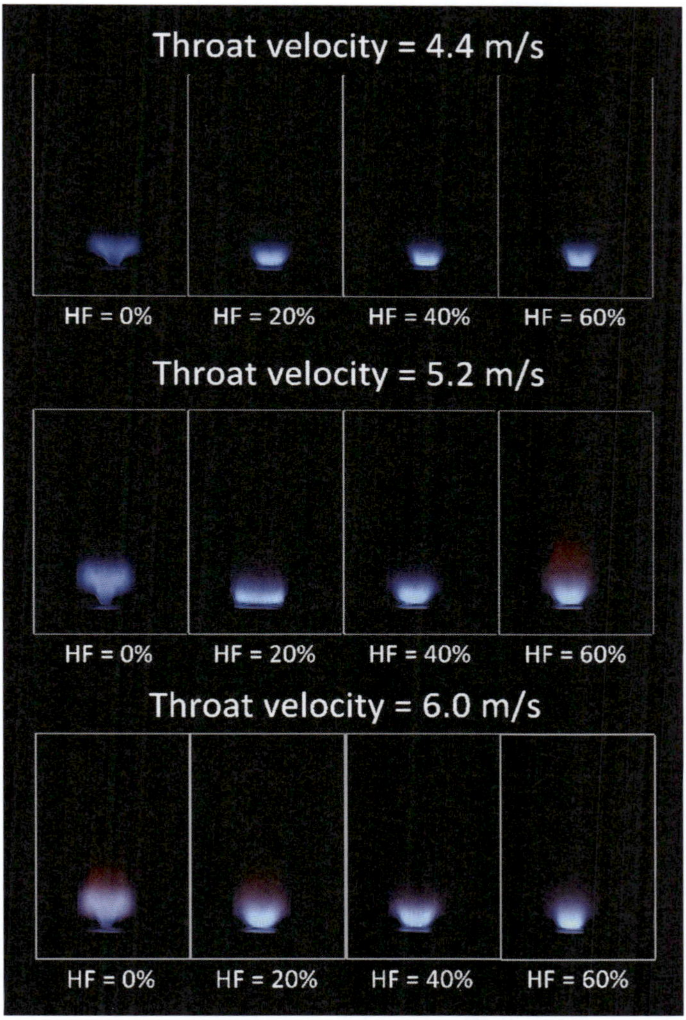


**Fig. 5.37** Flame images at *OF* value of 30% over a range of *HF* for different inlet throat velocities

in the combustion zone. At low *HF*, the flame tends to take the shape of the confinement by spreading over the reactor. However, the flames are more compact and intense at the bottom part of the combustion zone with the increase of *HF*. The visual flame length tends to decrease slightly while increasing *HF*. This can be explained by increased reaction rate and flame speed resulting in more confined flames. It has been observed that the audible combustion noise increases with the increase of *HF* until flashback occurrence. Moreover, no notable change was observed on the flame macrostructure with the increase of throat velocity; however, the flame length increases because of higher convective transport. Figure 5.38 depicts the effect of oxygen addition on flame macrostructure at fixed *HF* of 20% for different throat velocities. It can be observed that with the increase of *OF*, the brightness of the flame near the bottom part of the combustion zone increases. A transition from V-shape to U-shape was observed while increasing *OF* and was found to occur earlier

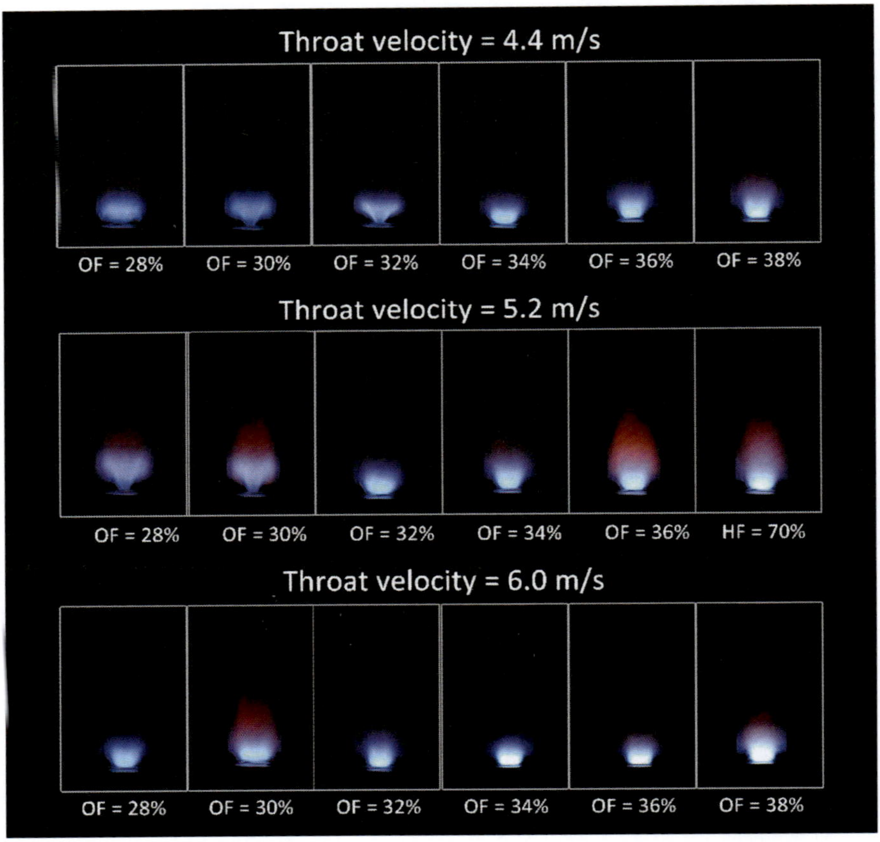


**Fig. 5.38** Flame images at *HF* of 20% over a range of *OF* for different inlet throat velocities

at higher inlet bulk velocities. Also, increasing the throat velocity results in more compact flames.

## 5.7 Combustor Operability of Premixed Oxy-propane Flames

The study in this section investigates the stability and combustion features of $C_3H_8/O_2/CO_2$ flames in a model dry low emissions (DLE) gas turbine burner adopting lean premixed (LPM) oxy-fuel combustion technology. The model combustor is typically the same one used in the above mentioned studies in this chapter. Relationship between the flame speed (*FS*) and the adiabatic flame temperature ($T_{ad}$) was premeditated for better analysis and prediction of the flame behavior. The stability of the $C_3H_8/O_2/CO_2$ flame is characterized by its blowout and flashback limits over ranges of equivalence ratio ($\phi$), from 0.1 to 1.0, and oxygen fraction (*OF*), from 15 to 70% volumetric concentration of $O_2$ in the $O_2/CO_2$ oxidizer mixture. The stability limits were plotted as function of $\phi$ and *OF*, and against the contours of $T_{ad}$.

### 5.7.1 *Test Conditions*

As mentioned earlier, the bulk flow velocity of the reacting mixture was kept constant at a value of 5.2 m/s at the burner throat. The mass flow rates of $C_3H_8$, $O_2$, and $CO_2$ are, therefore, changed with respect to each other while moving from one data point (based on a specific combination of $\phi$ and *OF*) to another. In other words, for any combination of $\phi$ and *OF*, these three flow rates were mutually adjusted. Equivalence ratio ($\phi$) was spanned from 0.1 to 1.0 while *OF* was scanned from 15 to 70%, taking one value in one data point in a step of 5%, for identifying the stability maps of the reactant mixture.

### 5.7.2 *Stability Map of Oxy-propane Flames*

The combustor stability map, while operated under oxy-propane combustion conditions, is presented in Fig. 5.39 in the form of a 2-D map on background contour plots of the adiabatic flame temperature ($T_{ad}$). The plot is drawn for investigating the correlations between *FS* and the adiabatic flame temperature. For a specific *OF*, blowout point is represented by a circular symbol whereas a diamond one indicates flashback point. The stable region between the two curves narrows down with increasing *OF* and emphasizes the presence of non-reacting dilute gas ($CO_2$ in our case) for stable combustion. The blowout curve shows consistent agreement with the $T_{ad}$ (~1200 K)

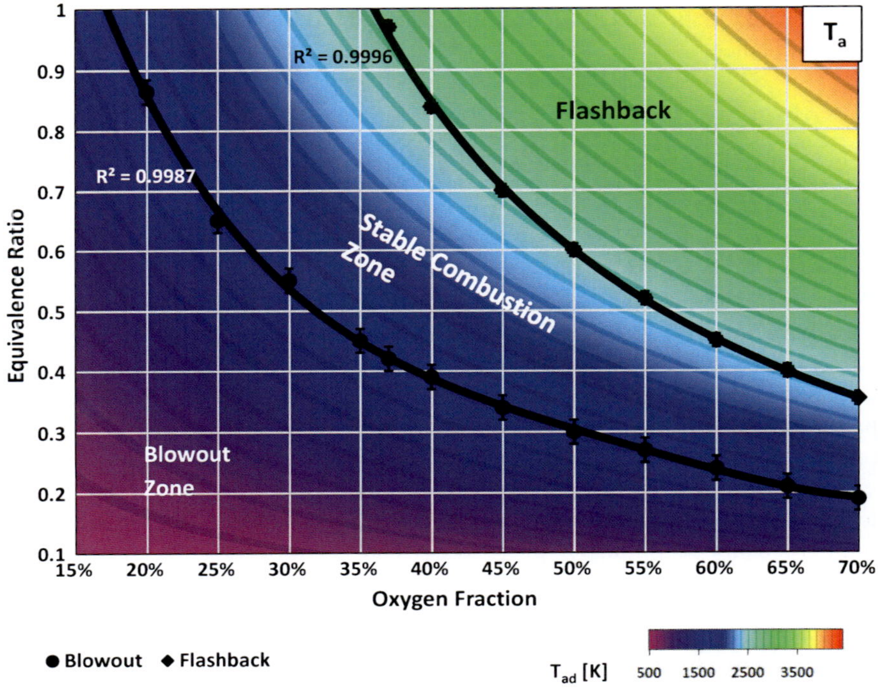


**Fig. 5.39** Combustor stability map at the fixed bulk throat velocity of 5.2 m/s plotted against the corresponding $T_{ad}$ curves

at constant bulk velocity of the reactant mixture of 5.2 m/s. Considering fixed burner geometry along with fixed inlet bulk velocity, one can assume that the flame blows out at fixed Damkohler number (*Da*). It is worthwhile to mention that $CH_4$–$O_2$–$CO_2$ premixed flames showed similar trend of the blowout at fixed $T_{ad}$ based on a previous study by the authors on the same swirl premixed combustor [121]. This shows that $T_{ad}$ is a more relevant controlling parameter choice of the combustor stability near the blowout limit.

### *5.7.3 Operability of Oxy-methane Versus Oxy-propane Flames*

The stability limits, in terms of flashback and blowout limits, of the investigated $C_3H_8/O_2/CO_2$ flames are compared in Fig. 5.40 with those of the $CH_4/O_2/CO_2$ flames obtained in a previous study [70] on the same premixed swirl combustor. It is observed that both the flashback and the blowout extinction limits of the methane flames are reached at higher $\phi$ as compared to those of the propane flames for the same *OF* in the

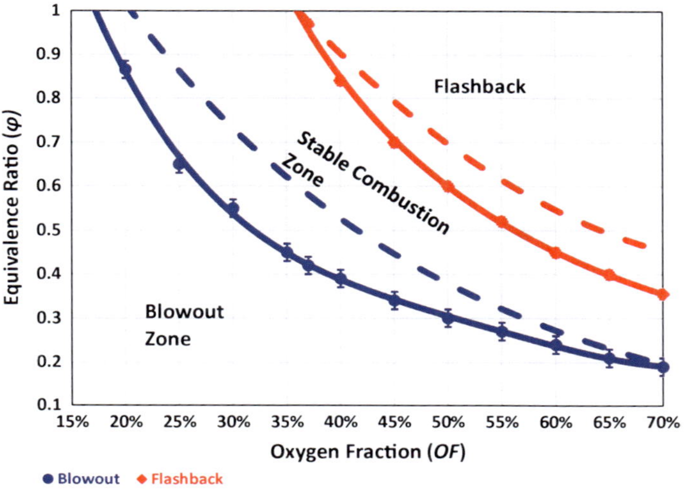


**Fig. 5.40** Comparison of static stability limits of $CH_4/O_2/CO_2$ and $CH_4/O_2/CO_2$ flames on the same premixed swirl combustor. The dashed line curves represent the limits for the methane flames, while the solid lines represent the limits for the propane flames

diluted oxidizer mixture. Since $T_{ad}$ of methane is higher (roughly about 200–250 K) than that of propane, this difference in $T_{ad}$ shifts the stability limits of methane to higher $\phi$, as compared to propane, for a given $OF$. In addition, it is observed that the propane flames have wider flame stability region than methane flames, and both zones narrow down towards larger $OF$ and smaller $\phi$. Even narrower stable flame region for oxy-propane flames than oxy-methane flames were obtained at higher $OF$ and lower $\phi$. It can be concluded that oxy-propane flames are more stable at lower $OF$ and higher $\phi$, whereas, oxy-methane flames are more stable at higher $OF$ and lower $\phi$.

## 5.8 Concluding Remarks

The operability of gas turbine combustors under fuel-flexible and oxidizer-flexible operating conditions is evaluated in this chapter based on experimental measurements and numerical calculations. The chapter started with a condensed literature review on the characteristics of fuel-flexible and oxidizer-flexible combustion, in terms of flame stability and emissions. It was concluded that the replacement of $N_2$ with $CO_2$ results in lower flame speeds and combustion temperatures in the oxy-fuel combustion when compared to the air-fuel one. Due to this reason, 21% $O_2$ in $O_2/CO_2$ oxidizer mixture is not achievable and therefore, $O_2$ composition should be more than this minimum threshold in the oxidizer mixture for achieving stable flames. The results of a set of studies based on experiments and numerical simulations performed on a premixed swirl-stabilized model gas turbine combustor are presented in this chapter. The studies examine the performance of the same gas turbine combustor under fuel-flexible and oxidizer-flexible operating conditions.

The first study of this chapter investigates experimentally the combustion behavior and stability map of hydrogen-enriched oxy-methane premixed flames in a model gas turbine combustor. Results indicated that both the flashback and blow-out limits tend to move towards the leaner side with increasing hydrogen fraction in the fuel stream indicating increased flame speed. Although, the stability limits are observed to follow the Reynolds number based on reactant gases velocity at the throat and the throat diameter. The findings can be comprehended from the fact that increasing hydrogen fraction in the fuel stream does not only reduce the Reynolds number but also enhance the kinetics of combustion. Reduction in Reynolds number is characterized by lower turbulence and hence, slower flame speed; however, increased kinetics tends to increase the flame speed. Moreover, an interesting correlation between the flammability limits and the throat velocities were noticed. While both the flashback and blow-out limits were observed to shift towards the leaner side with increasing throat velocities, blow-out limits tend to reach an asymptote at higher velocities. Increased turbulence level at higher velocities augments the flame front area resulting in higher flame speeds, which conforms the findings. However, blow-out limits are characterized by higher Reynolds numbers and it seems that turbulence effect reaches an asymptotic value for throat velocities higher than 5.2 m/s. Further, flame shapes were found to be affected by hydrogen fraction as well as equivalence ratios. Higher HF burning is characterized by compact and noisy flames due to increased flame speed. Equivalence ratio has a significant impact on AFT and flame shape. The effect of throat velocities on the flame shape was found to be prominent for $\phi < 0.80$, suggesting the flame speed to be a weaker function of turbulent intensity near stoichiometric combustion.

The second study of this chapter implemented the large eddy simulation (LES) methodology to evaluate the stability characteristics of hydrogen-enriched oxy-methane premixed flames. The simulations of the premixed turbulent flame were carried out to characterize the effect of hydrogen fraction in more detail under stoichiometric conditions. The results indicated wider operability range with hydrogen enrichment. The flames at higher hydrogen fractions have the characteristic *V*-shape. Unlike pure methane oxy-flames, the hydrogen-enriched ones do not follow constant Damkohler number at either the flashback or the blowout limits. At higher hydrogen fraction, the inner recirculation zone is characterized by more than one single eddy, and the number and size of eddies are mainly function of hydrogen fraction. The role of outer recirculation zone in controlling flame stability diminishes at higher hydrogen fraction. The rate of strain increases marginally with hydrogen fraction; however, the effect of hydrogen fraction on the critical strain rate is significant. Hence, the flames are more stable against quenching under lean combustion at higher hydrogen fractions. Moreover, the reaction rates enhance with hydrogen addition and carbon has to compete with hydrogen for oxygen; consequently, the flames at higher hydrogen fraction are characterized with higher $CO$ emissions at the combustor exit. The chemical time scale was found to be shorter than the smallest-eddy time scale, and hence, the inner flame structure is not affected by turbulence. Thus, the reactions are governed by the laminar burning rate for all the cases considered.

The third study of this chapter examines experimentally the effects of $H_2$-enrichment and inlet velocity on stability limits and shape of $CH_4/H_2$–$O_2/CO_2$ flames considering the same premixed swirl combustor and under stoichiometric conditions ($\varphi = 1.0$). The stability limits in terms of blowout and flashback were identified over wide ranges of hydrogen fraction (*HF*: %$H_2$ in $H_2$–$CH_4$ mixture) and oxygen fraction (*OF*: %$O_2$ in $O_2$–$CO_2$ mixture) at fixed inlet velocity of 6 m/s. The results show stable flame operation in the ranges 16% < OF < 44% and 0% < HF < 90%. At fixed throat inlet velocity, increasing *OF* at fixed *HF* results in an increase of the kinematic viscosity, which results in lower Re. This corresponds to lower Re at flashback than at that at blowout conditions. Moreover, increasing *HF* results in lower mixture Re values near stability limits. Furthermore, flame shapes were observed to be affected by the addition of hydrogen. At high values of *HF*, the flame shape tends to be shorter and more compact due to the increased reaction rate and flame speed. Higher *HF* burning is also characterized by noisy flames due to the unsteady heat release in the turbulent flame and the heat release rate fluctuations in the combustion zone. In the stable combustion zone, it was found that flame shapes are not significantly affected when Re is fixed. Slight changes in the flame shape were observed at fixed *AFT* due to the combined effect of increasing *HF* and decreasing *OF*. The flames become gradually shorter and more compact because of increased reaction rate in the combustion zone.

Finally, a study is presented at the end this chapter to identify static stability and combustion characteristics of oxy-propane flames in a premixed fuel-flexible swirl combustor. The stability of the $C_3H_8/O_2/CO_2$ flame is characterized by its blowout and flashback limits over ranges of equivalence ratio ($\phi$), from 0.1 to 1.0, and oxygen fraction (*OF*), from 15 to 70% volumetric concentration of $O_2$ in the $O_2/CO_2$ oxidizer mixture. The stability limits were plotted as function of $\phi$ and *OF*, and against the contours of $T_{ad}$. It was found that the blowout curve follows a fixed $T_{ad}$ curve on the stability map, whereas the flashback curve does not. This implies that $T_{ad}$ is a more relevant controlling parameter choice of the combustor stability near the blowout limit, whereas reaction rate is a suitable indicator of the flame stability limit near flashback. The stability map of $C_3H_8/O_2/CO_2$ flames was compared with that of $CH_4/O_2/CO_2$ flames obtained on the same combustor. It was found that at the same *OF*, the blowout and the flashback limits of the $C_3H_8/O_2/CO_2$ flame are achieved at a smaller $\phi$ as compared to those of $CH_4/O_2/CO_2$ flame.

**Acknowledgements** The authors wish to acknowledge the support received from King Fahd University of Petroleum & Minerals under Grant number BW191002 for the preparation of this book chapter.

## References


1. Priddle R (1998) IEA world energy outlook. IEA, Paris, France
2. Li H, Yan J, Yan J, Anheden M (2009) Impurity impacts on the purification process in oxyfuel combustion based $CO_2$ capture and storage system. Appl Energy 86(2):202–213
3. Kang Y, Wei S, Zhang P, Lu X, Wang Q, Gou X, Huang X, Peng S, Yang D, Ji X (2017) Energy 119:1195–1211
4. Xie Y, Wang J, Zhang M, Gong J, Jin W, Huang Z (2013) Experimental and numerical study on laminar flame characteristics of methane oxy-fuel mixtures highly diluted with $CO_2$. Energy Fuels 27:6231–6237. https://doi.org/10.1021/ef401220h
5. Oh J, Noh D (2012) Laminar burning velocity of oxy-methane flames in atmospheric condition. Energy 45(1):669–675. https://doi.org/10.1016/j.energy.2012.07.027
6. Konnov AA, Dyakov IV (2004) Measurement of propagation speeds in adiabatic flat and cellular premixed flames of $C_2H_6 + O_2 + CO_2$. Combust Flame 136:371–376
7. Konnov AA, Dyakov IV (2007) Experimental study of adiabatic cellular premixed flames of methane (ethane, propane) + oxygen + carbon dioxide mixtures. Combust Sci Technol 179(4):747–765
8. Abubakar Z, Shakeel MR, Mokheimer EMA (2018) Experimental and numerical analysis of non-premixed oxycombustion of hydrogen-enriched propane in a swirl stabilized combustor. Energy 165:1401–1414
9. Abubakar Z, Sanusi SY, Mokheimer EMA (2017) Stability of propane-air and oxyfuel diffusion flames in a swirlstabilized combustor; an experimental study. Energy Procedia
10. Konnov AA, Meuwissen RJ, De Goey LPH (2011) The temperature dependence of the laminar burning velocity of ethanol flames. Proc Combust Inst 33(1):1011–1019
11. Gu X, Huang Z, Wu S, Li Q (2010) Laminar burning velocities and flame instabilities of butanol isomers-air mixtures. Combust Flame 157(12):2318–2325
12. Van Lipzig JPJ, Nilsson EJK, De Goey LPH, Konnov AA (2011) Laminar burning velocities of n-heptane, iso-octane, ethanol and their binary and tertiary mixtures. Fuel 90(8):2773–2781
13. Vagelopoulos CM, Egolfopoulos FN (1998) Direct experimental determination of laminar flame speeds. Symp (Int) Combust 27(1):513–519
14. Tseng LK, Ismail MA, Faeth GM (1993) Laminar burning velocities and Markstein numbers of hydrocarbon/air flames. Combust Flame 95(4):410–426
15. Toporov D, Bocian P, Heil P, Kellermann A, Stadler H, Tschunko S et al (2008) Detailed investigation of a pulverized fuel swirl flame in $CO_2/O_2$ atmosphere. Combust Flame 155(4):605–618
16. Kaskan WE (1957) The dependence of flame temperature on mass burning velocity. Symp Combust 6(1):134–143
17. van Maaren A, Thung DS, de Goey LRH (1994) Measurement of flame temperature and adiabatic burning velocity of methane/air mixtures. Combust Sci Technol 96(4–6):327–344
18. Liao SY (2007) Determination of the laminar burning velocities for mixtures of ethanol and air at elevated temperatures. Appl Therm Eng 27:374–380
19. Metghalchi M, Keck JC (1980) Laminar burning velocity of propane-air mixtures at high temperature and pressure. Combust Flame 38:143–154
20. Yi B, Zhang L, Huang F, Mao Z, Zheng C (2014) Effect of $H_2O$ on the combustion characteristics of pulverized coal in $O_2$/ $CO_2$ atmosphere. Appl Energy 132:349–357
21. Galmiche B, Halter F, Foucher F, Dagaut P (2011) Effects of dilution on laminar burning velocity of premixed methane/air flames, pp. 948–954
22. Kim HK, Kim Y, Lee SM, Ahn KY (2007) NO reduction in 0.03–0.2 MW oxy-fuel combustor using flue gas recirculation technology. Proc Combust Inst 31(x):3377–3384
23. Habib MA, Badr HM, Ahmed SF, Ben-Mansour R, Mezghani K, Imashuku S, la O' GJ, Shao-Horn Y, Mancini ND, Mitsos A ()A review of recent developments in carbon capture utilizing oxy-fuel combustion in conventional and ion transport
24. Liu CY, Chen G, Sipöcz N, Assadi M, Bai XS (2012) Characteristics of oxy-fuel combustion in gas turbines. Appl Energy 89(1):387–394

25. Heil P, Toporov D, Stadler H, Tschunko S, Förster M, Kneer R (2009) Development of an oxycoal swirl burner operating at low $O_2$ concentrations. Fuel 88(7):1269–1274
26. Glarborg P, Bentzen LLB (2008) Chemical effects of a gigh $CO_2$ concentration in oxy-fuel combustion of methane. Energy Fuels 22(1):291–296
27. Wang J, Huang Z, Tang C, Miao H, Wang X (2009) Numerical study of the effect of hydrogen addition on methane-air mixtures combustion. Int J Hydrogen Energy 34(2):1084–1096. https://doi.org/10.1016/j.ijhydene.2008.11.010
28. Shy SS, Chen YC, Yang CH, Liu CC, Huang CM (2008) Effects of H2 or CO2 addition, equivalence ratio, and turbulent straining on turbulent burning velocities for lean premixed methane combustion. Combust Flame 153:510–524. https://doi.org/10.1016/j.combustflame.2008.03.014
29. Boushaki T, Dhué Y, Selle L, Ferret B, Poinsot T (2012) Effects of hydrogen and steam addition on laminar burning velocity of methane-air premixed flame: Experimental and numerical analysis. Int J Hydrogen Energy 37(11):9412–9422
30. Guo H, Smallwood GJ, Liu F, Ju Y, Gülder ÖL (2005) The effect of hydrogen addition on flammability limit and NOx emission in ultra-lean counterflow $CH_4$/air premixed flames. Proc Combust Inst 30(1):303–310
31. Tuncer O (2013) The effect of hydrogen enrichment of methane fuel on flame stability and emissions. In: International conference on Renewable Energy Research and Applications (ICRERA) 2013, pp 103 – 108
32. Kim HS, Arghode VK, Gupta AK (2009) Flame characteristics of hydrogen-enriched methane-air premixed swirling flames. Int J Hydrogen Energy 34(2):1063–1073. https://doi.org/10.1016/j.ijhydene.2008.10.035
33. Schefer RW (2002) Reduced turbine emissions using hydrogen-enriched fuels. In: Proceedings of the 2002 U.S. DOE Hydrogen Program Review 2002, pp 1–16
34. Yu G, Law CK, Wu CK (1986) Laminar flame speeds of hydrocarbon þ air mixtures with hydrogen addition. Combust Flame 63(3):339–347
35. Gauducheau JL, Denet B, Searby G (1998) A numerical study of lean $CH_4$/$H_2$/air premixed flames at high pressure. Combust Sci Technol 137(1–6):81–99
36. Aliyu Mansur, Nemitallah MA, Said Syed A, Habib Mohamed A (2016) Characteristics of $H_2$-enriched $CH_4$–$O_2$ diffusion flames in a swirl-stabilized gas turbine combustor: experimental and numerical study. Internation Journal of Hydrogen Energy 41:2041
37. Imteyaz BA, Nemitallah MA, Abdelhafez AA, Habib MA (2018) Combustion behavior and stability map of hydrogen-enriched oxy-methane premixed flames in a model gas turbine combustor. Int J Hydrogen Energy 43
38. Crocco L (1969) Symp Combust [Proc.] 12(1):85–99
39. Annaswamy M, Ghoniem F (2002) Control Syst IEEE 22(6):37–54
40. Lieuwen T, Torres H, Johnson C, Zinn BT (2001) J Eng Gas Turbines Power 123:182–189
41. Khare SP, Wall TF, Farida AZ, Liu Y, Moghtaderi B, Gupta RP (2008) Factors influencing the ignition of flames from air-fired swirl pf burners retrofitted to oxy-fuel. Fuel 87(7):1042–1049
42. Taamallah S, Chakroun NW, Watanabe H, Shanbhogue SJ, Ghoniem AF (2017) On the characteristic flow and flame times for scaling oxy and air flame stabilization modes in premixed swirl combustion. Proc Combust Inst 36(3):3799–3807. https://doi.org/10.1016/j.proci.2016.07.022
43. Shroll AP, Shanbhogue SJ, Ghoniem AF (2012) Dynamic-stability characteristics of premixed methane oxy-combustion. J Eng Gas Turbines Power 134(5):51504
44. Tang C, Huang Z, Jin C, He J, Wang J, Wang X et al (2008) Laminar burning velocities and combustion characteristics of propane–hydrogen–air premixed flames. Int J Hydrogen Energy 33(18):4906–4914. https://doi.org/10.1016/j.ijhydene.2008.06.063
45. Ishizuka S, Law CK (1982) An experimental study on extinction and stability of stretched premixed flames. In: Symposium (International) on combustion. Elsevier, Amsterdam
46. Kutne P, Kapadia BK, Meier W, Aigner M (2011) Experimental analysis of the combustion behaviour of oxyfuel flames in a gas turbine model combustor. Proc Combust Inst 33(2):3383–3390

47. Song Y, Zou C, He Y, Zheng C (2015) The chemical mechanism of the effect of $CO_2$ on the temperature in methane oxy-fuel combustion. Int J Heat Mass Transf 86:622–628
48. Saygin D, Kempener R, Wagner N, Ayuso M, Gielen D (2015)The implications for renewable energy innovation of doubling the share of renewables in the global energy mix between 2010 and 2030. Energies 8:5828–5865. https://doi.org/10.3390/en8065828
49. Deprez A, Colombier M, Iddri TS (2015) Transparency and the Paris Agreement : driving ambitious action in the new climate regime 2015
50. Wall T, Gupta R, Buhre B, Khare S (2005) Oxy-fuel ($O_2/CO_2$, $O_2$/RFG) technology for sequestration-ready $CO_2$ and emission compliance. In: 30th International technical conference on coal utilization and fuel systems, Cleanwater, USA
51. Al-Abbas AH, Naser J, Dodds D (2011) CFD modelling of air-fired and oxy-fuel combustion of lignite in a 100 KW furnace. Fuel 90:1778–1795. https://doi.org/10.1016/j.fuel.2011.01.014
52. Nemitallah MA, Habib MA, Badr HM, Said SA, Jamal A, Ben-Mansour R et al (2017) Oxy-fuel combustion technology: current status, applications, and trends. Int J Energy Res. https://doi.org/10.1002/er.3722
53. Nozaki T, Takano S, Kiga T (1997) Analysis of the flame formed during oxidation of pulverized coal by an $O_2$-$CO_2$ mixture. Energy 22:199–205
54. Chen L, Yong SZ, Ghoniem AF (2012) Oxy-fuel combustion of pulverized coal: Characterization, fundamentals, stabilization and CFD modeling. Prog Energy Combust Sci 38:156–214. https://doi.org/10.1016/j.pecs.2011.09.003
55. Zhang N, Lior N (2008) Two novel oxy-fuel power cycles integrated with natural gas reforming and $CO_2$ capture. Energy 33:340–351. https://doi.org/10.1016/j.energy.2007.09.006
56. Habib MA, Nemitallah MA, Ben-Mansour R (2013) Recent development in oxy-combustion technology and its applications to gas turbine combustors and ITM reactors. Energy Fuels 27:2–19. https://doi.org/10.1021/ef301266j
57. Amato A, Hudak B, D'Carlo P, Noble D, Scarborough D, Seitzman J et al (2011) Methane oxycombustion for low $CO_2$ cycles: blowoff measurements and analysis. J Eng Gas Turbines Power 133. https://doi.org/10.1115/1.4002296
58. Shroll AP, Shanbhogue SJ, Ghoniem AF (2012) Dynamics and stability limits of syngas combustion in a swirl-stabilized combustor. ASME J Eng Gas Turbines Power 5:51504
59. Watanabe H, Shanbhogue SJ, Ghoniem AF. Impact of equivalence ratio on the macrostructure of premixed swirling $CH_4$/air and $CH_4/O_2/CO_2$ flames. Vol 4B Combust Fuels Emiss V04BT04A014. https://doi.org/10.1115/gt2015-43224 (ASME; 2015)
60. Ditaranto M, Anantharaman R, Weydahl T (2013) Performance and NOx emissions of refinery fired heaters retrofitted to hydrogen combustion. Energy Procedia 37:7214–7220. https://doi.org/10.1016/j.egypro.2013.06.659
61. Riahi Z, Bounaouara H, Hraiech I, Ali Mergheni M, Sautet J-C, Ben Nasrallah S (2017) Combustion with mixed enrichment of oxygen and hydrogen in lean regime. https://doi.org/10.1016/j.ijhydene.2016.06.232
62. Rezgui Y, Guemini M (2017) Effect of hydrogen addition on equimolar dimethyl ether/iso-octane/oxygen/argon premixed flames. Int J Hydrogen Energy 42:29557–29573. https://doi.org/10.1016/J.IJHYDENE.2017.10.063
63. Shi B, Hu J, Ishizuka S (2015) Carbon dioxide diluted methane/oxygen combustion in a rapidly mixed tubular flame burner. Combust Flame 162:420–430. https://doi.org/10.1016/j.combustflame.2014.07.022
64. Shi B, Zhu Z, Wang N, Lu P (2015) An Experimental study on oxy-fuel combustion of methane under various oxygen mole fractions. In: 8th international symposium on coal combustion
65. Konnov AA, Dyakov IV (2005) Measurement of propagation speeds in adiabatic cellular premixed flames of $CH_4 + O_2 + CO_2$. Exp Therm Fluid Sci 29:901–907. https://doi.org/10.1016/j.expthermflusci.2005.01.005
66. Coppens FHV, Konnov AA (2008) The effects of enrichment by $H_2$ on propagation speeds in adiabatic flat and cellular premixed flames of $CH_4 + O_2 + CO_2$. Fuel 87:2866–2870. https://doi.org/10.1016/j.fuel.2008.04.009

67. Li J, Huang H, Kobayashi N, He Z, Nagai Y (2014) Study on using hydrogen and ammonia as fuels: combustion characteristics and NOx formation. Int J Energy Res 38:1214–1223. https://doi.org/10.1002/er.3141
68. Mokheimer EMA, Sanusi YS, Habib MA (2016) Numerical study of hydrogen-enriched methane-air combustion under ultra-lean conditions. Int J Energy Res 40:743–762. https://doi.org/10.1002/er.3477
69. İlbaş M, Yılmaz İ (2012) Experimental analysis of the effects of hydrogen addition on methane combustion. Int J Energy Res 36:643–647. https://doi.org/10.1002/er.1822
70. Abdelhafez AA, Rashwan SS, Nemitallah MA, Habib MA (2018) Stability map and shape of premixed CH4/$O_2$/$CO_2$ flames in a model gas-turbine combustor. Appl Energy 215:63–74
71. Mazas A, Lacoste D, Fiorina B, Schuller T (2009) Effects of water vapor addition on the laminar burning velocity of methane oxygen-enhanced flames at atmospheric pressure. In: Proceedings of European Combustion Meeting, pp 1–6
72. Li YH, Chen GB, Lin Y-C, Wu F-H, Chao YC (2013) Effects of flue gas addition on the premixed oxy-methane flames. In: 24th international colloquium on the dynamics of explosions and reactive systems. https://doi.org/10.1016/j.egypro.2015.07.623
73. Jourdaine P, Mirat C, Beaunier J, Joumani Y, Schuller T (2015) A comparison of the structure of $N_2$ and $CO_2$ diluted $CH_4$/$O_2$ premixed flames in a swirled combustor. In: Proceedings of European Combustion Meeting
74. Jourdaine P, Mirat C, Caudal J, Lo A, Schuller T (2016) A comparison between the stabilization of premixed swirling CO2-diluted methane oxy-flames and methane/air flames. Fuel. https://doi.org/10.1016/j.fuel.2016.11.017
75. Marsh R, Runyon J, Giles A, Morris S, Pugh D, Valera-Medina A et al (2016) Premixed methane oxycombustion in nitrogen and carbon dioxide atmospheres: measurement of operating limits, flame location and emissions. In: Proceedings of Combustion Institute, vol 0. Elsevier, Amsterdam. http://dx.doi.org/10.1016/j.proci.2016.06.057
76. Taamallah S, Shanbhogue SJ, Ghoniem AF (2016) Turbulent flame stabilization modes in premixed swirl combustion: physical mechanism and Karlovitz number-based criterion. Combust Flame 166:19–33. https://doi.org/10.1016/j.combustflame.2015.12.007
77. Halter F, Chauveau C, Gökalp I (2007) Characterization of the effects of hydrogen addition in premixed methane/air flames. Int J Hydrogen Energy 32:2585–2592. https://doi.org/10.1016/j.ijhydene.2006.11.033
78. Cozzi F, Coghe A (2006) Behavior of hydrogen-enriched non-premixed swirled natural gas flames. Int J Hydrogen Energy 31:669–677. https://doi.org/10.1016/j.ijhydene.2005.05.013
79. Tahtouh T, Halter F, Samson E, Mounaïm-Rousselle C (2009) Effects of hydrogen addition and nitrogen dilution on the laminar flame characteristics of premixed methane-air flames. Int J Hydrogen Energy 34:8329–8338. https://doi.org/10.1016/j.ijhydene.2009.07.071
80. Hu E, Huang Z, He J, Jin C, Zheng J (2009) Experimental and numerical study on laminar burning characteristics of premixed methane–hydrogen–air flames. Int J Hydrogen Energy 34:4876–4888. https://doi.org/10.1016/j.ijhydene.2009.03.058
81. Schefer RW (2003) Hydrogen enrichment for improved lean flame stability. Int J Hydrogen Energy 28:1131–1141
82. Di Sarli V, Di Benedetto A (2007) Laminar burning velocity of hydrogen–methane/air premixed flames. Int J Hydrogen Energy 32:637–46. https://doi.org/10.1016/j.ijhydene.2006.05.016
83. Aliyu M, Nemitallah MA, Said SA, Habib MA (2016) Characteristics of $H_2$-enriched $CH_4$-$O_2$ diffusion flames in a swirl-stabilized gas turbine combustor: experimental and numerical study. Int J Hydrogen Energy 41:20418–20432. https://doi.org/10.1016/j.ijhydene.2016.08.144
84. Gersen S, Anikin NB, Mokhov AV, Levinsky HB (2008) Ignition properties of methane/hydrogen mixtures in a rapid compression machine. Int J Hydrogen Energy 33:1957–1964. https://doi.org/10.1016/j.ijhydene.2008.01.017
85. Gülder ÖL, Snelling DR, Sawchuk RA (1996) Influence of hydrogen addition to fuel on temperature field and soot formation in diffusion flames. Symp Combust 26:2351–2358. https://doi.org/10.1016/S0082-0784(96)80064-6

86. Pandey P, Pundir BP, Panigrahi PK (2007) Hydrogen addition to acetylene-air laminar diffusion flames: studies on soot formation under different flow arrangements. Combust Flame 148:249–262. https://doi.org/10.1016/j.combustflame.2006.09.004
87. Kumar P, Mishra DP (2008) Experimental investigation of laminar LPG-$H_2$ jet diffusion flame. Int J Hydrogen Energy 33:225–231. https://doi.org/10.1016/j.ijhydene.2007.09.023
88. Park SH, Lee KM, Hwang CH (2011) Effects of hydrogen addition on soot formation and oxidation in laminar premixed $C_2H_2$/air flames. Int J Hydrogen Energy 36:9304–9311. https://doi.org/10.1016/j.ijhydene.2011.05.031
89. Choudhuri AR, Gollahalli SR (2000) Combustion characteristics of hydrogen-hydrocarbon hybrid fuels. Int J Hydrogen Energy 25:451–462. https://doi.org/10.1016/S0360-3199(99)00027-0
90. Nemitallah MA, Habib MA (2013) Experimental and numerical investigations of an atmospheric diffusion oxy-combustion flame in a gas turbine model combustor. Appl Energy 111:401–415. https://doi.org/10.1016/j.apenergy.2013.05.027
91. Vascellari M, Cau G (2009) Numerical simulation of pulverized coal oxy-combustion with exhaust gas recirculation. In: Proceeding CCT2009 Fourth International Conference on Clean Coal Technologies, Dresden, Germany
92. Andersson K, Johnsson F (2007) Flame and radiation characteristics of gas-fired $O_2/CO_2$ combustion. Fuel 86:656–668. https://doi.org/10.1016/j.fuel.2006.08.013
93. Imteyaz B, Habib MA, Ben-Mansour R (2017) The characteristics of oxycombustion of liquid fuel in a typical water-tube boiler. Energy Fuels 31:6305–6313. https://doi.org/10.1021/acs.energyfuels.7b00489
94. Joshi ND, Epstein MJ, Durlak S, Marakovits S (1994) Development of a fuel air premixer for aero-derivative dry low emissions combustors, vol 4. In: ASME Pap 94-GT 1994, p 253
95. Sullivan-Lewis E, McDonell V (2016) Predicting flameholding for hydrogen and natural gas flames at gas turbine premixer conditions. J Eng Gas Turbines Power 138:121502. https://doi.org/10.1115/1.4034000
96. Driscoll JF (2008) Turbulent premixed combustion: flamelet structure and its effect on turbulent burning velocities. Prog Energy Combust Sci 34:91–134. https://doi.org/10.1016/j.pecs.2007.04.002
97. Cheng RK, Littlejohn D (2008) Laboratory study of premixed $H_2$-Air and $H_2$–$N_2$–Air flames in a low-swirl injector for ultralow emissions gas turbines. J Eng Gas Turbines Power 130:31503. https://doi.org/10.1115/1.2836480
98. Cheng RK, Littlejohn D, Nazeer WA, Smith KO (2008) Laboratory studies of the flow field characteristics of low-swirl injectors for adaptation to fuel-flexible turbines. J Eng Gas Turbines Power 130:21501. https://doi.org/10.1115/1.2795786
99. Wietschel M, Ball M (2009) The future of hydrogen—opportunities and challenges. Hydrog Econ Oppor Challenges 9780521882:613–639. https://doi.org/10.1017/CBO9780511635359.021
100. Edwards PP, Kuznetsov VL, David WIF, Brandon NP (2008) Hydrogen and fuel cells: towards a sustainable energy future. Energy Policy 36:4356–4362. https://doi.org/10.1016/J.ENPOL.2008.09.036
101. European Commission (2008) HyWays: the European hydrogen roadmap: contract SES6-502596, EUR-OP, Luxembourg
102. Delattin F, Di Lorenzo G, Rizzo S, Bram S, De Ruyck J (2010) Combustion of syngas in a pressurized microturbine-like combustor: experimental results. Appl Energy 87(4):1441–1452
103. Khalil AE, Gupta AK (2013) Hydrogen addition effects on high intensity distributed combustion. Appl Energy 1(104):71–78
104. Gu M, Chu H, Liu F (2016) Effects of simultaneous hydrogen enrichment and carbon dioxide dilution of fuel on soot formation in an axisymmetric coflow laminar ethylene/air diffusion flame. Combust Flame 166:216–228. https://doi.org/10.1016/j.combustflame.2016.01.023
105. Imteyaz B, Nemitallah MA, Abdelhafez AA, Habib MA (2018) Combustion behavior and stability map of hydrogen-enriched oxy-methane premixed flames in a model gas turbine

combustor. Int J Hydrogen Energy 43:16652–16666. https://doi.org/10.1016/j.ijhydene.2018.07.087

106. Abdelwahid S, Nemitallah MA, Imteyaz B, Abdelhafez AA, Habib MA (2018) On the effects of $H_2$-enrichment and inlet velocity on stability limits and shape of $CH_4/H_2$-$O_2/CO_2$ flames in a premixed swirl combustor. Energy Fuels 32. https://doi.org/10.1021/acs.energyfuels.8b01958
107. Cheng RK, Oppenheim AK (1984) Autoignition in methanehydrogen mixtures. Combust Flame 58:125–139. https://doi.org/10.1016/0010-2180(84)90088-9
108. Sankaran R, Im HG (2006) Effects of hydrogen addition on the markstein length and flammability limit of stretched methane/air premixed flames. Combust Sci Technol 178:37–41. https://doi.org/10.1080/00102200500536217org/10.1080/00102200500536217
109. Nakahara M, Kido H (2008) Study on the turbulent burning velocity of hydrogen mixtures including hydrocarbons. AIAA J 46. https://doi.org/10.2514/1.23560
110. Daniele S, Jansohn P, Mantzaras J, Boulouchos K (2011) Turbulent flame speed for syngas at gas turbine relevant conditions. Proc Combust Inst 33:2937–2944. https://doi.org/10.1016/J.PROCI.2010.05.057
111. Daniele S, Jansohn P, Boulouchos K (2009) Experimental investigation of lean premixed syngas combustion at gas turbine relevant conditions: lean blow out limits, emissions and turbulent flame speed. Italian Section of the Combustion Institute
112. Syred N, Abdulsada M, Griffiths A, O'Doherty T, Bowen P (2012) The effect of hydrogen containing fuel blends upon flashback in swirl burners. Appl Energy 89(1):106–110
113. Khalil AE, Gupta AK (2013) Fuel flexible distributed combustion for efficient and clean gas turbine engines. Appl Energy 1(109):267–274
114. Zhang Q, Noble DR, Lieuwen T (2007) Characterization of fuel composition effects in $H_2/CO/CH_4$ mixtures upon lean blowout. J Eng Gas Turbines Power 129(3):688–694
115. Chen L, Ghoniem AF (2012) Simulation of oxy-coal combustion in a 100 kW th test facility using RANS and LES: a validation study. Energy Fuels 26:4783–4798. https://doi.org/10.1021/ef3006993
116. Rajhi MA, Ben-Mansour R, Habib MA, Nemitallah MA, Andersson K (2014) Evaluation of gas radiation models in CFD modeling of oxy-combustion. Energy Convers Manag 81:83–97. https://doi.org/10.1016/j.enconman.2014.02.019
117. Johansson R, Leckner B, Andersson K, Johnsson F (2011) Account for variations in the $H_2O$ to $CO_2$ molar ratio when modelling gaseous radiative heat transfer with the weighted-sum-of-grey-gases model. Combust Flame 158:893–901. https://doi.org/10.1016/j.combustflame.2011.02.001
118. Négishi N (1982) Lean premixture combustion on a coaxial burner. Symp Combust 19:441–447. https://doi.org/10.1016/S0082-0784(82)80216-6
119. Nemitallah MA, Kewlani G, Hong S, Shanbhogue SJ, Habib MA, Ghoniem AF (2015) Investigation of a turbulent premixed combustion flame in a backward-facing step combustor; effect of equivalence ratio. Energy 95:211–222. https://doi.org/10.1016/j.energy.2015.12.010
120. Abdelwahid Suliman, Nemitallah MA, Imteyaz Binash, Abdelhafez AA, Habib MA (2018) Effects of $H_2$ enrichment and inlet velocity on stability limits and shape of $CH_4/H_2$–$O_2/CO_2$ flames in a premixed swirl combustor. Energy Fuels 32(9):9916–9925
121. Liao SY, Jiang DM, Huang ZH, Zeng K, Cheng Q (2007) Determination of the laminar burning velocities for mixtures of ethanol and air at elevated temperatures. Appl Therm Eng 27(2–3):374–380

# Chapter 6
# Hybrid Membrane and Porous-Plates Reactors for Gas Turbine Applications

## 6.1 Introduction

Because of the high efficiency penalty associated with using cryogenic oxygen separation units in oxy-combustion systems, researchers were striving to find alternative solutions. One of these solutions is the use of ion transport membranes (ITMs) for oxygen production. These ITMs have the capability of extracting oxygen from air at elevated temperatures (above 650 °C). The permeation of oxygen through the ion transport membranes depends on the membrane type, thickness, operating temperature and the partial pressure of oxygen across the membrane. The integration of these membranes in oxy-combustion reactors led to the design of oxygen transport reactors (OTRs) where oxygen-air separation occurs at one side while the fuel burning occurs at the other side of the membrane. The idea is to replace the currently used conventional combustors by multi-membrane reactor system. However, such membrane reactors are suffering from low oxygen permeation and deterioration of membrane performance after a short period of time due to contamination caused by combustion products. Currently, research is focused on improving the membrane performance and chemical stability under more demanding operational conditions. In this regard, dense mixed-ionic-electronic conducting ceramic (MIEC) membrane has shown a good potential for use in oxy-fuel technology. One main challenge in designing an OTR to be used in power plants is the low oxygen flux that makes the required membrane surface area to be very large. Accordingly, improving oxygen permeation in ITMs is a necessary step towards the efficient design of OTRs. Another kind of membranes is the polymeric kind, which can separate gases under low temperatures. In polymeric membrane gas separation, there is no need for thermal heating of the membrane or the involvement of any sorbents to separate a gas mixture, as in the cases of cryogenic distillation and adsorption. The main driving force in polymeric membrane air separation is the partial pressure gradient across the membrane.



M. A. Nemitallah et al., *Approaches for Clean Combustion in Gas Turbines*,
Fluid Mechanics and Its Applications 122,
https://doi.org/10.1007/978-3-030-44077-0_6

The concept of gas separation using ceramic membranes is first introduced in this chapter, followed by describing the gas separation process using polymeric membranes and how the two kinds of membranes can be integrated together to improve the oxygen permeation flux for the combustion process. A condensed literature review is presented in this chapter to explore the possibility of the application of OTRs for clean power generation using oxy-fuel combustion technology. In addition, novel designs of OTRs are presented in this chapter and their performance is evaluated based on detailed numerical simulations supported by experimental measurements. In this regard, a detailed study is presented first on the characteristic of air separation in different hollow-fiber polymeric membranes for oxygen enriched air clean combustion applications. Targeting higher $O_2$ concentrations out of the polymeric membrane unit, multi-stage air separation unit is developed and presented in this study. Based on this study, a hybrid polymeric-ceramic membrane unit is developed and its performance is evaluated for carbon-free oxy-combustion applications. A ceramic oxygen transport membrane (OTM) unit is integrated with a polymeric membrane unit to form the hybrid polymeric-ceramic membrane reactor. Oxygen-enriched air out of the polymeric membrane unit is fed to a ceramic membrane oxy-fuel based reactor. A detailed numerical study is performed to investigate oxy-fuel combustion in a tubular oxygen transport reactor (OTR) that is fed by oxygen-enriched air.

Actually, the process of oxy-combustion inside the OTRs is difficult to be achieved at normal lab conditions and requires lot of precautions. In addition, the oxygen permeation flux across the membranes still is not enough for complete fuel conversion. Because of that, porous plates are used instead of membranes in a study that is presented in this chapter trying to explore the flame characteristics and its location with respect to the porous plate that should help in the development of better performance OTRs. The study investigates experimentally and numerically the laminar flow field and oxy-methane combustion characteristics in a face-to-face two-porous-plates reactor over wide range of operating global equivalence ratio. The reactor is of low power to mimic the operation of OTRs, but under sufficient oxygen permeation flux for combustion. This is achieved by using porous plates instead of low oxygen separation rate membranes. Another detailed study is presented in this chapter based on experimental measurements to investigate in detail the combustion characteristics and lean blow-out (LBO) limits of the generated non-premixed oxy-methane flames inside the porous-plates reactor.

## 6.2 Concept of Membrane Separation

The main difference between ITMs and polymeric membranes is that polymeric membranes can separate gases under low temperature; however, the produced gases are of lower purity when compared to ITMs. On the other hand, ITMs can only separate gases under high temperatures, a phenomenon that makes ITMs suitable for oxy-combustion applications in what is called OTRs for power generation. However, the oxygen permeation flux across the ITMs is low and needs to be improved for complete fuel conversion. In fact, oxygen permeation flux across the ITMs is mainly function of the difference in oxygen partial pressure between both sides of the membranes, and improving oxygen concentration in the feeding air in the feed side should result in improved oxygen permeation flux. This makes the polymeric membranes a good choice to be integrated with ITM unit in a preceding stage to supply the ITM unit with oxygen-enriched air in order to raise its oxygen permeation flux making it more convenient for oxy-combustion applications. In this section, the concepts of gas separation using ion transport membranes (ITMs) and polymeric membranes are introduced.

### 6.2.1 Ceramic Membranes

During last decades, especial care has been taken for the different applications of ion transport membranes (ITMs). These applications include pure oxygen separation from air (non-reacting medium), production of syngas (partial fuel conversion), and recently, oxy-fuel combustion technology [1]. All of these applications of ITMs are targeting lowered concentrations of the greenhouse gases in the ambient atmosphere as an action to reduce the globe temperature [2]. For the purpose of pure oxygen separation, cryogenic air separation units (CASU) are used and then the produced oxygen can be used in combustion in conventional combustion systems. However, the CASU is highly energy intensive, about 3–4% of the produced energy in oxy-combustion process is consumed in the CASU [3]. As an alternative to CASU, chemical looping air separation (CLAS) process can be used for pure oxygen separation for oxy-fuel thermal power plants. Due to its simplicity in terms of hardware, low cost, and the possibility of operation under atmospheric pressure; the CLAS process results in significantly low capital cost [4, 5]. Recently, a new idea has been appeared in order to combine both the oxygen separation and oxy-combustion processes in one unit; this unit is called ITM reactors. In such kind of reactors, membrane is used to selectively separate oxygen from the flowing air in the feed side to the permeate side where the oxy-combustion process occurs. The membrane is heated using portion of the heat released from the combustion process in the permeate side. Different ITM reactor designs using different ceramic membrane materials are considered in the literature. In all cases, the main goal is to maximize the oxygen permeation rate while keeping the membrane temperature within the operating limits of the considered membrane material. Different membrane materials were considered in the literature including the lanthanum cobaltite perovskite ceramics [6–9]. Due to the specific needs raised by different membrane applications, especially in ITM reactor technology, successive research is being done for the development of new ceramic membrane materials that can withstand higher operating temperatures. These new materials include ceramic–metal dual phase membranes, modified perovskite ceramics, thin dual phase membranes including the Pd phases and yttria-stabilized zirconia (YSZ), and structured ceramic [10–13]. Recently, Nemitallah et al. [14, 15] reported similar results at elevated membrane temperature. Using ultra-thin $La_{0.6}Sr_{0.4}Co_{0.2}Fe_{0.8}O_{3-\delta}$

ceramic membranes, Tan and Li [16] studied experimentally the influences of both the membrane surface temperature and the sweep gas flux on the oxygen permeation flux. Higher oxygen permeation fluxes were observed at higher membrane surface temperatures and higher sweep gas flow rates.

Other than pure oxygen separation, ITMs can be used in partial oxidation of methane to syngas. The separated oxygen is consumed in the sweep side of the membrane in the fuel conversion process [17]. Nowadays, most of the research works are focused on the application of the oxy-fuel combustion technology into ITM reactors. In ITM reactor technology, the separated oxygen is directly consumed in the combustion process with fuel in the sweep side of the membrane. In this case, the combustion products consist only of $H_2O$ and $CO_2$. $H_2O$ can be condensed easily through a simple condensation process, and then, $CO_2$ can be compressed for sequestration or industrial uses [14]. In case of ITM reactor applications, oxygen permeation rate is highly improved as compared to the non-reacting cases (pure oxygen separation). This is due to the reduced oxygen partial pressure in the sweep side of the membrane as a result of oxygen consumption in the combustion process. Also, due to elevated temperature as a result of exothermic reactions, the membrane total resistance for oxygen permeation is reduced, and then, the oxygen permeation flux is increased [15]. The characteristics of both oxygen permeation and oxy-combustion processes inside a stagnation flow ITM reactor were examined by Ben-Mansour et al. [18]. They reported significant increase in the oxygen permeation flux when the reactions are activated in the permeate side of the membrane as compared to the non-reactive case. Akin and Lin [19] conducted some experiments in order to investigate the influences of sweep gas flux and on the oxygen permeation flux. A reduction in the oxygen permeation flux was reported when the sweep gas flux was increased. Mancini and Mitsos [20] introduced a new design of an ITM reactor to be used for production of the required oxygen for the combustion process inside a power plant. Nemitallah et al. [15] presented an optimized monolith structure design of an ITM reactor in order to replace the combustor of a conventional gas turbine. The specifications of this reactor are 3.35 m height, 2700 $m^2$ total membrane surface area, 10 $m^3$ total reactor volume, and the reactor can deliver power ranging from 5.0 to 8.0 MWe, based on the cycle first law efficiency. Ahmed et al. [21] performed a CFD study of a novel ITM reactor design under oxy-combustion conditions. The novel reactor design in their work resulted in gradual burning of the fuel, and as a result, slow temperature rise and uniform temperature distribution are encountered along the membrane surface. Gunasekaran et al. [22] performed a detailed optimization study on the design and operating parameters of a membrane-based oxy-combustion power plant. The optimization resulted in considerable improvements in the efficiency and emissions of the power plant. Kotowicz and Michalski [23] conducted an efficiency analysis of a supercritical power plant fueled by hard coal utilizing high temperature membranes for air separation. The results showed significant improvement in the net efficiency of the analyzed supercritical power plant. Recently, Habib and Nemitallah [24] designed an ITM reactor for application in fire tube boilers based on three-dimensional simulations. They came up with a reactor design that can deliver power up to 8.0 MWe.

### 6.2.2 Polymeric Membranes

Fossil fuels are the dominant source of the global primary energy demand, and will likely remain so for the next decades. Carbon capture is essential to enable the use of fossil fuels while reducing the emissions of $CO_2$ into the atmosphere, and thereby mitigating global climate change. Carbon capture is generally grouped into three different strategies. These are post-combustion de-carbonization (separation of $CO_2$- from exhaust gas), oxy-fuel combustion and pre-combustion de-carbonization (production of carbon free fuel). The technology of membrane gas separation can be applied in both post-combustion and oxy-combustion carbon capture technologies. In post-combustion technology, $CO_2$ is separated from the exhaust stream using membranes with different techniques [25, 26]. One of the most common techniques for $CO_2$ capture is the membrane gas absorption (MGA) technique. This technique combines between membrane separation and absorption methods to provide effective removal of $CO_2$ [27]. Oxy-fuel combustion requires separation of air to provide high concentration of $O_2$. Due to the elevated cost of combustion using pure oxygen, oxygen enriched air combustion became an acceptable approach for controlling emissions [28]. Oxygen enriched air could be produced by three main air separation methods including; cryogenic distillation, pressure swing adsorption and membrane separation. Cryogenic distillation is the most common of these three technologies and produces high purities of oxygen (>99%) with large scale productions [29]. Pressure swing adsorption can reach up to 95% oxygen purities, but the requirement of solvents limits its size capacity, primarily due to capital costs [30, 31]. The third technology is the most recent of these three technologies, which is the membrane air separation technology through the use of high temperature ceramic ion transport membranes (ITMs) and polymeric membranes. ITMs can produce purities of close to 100% while polymeric membranes can produce different concentrations of oxygen enriched air [30].

In polymeric membrane gas separation, there is no need for thermal heating of the membrane or the involvement of any sorbents to separate a gas mixture, as in the cases of cryogenic distillation and adsorption. The main driving force in polymeric membrane air separation is the partial pressure gradient across the membrane. The permeability ($P$) of a polymeric membrane for a specific gas could be defined as follows [32]:

$$P = \frac{JL}{A(p' - p'')} \tag{6.1}$$

where $J$ is the permeation flux through the membrane for a specific gas, $p'$ and $p''$ are the gas partial pressure on the feed side and the permeate side, $A$ is the membrane area, and $L$ is the membrane thickness. The permeability coefficient is a specific property of a gas for a specific polymer. This chapter aims at investigating the performance of polymeric membranes for oxygen enriched air combustion applications. Based on that, a polymer with high oxygen permeability and low nitrogen permeability is needed for such applications. The selectivity ($\alpha$) is a commonly used parameter that characterizes the ability of a polymer to separate two gases, and it can be defined as follows:

$$\alpha_{a/b} = \frac{P_a}{P_b} \tag{6.2}$$

where $P_a$ and $P_b$ are the permeability of the two gases, gas $a$ and gas $b$. The permeability is a product of two coefficients; the solubility coefficient ($S$), which is the thermodynamic term, and the diffusion coefficient ($D$), the kinetic term. This relation can be written as follows [32]:

$$P = D * S \tag{6.3}$$

The values of these two terms ($D$ and $S$) depend on the penetrant gas and the material of the membrane. These coefficients are also influenced by temperature, pressure, gas combustion and other parameters depending on the penetrant gas and the polymer. Park and Pual [33] predicted and reported a correlation to calculate the permeability of glassy polymers using the fractional free volume (FFV), as follows:

$$P = A \exp\left(\frac{-B}{FFV}\right) \tag{6.4}$$

where, A and B are constants for a particular gas with specific conditions, and the fractional free volume (FFV) is defined by FFV = (V – $V_0$)/V, where $V_0$ is the dense volume and V is the measured volume. They determined the empirical factors of 105 different polymers data base. But this correlation still has a large degree of error, so the obtained experimental values of the permeability were used in the model.

A perfect polymer for air separation should have high oxygen permeability and high selectivity ($\alpha_{O_2/N_2}$). However, there is a limitation getting such polymer with high selectivity and high permeability that is represented in the "upper bound" relationship for membrane separation of gases. This limit, a log-log plot of permeability versus selectivity of the more permeable gas, demonstrated that almost all the data points were below this line [34]. Revisited upper bound was predicted by Robeson [35]. Many research works were carried out to develop high performance gas separation polymeric membranes. One of the recently developed groups of membranes is the thermally rearranged (TR) polymeric membranes group by Park et al. [36]. These polymeric membranes have high selectivity and high permeability for different gases [37]. Different studies were conducted to develop polymers from this family and to study the gas separation process through this type of membranes [38–41]. Sanders et al. [37] placed the TR polymers and other recent polymers in the upper bound in their review.

As a matter of fact, there are different parameters can affect the permeability of the polymers. Jiang and Kumar [42] conducted experiments to study the performance of hydrocarbons and nitrogen mixtures separation through silicone-coated polymeric membranes. It was found that permeability and selectivity are significantly lower in gas mixtures as compared to pure gases. Coombe and Nieh [43] measured the air separation performance of perfluorodioxole polymeric membranes and investigated the relationship between membrane performance and combustion parameters by experimentation. Pressure and temperature can also affect the permeability of the polymers, and the permeability can exhibit different behaviors with changes in the pressure. Alsari et al. [44] studied experimentally the effect of membrane thickness and pressure on the gases permeability in dense polyphenylene oxide (PPO) membranes. They used a sweep gas on the permeate side and air on the feed side; they found that the selectivity of $O_2$ ($\alpha_{O_2/N_2}$) increased with the feed pressure. This can be attributed to the increase in $O_2$ permeability and the decrease in $N_2$ permeability with the pressure. Similarly, different behavior was recorded for the temperature effect on the permeability. Feng et al. [45] found that the permeability increased with the temperature, following the Arrhenius type relationship, and the selectivity ($\alpha_{O_2/N_2}$) decreased. Dixon-Garrett [46] found that the ethylbenzene permeability decreased with increase in temperature and pressure.

Numerical investigation by computer programs became a very useful method to study air separation. Commercial and energy studies were conducted to compare different techniques of gas separation. Ahmad et al. [47] simulated $CO_2$ capture from natural gas to find the optimal design of the separation using Aspen HYSYS. They found that the minimum cost for the gas process can be achieved by the double stage separation with a permeate recycle system. A parametric study of energy and mass balances for a simulated single membrane process was carried out by Zhao et al. [48]. They used PRO II software to build a model for $CO_2/N_2$ mixture separation by cardo polyimide membrane. Kawachale et al. [49] used the Fluent program to study the mixing behavior of gases near the membrane, and the fluxes through it using 2D and 3D models. Binns et al. [50] used ASPEN HYSYS software to simulate gas separation by a multi-stage hollow fiber membrane. They compared different connections and arrangements of the separation system. Dong et al. [51] solved their model of gas separation in MATLAB using a TR-PBO hollow fiber membrane in a carbon capture application, and they proposed to include the shell-side pressure drop in the gas separation models. Zhang et al. [52] studied numerically the effect of membrane module design on the removal rate of $SO_2$. The results showed better permeation rates while increasing the number of fibers and decreasing the membrane thickness. Choi et al. [53] performed a study on $O_2$ removal from exhaust gas stream using polymeric membrane for complete capture of $CO_2$. They evaluated the convinced changes in polymer morphology and separation performance data considering single gas permeation.

## 6.3 Polymeric Membranes for Oxygen-Enriched Air Combustion Applications

In this section, a detailed study is presented on the characteristics of air separation in hollow-fiber polymeric membrane aiming at generation of highly oxygen-enriched air for clean combustion applications. A 2-D axisymmetric mesh was generated using Gambit software and, then, fed to Fluent 12.1 code. Detailed numerical study is presented for air separation using different polymers. Effects of sweep gas flow rate, feed gas flow rate, fed pressure, and membrane polymeric material (considering TR-PBO-co-PPL-28 and TR-PI-g-CD) on permeation fluxes and concentrations of $O_2$ and $N_2$ are investigated. Targeting higher exit $O_2$ concentrations, multi-stage air separation unit is developed.

### *6.3.1 Description of Membrane Unit and Ranges of Parameters*

The geometry and membrane properties of this model were made with reference to Feng et al. [45] shell-and-tube unit. A total number of 368 hollow fiber membranes were encased in a copper tubing of a shell and tube configuration. The current work utilizes an asymmetric hollow fiber polymeric membrane unit for air separation to perform the numerical simulations. This configuration of a single hollow fiber membrane unit and it shell is represented by the 2D schematic shown in Fig. 6.1. Bore-side feed counter-current flow configuration was considered in this study to give better separation results [45]. Air (20.5 and 79.5%, by volume) is fed to the fiber membrane cell in the bore-side and oxygen depleted air leaves as retentate at the end of this side. At the shell-side, methane is used as the sweep gas that has large molecular size to ensure no blocking in the membrane permeation [44]. In fact, the study presented in this section is mainly concerned with the application of oxygen-enriched air combustion technology for clean energy production. Oxygen enriched air-methane mixture leaves in the permeate outlet in the sweep cases, while oxygen enriched air leaves in the non-sweeping flow cases. The produced Oxygen enriched air-methane mixture can be introduced directly to a combustion system to generate energy at low emissions levels. This may justify the use of methane, a gaseous fuel, as sweep gas in the permeate side of the membrane. At sweep and feed inlets mass flow

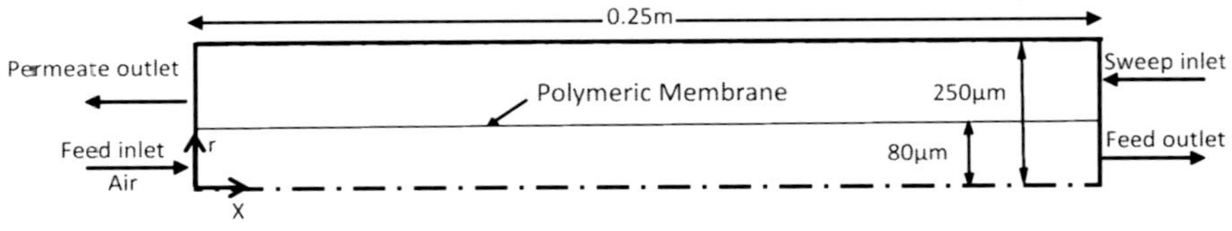


**Fig. 6.1** Hollow fiber membrane cell computational domain

inlet conditions have been specified while pressure outlet boundary conditions are specified at the exit sections of the feed and sweep sides (atmospheric pressure at the sweep side and specific pressure at the feed side as shown in Table 6.1). Membrane is modeled as a wall with coupled heat transfer mechanism to calculate its temperature based on the surrounding flows from both sides. The base case of this study has the same operating conditions as in Feng et al. [45] (no sweep flow, air is fed with varying flow rate at 23 °C, 100 psig feed pressure and 0.0 psig permeate pressure). The effect of sweep flow was investigated by feeding 100% $CH_4$ at the shell side with flow rate range from zero to $1.15 \times 10^{-7}$ kg/s. Feed air with flow rate ranging from $3.3 \times 10^{-9}$ to $5.89 \times 10^{-8}$ kg/s was included with feed pressure ranging from 2 to 10 atm. Oxygen enriched air was used as feed gas in the multi-stage configuration with oxygen mass fraction ranging from 0.228 to 0.8. Table 6.1 shows a summary of all operating conditions and ranges of parameters.

**Table 6.1** Summery of the operating conditions

| Study | Parameter | Range |
|---|---|---|
| Base case | Fed $O_2$ mass fraction | 22.8% |
| | Feed flow rate | $5.89 \times 10^{-8}$ kg/s |
| | Sweep flow rate | Zero |
| | Feed pressure | 100 psig |
| Multi-stage configuration | Fed $O_2$ mass fraction | 22.8–80.0% |
| | Feed flow rate | $7.4 \times 10^{-9} - 5.89 \times 10^{-8}$ kg/s |
| | Sweep flow rate | Zero |
| | Feed pressure | 100 psig |
| Effect of sweep flow rate | Fed $O_2$ mass fraction | 22.8% |
| | Feed flow rate | $5.89 \times 10^{-8}$ kg/s |
| | Sweep flow rate | $0 - 1.15 \times 10^{-7}$ kg/s |
| | Feed pressure | 100 psig |
| Effect of feed flow rate | Fed $O_2$ mass fraction | 22.8% |
| | Feed flow rate | $3.3 \times 10^{-9} - 5.89 \times 10^{-8}$ kg/s |
| | Sweep flow rate | Zero |
| | Feed pressure | 100 psig |
| Effect of feed pressure | Fed $O_2$ mass fraction | 22.8% |
| | Feed flow rate | $5.89 \times 10^{-8}$ kg/s |
| | Sweep flow rate | $1.15 \times 10^{-9}$ |
| | Feed pressure | 2 – 10 atm. |

### 6.3.2 CFD Modeling and Validation

The numerical model of the polymeric membrane gas separation unit was built and validated using Fluent software. The hollow fiber membrane cell, including the bore and shell sides, was simulated using 2D axisymmetric geometry due to the symmetry of this separation unit around its axis. Gambit software mesh builder software) was used to develop the mesh to be introduced to the Fluent 12.1 software [54]. Fluent software is a built in code that is available commercially, which contains the immense physical modeling competence required to model flow, heat transfer, and reactions [54]. The code solves the flow field through the solution of conservation equations including continuity, energy, momentum, and species. The numerical solution of the flow field includes the solution of the differential equations of continuity, energy, momentum, and species transport. The equations are presented as follows:

$$\nabla \cdot (\rho U) = S_i \tag{6.5}$$

$$\nabla \cdot (\rho U U) = -\nabla p + \mu \nabla^2 U \tag{6.6}$$

$$(\rho C_p)_f U \cdot \nabla T = \nabla \cdot (k_f \nabla T) \tag{6.7}$$

$$\nabla \cdot (\rho U Y_i) - \nabla \cdot (\rho D_{i,m} \nabla Y_i) = S_i \tag{6.8}$$

where ρ is the density, V is the velocity, P is the pressure, μ is the dynamic viscosity $D_{i,m}$ is the diffusion coefficient in the mixture, T is the temperature, and $Y_i$ is the mass fraction. Source/sink term, $S_i$, was used to consider the transfer of $O_2$ and $N_2$ across the polymer membrane. To include this source/sink term in the governing equations as presented above, user defined functions (UDF) were written in visual C++ and compiled and hooked to the Fluent software to close the governing equations system [15, 55, 56]. The law that controls the flow of a fluid through a porous medium which was used to solve for the permeation across the polymer membrane can be presented as follows [32]:

$$J_{O_2/N_2} = \frac{P\left(P' - P''\right)}{L} \tag{6.9}$$

where J is the permeation flux of $O_2$ or $N_2$ in mol/m$^2$/s and P is the permeability of the polymer membrane in mol m/m$^2$/s/Pa. p′ and p″ are the gas partial pressure on the feed side and the permeate side and L is the membrane thickness. Based on the flux equation, the equation for the source/sink term is developed for both $O_2$ and $N_2$ at each boundary membrane cell as follows:

$$S_i = \begin{cases} +\frac{J_{O_2/N_2} \cdot A_{cell} \cdot MW_{O_2/N_2}}{V_{cell}} & \text{at sweep side of the membrane} \\ -\frac{J_{O_2/N_2} \cdot A_{cell} \cdot MW_{O_2/N_2}}{V_{cell}} & \text{at feed side of the membrane} \end{cases} \tag{6.10}$$

where $MW_{O_2/N_2}$ is the molecular weight of $O_2$ or $N_2$ in kg/mol, $A_{cell}$ is the cell area in $m^2$ and $V_{cell}$ is the cell volume in $m^3$. This term has a positive sign in the sweep side of the membrane and a negative sign in the feed side of the membrane as the species of $O_2$ and $N_2$ transports from the feed side to the permeate side of the membrane. The term, $S_i$, is only calculated for the boundary cells, cells in direct contact with the membrane from both sides, and is zero elsewhere in the computational domain. In addition, this term has a value of zero for species other than $O_2$ and $N_2$. In all simulations, it was assumed that membrane permeability is independent of the feed gas composition, membrane permeability is not affected by the swept gas, and the temperature distribution is uniform along the membrane (ambient temperature). These governing equations are discretized by the finite volume method and then solved using algebraic approximations. Steady state pressure based solver was used. Second order upwind scheme was used to discretize the convective terms in the conservation equations. Calculations the semi-implicit method for pressure-linked equations (SIMPLE) algorithm was applied to account for coupling between pressure and velocity fields [57]. The calculations were performed for each running case until the summation of al residuals is less than $10^{-6}$.

#### 6.3.2.1 Grid Independence Study

It is important to show that the used grid in this model has enough resolution to correctly predict the results in the computational domain to make sure that the results are independent of the grid. To validate this, different grid size results were calculate and compared. Figure 6.2 shows the distribution of $N_2$ permeation flux through the

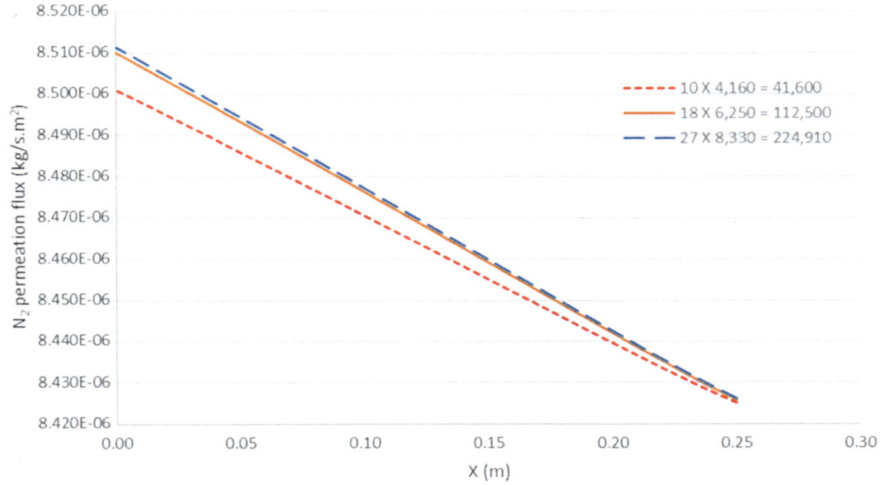


**Fig. 6.2** $N_2$ permeation flux through the membrane length for three different grids sizes at 100 psig feed pressure, 0.0 psig permeate pressure, $5.89 \times 10^{-8}$ kg/s feed flow rate and zero sweep

membrane length for different grid sizes. The results for the three grids are plotted for the same operating conditions. For all grids, the flux is maximum close to the feed inlet (x = 0 m), then it reduces in the axial direction as shown in the figure. This can be attributed to the reduction in mass fraction of $N_2$ in the feed side due to permeation and increase of it in the permeate side due to accumulation. This results in lower partial pressure driving force across the membrane and, as a result, $N_2$ flux is reduced. Based on the presented results, refining the grid above the size of 18 × 6250 cells resulted in insignificant changes in $N_2$ permeation flux. Based on that, all calculations were performed using grid with the size of 18 × 6250 cells.

#### 6.3.2.2 Model Validation

Experimental data by Feng et al. [45] were used to validate the model for polymer membrane system. In their experiment, hollow-fiber membranes were used to separate air and produce oxygen and nitrogen. A total number of 368 hollow fiber membranes were encased in a copper tubing of a shell and tube configuration. Shell-side feed and bore-side feed with countercurrent and co-current flow arrangements for different stage cuts were tested in this study. It was found that the bore-side feed countercurrent flow has the best performance. Table 6.2 represents the experimental data used in the validation study. Figure 6.1 shows the considered polymer membrane unit, which is a representation of one polymer membrane cell of the shell and tube configuration by Feng et al. [45]. So, the same mesh, details are mentioned above, is used in the validation study.

The model results and the experimental data by Feng et al. [45] are compared in terms of permeation flow rates of $O_2$ and $N_2$, and the gas concentrations at the permeated outlet as shown in Figs. 6.3 and 6.4. The results for gas concentrations at the permeate outlet are presented as function of the stage cut (θ) which is defined as:

$$\theta = \frac{permeate\ flow}{Feed\ flow} \tag{6.11}$$

As a result of increasing the feed flow rate of air, $N_2$ and $O_2$ retentate fluxes increased and the fraction of the penetrated $O_2$ to the fed $O_2$ reduced. This resulted in increase in $O_2$ mol fraction in the retentate as shown in Fig. 6.4 and reduction in $N_{2\ mol}$ fraction in the retentate as shown in Fig. 6.3. Stage cut decreases with increasing the feed flow rate while $N_2$ mol fraction decreases as presented in Fig. 6.3. Similarly,

**Table 6.2** The used experimental data for model validation

| Fiber outer diameter/length | Feed pressure | Permeate pressure | Temperature | Feed composition (% volume) | | Permeance (GPU) | |
|---|---|---|---|---|---|---|---|
| | | | | $O_2$ | $N_2$ | $O_2$ | $N_2$ |
| 160 μm/0.25 m | 100 psig | 0 psig | 23 °C | 20.5 | 79.5 | 9.3 | 1.8 |

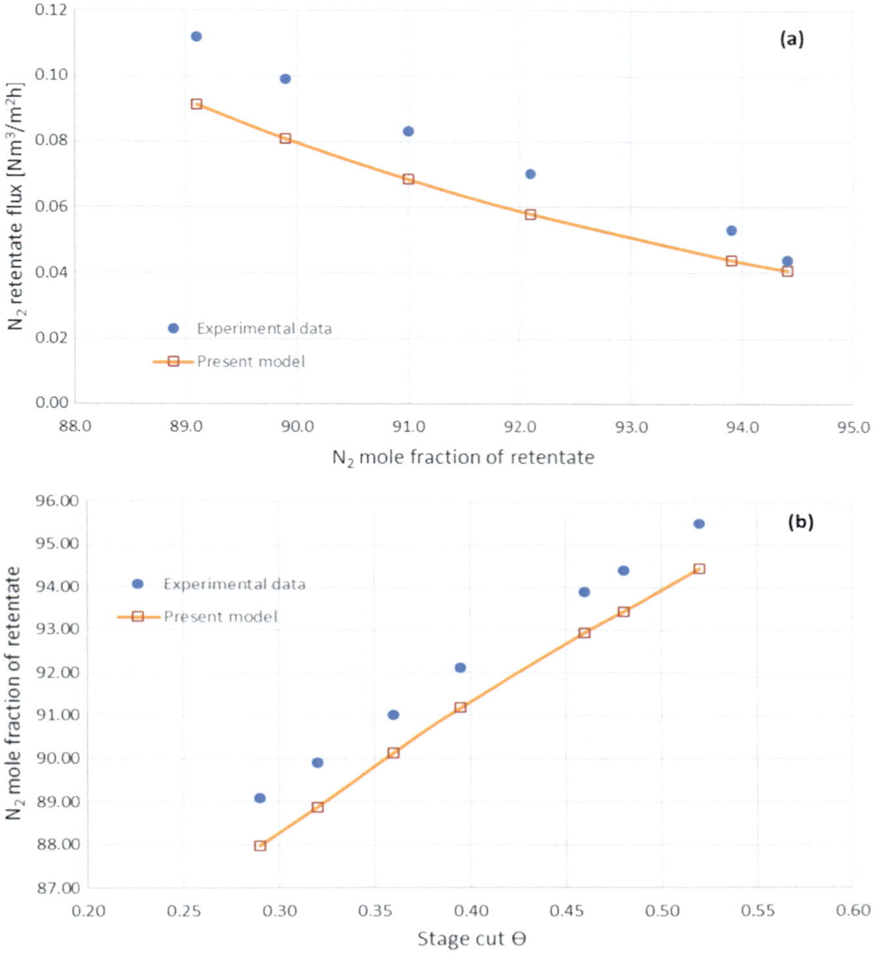


**Fig. 6.3** Model predictions against the experimental data by Feng et al. [45] at 100 psig feed pressure, 0.0 psig permeate pressure, varying feed flow rates and zero sweep; in terms of: **a** $N_2$ retentate flux and **b** $N_2$ mol fraction of retentate

increasing the feed flow rate resulted in increase in the permeation flux of oxygen and increase its mole fraction in the permeate side as shown in Fig. 6.4. This increase in the permeation flux can be attributed to the increase in the mass fraction of $O_2$ on the feed side as the feed flow rate increases. The results in Fig. 6.3 and Fig. 6.4 show good agreement with the experimental data.

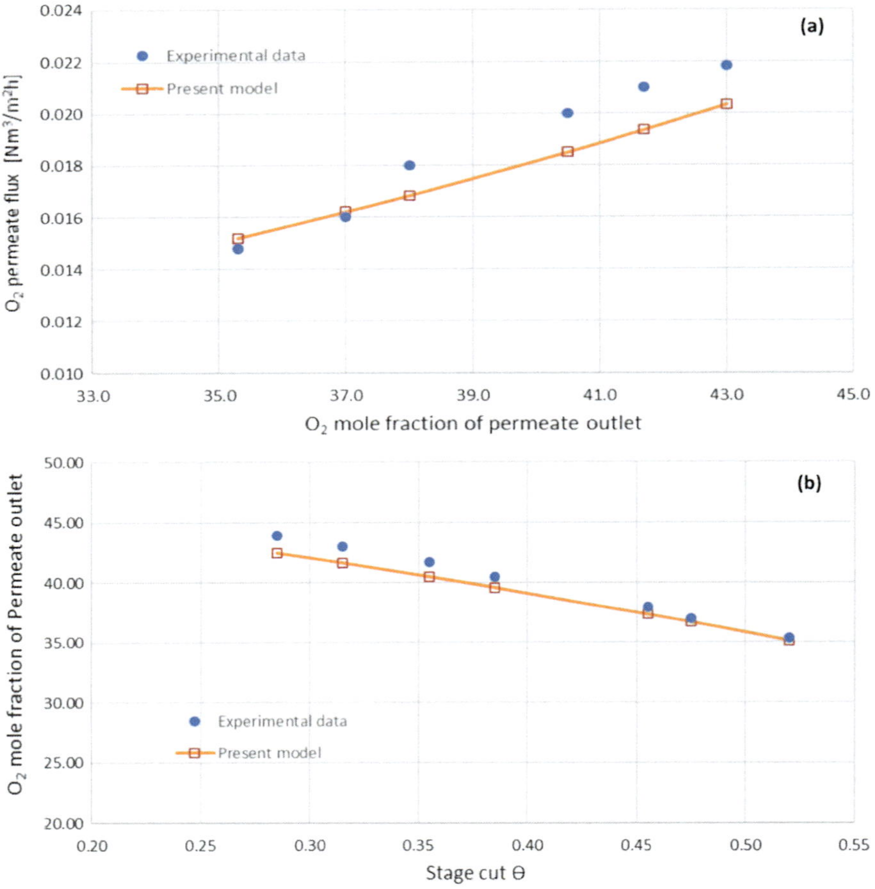


**Fig. 6.4** Model predictions against the experimental data by Feng et al. [45] at 100 psig feed pressure, 0.0 psig permeate pressure, varying feed flow rates and zero sweep; in terms of: **a** $O_2$ permeate flux and **b** $O_2$ mol fraction of permeate outlet

## 6.3.3 *Effect of Sweep Gas Flow Rate*

In this section, the effects of sweep gas flow rate on the permeation characteristics of the polymeric membrane unit are investigated. $CH_4$ is used as the sweeping gas under atmospheric pressure. Figure 6.5 shows how the change in sweep gas flow rate can affect the local distributions of permeation flow rates of $N_2$ and $O_2$. The results are also compared with the case of zero sweep gas flow rate. The plots show increase in the fluxes of $O_2$ and $N_2$ while increasing the sweep gas flow rate. This can be attributed to the reduction of $O_2$ and $N_2$ partial pressures in the permeate side due to the effective purging process of the permeated gases by the sweeping gas. Reductions in $O_2$ partial pressure are encountered in the permeate side while

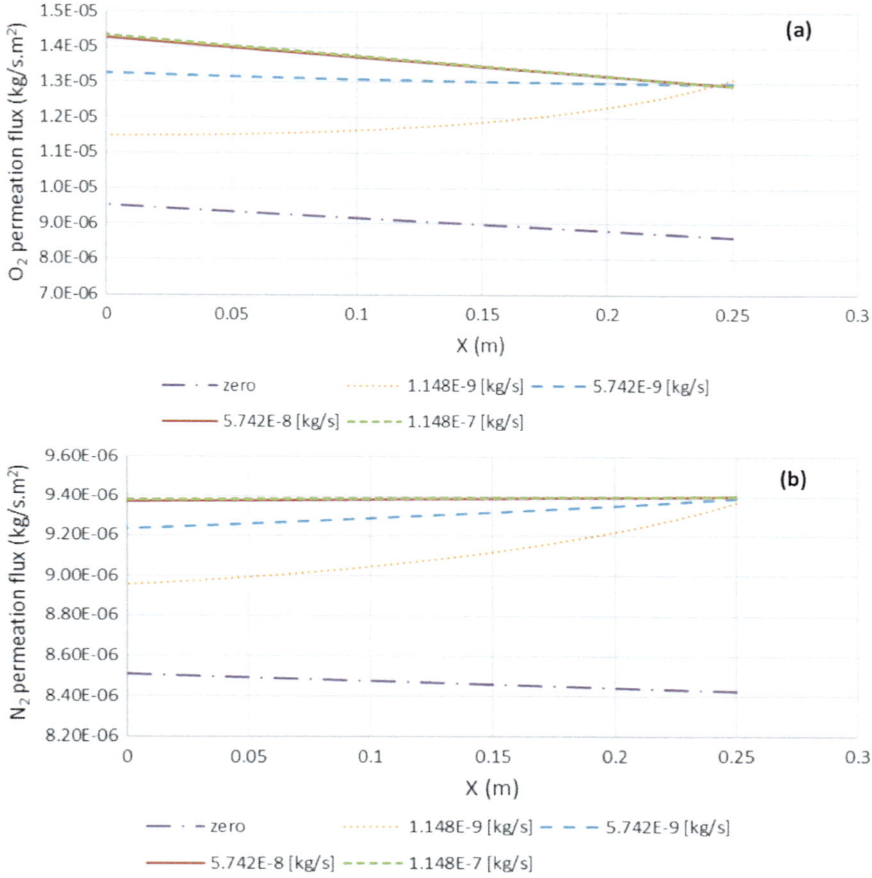


**Fig. 6.5** Distributions of: **a** $O_2$ permeate flux rate and **b** $N_2$ permeate flux rate, over a range of sweep gas flow rate at 100 psig feed pressure, 0.0 psig permeate pressure and 5.89e−8 kg/s feed flow rate

increasing the sweep gas flow rate as presented in Fig. 6.6. The improvements in the permeation fluxes of both $O_2$ and $N_2$ are significant when the flow sweep flow rate is increased from zero up to a value of 5.742E−8. Any further increase in the sweep flow rate does not result in any change neither in the permeation fluxes of $O_2$ and $N_2$, as shown in Fig. 6.5, nor in partial pressures of $O_2$ and $N_2$, as shown in Fig. 6.6. This can be attributed to the limit of permeability of the membrane, i.e. the maximum permeation flux is approached depending on the material of the membrane. At sweep flow rate of 1.148E−9 kg/s, the lowest considered sweep flow rate, sharp increases in the distributions of permeated oxygen and nitrogen flow rates are encountered. This may be attributed to the obtained sharp reductions in oxygen and nitrogen partial pressure distributions in the permeate side as presented in Fig. 6.6. At lower low

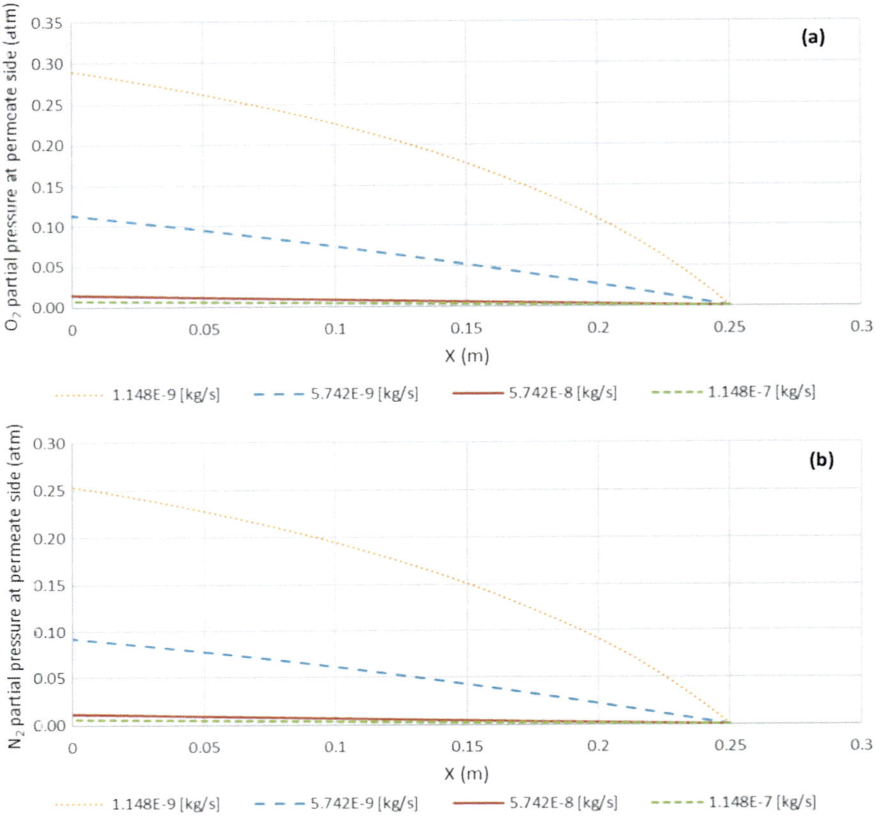


**Fig. 6.6** Distributions of: **a** $O_2$ partial pressure and **b** $N_2$ partial pressure, at the permeate side along the membrane surface over a range of sweep gas flow rate at 100 psig feed pressure, 0.0 psig permeate pressure and 5.89e−8 kg/s feed flow rate

rates, better diffusion rates of permeated oxygen and nitrogen in the sweep flow are obtained due to the reduced flow velocity and, as a result, better reductions in oxygen and nitrogen partial pressures in the permeate side are obtained. For the considered range of sweep flow rate except for the case of zero flow rate, identical values of the permeation fluxes are obtained at sweep gas inlet section, at x = 0.25 m. This can be attributed to the reduction in partial pressure driving force across the membrane at that section which corresponds to the retentate exit section of the feed flow.

Figure 6.7a, shows the overall permeation flow rates of $O_2$ and $N_2$ through the membrane over a rage of sweep gas flow rate. In addition, the corresponding values of mass fractions of $O_2$ and $N_2$ calculated at permeate outlet and retentate outlet are presented in Fig. 6.7b. The optimal value of the sweep flow rate could be observed from Fig. 6.7 where the permeation flow rates became unchanged. This optimal value could produce higher permeation and concentration of the oxygen-enriched air with

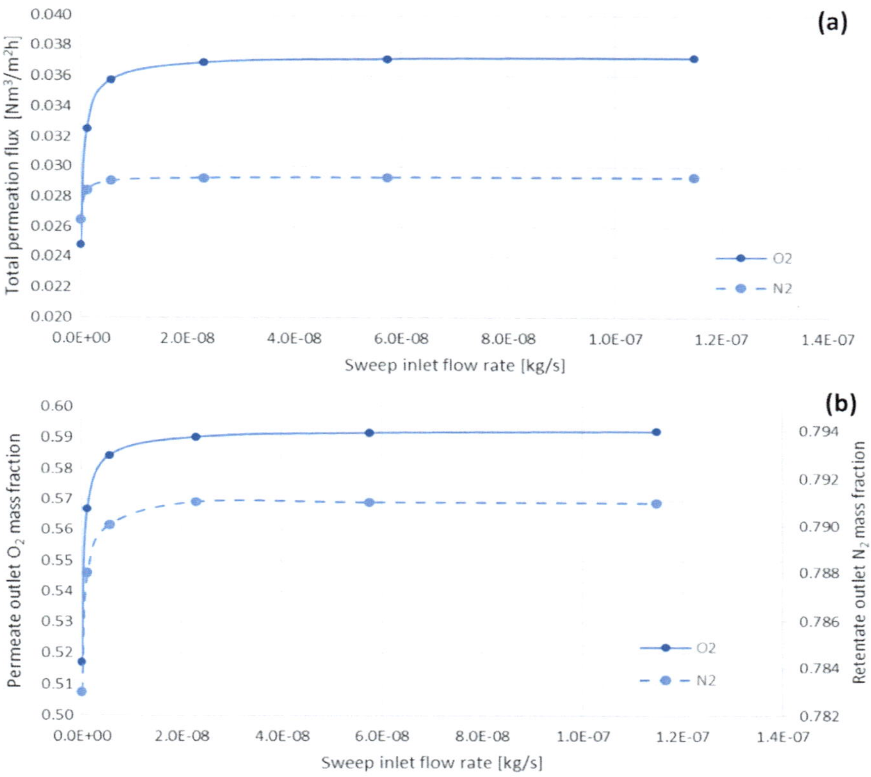


**Fig. 6.7** Distribution of: **a** $O_2$ and $N_2$ total permeate fluxes and **b** mass fractions of $O_2$ and $N_2$, calculated at permeate outlet and retentate outlet over a range of sweep gas flow rate at 100 psig feed pressure, 0.0 psig permeate pressure and 5.89e−8 kg/s feed flow rate

low energy consumption. As shown in the figure, the total permeation flow rates of $O_2$ and $N_2$ increase as the sweep flow rate increases. However, the increase in the permeation flux of $O_2$ is insignificant for sweep flow rates higher than 2.0E−8. The improvement in permeation flux of $N_2$ is insignificant at lower sweep flow rates as compared to $O_2$. Interestingly, at zero sweep gas flow rate, the permeation flux of $N_2$ is higher than that of $O_2$ as shown in Fig. 6.7a. So, sweep gas should be used in order to obtain higher oxygen permeation rates and higher concentrations of $O_2$ at permeate outlet. The value of sweep gas flow rate at which the fluxes of $O_2$ and $N_2$ are equal is 0.25 kg/s as shown in Fig. 6.7a. The plots of mass fractions of $O_2$ in Fig. 6.7b, show increase in the concentrations of $O_2$ at the permeate exit while increasing the sweep gas flow rate. Similar values of optimum sweep gas flow rate are obtained from plots in Fig. 6.7. The concentrations of $N_2$ at the retentate exit section also increased while increasing the sweep flow rate up to a certain optimum value. This increase in $N_2$ concentrations at the exit section of the feed side may be attributed to the increase of $O_2$ flux across the membrane while increasing the sweep gas flow. As a result, oxygen concentrations in the feed side are reduced and, accordingly, $N_2$ concentrations are increased. Similar trends were observed in the experiments by Li et al. [58].

### 6.3.4 *Effect of Feed Gas Flow Rate*

Actually, the change in sweep or feed flow rates results in modification in the flow field and, accordingly, change in partial pressure distributions of permeating gases on both sides of the membrane. This directly affects the rate of permeation across the membrane as it is a direct function of partial pressure of permeating gas. Increasing the sweep gas flow rate resulted in purging of the permeated gases and, accordingly, their partial pressures in the permeate side are reduced resulting in higher permeation rates. This happened up to a certain flow rate at which permeation fluxes are maximum and partial pressures in the permeate side are the lowest. Similar effect is obtained while increasing the feed gas flow rate as shown in Fig. 6.8 at zero sweep gas flow rate. Increasing the feed gas flow rate resulted in increase in the oxygen permeation flux. This can be attributed to the increase in flow velocity in the feed side in the vicinity of the membrane surface which results in purging of $N_2$-enriched air after $O_2$ and $N_2$ separation and, as a result, oxygen partial pressure is increased in the feed side and, accordingly, the permeation flux is increased. The increase in oxygen concentrations in the feed side in the vicinity of the membrane while increasing the feed gas flow rate is clearly presented in Fig. 6.9a. Interestingly, increasing the feed gas flow rate resulted in reduction in nitrogen permeation flux which helps getting higher oxygen concentrations at permeate outlet as shown in Fig. 6.10. Due to purging of $N_2$-enriched air after separation of $O_2$ and $N_2$ at certain location, the fresh

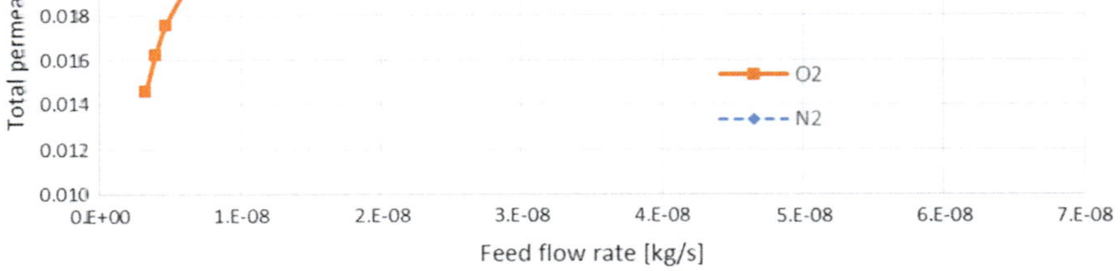


**Fig. 6.8** Effect of feed gas flow rate on total permeate fluxes of $O_2$ and $N_2$ at 100 psig feed pressure, 0.0 psig permeate pressure and zero sweep flow rate

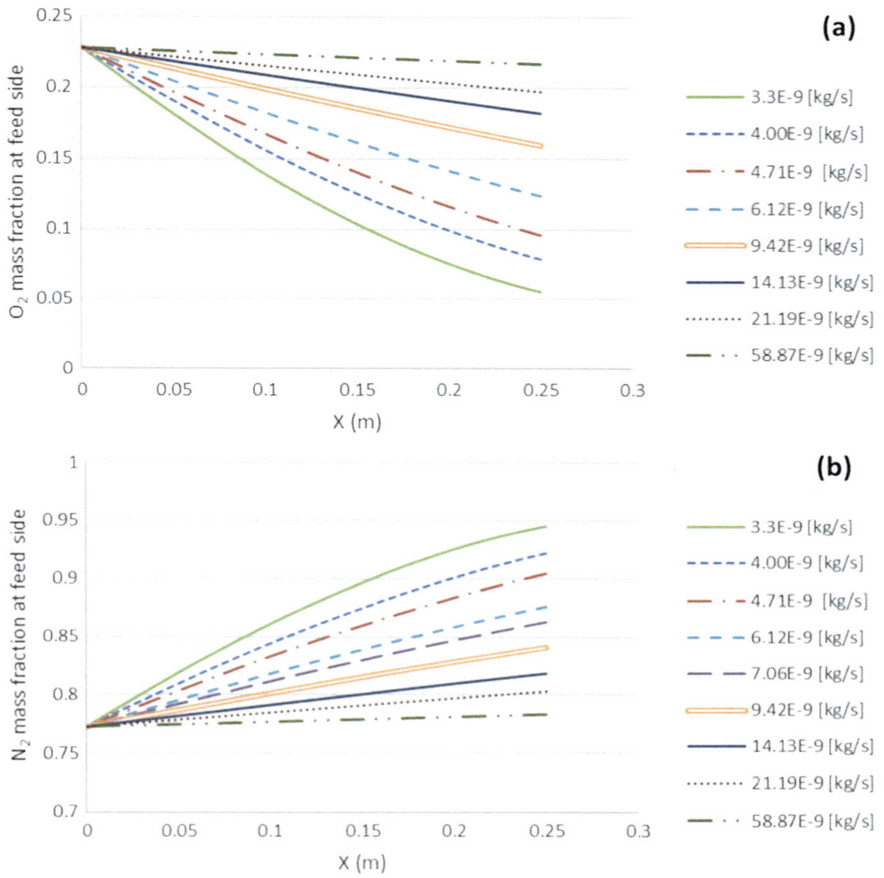


**Fig. 6.9** Distributions of: **a** mass fractions of $O_2$ and **b** mass fractions of $N_2$ along the membrane surface at feed side over a range of feed flow rate at 100 psig feed pressure, 0.0 psig permeate pressure, and zero sweep flow rate

incoming air has lower $N_2$ concentrations which results in reduction in $N_2$ partial pressure in the feed side. This results in reduction in partial pressure driving force for $N_2$ permeation and, accordingly, $N_2$ permeation flux is reduced as shown in Fig. 6.8. The distributions of $N_2$ mass fractions as function of feed gas flow rate are presented in Fig. 6.9, lower plot. Reductions in $N_2$ concentrations in the vicinity of membrane surface in feed side are encountered while increasing the feed gas flow rate. The increase in oxygen permeation flux is insignificant for feed flow rates higher than 4.0E−8 kg/s. This can be attributed to reaching the limit of oxygen permeation based on partial pressure differences and membrane permeability. In addition, increasing the feed gas flow rate more than 2.0E−8 kg/s resulted in controlling the $N_2$ partial pressure a fixed value and no significant change in $N_2$ permeation flux is obtained.

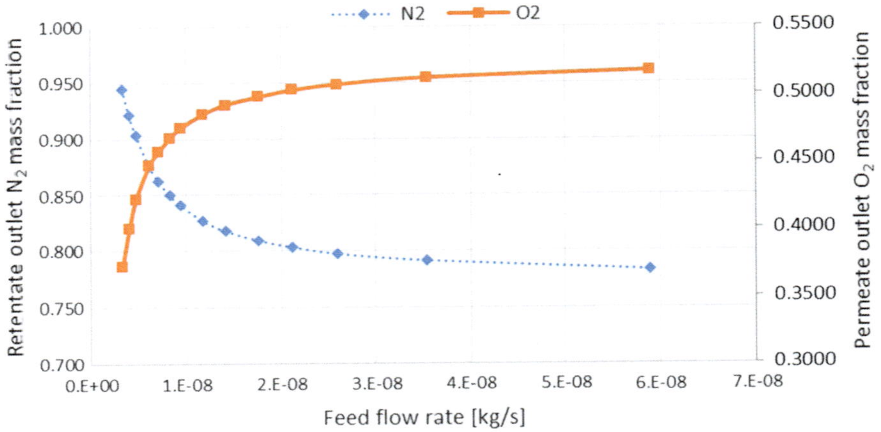


**Fig. 6.10** Distributions of mass fractions of $O_2$ and $N_2$ at permeate outlet and retentate over a range of feed flow rate at 100 psig feed pressure, 0.0 psig permeate pressure, and zero sweep flow rate

For the considered range of feed gas flow rate, $N_2$ permeation flux remained higher than that of $O_2$. As mentioned above (see Fig. 6.7), at zero sweep gas flow rate, $N_2$ flux is higher than $O_2$ flux. However, the flux of oxygen approaches that of $N_2$ at higher feed flow rates as shown in Fig. 6.8. The increase of $O_2$ permeation rate while increasing the feed flow rate resulted in higher $O_2$ mass fraction at permeate outlet as shown in Fig. 6.10. Also, due to permeation of $N_2$ at higher flow rates than $O_2$ from the feed side to the permeate side, the retentate $N_2$ mass fraction decreases while increasing the feed gas flow rate as shown in Fig. 6.10.

### 6.3.5 *Effect of Feed Pressure*

One of the main advantage of polymeric membranes over ceramic membranes is that the feed pressure can be raised with respect to the permeate pressure. This helps getting higher partial pressure of oxygen in the feed side and, as a result, oxygen permeation flux is increased. Figure 6.11 shows the effect of increasing the pressure in the feed side on permeation fluxes of $O_2$ and $N_2$ at sweep gas flow rate of 1.15E−9 kg/s. As shown in the figure, the permeation fluxes of $O_2$ and $N_2$ increase as the feed pressure increases. The relation between permeation flux and feed pressure seems to be linear, as the permeability is constant with the pressure. The feed pressure governs the permeation flux relation, and this could be clear if the permeation flux equation was rewritten as follows:

$$J = \frac{PA(p_1 y_1 - p_2 y_2)}{L} \tag{6.12}$$

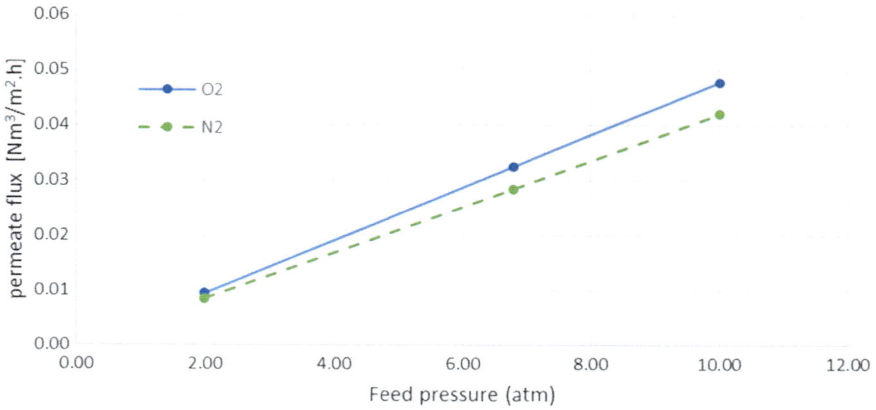


**Fig. 6.11** Total permeate flow rate for $O_2$ and $N_2$ as function of feed pressure at 0.0 psig permeate pressure, 5.89e−8 kg/s feed flow rate and 1.15e−9 kg/s sweep flow rate

where the partial pressure in feed and permeate sides are defined as $p' = y_1 * p_1$, and $p'' = y_2 * p_2$. Where, $p_1$ is the total pressure in the feed side and $y_1$ is the mole fraction of a specific gas in the feed gas and $p_2$ and $y_2$ are the pressure and the mole fraction at the permeate side. While P, A and L are constants, and for fixed pressure in the permeate side, then the permeation flux is mainly a linear function of feed pressure as presented in Fig. 6.11. As the sweep gas flux is greater than the value at which $O_2$ flux exceeds that value of $N_2$ flux, 0.25E−9 kg/s, $O_2$ flux is higher than $N_2$ flux at all pressure as shown in Fig. 6.11. Due to the increase of $O_2$ permeation flux while increasing the feed pressure, the concentrations of $O_2$ in the vicinity of membrane surface in the feed side are reduced as presented in Fig. 6.12. In addition, as shown

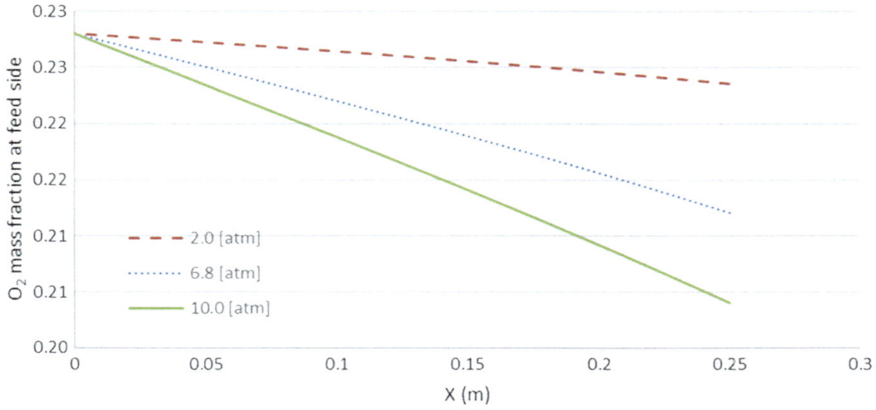


**Fig. 6.12** $O_2$ mass fraction at feed side along the membrane surface over a range of feed pressure at 0.0 psig permeate pressure, 5.89e−8 kg/s feed flow rate, and 1.15e−9 kg/s sweep flow rate

in the figure, the concentrations of $O_2$ are reduced toward the retentate exit section in the fed side due to permeation of $O_2$ through the whole length of the membrane.

### 6.3.6 Effect of Polymer Material

Different polymeric membranes can produce variant oxygen permeation fluxes depending on membrane material and feed gases. To study the effect of membrane material in terms of permeability and selectivity on $O_2$ permeation flux, three different polymeric membranes are considered in this study. The considered polymeric membrane materials s are TR-PBO-co-PPL 28, TR-PI-g-CD 450, in addition to the main studied hollow fiber membrane fabricated by Feng et al. [45]. The properties of the three polymeric membranes are presented in Table 6.3. For clarity, 1 GPU is equal to $3.35 \times 10^{-10}$ mol/(m$^2$ s Pa) or equal to $10^{-6}$ cm$^3$ (STP)/(cm$^2$ s cm Hg) and 1 Barrer is equal to $10^{-10}$ cm$^3$ (STP) cm/(cm$^2$ s cm Hg). Based on that, GPU is defined as the permeability (in Barrer) per unit membrane thickness (in μm). In Ref. [55], the permeability of $O_2$ is equal to 132 Barrer for a membrane having a thickness of 40 μm and, based on that, the permeance in GPU is calculated to be equal to 3.3 (132/40) GPU.

Figure 6.13 shows the distributions of stage cut (total permeation flow rate of $O_2$ plus $N_2$ divided by total feed flow rate) versus the separation factor, SF, (the ratio of $O_2$ to $N_2$ mass fraction in permeate-side divide by that ratio in feed-side) at zero sweep gas flow rate for the considered three polymeric membranes. Some of the results could be expected from those inputs in Table 6.3; however, for better understanding of the permeation characteristics for different membrane materials, the distributions of oxygen mass fractions at permeate outlet for different membranes are presented in Fig. 6.13. As shown in Fig. 6.13, oxygen mass fractions at permeate outlet for Feng et al. [45] fiber membrane and for TR-PI-g-CD 450 membrane are much different while they have the same selectivity. One can see that the highest $O_2$ mass fraction at permeate outlet (highest SF) and the lowest permeation flow rate (lowest stage cut) are obtained by the TR-PBO-co-PPL 28 membrane. This is because the TR-PBO-co-PPL 28 membrane has the highest selectivity and the lowest

**Table 6.3** Properties of selected polymeric membranes

| Membrane | TR-PBO-co-PPL 28 | TR-PI-g-CD 450 | Feng et al. [45] fiber membrane |
|---|---|---|---|
| Permeance ($O_2$) GPU | 3.3 | 67.67 | 9.3 |
| Permeance ($N_2$) GPU | 0.45 | 13.08 | 1.8 |
| Selectivity ($\alpha_{O_2/N_2}$) | 7.3 | 5.17 | 5.17 |
| Reference | [59] | [60] | [45] |

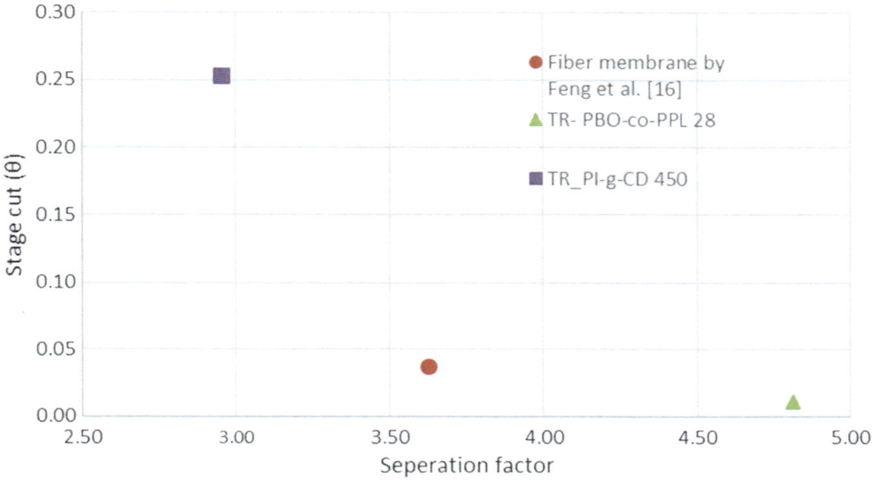


**Fig. 6.13** Stage cut against the separation factor at 100 psig feed pressure, 0.0 psig permeate pressure, 5.89e−8 kg/s feed flow rate and zero sweep flow rate

permeability among the considered three membranes. While the TR-PI-g-CD 450 membrane gives the highest permeation flow rate (highest stage cut) and the lowest $O_2$ mass fraction at permeate outlet (lowest SF). This can be attributed to the relatively high permeability and low selectivity of this membrane. The main considered hollow fiber membrane in this study, membrane by Feng et al. [45], gave moderate results in terms of permeation flow rate and $O_2$ mass fraction at permeate outlet as shown in Fig. 6.13. So, depending on the required gas to be separated at required flow rate and concentration, a suitable polymeric membrane material can be selected.

### 6.3.7 *Multi-stage Separation*

The aim of using polymeric membranes for oxygen separation is to increase oxygen concentration in air for oxygen-enriched air combustion application. If a certain high oxygen concentration is targeted for a specific application, multi-stage separation using polymeric membrane units is required to achieve the required high oxygen concentration. In this case, oxygen-enriched air mixture leaving the first unit from the permeate side is introduced to the second unit as feed flow. In this study, multi-stage gas separation is performed numerically considering three stages as presented in Fig. 6.14 at zero sweep gas flow rate. All considered fiber membranes in all stages are identical in terms of the size based on the design presented in Fig. 6.1. The first stage consists of two parallel multi-membranes units; each unit consists of 10 fiber membranes running in parallel. Air (with $O_2$ mass fraction of 0.228 and $N_2$ mass fraction of 0.772) is fed to the first stage. The feed flow rate for each fiber membrane

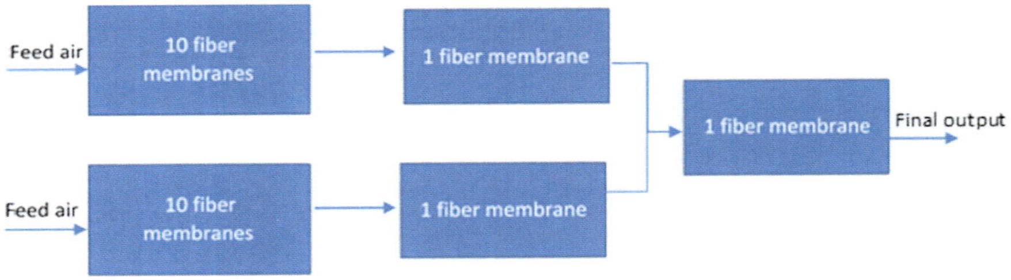


**Fig. 6.14** Configuration of the multi-stage separation unit

in the first stage is 5.89E−8 kg/s at zero sweep flow rate. Numerical simulations of oxygen permeation across one fiber membrane showed oxygen mass fraction of 0.517 at permeate outlet of the first stage. The permeate flow at the outlet of each unit in the first stage, with $O_2$ mass concentration of 0.517 and $N_2$ mass concentration of 0.483, is supplied as the feed flow to another membrane unit in the second stage as shown in Fig. 6.14. The second stage consists of two parallel membrane units; each unit consists of only one fiber membrane. The simulations at zero sweep gas flow rate of the second sage showed oxygen fraction of 0.8 at permeate outlet of the second stage. The final stage consists of one fiber membrane unit fed by the permeate flow out of the two parallel membrane units in the second stage. Calculations are performed considering zero sweep flow rate and the results showed oxygen fraction of 0.914 at permeate outlet of the third stage.

For all simulations, only one fiber membrane is considered in each stage. Figure 6.15 shows the stage cut of one fiber membrane cell against the gained $O_2$

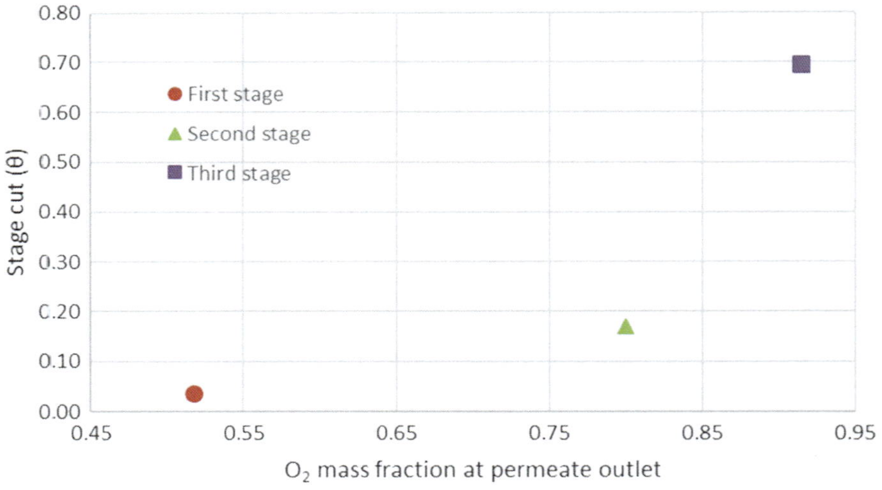


**Fig. 6.15** Stage cut versus $O_2$ mass fraction at permeate outlet for the series of the three stages at 100 psig feed pressure, 0.0 psig permeate pressure, varying feed flow rates and zero sweep flow ate

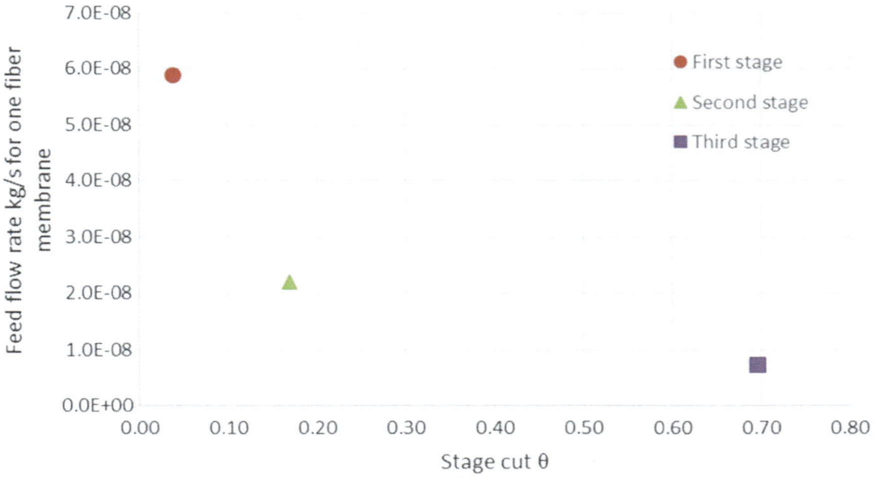


**Fig. 6.16** Stage cut versus feed flow rate for the series of the three stages at 100 psig feed pressure, 0.0 psig permeate pressure, varying feed flow rates and zero sweep flow rate

mass fraction at permeate outlet of one fiber membrane cell in each stage. In addition, Fig. 6.16 represents the feed flow rate against stage cut for one fiber membrane in each stage. However, each membrane cell in the first stage is fed by the highest feed flow rate as compared to the flow rates in the second and the third stages (see Fig. 6.16), the lowest $O_2$ permeation flux (lowest stage cut) is obtained in the first stage as shown in Fig. 6.15. This can be attributed to the relatively low $O_2$ mass fraction in the feed side in the first stage as compared to the second and the third stage. As a result, the driving force for $O_2$ permeation is reduced and, accordingly, oxygen permeation flux is reduced. So, higher stage cut is gained in the third stage as shown in Fig. 6.16 because of the high permeation flow rate relative to the feed flow. Better results can be obtained by recycling the retentate flows of the third and second stages to the feed side in the first stage as those retentate flows have higher $O_2$ mass fractions than normal air introduced in the first stage. Similar approach for recycling the retentate gases to the first stage was suggested by Katoh et al. [61].

## 6.4 Ceramic Membrane Reactors for Oxy-fuel Combustion Applications

Nowadays, the increased level of emissions of carbon dioxide as the main greenhouse gas represents one of the most critical problems facing our present and future lives. There are many running research works around the world trying to reduce the emissions of carbon dioxide targeting the control of the globe temperature. Some focus is made on the use of renewable and clean energies; however, for all of these natural sources of energy, the economic cost is the main barrier toward their intensive applications [1]. At present, the whole world is forced to use the hydrocarbons fossil fuels in order to cover the tremendously increasing demand of energy. As a result, carbon dioxide emissions will be continuously increasing. This necessitates the treatment of carbon dioxide through its capture or storage [62]. Based on that, the research work is directed toward technologies for carbon capture and sequestration. These technologies include pre-combustion, post-combustion and oxy-combustion carbon capture technologies. In the pre-combustion carbon capture technology, carbon dioxide is captured before the combustion process through gasification and shift conversion reaction processes of the hydrocarbon fuel. In the post-combustion capture, carbon dioxide is captured after the combustion process from the exhaust gases through scrubbing process. Chemically active agents like mono-ethanolamine (MEA) and methyldiethanolamine (MDEA) are primarily considered in the scrubbing process [2]. In the above mentioned two technologies for carbon capture, additional devices and modifications in the existing systems are required which will result in additional big cost. As one of the most promising techniques for $CO_2$ capture, oxy-combustion technology can be applied into the existing systems with slight modifications. In oxy-combustion technology, the hydrocarbon fuel is burned in a medium of pure oxygen and some recycled exhaust gases instead of air (nitrogen is not introduced into the combustion chamber). In this process, the combustion products consist of a mixture of only carbon dioxide and water vapor [63]. Water vapor can be easily condensed and accordingly carbon dioxide can be captured for industrial use or storage.

Oxy-combustion technology can be applied in the conventional combustion systems; however, this requires the separation of oxygen in another unit called air separation unit (ASU). This unit requires a part of the output power of the combustion system, and as a result, the overall system efficiency is reduced [64]. Recently, another approach for the application of oxy-combustion technology was proposed through the use of what's called oxygen transport membrane reactors (OTMR). In such kind of reactors, oxygen is being separated inside the combustion system through the use of ion transport membranes (ITM). At temperatures ranging from 650 to 950 °C, these membranes are activated for oxygen separation from the feed side to the permeate side of the membrane [21]. In the permeate side of the membrane, fuel is being burned with the separated oxygen in a medium of recycled carbon dioxide. There are many membrane materials which can be used in such applications including ceramic membranes of the same perovskite crystal structure of lanthanum cobaltite, perovskite ceramics, modified proviskite ceramics [65], brownmillerite structured ceramics [66], ceramic metal dual phase membranes [67], in addition to, thin duel phase membranes which consists of chemically stable yttria-stabilized zirconia (YSZ) [68].

Xu and Thomson [69] developed an explicit oxygen permeation model for ion conducting membranes while emphasizing the surface exchange kinetics at both sides of a $La_{0.6}Sr_{0.4}Co_{0.2}Fe_{0.8}O_{3-\delta}$ (LSCF) membrane. They performed a series of experimental measurements over a broad range of oxygen partial pressure and temperature. They reported increase in the oxygen permeation flux while increasing the temperature and the oxygen flux is controlled by bulk diffusion at high temperatures.

Rui et al. [13] studied the permeation of oxygen across ceramic oxygen ionic conducting membranes under finite rate reactions conditions. They reported drop in the partial pressure of oxygen in the sweep side due to combustion, and accordingly, an increase in the oxygen permeation flux was obtained. In addition, both the oxygen partial pressure in the permeate side and the reaction rate were lowered as a result of increasing the inlet inert gas in the permeate side. The permeation of oxygen across ceramic oxygen ionic conducting membranes was investigated by Akin and Lin [19] using a simple mathematical model for different oxidation reaction kinetics mechanisms including both extremely fast reaction and no reaction conditions. The results were compared with the experimental oxygen permeation data for $Bi_{1.5}Y_{0.3}Sm_{0.2}O_{3-\delta}$ (BYS) for different conditions of reaction using ethane and methane. They showed that the flux of oxygen is strongly affected by the reducing gas flux and oxidation reaction. Their results showed high oxygen permeation flux by one order of magnitude in case of ethane than with methane due to the faster reaction rate of ethane as compared to methane. Habib et al. [70] examined the influence of reactivity of methane in the permeate side of a stagnation flow ITM reactor on the permeation flux of oxygen across a $La_{0.6}Sr_{0.4}Co_{0.2}Fe_{0.8}O_{3-\delta}$ (LSCF-6428) ionic conducting ceramic membrane. They attributed the sharp rise in the flux of oxygen while the reactions are activated in the permeate side mainly to the rise in the combustion temperature which enhances the diffusion of oxygen and reduces the bulk diffusion resistance of the membrane.

Hong et al. [71] studied numerically the operation of a LSCF ITM reactor under reactions conditions while incorporating a detailed gas-phase model of methane. Their results showed the dependence of oxygen permeation flux on reactor geometry, membrane temperature and feed and sweep flux rates. Ben-Mansour et al. [72] and Nemitallah et al. [55] studied the performance of a $La_{0.1}Sr_{0.9}Co_{0.9}Fe_{0.1}O_{3-\delta}$ (LSCF-1991) ITM reactor under non-reacting and reacting conditions considering different operating conditions. They reported significant effects of a set of parameters including inlet temperature, inlet methane concentration, and reactor geometry on the operation of the IT reactor. Hong et al. [73] studied the interactions between oxygen permeation and homogeneous-phase methane combustion in the permeate side of a LSCF membrane. They found enhancement of the oxygen permeation flux while the inlet temperature and fuel concentration are increased. Hunt et al. [74] measured the oxygen profiles normal to the membrane surface and oxygen permeation flux across a $La_{0.9}Ca_{0.1}FeO_{3-\delta}$ (LCF) membrane and they developed a multi-step surface exchange model. Kirchen et al. [75] presented a novel ITM reactor for fundamental investigations of oxygen permeation and oxy-fuel combustion under reactive flow conditions. They studied experimentally the influence of inlet methane fraction on the oxygen permeation and fuel conversion in the sweep side of the membrane. They reported that in order to use an ITM reactor for complete fuel conversion (i.e. oxy-combustion application) the inlet fuel concentration should be kept below a certain limit to satisfy the stoichiometric conditions for combustion.

Recently, Mancini and Mitsos [20] presented a study on the application of ITM reactors for large scale power generation. They proposed a design of a monolith structure ITM reactor to be used for power generation based on detailed numerical analysis. Their reactor has a total volume of 1000 $m^3$, 100,000 total feed and permeate channels, 266,700 $m^2$ surface area, a height of 4.75 m, length of 44.44 m and the reactor is able to deliver power in the range of 300 to 500 MWe based on the cycle first law efficiency. Nemitallah et al. [15] developed a multi-parallel short channels compact ITM reactor design for gas turbine combustion application based on detailed 3-D numerical analysis. Their ITM reactor has a length of only 0.9 m, height of 3.35 m, volume of 10 $m^3$, surface area of 2700 $m^2$, 25,000 channels per stream, and the reactor is able to produce power ranging from 5 to 8 MWe based on the cycle first law efficiency. However of the extensive work done on the analysis of ITM reactors, still more research work is required on membrane materials to increase their oxygen permeability and to make them able to withstand high temperature under reacting conditions. In addition, there are some problems regarding to the sustainability of the flame inside the ITM reactor under such difficult conditions for combustion.

## 6.5 Hybrid Polymeric-Ceramic Membrane Reactor for Gas Turbine Applications

Oxy-fuel combustion is considered to be one of the most promising carbon capture technologies toward the control of global warming issue. Pure oxygen or oxygen-enriched air could be generated by three main air separation methods including; cryogenic distillation, pressure swing adsorption, and membrane separation. Cryogenic distillation is the most common of these three technologies and produces oxygen of more than 99% purity with large scale production [29]. Pressure swing adsorption can reach up to 95% oxygen purity, but the requirement of solvents limits its size capacity. This is primarily due to capital costs [30, 31]. The membrane separation technology is the most recent of these three technologies. This technology utilizes high temperature ceramic ion transport membranes (ITMs) and polymeric membranes. Polymeric membranes for gas separation have been studied extensively in the literature targeting the development of high performance membranes for different applications [38, 39, 76]. One of the recently developed groups of membranes is the thermally rearranged (TR) polymeric membranes group by Park et al. [76]. These polymeric membranes have high selectivity and high permeability for different gases. Various studies were conducted to develop polymers from this family and to explore the gas separation process through this type of membranes [38–40, 77]. Dong et al. [51] solved a model for gas separation in a TR-PBO hollow fiber membrane using MATLAB, and they proposed to include the shell-side pressure drop while modeling gas separation. Choi et al. [53] performed a study on $O_2$ removal from exhaust gas streams using polymeric membrane for complete capture of $CO_2$. They evaluated the permanent changes in polymer morphology and separation performance data considering single gas permeation.

The use of ceramic ion transport membranes (ITMs) allows production of highly pure oxygen through air separation at high temperature (more than 973 K) and high pressure and at reduced costs [78]. ITMs have high oxygen selectivity compared to polymeric membranes, but they only operate at high temperatures that could eliminate their utility for high temperature applications [79]. However, ITMs can be a good choice for oxy-fuel combustion applications, as the combustion process provides the required heat to raise the membrane temperature for air separation [80]. Increasing $O_2$ concentration on the feed side through using oxygen-enriched air instead of atmospheric air enhances the partial pressure difference across the membrane resulting in higher oxygen permeation rate. This oxygen-enriched air stream can be generated through the use of polymeric membranes. In this case, atmospheric air is fed to the polymeric membrane in the feed side to produce oxygen-enriched air mixture in the permeate side, which can be fed to the ceramic membrane unit [81]. Polymeric membranes can produce oxygen-enriched air with more than 50% mole fraction, depending on the polymer type. The study presented in this section aims at developing a hybrid polymeric-ceramic membrane unit for oxy-fuel combustion applications. In these hybrid membranes, the oxygen-enriched air out of the polymeric membrane unit is fed to an oxygen transport reactor (OTR) unit for oxygen separation and oxy-combustion of gaseous fuel.

The performance of different types of ceramic membranes for pure oxygen separation has been investigated extensively in the literature with little emphasis on oxygen permeation under oxy-combustion conditions [55, 80–83]. Yantovski et al. [82] used ITM to feed oxygen for an oxy-fuel combustion power plant. They found that the corresponding turbine inlet temperatures ranged between 1573 K and 1773 K and the thermal efficiencies ranged from 46 to 55%. Habib et al. [80] modeled an isothermal multi-channel reactor for oxy-fuel combustion to study the permeation and combustion behavior at different operation conditions. It was found that the reactions in the permeate side lead to an increase in the oxygen partial pressure gradient across the membrane as well as the membrane temperature, thereby, oxygen permeation flux increased. Nemitallah et al. [55] initiated a numerical study to investigate the performance of an ITM reactor under oxy-combustion conditions and the effect of various parameters on the reactor operation. Hong et al. [83] numerically investigated laminar oxy-fuel non-premixed flames in the permeate side of an ITM. They found that the flame location affects the heat transfer and species diffusion from the reaction zone to the ITM, which affects the flame temperature and the oxygen permeation flux.

Cryogenic distillation, the most conventional method of air separation, is an expensive method and does reduce the plant's efficiency [84]. Burdyny and Struchtrup [85] studied the air separation using a hybrid membrane-cryogenic system for small and medium oxy-fuel systems. The results showed that the process efficiency could be improved by 0.9% in this hybrid system. In the study presented in this section, Fluent software was utilized to perform a detailed numerical study on the characteristics of air separation and oxy-fuel combustion in a hybrid polymeric-ceramic membrane unit. The produced oxygen-enriched air from the polymeric membrane is fed to the ceramic oxygen transport reactor (OTR) to build the hybrid system. User defined functions (UDFs) written in visual C++ were coupled and hooked to the Fluent software to account for the transfer of both oxygen and nitrogen across the polymeric membrane and the transfer of oxygen only across the ceramic membrane. Effects of different parameters on the permeation characteristics and the oxy-fuel combustion process in the OTR were investigated. A proposed design of the hybrid unit is presented based on this study.

### 6.5.1 Design of Polymeric Membrane Unit

Recently, a detailed numerical study was performed to understand the characteristics of air separation in hollow-fiber polymeric membrane unit targeting the production of oxygen-enriched air for clean combustion applications [86]. In this study, a shell and tube design of a polymeric membrane unit is considered, and 2-D numerical simulations were performed considering asymmetric single hollow fiber membrane unit with its surrounding shell. The same polymeric membrane unit is used in the study presented in this section with the same configuration as in [86], and the simulation results are utilized for the development of a hybrid polymeric-ceramic membrane unit for oxy-combustion applications. Oxygen-enriched air stream leaving the polymeric membrane unit is utilized as feed flow for the oxygen transport reactor (OTR) ceramic membrane unit. Counter-current flow bore-side feed configuration was considered in this model to give better separation results [45]. Air (20.5% $O_2$ and 79.5% $N_2$, by volume) is fed to the fiber membrane cell in the bore-side, and oxygen depleted air leaves as retentate at the end of the feed side. While at the shell-side, oxygen-enriched air is produced. This is utilized to feed the oxygen transport reactor (OTR) in the hybrid system. Pressure outlet conditions have been specified at the permeate outlet and feed outlet (atmospheric pressure at the permeate outlet and 10.13 bar at the feed outlet). Mass flow rate inlets boundary conditions have been assigned for the sweep and the feed entries.

The numerical model of the polymeric membrane gas separation unit was established and validated using Fluent software. The numerical solution of the flow field includes the solution of the partial differential equations of continuity, energy, momentum, and species transport. The equations are presented as follows [55]:

$$\nabla \cdot (\rho U) = S_i \tag{6.13}$$

$$\nabla \cdot (\rho U U) = -\nabla p + \mu \nabla^2 U \tag{6.14}$$

$$(\rho C_p)_f U \cdot \nabla T = \nabla \cdot (k_f \nabla T) \tag{6.15}$$

$$\nabla \cdot (\rho U X_i) - \nabla \cdot (\rho D_{i,m} \nabla X_i) = S_i \tag{6.16}$$

where $U$ is the velocity, $\mu$ is the dynamic viscosity, $\rho$ is the density, p is the pressure, $D_{i,m}$ is the diffusion coefficient in the mixture, $T$ is the temperature, and $X_i$ is the mass fraction. Source/sink term, $S_i$, was used to consider the transfer of $O_2$ and $N_2$ across the polymer membrane. To include this source/sink term in the governing equations as presented above, user defined functions (UDFs) were written in Visual C++ and compiled and hooked to the Fluent software to close the set of the governing equations [55]. The law that controls the fluid flow through a porous medium, which was used to solve for the permeation across the polymer membrane, can be expressed as follows [87]:

$$J_{O_2/N_2} = \frac{P(p' - p'')}{L} \tag{6.17}$$

where $J$ is the permeation flux of $O_2$ or $N_2$ in mol/m$^2$/s and $P$ is the permeability of the polymer membrane in mol. m/m$^2$/s/Pa. $p'$ and $p''$ are the gas partial pressure in the feed and the permeate sides, and $L$ is the membrane thickness. Based on the flux equation, the equation of the source/sink term is developed for both $O_2$ and $N_2$ at each boundary membrane cell. A grid independence study has been performed to select the appropriate mesh size and to make sure that the results are independent of the mesh size [86]. Experimental data by Feng et al. [45] were used to validate the implemented model for polymeric membrane system. More details about the polymeric model and its validation are illustrated in detail in our previous work [86].

### 6.5.2 *Design of Oxygen Transport Membrane Reactor (OTMR) Unit*

The current work utilizes an oxygen transport reactor (OTR) unit that consists of tubular oxygen transport membrane (OTM) surrounded by a shell to perform the numerical simulations of the ceramic membrane unit. This configuration of an OTR unit is represented by the 2D schematic shown in Fig. 6.17. The used geometry considers 2 mm radius membrane tube with a thickness of 0.9 mm surrounded by 10 mm radius shell, and the unit has a length of 100 mm. LCF membrane material is used as the OTM in this study having properties as described in the study of Hunt et al. [74].

$CH_4/CO_2$ mixture is supplied at the sweep side (inside the OTM tube) as diluted fuel by $CO_2$ with various $CH_4/CO_2$ ratios. While oxygen-enriched air with different oxygen concentrations or atmospheric air (23.3% $O_2$ and 76.7% $N_2$ by mass) is fed

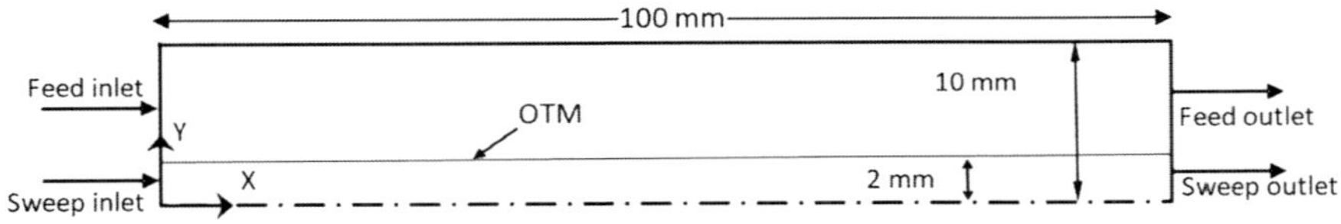


**Fig. 6.17** Computational domain of the considered oxygen transport reactor (OTR)

to the shell side. $O_2$-depleted air leaves as retentate at the feed outlet as a result of the oxygen permeation through the OTM, while a highly $CO_2$-concentrated mixture resulting from the combustion process leaves the sweep side as an exhaust stream. The sweep and feed inlet streams are to be preheated before the OTR entries from the atmospheric temperature to temperatures of 1173 K or 1123 K. High entrance temperature to the OTR is required to ensure the activation process for oxygen permeation. Mass flow rate inlet boundary conditions are specified for the feed and the sweep entries, while atmospheric pressure outlet boundary conditions have been specified at the sweep and feed outlets. Co-current flow configuration was considered in this model to ensure better separation results [88]. Table 6.4 shows a summary of ranges of parameters considered in this study.

**Table 6.4** Summary of the ranges of parameters considered in OTR simulations

| Study | Parameter | Range |
|---|---|---|
| Effect of fed $O_2$ mass fraction | Fed $O_2$ mass fraction | 23.3–85.0% |
| | Feed flow rate | $5.0 \times 10^{-6}$ kg/s |
| | Sweep flow rate | $2.0 \times 10^{-7}$ kg/s |
| | Swept $CH_4$ mass fraction | 0.5 |
| | Streams inlet temperature | 1173 K |
| Effect of sweep flow rate | Fed $O_2$ mass fraction | 52.25% |
| | Feed flow rate | $5.0 \times 10^{-6}$ kg/s |
| | Sweep flow rate | $8.0 \times 10^{-8} - 8.0 \times 10^{-7}$ kg/s |
| | Swept $CH_4$ mass fraction | 0.5 |
| | Streams inlet temperature | 1173 K |
| Effect of feed flow rate | Fed $O_2$ mass fraction | 52.25% |
| | Feed flow rate | $2.0 \times 10^{-6} - 1.0 \times 10^{-5}$ kg/s |
| | Sweep flow rate | $2.0 \times 10^{-7}$ kg/s |
| | Swept $CH_4$ mass fraction | 0.5 |
| | Streams inlet temperature | 1173 K |
| Effect of swept $CH_4$ mass fraction | Fed $O_2$ mass fraction | 52.25% |
| | Feed flow rate | $5.0 \times 10^{-6}$ kg/s |
| | Sweep flow rate | $2.0 \times 10^{-7}$ kg/s |
| | Swept $CH_4$ mass fraction | 0.05–1.00 |
| | Streams inlet temperature | 1173 K |
| Stoichiometric cases | Fed $O_2$ mass fraction | 52.25% |
| | Feed flow rate | $1.14 \times 10^{-6} - 3.81 \times 10^{-6}$ kg//s |
| | Sweep flow rate | $1.5 \times 10^{-6} - 2.0 \times 10^{-6}$ kg/s |
| | Swept $CH_4$ mass fraction | 0.1 |
| | Streams inlet temperature | 1123 and 1173 K |

The simulations of the OTR were performed using 2D axisymmetric geometry due to the symmetry of the unit as shown in Fig. 6.17. Gambit mesh builder software was used to develop the mesh that is used by Fluent 16.1 software. The numerical solution of the flow field includes the solution of the differential equations of continuity, energy, momentum, and species transport (Eqs. 6.13–6.16). Regarding the methane reaction kinetics model, the study utilizes the single-step kinetics model of finite rate reaction as follows:

$$CH_4 + 2O_2 \rightarrow CO_2 + 2H_2O \tag{6.18}$$

The reaction rates are calculated based on the default values of the activation energy and the pre-exponential coefficient available in the Fluent software, 2.027 × $10^8$ J/kg mol, and 2.119 × $10^{11}$ respectively. Radiation modeling is included in the computational work to accurately predict the temperature within the reactor. Discrete ordinates (DO) radiation model is adapted in the numerical model for the calculations of radiation parameters under oxy-combustion conditions [89]. The DO model solves the radiative transfer equation for absorbing, emitting, and scattering mediums. The radiative transfer equation in the direction $\overrightarrow{S}$ is given as follows [89]:

$$\frac{dI(\vec{r},\vec{s})}{ds} + (a+\sigma_s)I(\vec{r},\vec{s}) = \frac{an^2(\sigma T^4)}{\pi} + \frac{\sigma_s}{4\pi}\int_0^{4\pi} I(\vec{r},\vec{s}')\phi(\vec{s}\cdot\vec{s}')\,d\Omega' \tag{6.19}$$

where $n$ is the refractive index, $a$ is the absorption coefficient, $I$ is the intensity of radiation and $s$ is the path length. Oxygen separation through the dense non-porous LCF OTM is achieved using oxygen permeation flux equation as presented by Xu and Thomson [79]:

$$J_{O_2,OTM} = \frac{D_V K_r\left(p'^{0.5}_{O_2} - p''^{0.5}_{O_2}\right)}{2LK_f\left(p''_{O_2}p'_{O_2}\right)^{0.5} + D_V\left(p'^{0.5}_{O_2} + p''^{0.5}_{O_2}\right)} \tag{6.20}$$

where $D_v$ is the diffusion coefficient of oxygen vacancy, $K_r$ is the reverse reaction rate, $K_f$ is the forward reaction rate, $p'$ and $p''$ are the gas partial pressures in feed and permeate sides respectively, and L is the membrane thickness. $D_V$, $K_r$, and $K_f$ can be expressed as follows:

$$D_V = D_V^o \exp\left(-\frac{E_D}{RT}\right) \tag{6.21}$$

$$K_r = K_r^o \exp\left(-\frac{E_r}{RT}\right) \tag{6.22}$$

$$K_f = K_f^o \exp\left(-\frac{E_f}{RT}\right) \tag{6.23}$$

**Table 6.5** Experimentally fitted values of activation energies and pre-exponential factors for the LCF membrane [74]

| Material constant | Pre-exponential factor | | Activation energy (J/mol) |
|---|---|---|---|
| | Units | Value | |
| $k_f$ | cm/atm$^{0.5}$ s | $4.2 \times 10^{11}$ | $3.37 \times 10^5$ |
| $K_r$ | mol/cm$^2$ s | $1.34 \times 10^{10}$ | $3.79 \times 10^5$ |
| $D_v$ | cm$^2$/s | $4.98 \times 10^{-3}$ | $5.96 \times 10^4$ |

where, $E_D$, $E_r$, and $E_f$ are the activation energies and $K_r^o$, $K_f^o$, and $D_V^o$ are the pre-exponential factors. Table 6.5 shows the values of the activation energies and the pre-exponential factors that are fitted experimentally by Hunt et al. [74] for the LCF membrane.

Based on the permeation Eq. (6.20), the equation for the source/sink term is developed to be applied at membrane boundary cells as follows:

$$S_i = \begin{cases} +\frac{J_{O_2.OTM} \cdot A_{cell} \cdot MW_{O_2}}{V_{cell}}, & \text{at the sweep side of the OTR} \\ -\frac{J_{O_2.OTM} \cdot A_{cell} \cdot MW_{O_2}}{V_{cell}}, & \text{at the feed side of the OTR} \end{cases} \tag{6.24}$$

where $MW_{O2}$ is the molecular weight of $O_2$, $A_{cell}$ is the cell area and $V_{cell}$ is the cell volume. This term has a positive sign in the sweep side of the membrane and a negative sign on the feed side of the membrane as the species of $O_2$ transport from the feed side to the permeate side of the membrane. The term, $S_i$, is only calculated for the boundary cells, cells in direct contact with the membrane from both sides, and it has a value of zero elsewhere inside the computational domain. In addition, this term has a value of zero for species other than $O_2$. User defined functions (UDFs) written in visual C++ are coupled and hooked to the Fluent software to account for the transfer of $O_2$ across the membrane. UDFs are attached to the software through a source/sink term in mass and momentum equations.

To ensure that the used grid in this model has enough resolution to accurately predict the results in the computational domain and to make sure that the results are grid independent, oxygen permeation flux was calculated using different grids and the values were compared. Four different mesh sizes with total number of cells of 5500, 13,000, 23,600 and 36,000 were considered at $1.0 \times 10^{-6}$ kg/s feed flow rate with 0.525 $O_2$ mass fraction and at $2.0 \times 10^{-7}$ kg/s sweep flow rate with 0.50 $CH_4$ swept mass fraction. Based on the calculated results, refining the grid above the size of 23,600 grid cells resulted in insignificant changes in oxygen permeation flux. Based on that, all calculations were performed using a grid size of 23,600 cells.

Experimental data by Hunt et al. [74] were used to validate the numerical model for oxygen permeation across an LCF membrane. In their experiments, bottom cell reactor was used to permeate the oxygen through the LCF membrane. Different membrane thicknesses were included in their study to provide higher accuracy model. Table 6.6 represents the experimental data used in the validation study. The validation study is performed to calculate the oxygen permeation flux for two experimental sets

**Table 6.6** Experimental data by Hunt et al. [74] used for the validation of the numerical model

| OTM thickness (mm) | Feed gas | Sweep gas | Sweep gas flow rate | Feed flow rate | Configuration | Membrane effective area |
|---|---|---|---|---|---|---|
| 0.9–1.4 | Air | $CO_2$ | 1500–2400 sccm | 6000 sccm | Flat membrane | $4.76 \times 10^{-3}$ $m^2$ |

at different membrane thicknesses and different sweep flow rates over a range of operating temperature as presented in Fig. 6.18. The results show that the total oxygen permeation is affected by the membrane temperature. The total oxygen permeation

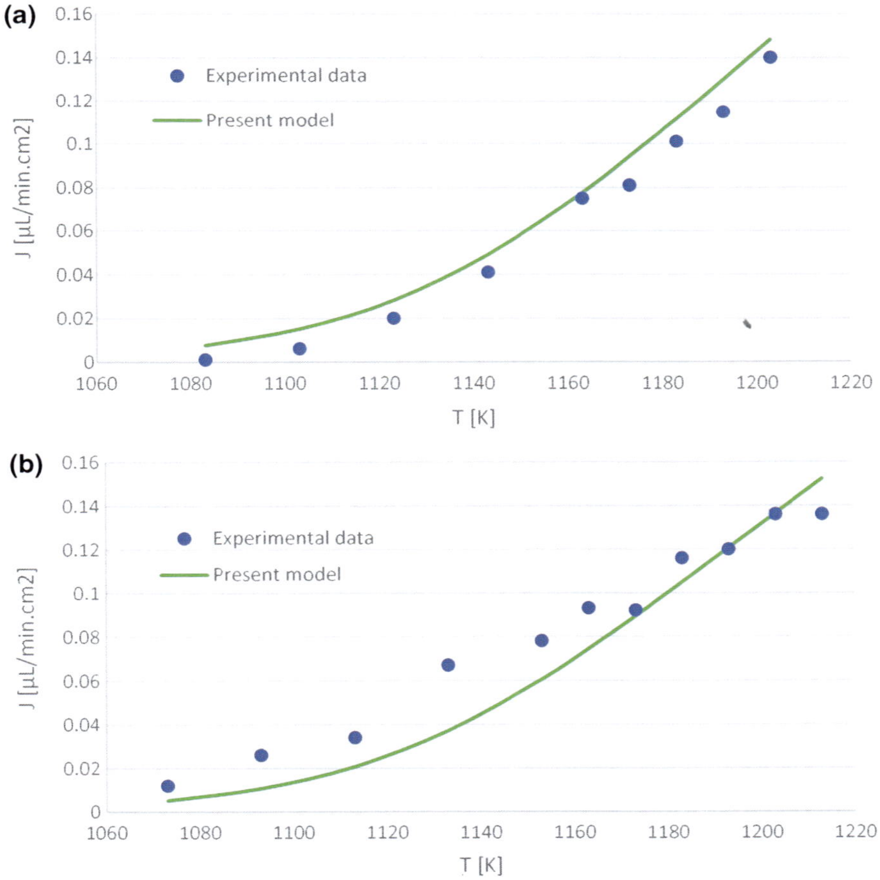


**Fig. 6.18** Model predictions in terms of $O_2$ permeation flux against the experimental data by Hunt et al. [74]: **a** at 0.9 mm membrane thickness, and 1500 sccm sweep flow rate, and **b** at 1.4 mm membrane thickness, and 2400 sccm sweep flow rate

flux through the membrane increases as a result of the temperature increase. This can be attributed to the increase in the diffusion coefficient and the reduction of permeation resistance while increasing the membrane temperature as indicated by oxygen permeation equation, Eq. (6.20). In addition, Fig. 6.18 shows good agreement between the numerical model results and the experimental data by Hunt et al. [74].

### 6.5.3 Analysis of the Hybrid Unit

The outputs of the polymeric membrane model were used as inputs to the OTR model to build the hybrid system. The thermally rearranged polymeric membrane, TR-PI-g-CD 450, with high permeation and selectivity of oxygen was used [90]. Based on the detailed numerical simulations of the polymeric membrane unit conducted by Habib et al. [86], Table 6.7 shows the inputs and the outputs of one fiber polymeric membrane unit fed by atmospheric air. Figure 6.19 shows the permeation flux of $O_2$ and $O_2$ mass fraction at permeate outlet over range of feed flow rates for the polymeric membrane unit. The plots show increase of $O_2$ mass fraction and $O_2$ permeation flux with increasing the feed flow rate. First, both $O_2$ mass fraction and $O_2$ permeation flux are increased sharply while increasing the feed flow rate above zero, then, they are linearly increased [86]. This can be attributed to the increase of flow velocity in the feed side resulting in better purging of the $O_2$-depleted mixture from the membrane surface in the feed side by the incoming fresh air. This should sustain high partial pressure differences across the membrane. Due to the raised pressure drop with the increased flow rate in the feed side of the membrane, medium feed flow rate has been selected for the operation of the polymeric unit in the hybrid system. Thus, sustaining high $O_2$ partial pressure at the feed side while keeping feasible pressure drop (near to 0.5 bar).

The outputs (oxygen-enriched air) have been used as inputs to the OTR. Figure 6.20 shows the contours of temperature, $O_2$ mass fraction, and $CH_4$ mass fraction of the OTR fed by oxygen-enriched air ($O_2$ mass fraction of 0.525) at $5.0 \times 10^{-6}$ kg/s feed flow rate. The temperature contours (Fig. 6.20a) show that the flame temperature reaches 2050 K, with an increase of 800 K above the inlet temperature. Moreover, because of the high thermal conductivity of the OTM, the outlet temperature at feed and sweep outlets are similar (around 1925 K). Lower temperature could be achieved by reducing the burned fuel amount as will be illustrated subsequently. $O_2$ mass fractions (Fig. 6.20b) show the oxygen depletion at the feed side, where $O_2$ mass fraction is reduced from the oxygen-enriched level (0.525 $O_2$ mass fractions) at inlet to the oxygen depleted air (lower than 0.10 $O_2$ mass fractions) at outlet. The reduction of $O_2$ mass fraction is a result of the oxygen permeation through the membrane, which increases with the increase of both $O_2$ partial pressure difference across the membrane and the membrane temperature. At the sweep side, $CH_4$ is consumed in the combustion zone and, as a result, its mass fraction is reduced from 0.25 at inlet to zero at outlet as shown in Fig. 6.20c. Downstream of the flame zone, there still permeation of $O_2$ through the membrane. This can be attributed to high membrane

**Table 6.7** Inputs and outputs of one fiber polymeric membrane unit

| Polymeric membrane | Area ($cm^2$) | Selectivity $\alpha_{O_2/N_2}$ | $O_2$ Permeance (GPU) | $N_2$ Permeance (GPU) | Feed flow rate (kg/s) | Feed pressure | $O_2$ mass fraction at permeate outlet | Permeate flow rate (kg/s) |
|---|---|---|---|---|---|---|---|---|
| TR-PI-g-CD 450 | 1.26 | 5.17 | 67.67 | 13.08 | $2.35 \times 10^{-7}$ | 10.13 bar | 0.525 | $2.56 \times 10^{-8}$ |

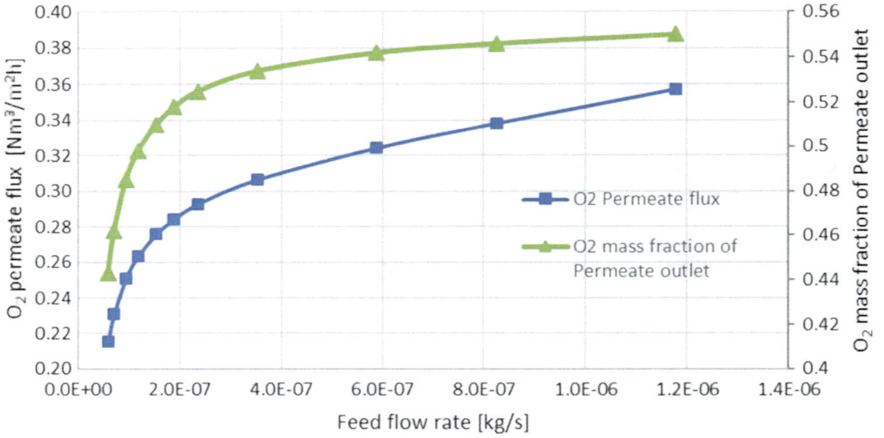


**Fig. 6.19** $O_2$ permeation flux and $O_2$ mass fraction at permeate outlet against the feed flow rate of the TR-PI-g-CD 450 polymeric membrane at 10.13 bar feed pressure, atmospheric permeate pressure and at zero sweep flow rate

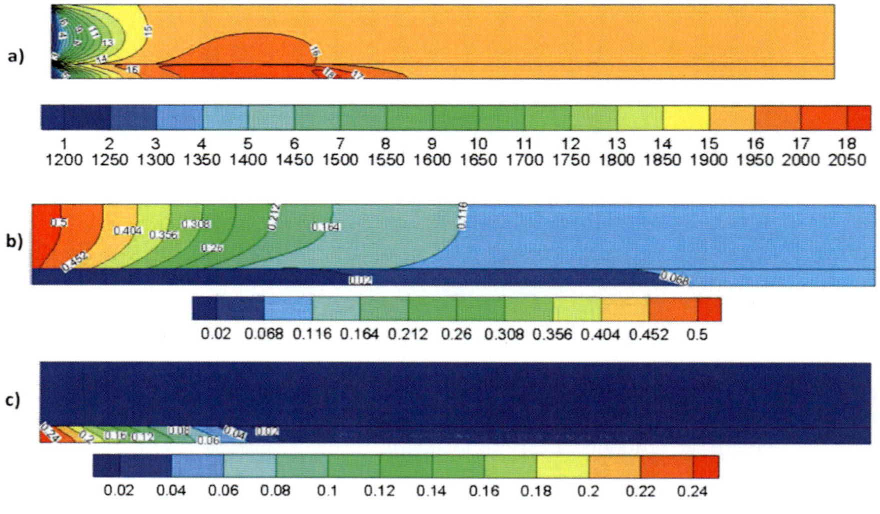


**Fig. 6.20** Contours of: **a** Temperature (K), **b** $O_2$ mass fraction, **c** $CH_4$ mass fraction, at $5.0 \times 10^{-6}$ kg/s feed flow rate with 0.525 $O_2$ mass fraction, $2.0 \times 10^{-6}$ kg/s sweep flow rate with 0.25 $CH_4$ mass fraction and at $\phi = 1.31$

temperature and high $O_2$ partial pressure difference across the membrane. As all $CH_4$ has been consumed by the reaction, the permeated $O_2$ will cause an increase of $O_2$ mass fraction at the sweep side (Fig. 6.20b). The existence of oxygen at the sweep outlet is not preferable for cost effectiveness of the oxy-combustion process and ease capturing of $CO_2$. To solve this issue, just the stoichiometric amount of $O_2$ should

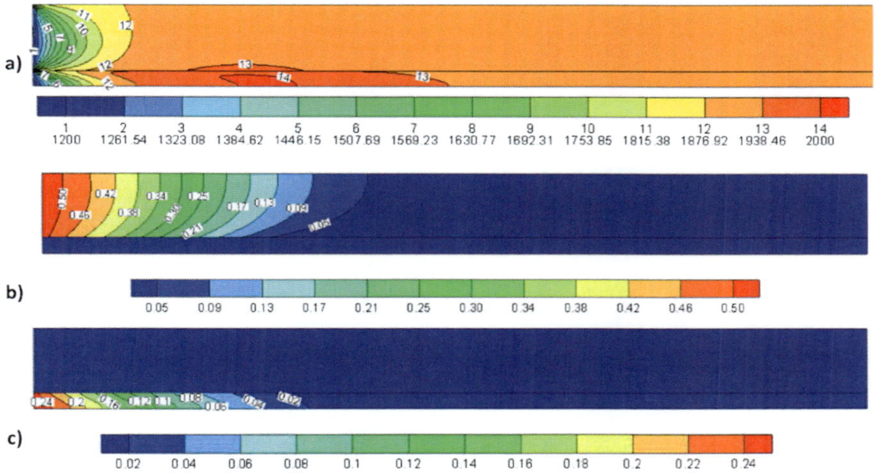


**Fig. 6.21** Contours of: **a** Temperature (K), **b** $O_2$ mass fraction, **c** $CH_4$ mass fraction, at $3.81 \times 10^{-6}$ kg/s feed flow rate with 0.525 $O_2$ mass fraction, $2.0 \times 10^{-6}$ kg/s sweep flow rate with 0.25 $CH_4$ mass fraction, and at $\varphi = 1.0$

permeate to the sweep side, to establish for a complete combustion without excess oxygen.

Figure 6.21 shows the contours of temperature, $O_2$ mass fraction and $CH_4$ mass fraction, at $3.81 \times 10^{-6}$ kg/s feed flow rate with 0.525 $O_2$ mass fraction, $2.0 \times 10^{-6}$ kg/s sweep flow rate with 0.25 $CH_4$ mass fraction, and at stoichiometric conditions ($\varphi = 1.0$). Stretched and lower temperature flame could be observed in this case. This can be attributed to the lower permeation across the membrane comparing with the previous case at $\phi = 1.31$. Figure 6.21b, shows no excess of $O_2$ at sweep outlet (less than 0.01 mass fraction). Also, complete combustion is achieved, where all the $CH_4$ is burned without excess oxygen at the outlet. So, the solution of the problem of excess oxygen at the sweep outlet can be solved by using stoichiometric feed flow rate ($\varphi = 1.0$). Another advantage of using this condition is the production of highly oxygen-depleted air (higher than 0.95 $N_2$ mass fraction) at the feed outlet, which can be used to produce high purity $N_2$.

### 6.5.4 *Effect of Feed $O_2$ Mass Fraction*

The local permeations through the membrane length for different fed $O_2$ mass fractions are shown in Fig. 6.22a. The case of air ($O_2$ mass fraction of 0.233) showed lower permeation level but with longer distribution along the membrane. While in case of $O_2$ mass fraction of 0.85, higher permeation is observed but with shorter distribution. The increment decrease of permeation at a specific region of the membrane

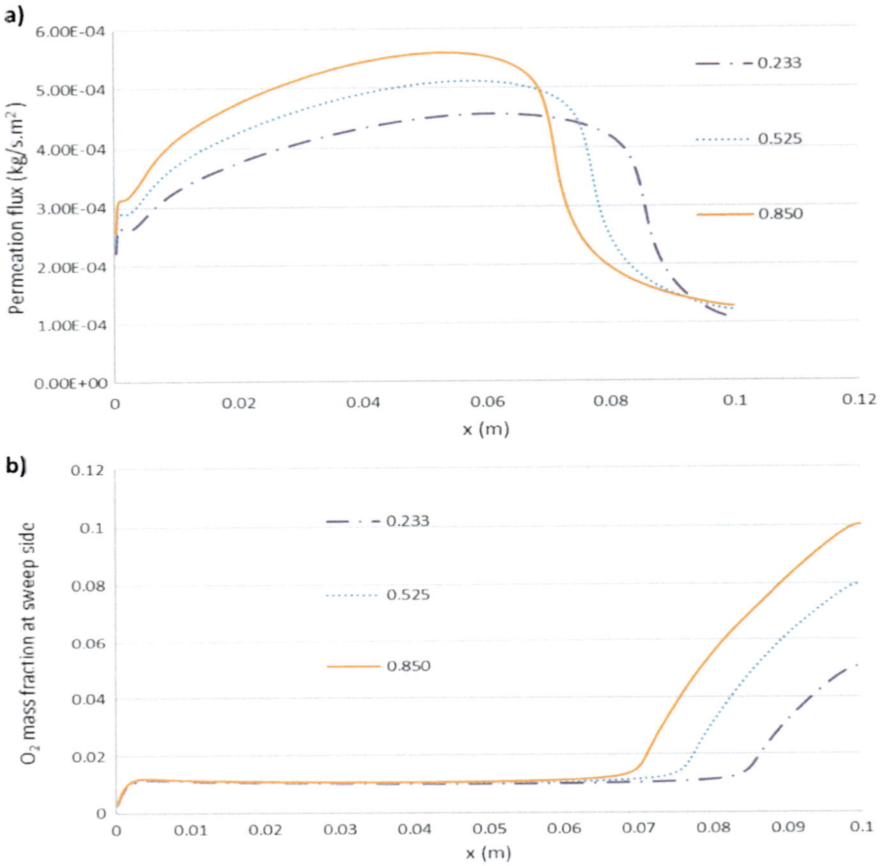


**Fig. 6.22** Distributions of **a** permeate flux rates of $O_2$, and **b** $O_2$ mass fraction at the sweep side, over range of fed $O_2$ mass fractions at $5.0 \times 10^{-6}$ kg/s feed flow rate, $2.0 \times 10^{-7}$ sweep flow rate with 0.50 $CH_4$ mass fraction, and at $\varphi > 1.0$

can be attributed to the increase of the $O_2$ mass fraction at the sweep side after complete fuel consumption. Figure 6.22b, shows the distributions of $O_2$ mass fraction at the sweep side for different feed oxygen concentrations. The case of 0.85 feed $O_2$ mass fraction has higher permeation of $O_2$ through the membrane and faster fuel consumption (see Fig. 6.23a) and, therefore, this case has faster increase of $O_2$ mass fraction at the sweep side as shown in Fig. 6.22b. This increase of $O_2$ in the sweep side increases the partial pressure at the sweep side, then, reduces the permeation through the membrane. The Higher permeation rate has been achieved by increasing the fed $O_2$ mass fraction. This high permeation has two main reasons; first, the higher $O_2$ partial pressure at the feed side and second, the higher membrane temperature as shown in Fig. 6.23b. The increased membrane temperature as the fed $O_2$ fraction increases can be attributed to faster burning rate of the fuel due to increased $O_2$

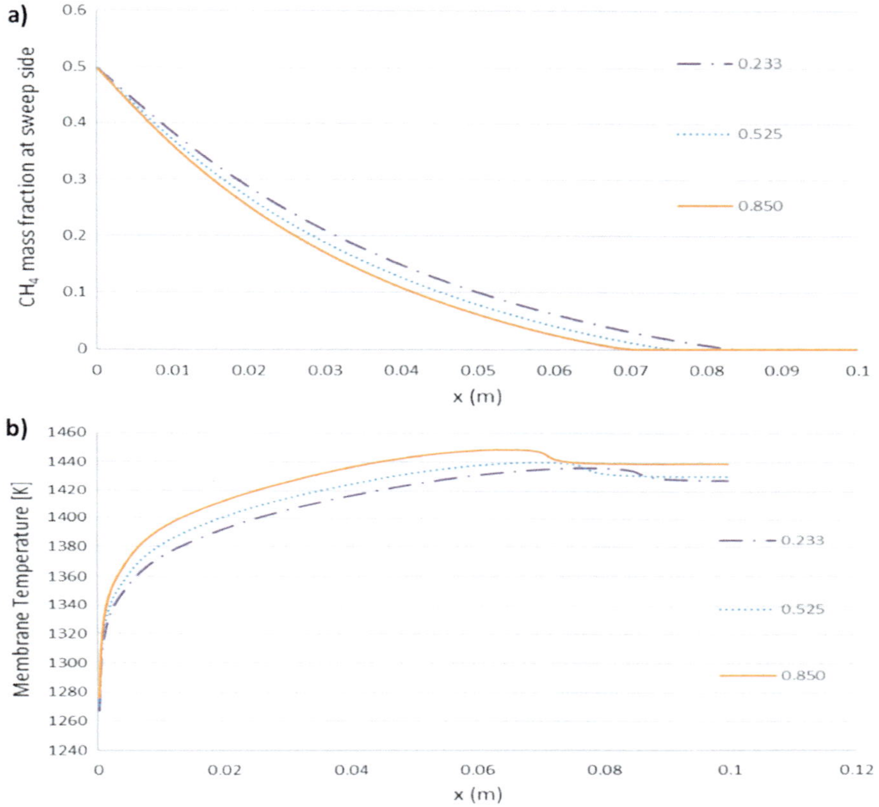


**Fig. 6.23** Distributions of **a** $CH_4$ mass fraction at sweep side, and **b** membrane temperature (K), over range of fed $O_2$ mass fractions at $5.0 \times 10^{-6}$ kg/s feed flow rate, $2.0 \times 10^{-7}$ sweep flow rate with 0.50 $CH_4$ mass fraction, and at $\varphi > 1.0$

permeation. Another reason for this higher temperature is the relatively high specific heat of $N_2$ as compared to $O_2$, which results in higher heat absorption in case of air (with higher $N_2$ concertation) as compared to the cases of oxygen-enriched air.

Comparing stoichiometric feed flow rate for both fed oxygen-enriched air case and fed atmospheric air case shows an advantage of using the hybrid system. In the air fed case, the feed flow should be higher than the oxygen-enriched case of 0.525 fed $O_2$ mass fraction by the ratio 0.525/0.233 to achieve the stoichiometric condition at the same $CH_4$ flow rate. This shows a major advantage of using oxygen-enriched air in the feed side instead of using atmospheric air. Reducing the feed flow rate by using the oxygen-enriched air reduces the required pumping power of the feed stream, and reduces the required amount of preheating of the feed stream. Therefore, the use of the hybrid system increases the cycle efficiency.

### 6.5.5 *Effect of Swept Fuel Mass Fraction*

In the absence of $N_2$, fuel in oxy-fuel combustion should be diluted by $CO_2$ to control the combustion temperature. Figure 6.24a, shows how the permeation flux increases with the increase of $CH_4$ mass fraction (decreasing $CO_2$ mass fraction) at constant total sweep flow rate. This increase in the permeation flux with the $CH_4$ fraction is increased can be attributed to the higher membrane temperature as shown in Fig. 6.24b. Higher $CH_4$ flow rate results in reduction of flame length and increase of flame temperature. The reduction in the flame length can be attributed to the higher oxygen permeation which results in fully burning of the fuel early as shown in Fig. 6.25a. The rapid reduction of the permeation flux of oxygen at higher $CH_4$

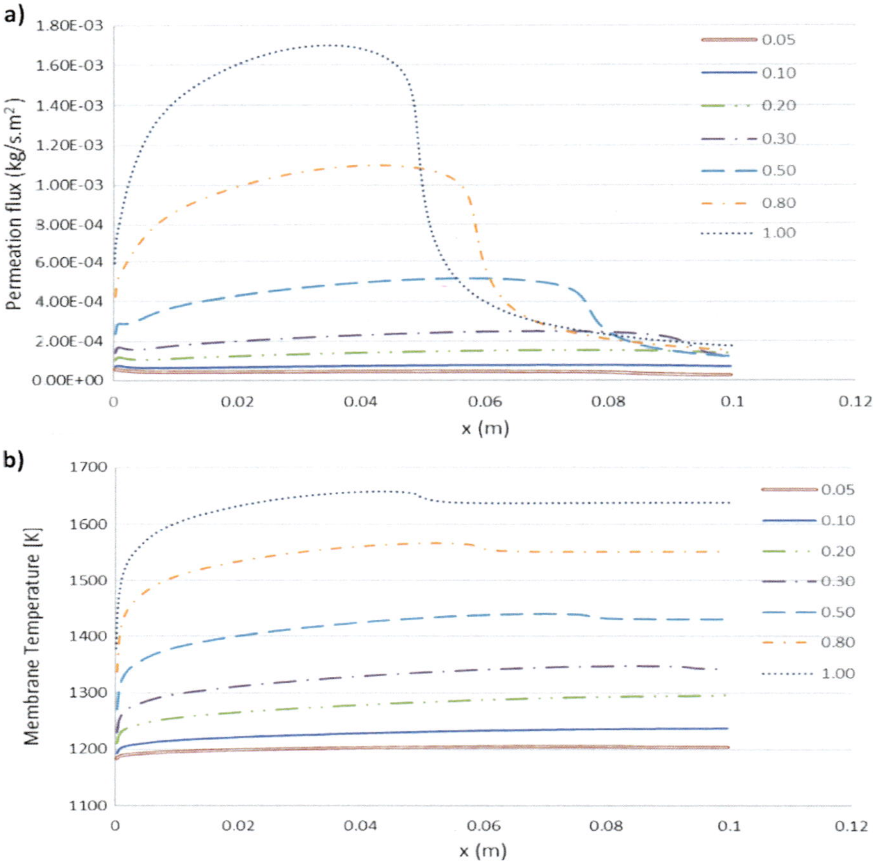


**Fig. 6.24** Distributions of **a** $O_2$ permeation flux and **b** membrane temperature (K), over range of swept $CH_4$ mass fractions at $5.0 \times 10^{-6}$ kg/s feed flow rate with 0.525 $O_2$ mass fraction, $2.0 \times 10^{-7}$ kg/s sweep flow rate, and at $\varphi > 1.0$

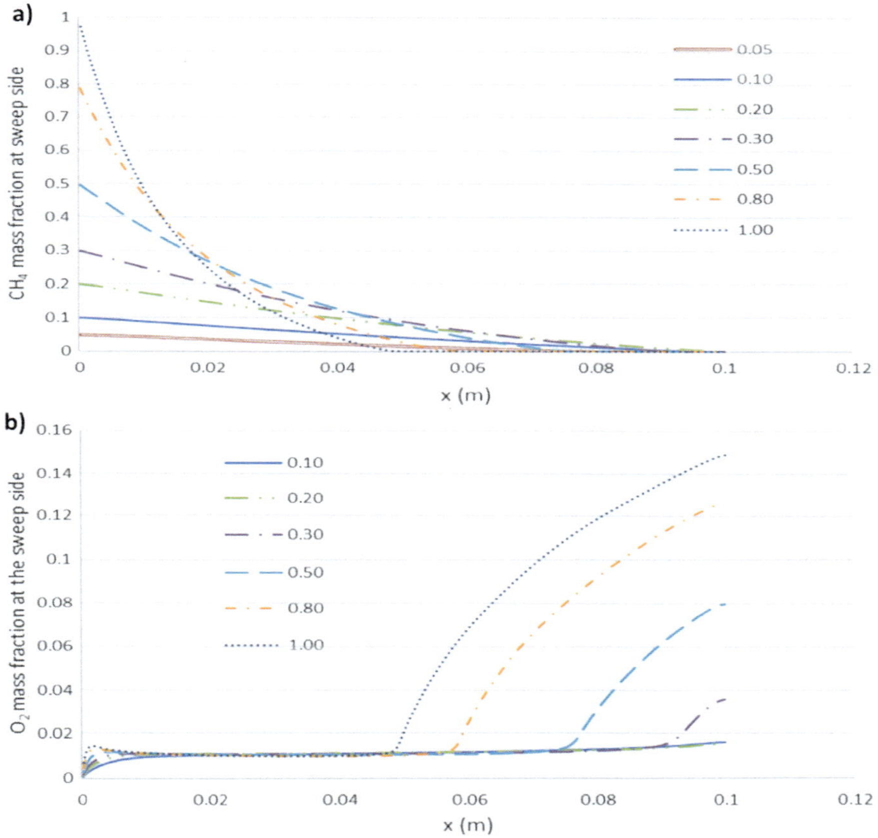


**Fig. 6.25** Distributions of **a** permeate flux rates of $O_2$, and **b** membrane temperature, over range of feed flow rates at 0.525 fed $O_2$ mass fraction, $2.0 \times 10^{-7}$ kg/s sweep flow rate with 0.5 $CH_4$ mass fraction, and at $\varphi > 1.0$

mass fraction can be attributed to the increase in the $O_2$ partial pressure at the sweep side due to oxygen permeation downstream of the flame zone.

### 6.5.6 *Effect of Feed Flow Rate*

At lower feed flow rates, fuel rich combustion takes place in the sweep side, and excess fuel appears with the combustion products. On the other hand, at higher feed flow rates, fuel lean combustion takes place, and excess permeated oxygen is lost at the sweep outlet, which is economically undesirable in oxy-fuel combustion. To study the effect of lean cases, and effect of excessive feed flow rates, Fig. 6.26 shows

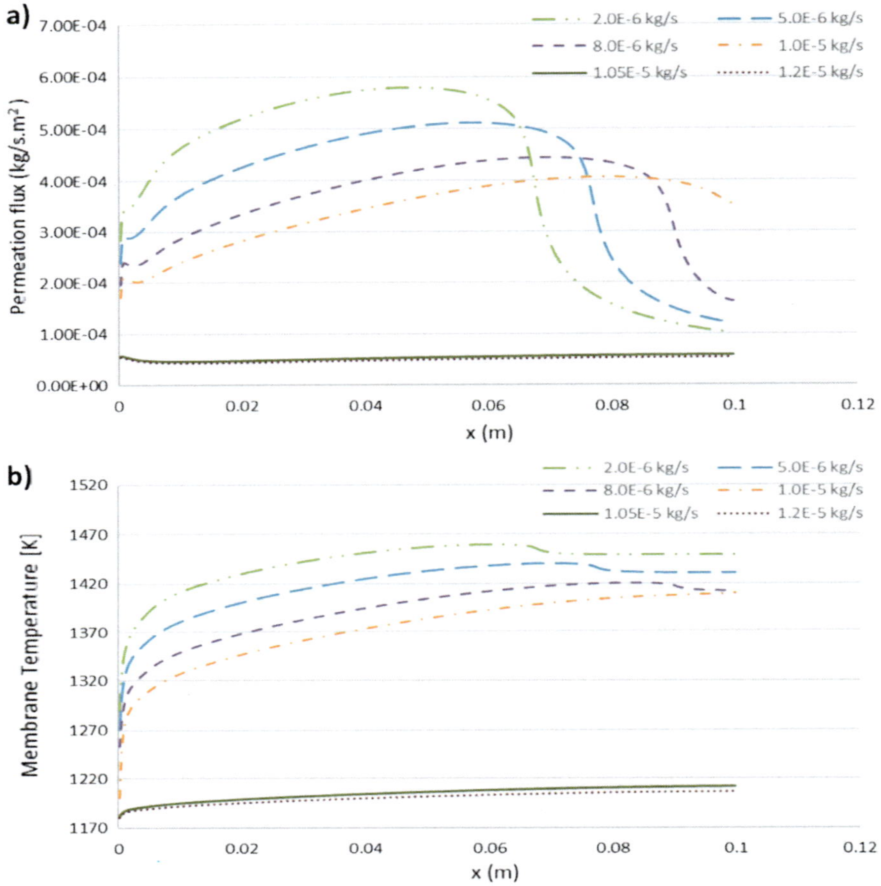


**Fig. 6.26** Distributions of **a** permeate flux rates of $O_2$, and **b** membrane temperature (K), over range of feed flow rates at 0.525 fed $O_2$ mass fraction, $2.0 \times 10^{-7}$ kg/s sweep flow rate with 0.5 $CH_4$ mass fraction, and at $\varphi > 1.0$

the local permeation flux through the membrane and the local membrane temperature for fuel lean cases. Higher temperatures are observed with reducing the feed flow rates resulting in increase of the local permeation flux. This higher temperature can be attributed to the decrease of heat losses as the feed flow rate is reduced. Figure 6.26a, shows that the higher permeation occurs at lower feed flow rates upstream of the flame core, then, it decreases sharply after combustion is completed at the sweep side. Incomplete combustion of the fuel at the cases of $1.05 \times 10^{-5}$ kg/s and $1.2 \times 10^{-5}$ kg/s feed flow rates, as shown in Fig. 6.27, is observed. This can be attributed to the significant heat losses from the sweep side to the feed side, which causes reduction in the membrane temperature (see Fig. 6.26a) and decrease in oxygen permeation flux. Therefore, the lean combustion will persist while increasing the

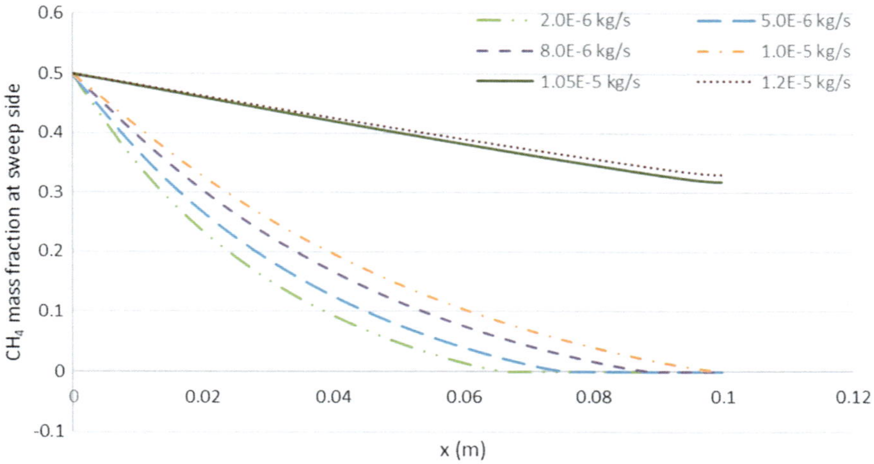


**Fig. 6.27** Distributions of $CH_4$ mass fraction, over range of feed flow rates at 0.525 fed $O_2$ mass fraction, $2.0 \times 10^{-7}$ kg/s sweep flow rate with 0.5 $CH_4$ mass fraction, and at $\varphi > 1.0$

feed flow rate until reaching a point where the heat losses to the feed side leads to incomplete combustion.

As the produced flow rate from the polymer membrane is known (Table 6.7), and for specific feed flow rate, the total number of the fiber polymer membranes required to feed one tubular OTM could be calculated as follows:

$$N = \frac{\dot{m}_{Feed}}{2.56 \times 10^{-8}} \tag{6.25}$$

where N is the required number of fiber polymer membrane to feed one tubular OTM, and one fiber polymer membrane production equal to $2.56 \times 10^{-8}$ kg/s of oxygen-enriched air (0.525 $O_2$ mass fraction). The ratio between the required fiber polymer membranes area to the OTM's area (R) could be defined as follows:

$$R = \frac{N \times A_{polymer}}{A_{LCF}} \tag{6.26}$$

where the area of one fiber polymer membrane is $1.26 \times 10^{-4}$ m$^2$ and the area of one tubular OTM membrane is $1.26 \times 10^{-3}$ m$^2$. Therefore, the total permeation of oxygen through the OTM could be distributed over a range of feed flow rates as shown in Fig. 6.28. This distribution indicates that increasing the feed flow rate results in gradual reduction of the total permeation flux due to the heat losses until the permeation flux drops abruptly at a flow rate of $1.0 \times 10^{-5}$ kg/s. This abrupt drop of the oxygen permeation at high feed flow rates can be attributed to the incomplete combustion at these higher flow rates. Therefore, using high feed flow rate reduces the system efficiency by requiring more fiber polymer membranes and more preheating

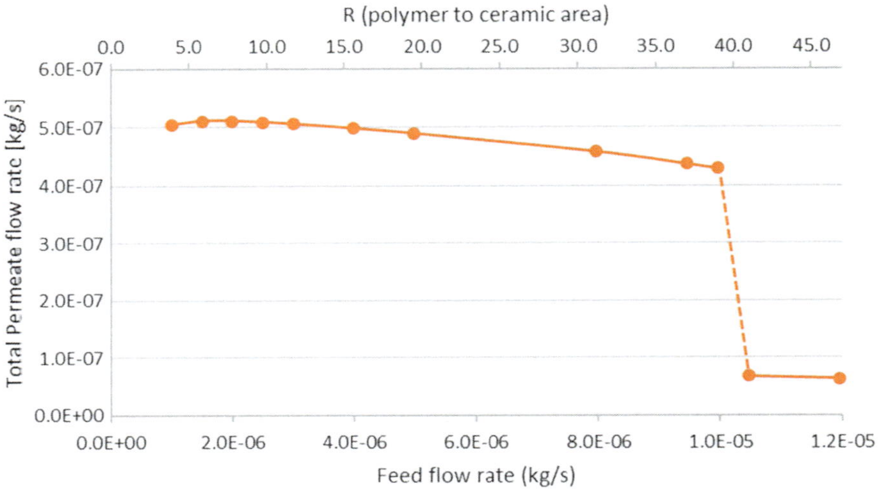


**Fig. 6.28** Distributions of total permeate flow rates of $O_2$ over a range of feed flow rate at 0.525 fed $O_2$ mass fraction, $2.0 \times 10^{-7}$ kg/s sweep flow rate with 0.5 $CH_4$ mass fraction, and at $\varphi > 1.0$

for the feed stream. Therefore, using the stoichiometric feed flow rate seems to be the best choice for minimum system requirements and high efficiency.

### 6.5.7 *Effect of Sweep Flow Rate*

The local permeations along the membrane length for different sweep flow rates while keeping the mass fraction of $CH_4$ and the other operation conditions unchanged are shown in Fig. 6.29a. At low sweep flow rate, lower permeation level is achieved with longer distribution along the membrane. While increasing the sweep flow rate, higher permeation is observed accompanied by shorter distribution along the membrane. The increase in permeation flux while increasing the sweep flow rate can be attributed to the increase in the $CH_4$ flow rate. As the swept $CH_4$ flow rate increases, the flame temperature increases as shown in Fig. 6.29b resulting in higher $O_2$ permeation flux and, accordingly, earlier burning of the fuel as shown in Fig. 6.30a. Also, as the sweep flow rate increases, the partial pressure of oxygen at the sweep side decreases, resulting in increase in the $O_2$ permeation flux as per the permeation Eq. (6.20).

The sharp decrease of permeation at a specific region of the membrane (see Fig. 6.29a) can be attributed to the increased $O_2$ mass fraction at the sweep side as shown in Fig. 6.29b, after consuming all $CH_4$. One can see from Fig. 6.29b, that higher sweep flow rates have higher $O_2$ mass fraction at the sweep side. This can be attributed to the greater membrane temperature that allows for higher rate of $O_2$ permeation to the sweep side. The lower $O_2$ mass fraction at the sweep side is achieved at the smaller sweep flow rate. However, as the sweep flow rate is reduced, the power

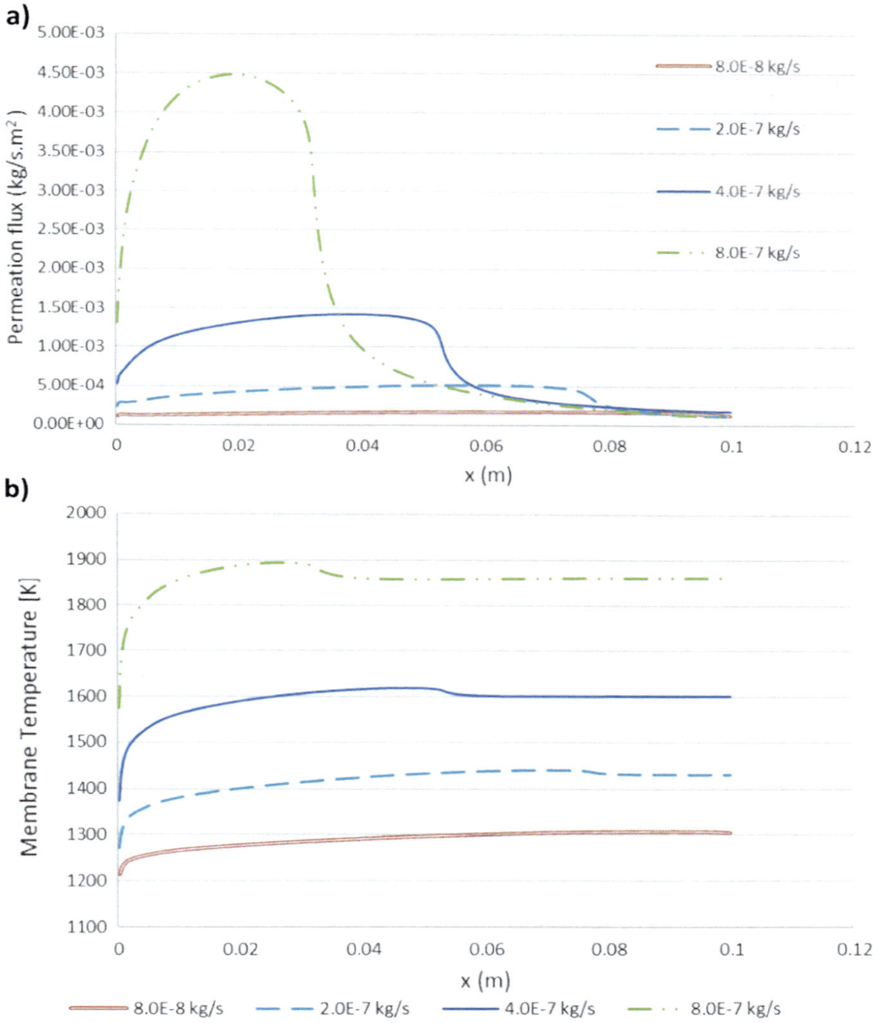


**Fig. 6.29** Distributions of: **a** $O_2$ permeation flux, and **b** membrane temperature (K), over range of sweep flow rates at $5.0 \times 10^{-6}$ kg/s feed flow rate with 0.525 $O_2$ mass fraction, 0.50 $CH_4$ swept mass fraction, and at $\varphi > 1.0$

output is reduced because of decreasing the fuel flow rate. Thus, reducing the sweep flow rate is not a solution of the excess $O_2$ at the sweep outlet. The solution of this problem is by reducing the feed flow rate to the stoichiometric conditions while using the higher possible sweep flow rate.

Figure 6.31 shows the effect of the sweep gas flow rate on the oxygen permeation for different swept $CH_4$ mass fractions. Smaller swept $CH_4$ mass fractions produce lower flame temperature and increase the ability of using higher sweep flow rates.

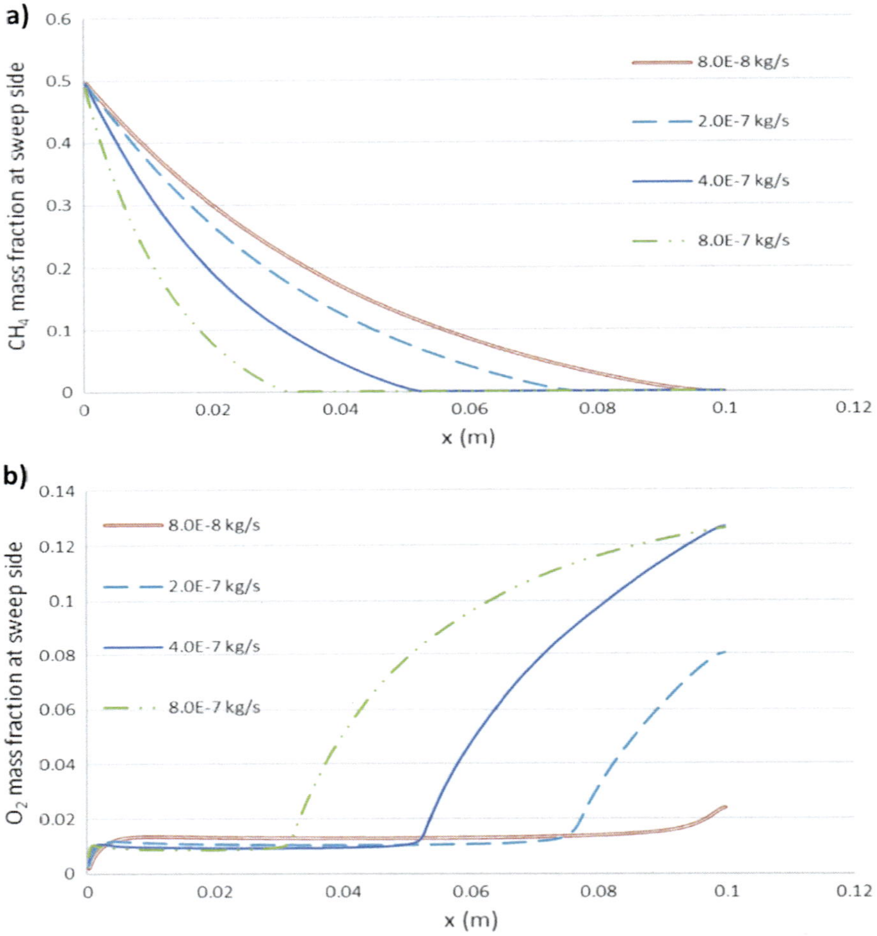


**Fig. 6.30** Distributions of: **a** $CH_4$ mass fraction, and **b** $O_2$ mass fraction, at the sweep side, over range of sweep flow rates at $5.0 \times 10^{-6}$ kg/s feed flow rate with 0.525 $O_2$ mass fraction, 0.50 $CH_4$ swept mass fraction, and at $\phi > 1.0$

Therefore, lower permeation could be observed at the lower swept $CH_4$ flow rates. Thus, lower membrane temperature can be realized by reducing the swept $CH_4$ mass fraction and increasing the sweep flow rate, in another word, keeping the same fuel flow rate and increasing the dilute flow rate.

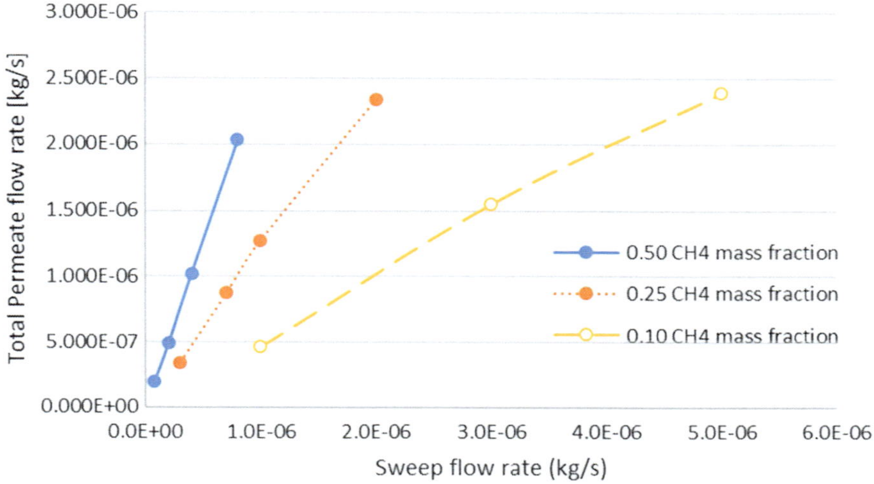


**Fig. 6.31** Distributions of total permeate flow rates of $O_2$ over range of sweep flow rates for different swept $CH_4$ mass fractions at $5.0 \times 10^{-6}$ kg/s feed flow rate with 0.525 $O_2$ mass fraction, 0.50 $CH_4$ swept mass fraction, and at $\phi > 1.0$

### *6.5.8 Design and Energy Analysis of the Hybrid Unit*

For stoichiometric air-to-fuel cases, the ratio between the required polymer membrane area and the OTM membrane (R) could be calculated using Eqs. (6.25) and (6.26) as follows:

$$R_{stoich} = \frac{4X_{CH_4}\dot{m}_{Sweep}}{X_{O_2}2.56 \times 10^{-7}} \tag{6.27}$$

This relation is function of the sweep gas-flow rate, the swept $CH_4$ mass fraction, and the fed $O_2$ mass fraction. Figure 6.32 illustrates the relation between the ratio (R) and the sweep gas-flow rate at a mass fraction of fed $O_2$ of 0.525 and 0.2 swept $CH_4$ mass fraction. This linear relation shows how the area ratio increases with increasing the sweep flow rate. As the sweep flow rate increases, the stoichiometric feed flow rate increases requiring more polymer membrane, and resulting in increase of the area ratio (R).

The power output based on thermal efficiency of 0.35 and including the preheat requirements can be calculated as follows:

$$Power\ output = 0.35HV_{CH_4}\dot{m}_{CH_4} - preheat\_load \tag{6.28}$$

where $HV_{CH4}$ is the heating value of methane and is equal to 55.688 MJ/kg. Moreover, the required power to heat the feed and the sweep streams to the required OTM inlet temperature (1123 K) can be calculated as follows:

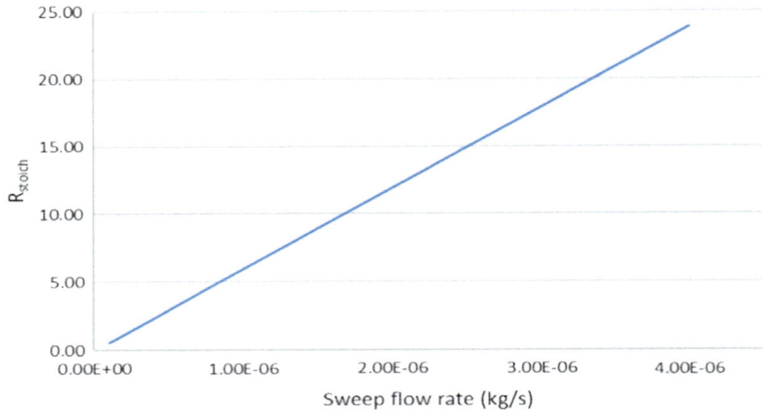


**Fig. 6.32** The relation between area ratio (R) and the sweep flow rate at 0.525 fed $O_2$ mass fraction, 0.2 swept $CH_4$ mass fraction, and at $\phi = 1.0$

$$preheat\_load = \Delta T\,(\dot{m}_{CH_4} c_{p,CH_4} + \dot{m}_{O_2} c_{p,O_2} + \dot{m}_{N_2} c_{p,N_2} + \dot{m}_{CO_2} c_{p,CO_2}) \tag{6.29}$$

where $\Delta T = 1123\ K - 673\ K = 450\ K$ and the specific heat ($c_p$) have been calculated at an average temperature of 900 K. The preheat load is the required power to heat the sweep and the feed streams from 673 K to the inlet streams operation temperature 1123 K. The 673 K is the temperature of the preheated streams (feed and sweep streams) at the inlet of the second preheating stage, where the first stage is for heat recovery from the exhaust of the system in which the temperature of the sweep and feed streams is raised from the ambient temperature to 673 K. The power output of this system could be expressed in terms of the sweep flow rate ($\dot{m}_{sweep}$), the fed $O_2$ mass fractions, and the swept $CH_4$ mass fraction as follows:

$$Power\ output = \dot{m}_{sweep}\Big[0.35HV_{CH_4}X_{CH_4} - \Delta T\Big(X_{CH_4}c_{p,CH_4} + (1 - X_{CH_4})c_{p,CO_2} + c_{p,N_2}\frac{4(1 - X_{O_2})X_{CH_4}}{X_{O_2}} + 4X_{CH_4}c_{p,O_2}\Big)\Big] \tag{6.30}$$

The output power, preheat load, and stoichiometric feed flow rate have been calculated for both oxygen-enriched air case and for the atmospheric air case at 7.5 $\times$ $10^{-7}$ kg/s sweep flow rate with 0.2 swept $CH_4$ mass fraction and the values are listed in Table 6.8.

Based on the values in Table 6.8, one can observe that using oxygen-enriched air as feed gas enhances the power output from 1.00 W to 1.74 Was a result of the reduction of the preheat load. Figure 6.33 shows the effect of sweep flow rate on the power output of one OTM for both 0.525 and 0.233 fed $O_2$ mass fractions. The linear relation in the previous Eq. (6.30) appears in this plot. The ratio of the power output

**Table 6.8** Power output calculations for both oxygen enriched air case and the atmospheric air case for one OTR at $7.5 \times 10^{-7}$ kg/s sweep flow rate with 0.2 swept $CH_4$ mass fraction and at $\phi = 1.0$

| Fed $O_2$ mass fraction | Stoichiometric feed flow rate (kg/s) | Number of fiber polymers | $R_{stoich}$ | Preheat load (W) | $0.35 HV_{CH_4} \dot{m}_{CH_4}$ (W) | Power output (W) |
|---|---|---|---|---|---|---|
| 0.525 | $1.14 \times 10^{-6}$ | 44.6 | 4.46 | 1.18 | 2.92 | 1.74 |
| 0.233 | $2.58 \times 10^{-6}$ | – | – | 1.92 | 2.92 | 1.00 |

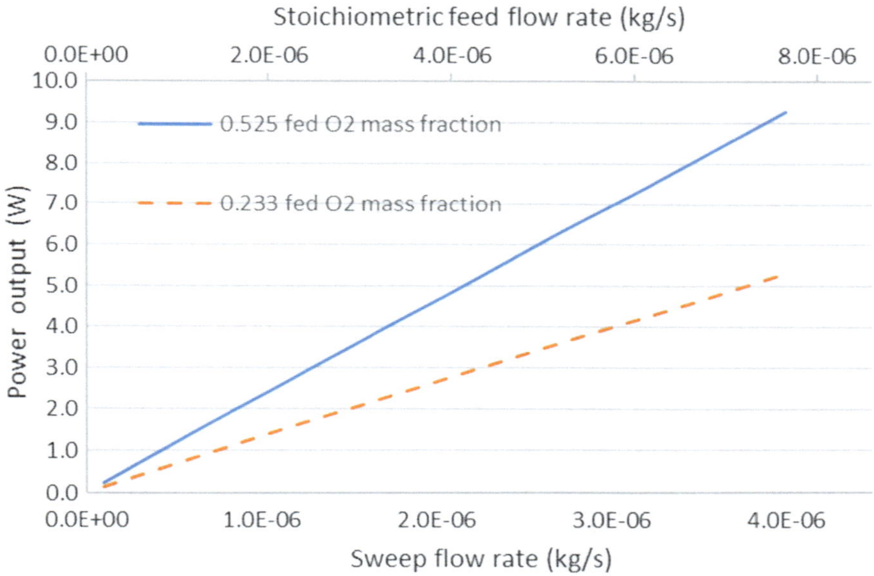


**Fig. 6.33** Effect of sweep flow rate on the power output of one OTR at 0.2 swept $CH_4$ mass fraction, and at $\phi = 1.0$

for the two cases of 0.525 and 0.233 fed $O_2$ mass fractions remains unchanged at a value of 1.74 while increasing the sweep flow rate.

The effect of the fed $O_2$ mass fraction on the power output is shown in Fig. 6.34. The figure shows nonlinear relation between the fed $O_2$ mass fraction and the output power as per Eq. (6.30). The increase in the power output would recommend the use of the higher possible feed $O_2$ mass fraction. Figure 6.35 shows linear relation between the swept $CH_4$ mass fraction and the power output. The increase of the swept $CH_4$ mass fraction increases the power output. The power enhancement ratio of using 0.525 fed $O_2$ mass fraction as compared to 0.233 fed $O_2$ mass fraction (atmospheric air) is more than 1.5 over the range of swept $CH_4$ mass fractions. The limitation of using higher $CH_4$ mass fraction is the higher temperature associated with rising $CH_4$ concentrations.

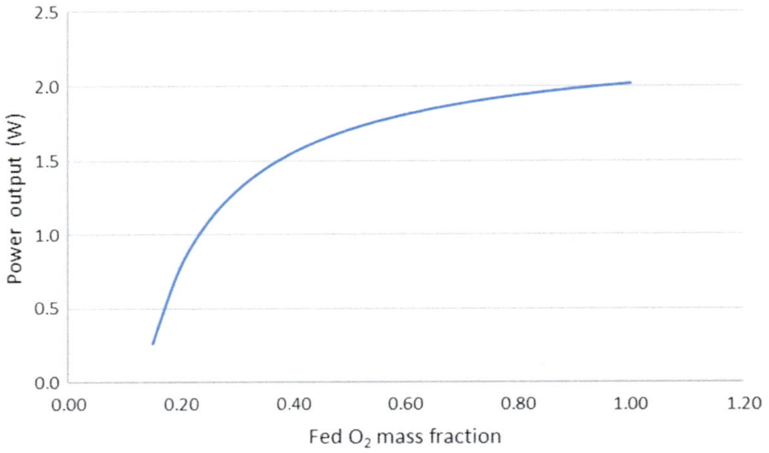


**Fig. 6.34** Effect of the fed $O_2$ mass fraction on the power output of one OTR at $7.5 \times 10^{-7}$ kg/s sweep flow rate and 0.2 swept $CH_4$ mass fraction, and at $\phi = 1.0$

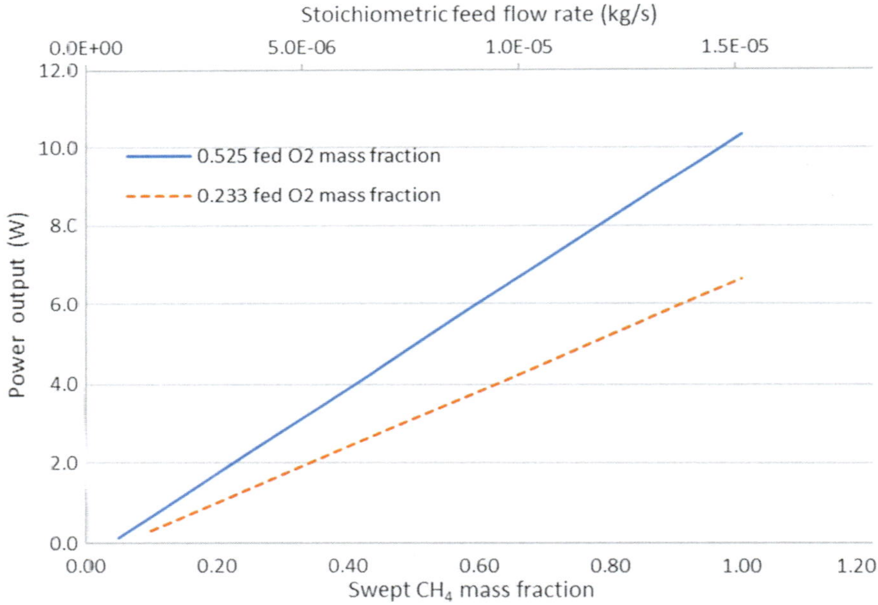


**Fig. 6.35** Effect of the swept $CH_4$ mass fraction on the power output of one OTR at $7.5 \times 10^{-7}$ kg/s sweep flow rate, and at $\phi = 1.0$

Based on the above analyses, a proposed hybrid system is introduced. The hybrid system considers the case of $7.5 \times 10^{-7}$ kg/s sweep flow rate with 0.2 swept $CH_4$ mass fraction, $1.14 \times 10^{-6}$ kg/s stoichiometric feed flow rate with 0.525 fed $O_2$ mass fraction, 1123 K streams inlet temperatures, and $\phi = 1.0$ to be the optimum design of the hybrid system. The membrane temperature and the permeation flux of this case are shown in Fig. 6.36. Higher temperature is observed in this case. The flame is elongated through the length of the membrane with uniform membrane surface temperature. This reduced the possibility of thermal fracture of the membrane. Inputs and outputs of the proposed case of the hybrid system are shown in Table 6.9. The illustrated geometry of the fiber polymeric membrane and the OTM are taken up in the proposed design. 29 OTM's are proposed to be collected in one shell, while 325 fiber polymer membranes are suggested to be collected in one shell [45]. The outputs of a single OTM shell system are illustrated in Table 6.9. A schematic diagram is presented in Fig. 6.37 to brief the configuration of the proposed hybrid polymer/ceramic oxy-fuel combustor system. This figure shows the inputs and the outputs of the designed hybrid system, and it shows a cross sectional view of the OTM shell.

The species mass fractions along the OTM for the optimum design are shown in Fig. 6.38. $CH_4$ is consumed totally in the stoichiometric conditions. A slight amount

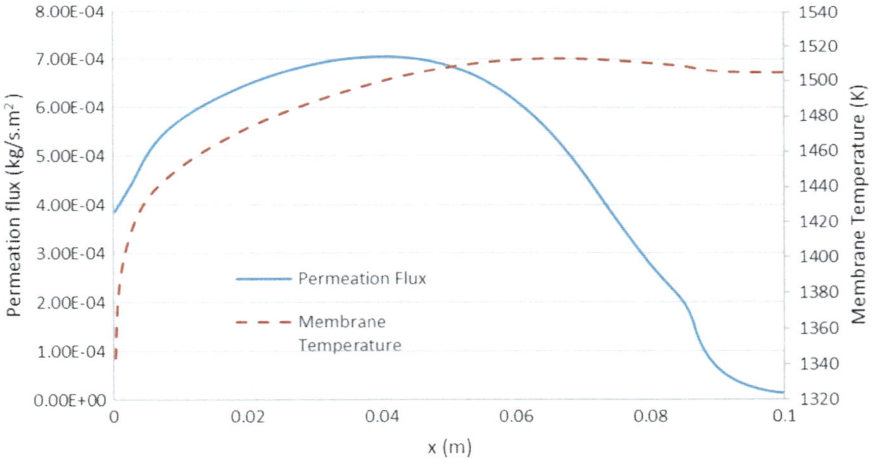


**Fig. 6.36** Permeation flux along the membrane length and membrane temperature for the proposed design case at $1.14 \times 10^{-6}$ kg/s feed flow rate with 0.525 $O_2$ mass fraction, $7.5 \times 10^{-7}$ kg/s sweep flow rate with 0.2 $CH_4$ mass fraction, and at $\phi = 1.0$

**Table 6.9** Proposed single OTM shell and its requirements and power

| Number of OTM's per shell | Number of fiber polymers | Number of fiber polymers per polymer shell | Number of fiber polymeric shells | Power output (W) |
|---|---|---|---|---|
| 29 | 1300 | 325 | 4 | 50.46 |

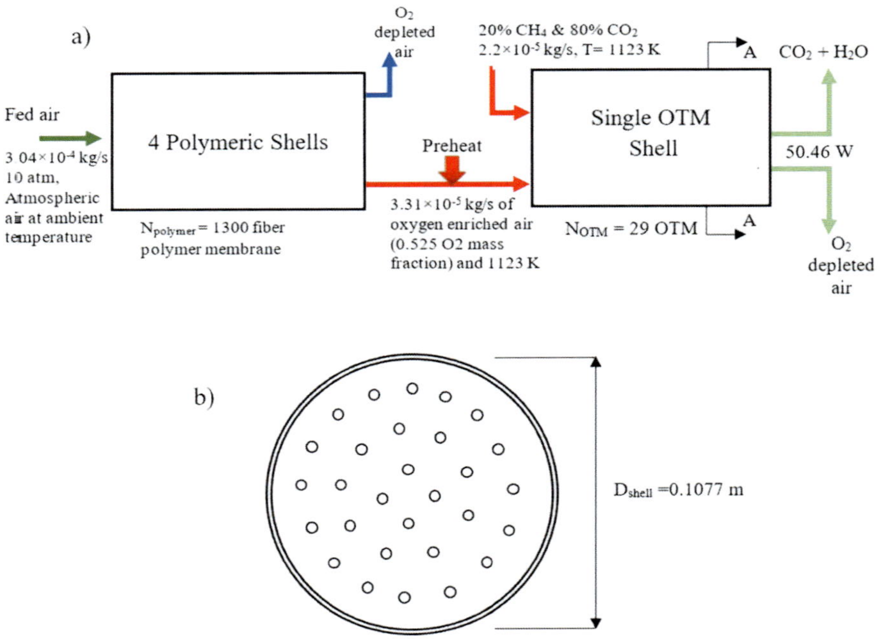


**Fig. 6.37** Schematic diagram showing: **a** configuration of the proposed hybrid system and, **b** sectional view of the OTM shell (A-A)

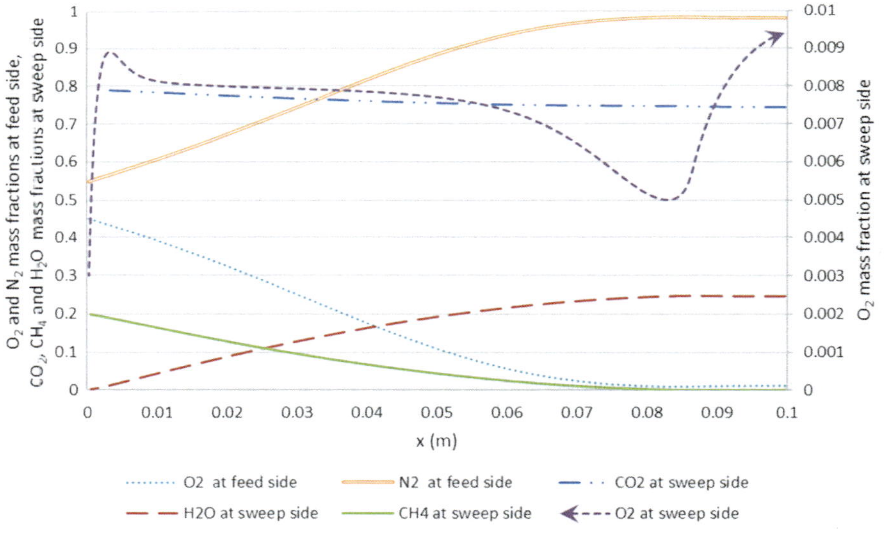


**Fig. 6.38** Distributions of species mass fractions at sweep and feed sides along the OTR length. At ($1.14 \times 10^{-6}$ kg/s feed flow rate with 0.525 $O_2$ mass fraction, $7.5 \times 10^{-7}$ kg/s sweep flow rate with 0.2 $CH_4$ mass fraction, and $\phi = 1.0$)

**Table 6.10** Feed and sweep outlets compositions and flow rates for the proposed design case

| $CH_4$ at sweep outlet | $O_2$ at sweep outlet | $H_2O$ at sweep outlet | $CO_2$ at sweep outlet | $O_2$ at feed outlet | $N_2$ at feed outlet | Sweep outlet flow rate (kg/s) | Feed outlet flow rate (kg/s) |
|---|---|---|---|---|---|---|---|
| Zero | 0.009 | 0.25 | 0.74 | 0.01 | 0.99 | $1.38 \times 10^{-6}$ | $5.177 \times 10^{-7}$ |

of $O_2$ is observed at the sweep and feed outlets as a result of higher feed flow rate (above the stoichiometric feed flow rate). The mass fraction of $CO_2$ decreases in the axial direction with the combustion is proceeding due to the increase in $H_2O$ mass fraction. $N_2$ mass fraction at the feed side increases because of the oxygen permeation that causes a reduction of $O_2$ mass fraction on the feed side. The growth of $O_2$ mass fraction in the sweep side at $x = 0.085$ m is attributed to the terminated reaction while the permeation of the oxygen from the feed side is persisting. Table 6.10 shows compositions and flow rates of the feed and sweep outlets. The high $N_2$ mass fraction at the feed outlet can be used to produce higher $N_2$ purities. Sweep flow rate outlet records an increase while feed flow rate decreased due to the high permeation of oxygen from the feed side to the sweep side.

## 6.6 Low-Power Porous-Plate Reactors

Greenhouse gases (GHG) anthropogenic emission in the atmosphere has been the ultimate cause of the current climatic change [91]. Among the main sources of anthropogenic greenhouse gases emissions, burning of fossil fuels has been identified as the main concern in the current century [92]. As reported by the International Energy Agency [93], the global energy consumption based on fossil fuel amounts to about 80% of the total global energy demand. This resulted in the emission of 32.3 Gt of $CO_2$ to the atmosphere in the year 2014 [93]. Recent findings indicated that about 40% of the global $CO_2$ emission is a direct result of electricity generation, with more than 30% coming from fossil fuels [94]. Several routes for lowering $CO_2$ emissions can be applied including increasing of plant efficiency (provides reduction of $CO_2$ emission by 2–3% for increase of plant efficiency of 1%), decreasing of carbon content in the fuel by utilizing less carbon fuels, reducing unnecessary fuel consumption, and employing carbon capture and storage technologies [92, 95]. CCS is the process of capturing, purifying, compressing, transporting and storing of unwanted $CO_2$ emitted from the above-mentioned sources. The CCS technologies have been categorized as post-combustion, pre-combustion and oxy-fuel combustion [96, 97]. In pre-combustion process, the fuel carbon is converted into syngas at initial stage before the combustion process. In post-combustion process, carbon dioxide emission is mitigated by separation and removal of $CO_2$ from the flue gases via absorption using solvents (basically amines), adsorption using metal organic framework, as well as $CO_2$ selective membrane separation. In oxy-fuel combustion process, pure oxygen is being used for the combustion of fuel resulting in only $H_2O$ and $CO_2$ as products of combustion. Oxy-fuel combustion process results in high combustion temperature. Part of the flue gas is usually recycled into the combustion zone to lower the combustion temperature [98].

To maintain stable flame under oxy-combustion conditions, there should be certain minimum amount of oxygen concentration in the oxidizer ($O_2 + CO_2$) to enable sufficient temperature levels within the combustor to sustain the combustion reactions [99]. Carbon dioxide addition, as a diluent, affects the oxy-combustion characteristics of natural gas flames in terms of variations in the following parameters within the combustor [100, 101]: (a) adiabatic flame temperature of the mixture, (b) radiative heat transfer, (c) transport properties (viscosity, mass diffusivity, and thermal conductivity), (d) specific heat of the mixture, (e) chemical kinetics, and (f) flame structure. Therefore, oxy-methane combustion in an environment of high $CO_2$ concentrations provides new combustion challenges and opportunities for the study of the above parameters. However, burning fuel using oxygen requires the separation of oxygen from air using air separation unit (ASU), which consumes about 15% of the total output power of the system, making the process of oxy-combustion cost ineffective [102]. Successful implementation of oxy-fuel combustion on mixed ionic-electronic conducting (MIEC) membranes is the ultimate goal that will allow conventional combustors to be replaced with oxygen transfer membrane reactors (OTMRs), where oxygen separation and oxy-combustion are co-located at the membrane instead of being performed separately. Numerical investigations of the OTMR concept are abundant [103, 104], but experimental studies are very rare [105]. Combining the process of oxygen separation (using ceramic membrane) with the oxy-fuel combustion of fuel in the sweep side of the membrane (using the separated oxygen) results in a compact unit for power production. This unit is called as high-temperature membrane reactor (HTMR). One of the goals of the study presented under this section is to mimic the process of oxygen permeation and oxy-combustion inside HTMR using porous-plate reactor. However, the process of oxygen separation is not investigated considering the current reactor design, only the process of oxy-fuel combustion is investigated.

### 6.6.1 Combustion Characteristics in Porous-Plate Reactors

The use of porous plate reactors has been increased significantly due to its flexibility of introducing the oxidizer mixture through the whole combustion chamber. This is of immense importance in case of oxy-fuel combustion to uniformly distribute oxygen through the whole combustion chamber and, accordingly, reduce the creation of elevated temperature spots within the flame zone. Habib et al. [106] studied the characteristics of oxy-fuel combustion in a porous plate reactor experimentally and numerically for the reactive and non-reactive cases. The results showed uniform temperature distribution through the porous plates due to the uniformity of oxygen distribution through the combustion zone. The effect of oxy-fuel mixing is significant especially for porous plate reactors as it affects directly the combustion chemistry within the combustor. Mixture fraction is defined as the fraction of the fluid mass that originates as fuel. Mass fractions for all other species can be derived based on state relationships from the mixture fraction [107]. Combustion models based on the mixture fraction concept has been commonly used to simulate and evaluate the combustion characteristics in combustors and fire models, including the Fire Dynamics Simulator (FDS) developed at the National Institute of Standards and Technology (NIST) [108]. Numerical simulations, using the Commercial CFD software, CFX-11, were performed by Dharavath et al. [109] to investigate the effect of equivalence-ratio on the combustion behavior and air-fuel mixing on the full-scale scramjet combustor of a hypersonic air-breathing vehicle with ethylene fuel at ground test conditions by solving species transport equations, k-ω turbulence model as well as the three-dimensional RANS equations. They validated the computational results against experimental data available in the literature and, then, performed parametric study of the full-scale combustor performance parameters, namely: total pressure loss, combustion efficiency, and thrust [109]. A similar work was conducted by Han et al. [110] to investigate the effects of air–fuel mixing quality on combustion parameters for two low temperature diesel combustion regimes using conventional diesel combustors. They found that the intake and injection pressures are the most crucial factors influencing the air–fuel mixing process. To achieve the different combustion regimes, they varied the concentration of intake oxygen. The improvement of air–fuel mixing with enhanced combustion process in the two cases was studied [110].

The ultimate goal of the study presented in this section is the application of oxy-fuel combustion technology as one of the carbon capture technologies for clean and sustainable environment. In the study presented in this section, both the flow field and oxy-methane combustion characteristics are investigated experimentally and numerically inside a face-to-face two-porous-plates reactor over wide ranges of mixing conditions and equivalence ratio under laminar inlet flow conditions. The oxy-combustion cold/non-reactive flow field was investigated both experimentally and numerically in order to understand the flow field characteristics, while the hot oxy-combustion was studied numerically. As per the open literature, there is a lack of data about the operation of the HTMR under oxy-combustion conditions. The porous plate reactor mimics the process of oxygen permeation and oxy-fuel combustion inside the HTMR. The process of oxy-combustion inside the HTMRs is difficult to be achieved at normal lab conditions and the oxygen permeation flux across the membrane still not enough for complete fuel combustion. Because of that, porous plates are used instead of membranes trying to see the flame characteristics and location with respect to the porous plate that should help in the development of HTMRs. Firstly, the cold flow characteristics are investigated experimentally and numerically to predict the flow mixing behavior and the location of the flame zone. Particle Imaging Velocimetry (*PIV*) technique is used to visualize the cold flow in the main channel under oxygen permeation in terms of average axial velocity of the main stream flow, velocity vectors and streamlines. Secondly, flow field and oxy-combustion characteristics of methane are investigated numerically over ranges of operating conditions.

### 6.6.2 Operating Conditions

Figure 6.39 shows a schematic diagram of the experimental setup of the porous-plate reactor. The reactor comprises three sections as shown in Fig. 6.40; the central combustion chamber where fuel and $CO_2$ are being supplied and the upper and the lower chambers where oxygen is supplied. Pure oxygen is supplied to the two oxygen chambers from a pressurized oxygen gas cylinder. Meanwhile, the fuel, together with the carbon dioxide, which is used as diluent in the reactive case, are also supplied in their pure form from pressurized cylinders, as shown in Fig. 6.39. Each pressurized cylinder is connected to a mass flow controller, which regulates the gas flow rate. Fuel and diluent are mixed upstream of the reactor in a mixing unit before being introduced to the reactor. The length of the mixing chamber and the reactor cover are 800 and 440 mm, respectively. In the reaction zone (central zone), both fuel and diluent are supplied axially, and mix with the incoming oxygen from the upper and the lower oxygen chambers through the porous plates. Particle Imaging Velocimetry (PIV) technique was employed in this work to characterize the non-reactive flow field under oxygen permeation through the porous plates. The PIV system was setup for flow visualization, velocity measurement and flow characterization of an interrogation (Imaged) area (IA) of 750 × 25.3 mm$^2$. Aluminum-oxide active seeding particles

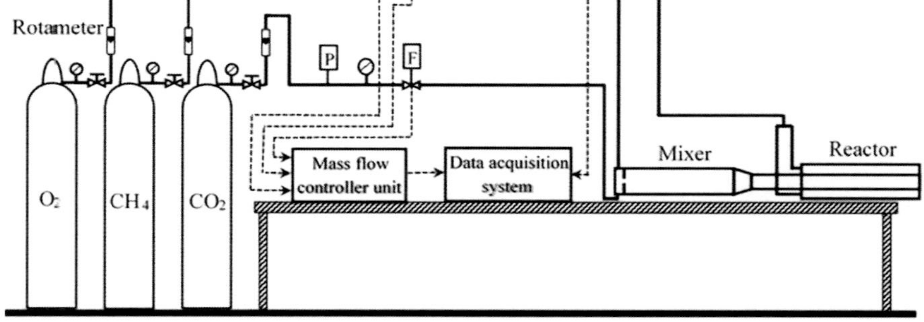


**Fig. 6.39** Schematic diagram of the experimental setup

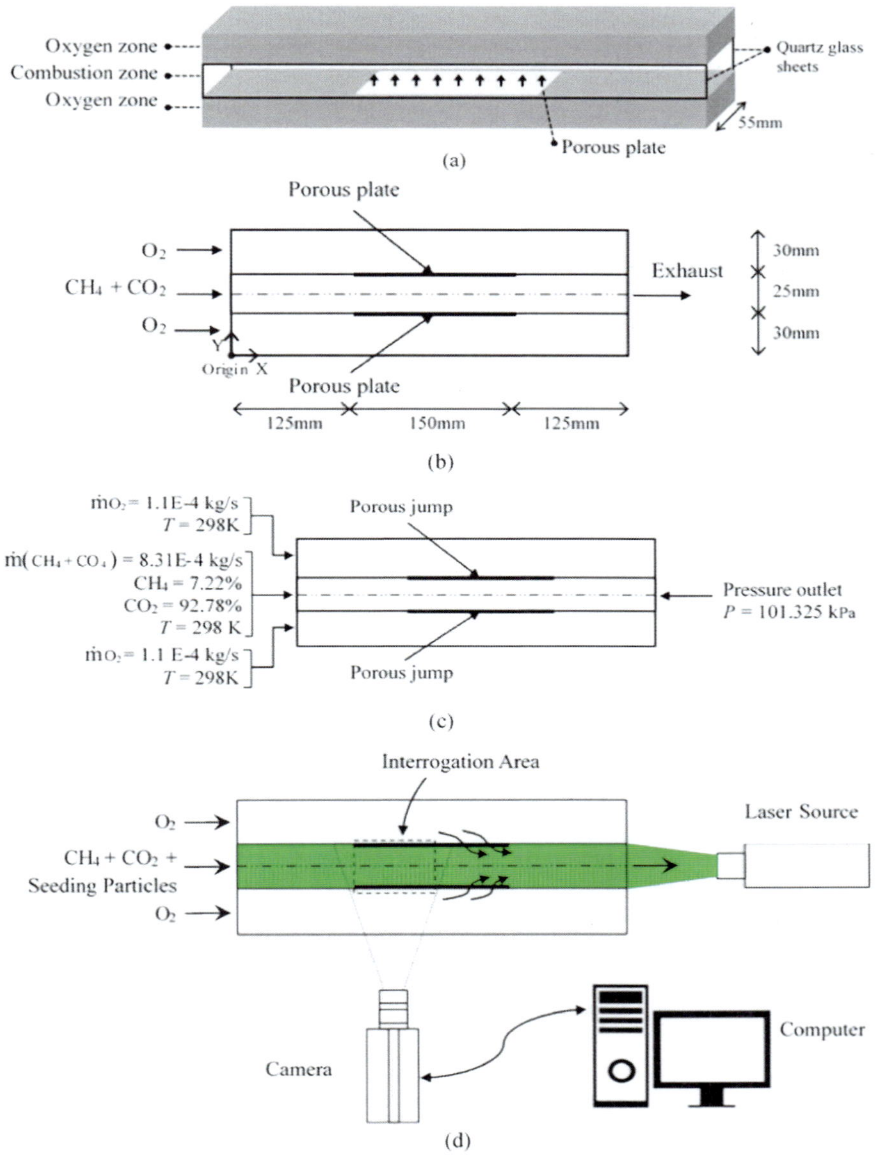


**Fig. 6.40** Schematic diagram of the porous plate reactor: **a** reactor configuration, **b** 2-D representation of the reactor and **c** boundary conditions for the combustion simulation at $\Phi = 1$, **d** PIV setup

having ~2 microns aerodynamic diameter was used ($Al_2O_3$, 0.9 g/mol weight, density of 4 g/cm$^3$, 70–230 mesh size) to ensure that the flow fluctuations are captured effectively. A high speed digital camera, facing the quartz sheet and normal to the flow direction, was used to capture the images of the flow of the illuminated particles due to the reflection of the laser light at selected time intervals (micro-seconds). The images are pre-processed and analyzed using different correlation analysis methods depending on the nature of the fluid flow in question. The velocity distribution and other related parameters can be obtained by post-processing the capture flow images [111–113].

A manual RayPower 2000 Laser from DANTEC Dynamics [113], was used to supply the laser light sheet of about 45 mm from the base plate with a wave length of 532 nm and output power of about 2000 mW. This light beam illuminates the seeding particles supplied with the fuel ($CH_4$). Phantom high-speed camera, SpeedSense 9040 series, from Vision Research—AMETEK Material Analysis Division, with maximum resolution of 1632 × 1200 *pixels* was calibrated and used to acquire stream of images at a trigger rate of 600 f/s. Detailed specifications of the Laser, Camera and the Lens are shown in Table 6.11a–c. The detailed procedure of the PIV experiments runs started with setting up of the required gas flows for an equivalence ratio from the experimental plan for the reactor at 1.36 kW power (Table 6.12) using the overall flow setup shown in Fig. 6.39. The seeding particles were injected along the $CO_2$ supply line after the mass flow controller before the mixer. The fuel plus seeded $CO_2$ then mixed in the mixer before the combustor inlet. The oxygen supply line was not seeded to avoid clogging of the porous plates. Hence, the effect of the oxygen permeation in the flow mixing in the reaction zone can be captured near the $O_2$-fuel stream boundaries. As suggested by [114, 115], the camera intensity calibration was conducted by keeping the cap on the camera assembly in order to obtain a reference of the intensity. Also, the optimization of the laser repetition rates and current follow the technique enlisted by [114]. The laser was receiving continuous trigger signal matching the set-frequency before it was switched to external mode. The camera mode was set to continuous grab and the particle images were well focused. The camera was being controlled by a computer program Phantom Camera Control (PCC). Steady state videos were recorded in cine format and then converted into 1000 images and saved. The same procedure was repeated for all the three runs of the experimental plan. The flow images were acquired as single frames with the camera using the PCC studio. Proceeding to the PIV analysis, the images were exported to Dantec Dynamic Studio and stored in the database. The software was designed to acquire, analyze and post-process data. The field of view method of calibration was used to inspect and modify origin and scale factor used in converting image plane (pixels) to object plane (metric units) by specifying the height which is equivalent to the height of the reaction zone. An analysis sequence for the gas flow was build-up and utilized to obtain the velocity vectors, velocity profiles, and the streamlines from the images. The following steps show the details of the analysis sequences:

**Table 6.11** (a) Laser specifications, (b) camera specifications, (c) lens specifications

(a)

| Specification | Raypower 2000 |
|---|---|
| Lasing medium | Gas |
| Wavelength | 532 ± 1 nm |
| Output power | >2000 mW |
| Beam diameter at the aperture | 3.0 nm |
| Polarization ratio | >100:1 |
| Operating temperature | 10–35 °C |
| Mains supply | 100–240 V AC, 3 A max, 50–60 Hz |
| TTL modulation frequency | Max 10 kHz |

(b)

| Specifications | Speedsense Camera 9040 |
|---|---|
| Sensor type | CCD progressive scan monochrome |
| Chip size | 18.77 × 13.8 mm |
| Resolution | 1632 × 1200 pixels |
| Pixel size | 11.5 μm |
| Bit depth | 8, 12, 14 |
| Maximum exposure time | 2 μs |
| Fps (full frame) | 1016/508 f/s |

(c)

| Specifications | Nikon AF Micro Nikkor 60 mm f/2.8 *D* |
|---|---|
| Focal length | 60 mm |
| f-number | 2.8–32 |
| Distance scale | 0.219–∞ m |

**Table 6.12** Non-reactive experimental plan for the porous plate reactor of 1.36 kW power

| Experiment No. | Volume flowrate (lit/min) | | | $O_2/CO_2$ | φ | Re |
|---|---|---|---|---|---|---|
| | $CH_4$ | $CO_2$ | $O_2$ | | | |
| Case 1 | 2.5 | 8.3 | 8.3 | 1 | 0.6 | 425.53 |
| Case 2 | 2.5 | 10.0 | 10.0 | 1 | 0.5 | 493.62 |
| Case 3 | 2.5 | 12.5 | 12.5 | 1 | 0.4 | 595.75 |

1. Making Double Frame; this process is applied to the raw images to convert them from single frames to double frames by combining them in twos. This was done by specifying N/2 double images in the option.
2. Image Min and Max; this method is a subset of Image Processing category and it is used to determine the field of minima/maxima over a series of images. Here,

the method was used to extract the field of minima from the paired raw images obtained from the previous conversion and returns one image.

3. Image Arithmetic; this method allows performing arithmetic (addition, subtraction, multiplication and division) on pixel values of the image. It was used to subtract the minimum image from step (2) from the raw images in step (1). This is a form of background subtraction technique to remove background noises from the raw images.
4. Image Mean; this calculates the average intensity of corresponding pixels (pixels with identical coordinates) in all selected images. The images must be compatible in terms of dimensions and grayscales. The inputs for this step are images from step (3) and the output is a single map.
5. Image Processing Library (IPL)-Blur; this module contains filters like low-pass and high-pass filters, Morphology, signal processing, utility and threshold that can be used to smooth images, detect edges, enhance image contrast, carryout non-linear calculations. The image from step (4) was processed to reduce the particle image intensity by blurring.
6. Image Arithmetic; this is used to subtract output image of step (5) from the raw images obtained in step (3). The aim is to reduce the intensity of each image in the ensemble.
7. Image Processing Library (IPL); here Gaussian filters (5 × 5) were applied twice to images from step (6) to smooth the images. This step makes the images of particles more visible. The Gaussian filter is a linear low-pass filter like mean filters but differs in the sense that it weighs grayscale at the center of kernels higher than those at the edges.
8. Image Min and Max; this step selects the minimum field form images from the output of step (7).
9. Image Arithmetic; this is used to subtract output image of step (8) from the raw images obtained in step (7). The aim is to check the numeric diffusion that may have been introduced by applying filters.
10. Adaptive Correlation; Correlation algorithm was applied to the output images from step (9) to generate vector maps. Adaptive correlation is a form cross-correlation with additional post-processing algorithms like local validation. Additionally, processes like local validation to detect bad vectors, the overlap of IA to compensate for the loss of vector field resolution, high accuracy module, and window deformation were applied to the images using this correlation type. Local validation was achieved through Peak Validation and Local Neighborhood Validation. The overlap of IA was achieved with 50% overlap between IA pairs in both horizontal and vertical directions. The velocity vectors were obtained from displacement vectors and time interval using the Central Difference Scheme which was the most accurate methodology suitable for PIV measurement. The high accuracy sub-pixel interpolation was part of the adaptive correlation algorithm that was used to determine correlation peak.
11. Vector Statistics: this calculates statistics from multiple vector maps and presents the results of mean velocity vectors in a vector map. Other parameters obtained were the streamlines and the velocity profiles.

The porous plates are fired alumina ceramic plates having length of 150 mm, width of 50 mm, thickness of 1 mm, porosity of 40%, effective thermal conductivity of 3.85 W/m K, and maximum operating temperature of 1550 °C (recommended temperature of 1450 °C) (42520-X) made by ESL Electro-Science, USA. The upper and lower oxygen chambers are closed; hence the pressurized oxygen supply is passed through the porous plates into the central combustion zone. The sides of the combustion zone are made of two quartz plates for flow/combustion visualization. The quartz glass plates have a thickness of 3 mm each. The nominal height of the two-oxygen chambers is 30 mm each. The reaction zone is 25.5 mm in height with length of 400 mm, and width of 55 mm, as shown in Fig. 6.40b. D-type high-tech mass flow controllers by Bronkhorst manufacturers, with accuracy of ±0.5%, were utilized to supply and maintain the required volume flowrates of the gases. The experimental measurements were conducted considering the cold flow (non-reactive) conditions. The operating equivalence ratio (Φ) was calculated based on the flow rates of methane and oxygen to be in the range from 0.4 to 0.6 as presented in Table 6.12. The experiments were conducted at low reactor power of 1.36 kW to effectively mimic the HTMR operation. The volume flow rate ratio of $O_2/CO_2$ of 1.0 was considered throughout the experiments. The mean flow velocities were calculated from the volumetric flow rates based on the cross-sectional area of the flow field and flow properties.

### 6.6.3 Modeling Non-reactive Flow in Porous-Plate Reactor

Computational fluid dynamics (CFD) was employed to simulate the non-reactive flow field characteristics inside a porous-plate reactor, based on the CFD method devised by Habib et al. [106]. Mesh was constructed and the flow field and oxygen permeation across the porous plates were solved, considering a two-dimensional (2-D) domain as per Fig. 6.40b, using the commercial ANSYS 15.0 Workbench software. The considered mesh is of 144,000 cells, 289,980 faces and 145,977 nodes with maximum aspect ratio of 1.7. Finer mesh grids were considered at the relevant boundaries and at the combustion zone, where higher gradients are expected. The general laminar flow conservation equations including continuity, momentum and species transport were solved for the fields of velocity, pressure, and species concentration [80, 116]:

$$\nabla \cdot (\rho \mathrm{U}) = 0 \tag{6.31}$$

$$\nabla_1(\rho \mathrm{UU}) = -\nabla \mathrm{P} + \mathrm{u}\nabla^2 \mathrm{U} \tag{6.32}$$

$$\nabla \cdot (\rho \mathrm{UY_i}) - \nabla \cdot \left(\rho \mathrm{D_{i,m}} \nabla \mathrm{Y_i}\right) = 0 \tag{6.33}$$

where, ρ is density, $U$ is velocity, $P$ is pressure, μ is dynamic viscosity, $Y_i$ is the mass fraction of $i$th specie and $D_{i,m}$ is the diffusion coefficient of the $i$th specie.

ANSYS-Fluent 16.5 software based on finite volume approach was employed for the discretization of the above conservation equations for the numerical calculations using the boundary conditions indicated in Fig. 6.40c. Pressure-velocity coupling was achieved using a built-in 2-D double precision solver semi-implicitly. Second-order upwind schemes were used for the species and momentum discretization while standard scheme was applied for the pressure. Advanced Multi Grid solver (AMG) was used to ensure convergence using the following values for under-relaxation factor: 0.3 for pressure; 1.0 for density; 0.7 for momentum; 1.0 for body force; 0.8 for energy and 1.0 for species mass fractions. Recursive f-cycle process was set, and gradient stabilized bi-conjugate method was used to prevent convergence irregularities. An absolute convergence check criterion was set to $10^{-6}$ for all the residuals, which implies convergence after all the residuals falls below the set criterion. Velocity field and species mass fractions were obtained and were utilized for the calculation of the mixture fractions and the average local equivalence ratios inside the combustor. The mixture fraction is a non-dimensional conserved quantity that indicate the local fuel-oxygen ratio in non-premixed mixture as given by Eq. (6.34) and the stoichiometric mixture fraction as given by Eq. (6.35) [117–119]. A unique relation between the localized equivalence ratio and the mixture fraction has been developed by [117, 118] as shown in Eq. (6.36).

$$Z = \frac{\upsilon Y_f - Y_O + Y_{O,2}}{\upsilon Y_{f,1} + Y_{O,2}} \tag{6.34}$$

$$Z_{st} = \frac{Y_{O,2}}{\upsilon} Y_{f,1} + Y_{O,2} \tag{6.35}$$

$$\phi = \frac{Z(1 - Z_{st})}{Z_{st}(1 - Z)} \tag{6.36}$$

The cases of $Z < Z_{st}$, $Z > Z_{st}$ and $Z = Z_{st}$, represent lean, rich and stoichiometric mixtures, respectively.

The thin thickness of the porous plates used in this work enabled the use of Porous jump as the pressure drop boundary condition as a reduced model of porous media to one-dimension [116]. This simpler model is more robust with better convergence compared to the full porous media model. The pressure-drop in porous media having a finite thickness uses the combination of Darcy's Law and inertia loss term [116]:

$$\Delta p = \left[\mu \cdot \upsilon/\alpha + \left(C_2 \cdot \rho \cdot \upsilon^2\right)/2\right] \cdot \Delta m \tag{6.37}$$

where $\mu$ is the dynamic viscosity, $\nu$ is the normal velocity to the porous face, $\alpha$ is the permeability of the media, $C_2$ is the pressure-jump coefficient, $\Delta m$ is the thickness of the media.

The porous plates were modeled using Porous Jump as boundary condition due to the small thickness of 1 mm using Eq. (6.37). This required only the thickness, porosity and pressure-jump coefficient. The pressure-jump coefficient was calculated based on the procedure recommended by [116].

### 6.6.4 Modeling Reactive Flow in Porous-Plate Reactor

The CFD approach was employed for the reactive flow modeling via solving the same governing Eqs. (6.31)–(6.33) in addition to energy and radiative transfer equations as presented in Eqs. (6.38)–(6.39). The inlet characteristics of the flow are laminar fully developed due to low inlet Reynolds number as shown in Table 6.12. The thermal boundary conditions applied to the porous plates and the inner walls supporting them was 'coupled'. Under this condition the solver will directly compute the heat transfer from the adjacent cells solutions and hence, no additional thermal boundary conditions are required. For the other external walls 'mixed' thermal boundary condition was applied which accounts for both convective and radiative heat transfer to the ambient at temperature of 298 K, external emissivity of 1 and heat transfer coefficient of 20 $W/m^2K$. All the walls were modeled as opaque while the quartz glass was modelled as semi-transparent. The outlet boundary condition was constant atmospheric pressure at temperature of 298 K. The energy equation accounts for energy transfer due to conduction, viscous dissipation, species diffusion, as well as energy source terms [120]. Discrete Ordinates (DO) radiation model is used for computing radiative heat transfer for both the exhaust gases, which are mostly participating media, and surface-to-surface radiation. To determine the absorption coefficient of the exhaust gas mixture, the Weighted-Sum-of-Gray-Gases Model (WSGGM) was used for the computation of a variable absorption coefficient in the DO model. This model predicts the thermal radiation heat transfer between the flue gases, the walls of the combustor and the surrounding.

$$(\rho C_p)U \cdot \nabla T = \nabla \cdot (k \nabla T) \tag{6.38}$$

$$\frac{dI(r,s)}{ds} = KI_b - (K + \sigma_s)I(r,s) \tag{6.39}$$

As a matter of fact, most of the detailed mechanisms are validated for normal air combustion case not for oxy-combustion case. The modified two-step reaction kinetics model by Andersen et al. [121] for methane combustion in highly $CO_2$-concentrated medium, oxy-combustion conditions was applied. This model proved high accuracy based on the validation studies performed in references [106, 122]. The model considers the following reactions:

$$CH_4 + 1.5O_2 \rightarrow CO + 2H_2O \tag{6.40}$$

$$CO + 0.5O_2 \rightarrow CO_2 \tag{6.41}$$

$$CO_2 \rightarrow CO + 0.5O_2 \tag{6.42}$$

Reaction rates were calculated based on the Arrhenius equation as per Nemitallah and Habib [122]. The system used 4 processor with 4 GB RAM. The average run-time for convergence per case was 11 h.

Numerical calculations were performed considering the 2-D porous plate reactor geometry presented in Fig. 6.40b at reactor power of 3 kW. Grid independence study was performed, and it was found that there were no significant variations of the results (<1%) beyond the grid points of 29,520 which was used in the simulations. The mass flow inlet boundary condition was selected for all gas flows with inlet temperature of 300 K, while pressure outlet boundary condition was set at the common exit section of the reactor. The energy, momentum and species transport as well as the radiative transfer equations were solved using steady pressure-based solver in the ANSYS Fluent 16.5 software. The energy and radiation models are implemented one after the other (uncoupled) sequentially using the conservative finite-volume scheme [123]. The used pressure-velocity segregate algorithm was the semi-implicit method for pressure linked equations (SIMPLE) with higher order of the relaxation terms. Table 6.13 shows the seven cases considered for the analysis of oxy-combustion characteristics inside the porous-plate reactor over a range of equivalence ratio from the stoichiometric condition down to very lean operation at equivalence ratio of 0.4. In all cases, fuel mass flow rate is kept constant at $0.6 \times 10^{-4}$ kg/s (reactor power of 3 kW), and the volumetric percentage of $O_2$ in the $CO_2$–$O_2$ oxidizer is kept fixed at 30% in order to obtain similar adiabatic temperature for the case of normal air combustion [124]. The flow rates of fuel and oxidizer are kept within the low range to mimic the processes of oxygen permeation and oxy-combustion in side HTMRs for clean power production applications.

**Table 6.13** Sets of numerical cases for the analysis of oxy-combustion characteristics within the porous-plate reactor at various equivalence ratios

| Case No. | Mass Flow rates $\times 10^{-4}$ (kg/s) | | | Equivalence ratio ($\Phi$) |
|---|---|---|---|---|
| | $CH_4$ | $O_2$ | $CO_2$ | |
| 1 | 0.60 | 2.41 | 7.71 | 1.0 |
| 2 | | 2.67 | 8.56 | 0.9 |
| 3 | | 3.01 | 9.63 | 0.8 |
| 4 | | 3.44 | 11.0 | 0.7 |
| 5 | | 4.01 | 12.8 | 0.6 |
| 6 | | 4.81 | 15.4 | 0.5 |
| 7 | | 6.02 | 19.3 | 0.4 |

### 6.6.5 Non-reactive Flow Field Characteristics

Three sets of experiments were conducted to determine the effect of the oxygen permeation from the two porous plates on the velocity profiles within the central (mixing) zone of the reactor under non-reactive flow conditions. In all the three cases, as per Table 6.12, low fixed fuel flow rate of 2.5 l/min and fixed volumetric flow rate ratio of $CO_2/O_2$ of 1.0 were considered. This low fuel flow rate was considered to mimic the operation of high-temperature membrane reactors (HTMRs) combining both oxygen separation and oxy-combustion of the fuel using the permeated oxygen in the permeate side of the membrane. Still oxygen separation membranes are under development to produce higher oxygen permeation flux for complete combustion of the fuel for HTMR applications. In this study, two face-to-face porous-plates are used instead of membranes to allow ease control of oxygen supply enabling the study of flow field and oxy-combustion characteristics of HTMRs. The experiments were conducted at different $O_2$ flow rates (typically 8.3, 10.0 and 12.5 l/min) and equivalence ratio (typically $\Phi = 0.6$, 0.5 and 0.4).

From the PIV analyses, the averaged velocity vectors and the streamlines plots for the first part of the IA (sections a–c, Fig. 6.43d) are presented for the considered three cases of Table 6.12 in Figs. 6.41 and 6.42, respectively. Actually, the flow is laminar and fluctuations in the flow velocity are very low to make significant difference between the average velocity and the velocity profile. That's why the results are presented based on the average velocity profile. Near the porous plates, the velocity vectors are scarce and shorter in length indicating low velocity in the mixing boundary layer and deflected because of the oxygen gas permeation through the porous plates from the oxygen zone into the central zone. As per Fig. 6.41, the level of mixing increases with the decrease of equivalence ratios. This can be attributed to the increase in oxygen flow rate associated with the increase in equivalence ratio. This behavior can be seen more evidently from the contours of streamlines presented in Fig. 6.42. The third case, having the highest oxygen flow rate, showed increased in fluctuations near the porous plates due to the enhanced flow mixing. If the results are projected on HTMR operation, this enhanced flow mixing should help reducing oxygen partial pressure at the permeate side of the membrane resulting in increased oxygen permeation flux. Also, this enhanced mixing zone beside the membrane protects the membrane against the flame elevated temperature and helps in flame anchoring and stabilization within the HTMR. Figure 6.43 shows the traverse (vertical) velocity profiles of the fuel-diluent and oxygen mixture obtained at three different sections of the IA for the considered range of experiments. Sections *a, b* and *c* are located 10 mm, 35 mm and 70 mm from the leading edge of the porous plate axially, as presented in Fig. 6.43d. As per Fig. 6.43, the velocity profiles tend to deviate from the known parabolic profile of internal flows along the length of the porous plate due to permeation of oxygen across the porous plates. At a given equivalence ratio, the total permeation flow rate of oxygen increases along the reactor in the axial direction, from section *a* to section *c*, resulting in an increase

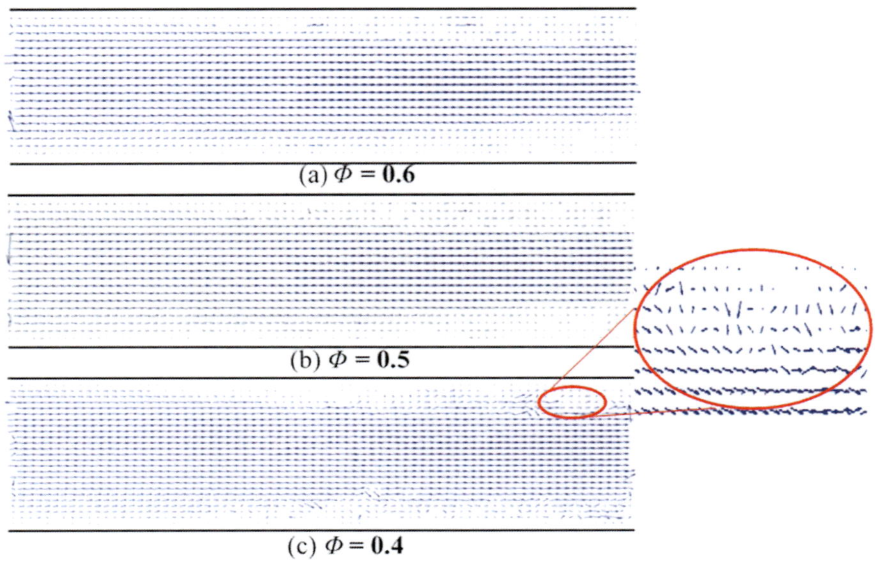


**Fig. 6.41** Experimental velocity vectors at different equivalence ratios of: **a** 0.6, **b** 0.5 and **c** 0.4, $\chi = 125$–200 mm, Y = 30–55 mm

in flow velocity and more distortion of the profiles at the end of the IA, which corresponds to the middle zone of the porous plates. Whereas, the distortion of the velocity profiles due to oxygen mixing also increases with the decrease in equivalence ratio. While decreasing the equivalence ratio, the maximum velocity increases due to the increase in compression of the fuel-diluent stream by the increase in the pressure of the oxygen cross-flow.

The images of the flow field were analyzed using the PIV system software for the entire length of the porous plates at equivalence ratio of 0.4. Based on the analysis of those images, the mean velocities of the fuel and diluent mixture were calculated at different axial locations in the central zone between the two porous plates. When the experimental conditions were simulated numerically using the CFD model at the corresponding equivalence ratios, the corresponding average velocities of the flow in the central zone between the two porous plates were calculated at different axial locations using the mass-weighted average values based on surface integrals. Both experimental and numerical values of average velocities are compared as shown in Fig. 6.44. The comparison of average velocities shows good agreement between the experimental and simulated results with absolute error of less than 4%. Due to continuous oxygen permeation axially through the porous plates to the central channel, the total volume flow rate is increased, resulting in increase in the average velocity in the axial direction downstream of the porous plate. Projecting the analysis to the HTMR operation, the excess oxygen permeated from the end parts of the membranes in the axial direction cools down the flame temperature and protects the membrane against

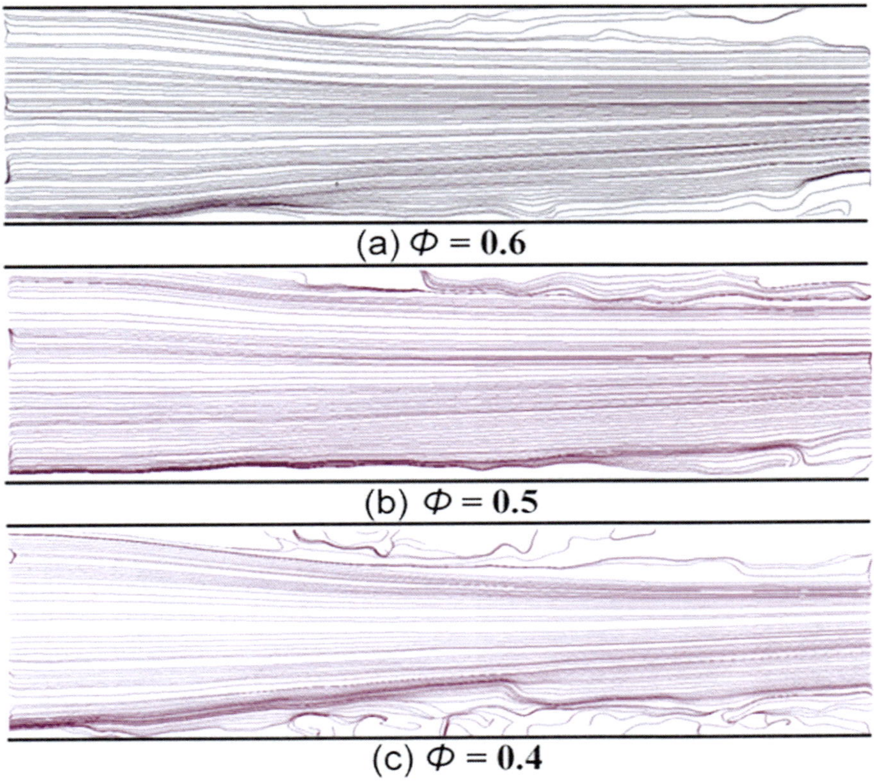


**Fig. 6.42** Experimental streamlines at different equivalence ratios of: **a** 0.6, **b** 0.5 and **c** 0.4, $\chi$ = 125–200 mm, Y = 30–55 mm

excessive temperature rise. Also, the enhanced velocity downstream of the membrane helps better flame anchoring and stabilization downstream of the membrane zone.

The mass fractions of $CH_4$, $CO_2$ and $O_2$ were obtained numerically at different axial positions, and Eqs. 6.34 through 6.36 were applied to compute the mixture fraction and the local equivalence ratio through the reactor as presented in Figs. 6.45 and 6.46. As per Fig. 6.45, the mixture fraction decreases axially along the porous plate section and remain constant downstream of the porous plate at all operating equivalence ratios. The results show no oxygen at the beginning of the porous plate. The mass flux of the oxygen increases almost linearly up to about three-quarters of the porous section and, then, increases exponentially toward the end of the porous plate. The mixture fraction reaches its lowest value at the end of the porous plate and remains constant up to the exit section of the reactor since there is no more oxygen inflow. These results coincide with the findings by Farooqui et al. [125] for oxygen permeation inside a HTMR. The results show insignificant increase of

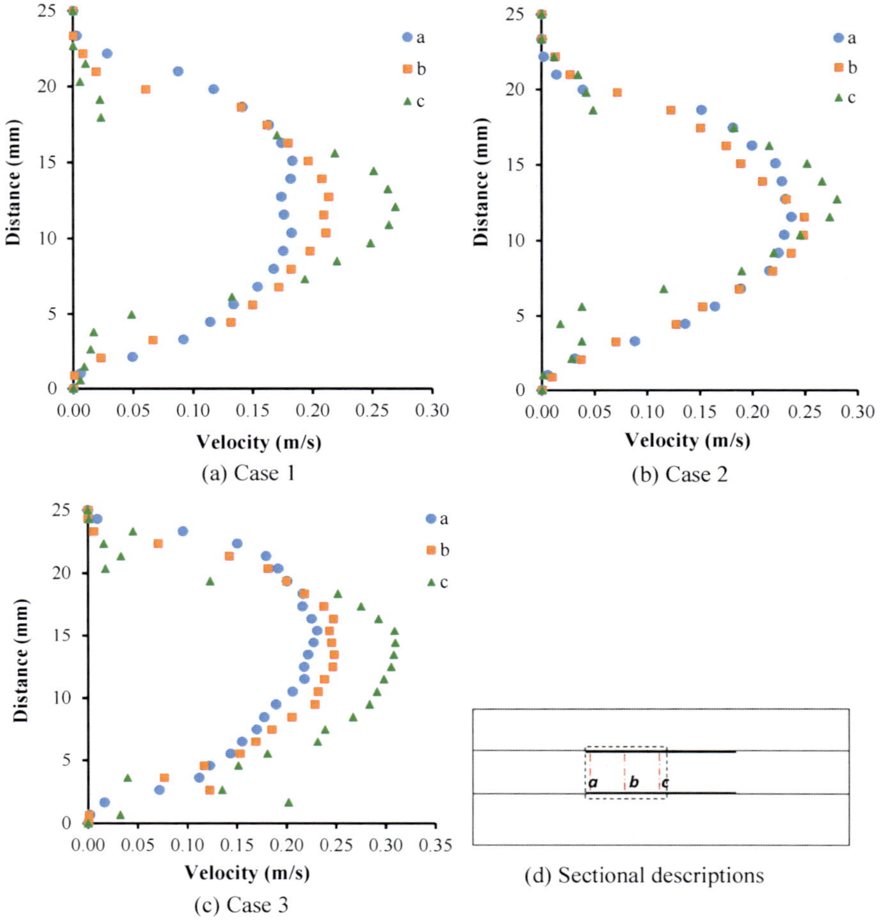


**Fig. 6.43** Experimental traverse velocity profiles at three different sections (a = 135, b = 160 and c = 195 mm) within the porous-plate reactor for different equivalence ratios of: **a** 0.6, **b** 0.5 and **c** 0.4

mixture fraction with the increase in the overall operating equivalence ratio through the center of the reactor as the oxygen flow is introduced from top and bottom porous plates. Figure 6.46 shows the variation of average local equivalence ratio along the length of the porous plate reactor. Since there is little amount of oxygen at the beginning of the porous plate, the average equivalence ratio is the highest, and it decreases rapidly to the stoichiometric value towards the end of the porous plates due to oxygen permeation across the porous plates. Downstream of the porous plates, the local equivalence ratio decreases to the set value (global equivalence ratio) for the three cases. This shows that the reactor attains an averaged equivalence ratio equal to the set value, based on the inlet flow rates, just downstream of the porous plate.

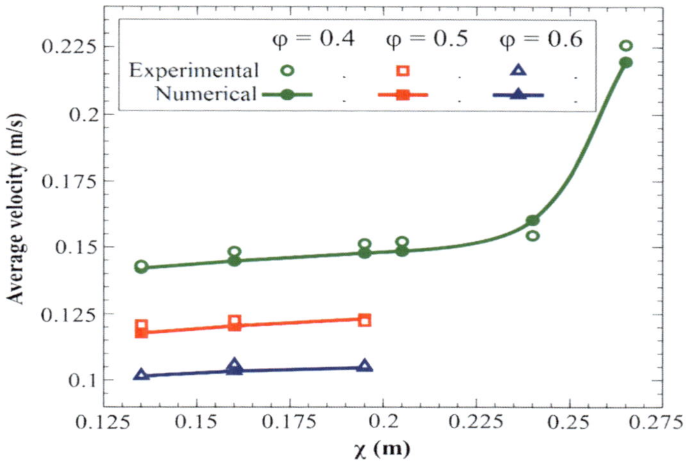


**Fig. 6.44** Comparison of experimental and numerical values of average velocities along the combustor at equivalence ratio of 0.4

**Fig. 6.45** Averaged axial variations of fuel mixture fraction for different equivalence ratios based on numerical calculations

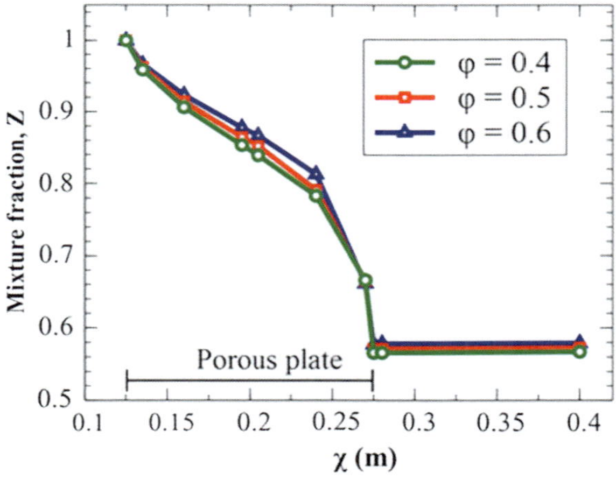


This gives an idea about the position in the reactor where the set equivalence ratio is achieved and, accordingly, the location of the flame downstream of the porous plates. The interaction between the permeated oxygen and the main flow creates recirculation zone through the entire length of the membrane which helps stabilizing the flame and acts as flame anchoring point for the flame downstream of the porous plates. The location of the flame downstream of the membrane for the operation of HTMRs protects the membrane against high flame temperature and allows for complete fuel conversion.

The analysis of the average equivalence ratio paved the way for establishing a general relationship between the set (global) equivalence ratio and the position

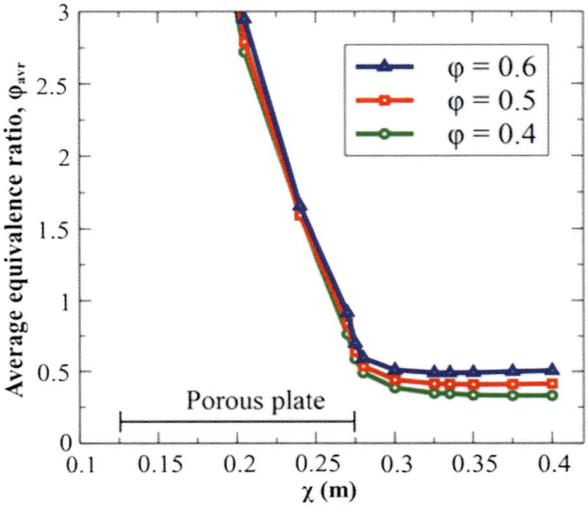


**Fig. 6.46** Axial variations of the average local equivalence ratio for different equivalence ratios based on numerical calculations

within the reactor where this value is achieved. This has been depicted in Fig. 6.47. A linear relationship between the global equivalence ratio and the axial position where it occurs was obtained to be as follows:

$$\phi_g = -12.857\chi + 4.1857 \tag{6.43}$$

The plot shows that the higher the set (global) equivalence ratio, the lower the mixture fraction and the shorter the axial distance at which it is achieved. This means that for operation toward stoichiometric combustion, the flame will be shifted upstream. This may be attributed to the reduced flow rate of $O_2$ and $CO_2$ and reduction in total axial flow velocity. For more insight, the traverse variations of mixture fraction and local equivalence ratio within the reactor, in the zone between the two porous plates, are presented in Figs. 6.48 and 6.49, respectively. The results in Fig. 6.48

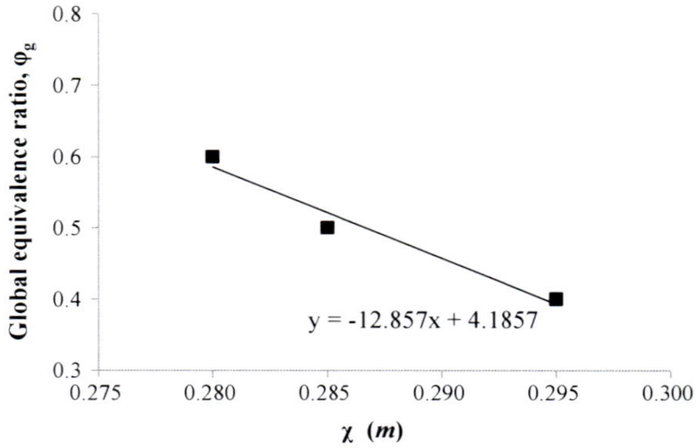


**Fig. 6.47** Axial position at which the set (global) equivalence ratio is achieved for different set values of the equivalence ratio based on numerical calculations

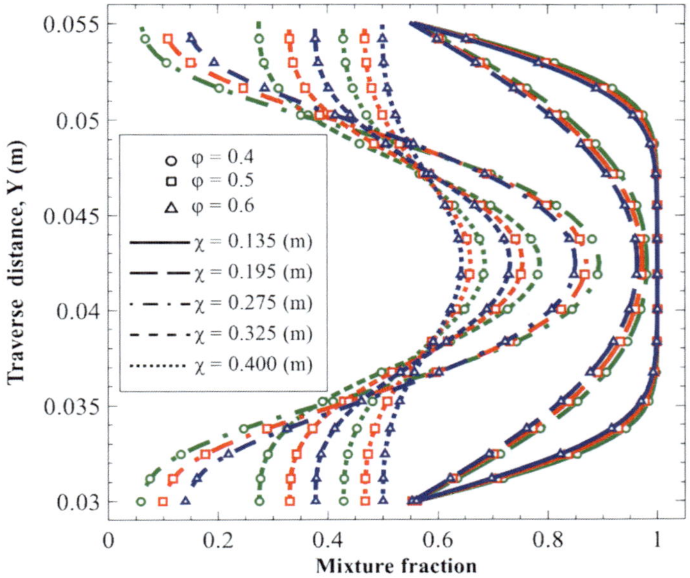


**Fig. 6.48** Traverse variations of fuel mixture fraction at different axial locations for different set equivalence ratios based on numerical calculations

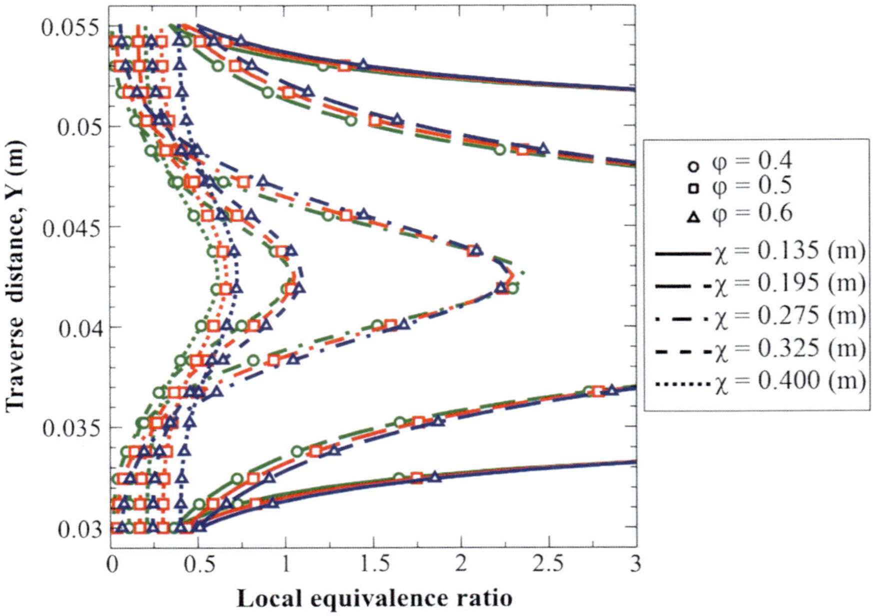


**Fig. 6.49** Traverse variations of local equivalence ratio at different axial locations for different set equivalence ratios based on numerical calculations

indicate that higher oxygen mass fractions are obtained at the end of the porous plates, and the profile is sharper at the centerline of reactor with a steep decrease towards the porous plates. As per Fig. 6.48, the variations in the profile of mixture fraction in the traverse direction are reduced downstream of the porous plates indicating better mixing. The improved mixing downstream of the porous plates is assured by the distributions of local equivalence ratio as per Fig. 6.49. The values of mixture fraction and the equivalence ratio increases in the traverse direction, from the surface of the porous plates toward the center of the reactor where their maximum values are attained, for all operating set equivalence ratios.

### 6.6.6 Reactive Flow Field Characteristics

This section investigates the characteristics of oxy-combustion with the porous plate reactor numerically to determine the location of the flame with respect to the porous plate and the distributions of temperature and species concentrations. This should give an idea about the generated flame when the case is projected on HTMR applications. Figure 6.50 shows the effect of equivalence ratio, over a range from 0.4 up to 1.0, on the distributions of temperature in the traverse and axial directions. As per Fig. 6.50a, the traverse distributions of temperature (at the trailing edge of the porous plate) show increase in the temperature level within the reactor while increasing the equivalence ratio toward stoichiometry. This may be attributed to improved convective heat transfer within the rector at lower equivalence ratios due to the associated increase in the total oxidizer ($O_2$ plus $CO_2$) flow rate. The high flow of $CO_2$ at lower

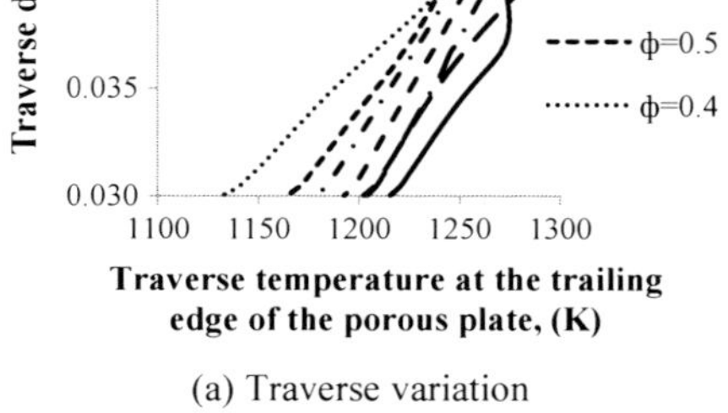


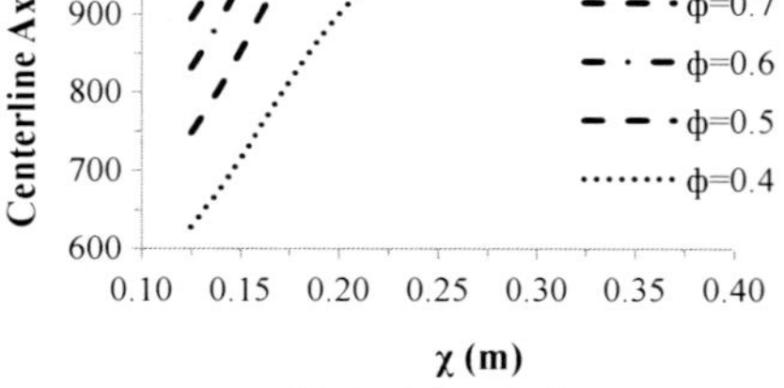


**Fig. 6.50** Effect of equivalence ratio at reactor power of 3 kW on the temperature distributions based on numerical calculations: **a** traverse distributions at the trailing (end) edge of the porous plate, and **b** axial distributions

equivalence ratio slows down the reaction kinetic rates and, as a result, the combustion temperature is reduced. The excess amount of permeated oxygen from the trailing part of the porous plate cools down the combustion products and reduces the combustion temperature at lower equivalence ratio. However, in HTMRs, oxygen is permeated through a membrane which consumed portion of the heat of combustion for oxygen separation. Based on that, HTMRs are preferred to work under stoichiometric conditions ($\Phi = 1.0$). At $\Phi = 0.9$ in Fig. 6.50a, the main stream temperature at the center of the reactor is higher than the obtained temperature at $\Phi = 1.0$. This can be attributed to the reduced level of mixing between the permeated oxygen and the main flow stream containing the fuel due to the reduction of the amount of permeated oxygen at $\Phi = 1.0$. As per Fig. 6.50a, the temperature on the trailing edge of the porous plate ranges from 1130 K (at $\Phi = 0.4$) to 1240 K ($\Phi = 1.0$). This temperature range lies within the operating limit of operation of the ceramic membranes when the cased is projected on HTMR operation. At all equivalence ratios, the intense reactions are delayed in the main flow to the downstream region of the porous plate as per the temperature plots in Fig. 6.50b and the temperature contours in Fig. 6.51. The improved mixing in the downstream region of the porous plate results in improvement in the combustion temperature. This is shown in Figs. 6.50b and 6.51. As per the results in Figs. 6.50b and 6.51, lean combustion operation resulted in better mixing and higher flame temperature as compared to stoichiometric combustion. However, the excess amount of oxidizer under lean operation cools the exhaust stream down to reach similar temperature as the stoichiometric combustion at the exit section of the reactor. These results support the operation of the reactor under stoichiometric condition from performance points of view, especially when the case is projected on the HTMR operation. As per Fig. 6.51, the reactor design keeps the zone beside the porous plate hot due to initiation of reactions once oxygen appears in the central zone. This feature of the reactor design is very crucial for the operation when the porous plate is replaced by ceramic membrane for HTMR applications, as the membrane should be kept at elevated temperature to be activated for oxygen permeation. The lower amount of permeated oxygen and reduced level of mixing with the main stream results in elongated flame at stoichiometric condition when compared to lean operation as shown in Fig. 6.51. For same reasons, methane depletion rate within the reactor is lower at stoichiometric condition compared to lean operation as shown from the contour plots of methane concentrations presented in Fig. 6.52. Also, full conversion of the fuel is delayed in the downstream zone close to the exit section at higher equivalence ratios as shown in Fig. 6.52.

## 6.7 Operability Limits of Porous-Plate Reactor Mimicking OTMR Operation

Most of the existing power plants utilize air as the oxidizer for combustion; meanwhile oxy-fuel combustion concept requires the use of pure oxygen. This necessitates the oxygen separation from air using various techniques to fit into the conventional as well as the future power plants. The three main oxygen/air separation processes are

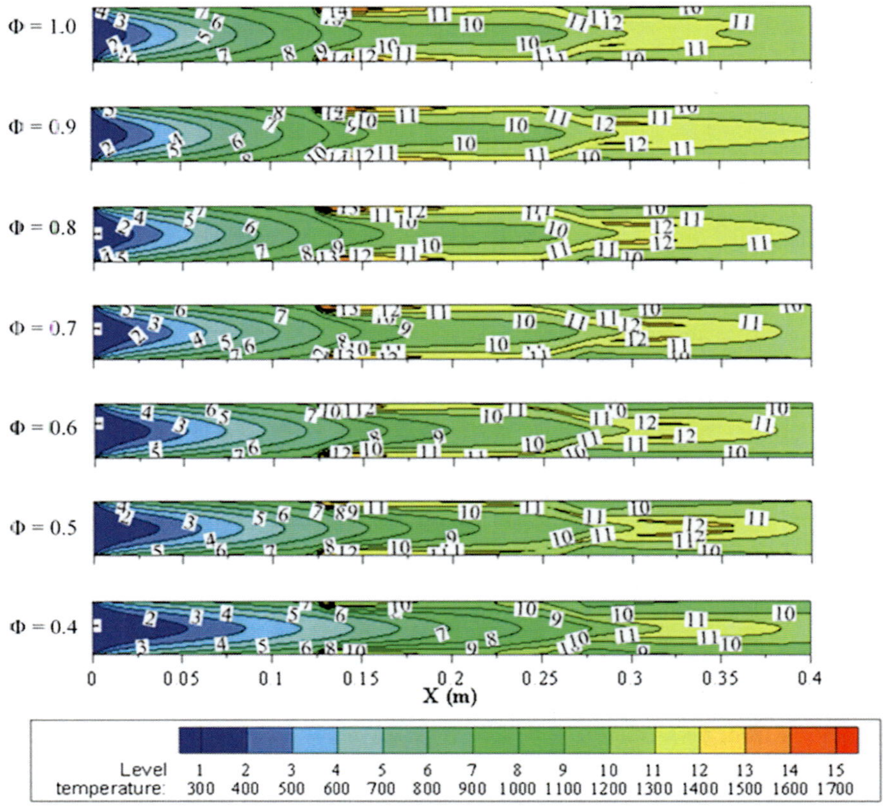


**Fig. 6.51** Numerical contour plots of temperature (K) over a range of equivalence ratio and at reactor power of 3 kW

cryogenic, membrane and solid sorbent processes [126, 127]. The most established technology among the three is the cryogenic distillation that allows for large-scale production of $O_2$ at very high purities (up to 99%) at low temperature [127, 128]. Despite the production of high purity of $O_2$ by this technique, the design complexity, high energy demand as well as low exergetic efficiencies (15–24%) necessitate the investigation of alternative air separation concepts [129, 130]. The solid sorbent (adsorption) process, despite having $O_2$ purity of up to 95%, is not fully mutualized due to excessive capital requirements of the solvent [128, 131]. The most recent separation technology is the membrane separation, which includes the use of high-temperature ion transport ceramic membrane (ITM) [132]. Combining the process of oxygen separation (using ceramic membrane) with the oxy-fuel combustion of fuel in the sweep side of the membrane (using the separated oxygen) results in a compact unit for power production. This is called high-temperature membrane reactor

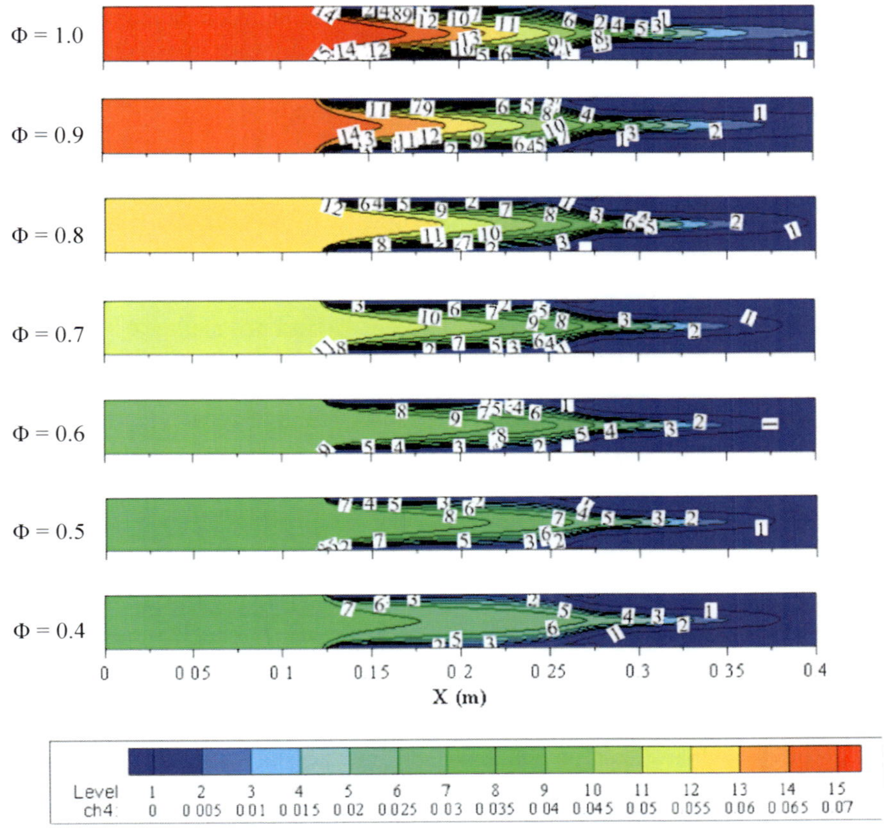


**Fig. 6.52** Numerical contour plots of mass fractions of $CH_4$ over a range of equivalence ratio and at reactor power of 3 kW

(HTMR). The use of porous plate reactors is increased significantly due to its flexibility of introducing the oxidizer mixture through the whole combustion chamber. This is of immense importance in case of oxy-fuel combustion to uniformly distribute oxygen through the whole combustion chamber and, accordingly, reduce the creation of hot temperature spots within the flame zone. Habib et al. [133] studied the characteristics of oxy-fuel combustion in a porous plate reactor experimentally and numerically for the reactive and non-reactive cases. The results showed significant effect of oxy-fuel mixing as it directly affects the combustion chemistry within the combustor.

Flame blow-out occurs when the flame is incapable of sustaining combustion at locations downstream from the exit of the combustor [134, 135]. Karbasi and Wierzba [136] studied the effect of dilution of fuel jet for a co-flow arrangement in order to determine the velocities at the flame blow-out. The diluents used were Nitrogen ($N_2$), Helium (He), and carbon dioxide ($CO_2$). It was found that the addition of the diluent in the co-flow has a greater influence on the blow-out limit in comparison to the case when the diluent was premixed within the fuel jet. The diluents, in descending order of their influence, are $CO_2$, $N_2$ and He. Several models were developed based on flow exit velocity and fuel properties to estimate the blow-out limits. The method relies heavily on how the fuel velocity and diluent, or co-flow is measured. Mostly, the blow-out limit is found to decrease with increase in the flow velocity [134]. The LBO limit was theoretically characterized using two different approaches [137–139]. The first approach is based on the Damköhler (Da) number concept, defined as the ratio of characteristic mixing time to the characteristic chemical reaction time. It is based on the postulation that lean blow-out occurs when the residence time is not sufficient to allow the chemical reactions to occur. The second method considers the ratio of the flame speed to the flow speed based on the assumption of flamelet-like combustion, in that, the lean blow-out occurs when the flow speed exceeds the flame speed at critical positions within the combustor [140]. In both the approaches, lean blow-out is influenced by heat content, fuel kinetic characteristics, diffusivity, and the flow field of the combustor [141].

The study presented under this section focuses on experimental testing of oxy-fuel combustion characteristics in a face-to-face two-porous-plates reactor over wide ranges of mixing conditions and equivalence ratio. This reactor mimics the process of oxygen permeation and oxy-fuel combustion inside the HTMR. The process of oxy-combustion inside the HTMRs is difficult to be achieved at normal lab conditions and the oxygen permeation flux across the membrane still not enough for complete fuel combustion. Because of that, porous plates are used in the presented study instead of membranes trying to see the flame characteristics and location with respect to the porous plate that should help in the development of HTMRs. The experimental study aims at investigating the combustion characteristics and lean blow-out limits of oxy-methane non-premixed flames over wide ranges of operating conditions.

### *6.7.1 Combustor Setup*

The setup used in investigating the combustion characteristics and lean blow-out limits of non-premixed oxy-combustion flames in porous plate reactor is shown in Fig. 6.53. Pure oxygen was supplied into the two oxygen chambers from a pressurized oxygen gas cylinder while the fuel (methane), together with the diluent (carbon-dioxide) were also supplied in their pure forms from their respective pressurized cylinders. Each pressurized cylinder was connected to a mass flow controller, which regulates the gas flowrate entering the sections. All controllers are Bronkhorest HIGH-TECH type with uncertainty of $\pm 0.5\%$ of full scale. The fuel and diluent were mixed upstream of the reactor in a stainless-steel mixing chamber. The length of the mixing chamber is 80 cm. The fuel and $CO_2$ were supplied through a rectangular cross-sectional duct while the pure oxygen was supplied via the upper and lower porous steel chambers. The uniformity of the oxygen flow along the porous plate is controlled by the applied uniform pressure in the oxygen chambers, which

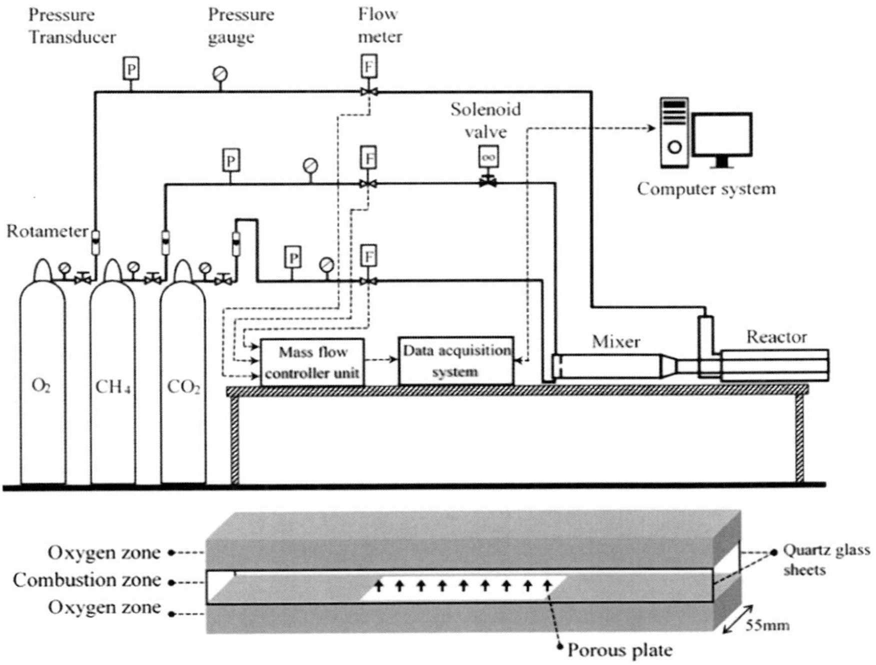


**Fig. 6.53** Schematic diagram of the experimental setup and the porous-plate reactor

are closed with one inlet and no exit except through the porous plates. In reaction zone (central zone), both the fuel and the supplied diluent mix with the oxygen permeating through the porous plate from both the upper and the lower sides of oxygen chambers, forming a non-premixed flame attached to the fuel inlet side of combustor. Two quartz glass plates of 2 mm thick each were used to form the rectangular combustion zone sides and permitted for the optical access to the flame. A high-speed camera (80 frames/second) was employed for flame structures imagining. The camera was mounted at the same height of the horizontal combustion zone about 1000 mm away. The size of the oxygen chamber is 260 mm in x-direction, 50 mm in y-direction, and 50 mm in z-direction. The dimensions of the combustion zone are 260 mm in x-direction from the fuel-diluent inlet of the combustor, 20 mm in y-direction, and 50 mm in z-direction. The cross-section of the fuel-diluent inlet (y–z) is 10 × 20 mm, while the porous oxygen inlet cross-section (x–z) is 120 × 25 mm as shown in Fig. 6.54. The two porous plates used for oxygen permeation have are 1 mm thick, fired alumina ceramic plates, with porosity of 40% and maximum operating temperature of 1550 °C (model 42520-X).

The objectives of the study were achieved by executing the experimental plans as shown in Tables 6.14, 6.15, 6.16 and 6.17. In Tables 6.14 and 6.15, the fuel flowrate was fixed at 2.76 l/min (equivalent to 1.5 kW calorific value of the fuel) while varying the percentage of $CO_2$ in the oxidizer at a fixed equivalence ratio and varying the

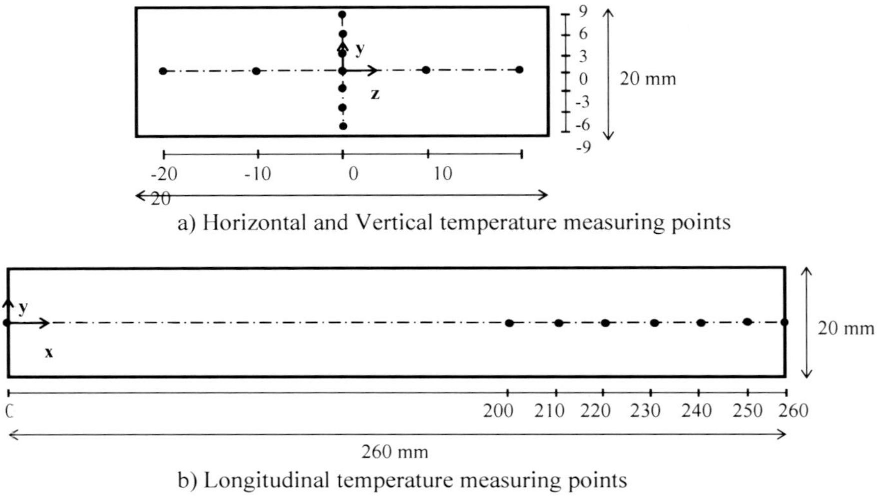


**Fig. 6.54** Nodalization of temperature measurements

**Table 6.14** Operating volume flow rates of $CH_4$, $O_2$ and $CO_2$ fed to the reactor at different concentrations of $CO_2$ and at fixed fuel firing rate of 1.5 kW for a range of equivalence ratio

| % $CO_2\|_{oxidizer}$ | $\dot{Q}_{CH_4}$ (L/min) | $\phi = 1.0$ | | | $\phi = 0.7$ | | $\phi = 0.5$ | |
|---|---|---|---|---|---|---|---|---|
| | | $\dot{Q}_{O_2}$ (L/min) | $\dot{Q}_{CO_2}$ (L/min) | Re | $\dot{Q}_{O_2}$ (L/min) | $\dot{Q}_{CO_2}$ (L/min) | $\dot{Q}_{O_2}$ (L/min) | $\dot{Q}_{CO_2}$ (L/min) |
| 70 | 2.76 | 5.52 | 12.87 | 2466 | 7.88 | 18.39 | 11.03 | 25.75 |
| 55 | 2.76 | 5.52 | 6.74 | 1550 | 7.88 | 9.63 | 11.03 | 13.49 |
| 45 | 2.76 | 5.52 | 4.51 | 1092 | 7.88 | 6.45 | 11.03 | 9.03 |
| 35 | 2.76 | 5.52 | 2.97 | 817 | 7.88 | 4.24 | 11.03 | 5.94 |
| 27 | 2.76 | 5.52 | 2.04 | 633 | 7.88 | 2.92 | 11.03 | 4.08 |
| 0 | 2.76 | 5.52 | 0.00 | 267 | 7.88 | 0.00 | 11.03 | 0.00 |

**Table 6.15** Operating volume flow rates of $CH_4$, $O_2$ fed to the reactor at different equivalence ratios and at fixed fuel firing rate of 1.5 kW

| $\phi$ | $\dot{Q}_{CH_4}$ (L/min) | $\dot{Q}_{O_2}$ (L/min) |
|---|---|---|
| 1.0 | 2.76 | 5.52 |
| 0.9 | 2.76 | 6.13 |
| 0.8 | 2.76 | 6.90 |
| 0.7 | 2.76 | 7.88 |
| 0.6 | 2.76 | 9.20 |
| 0.5 | 2.76 | 11.03 |

**Table 6.16** Operating volume flow rates of $CH_4$, $O_2$ fed to the reactor at different equivalence ratios and different fuel firing rates

| ϕ | 1.5 kW | | 1.0 kW | | 0.75 kW | | 0.5 kW | |
|---|---|---|---|---|---|---|---|---|
| | $\dot{Q}_{CH_4}$ (L/min) | $\dot{Q}_{O_2}$ (L/min) | $\dot{Q}_{CH_4}$ (L/min) | $\dot{Q}_{O_2}$ (L/min) | $\dot{Q}_{CH_4}$ (L/min) | $\dot{Q}_{O_2}$ (L/min) | $\dot{Q}_{CH_4}$ (L/min) | $\dot{Q}_{O_2}$ (L/min) |
| 1.0 | 2.76 | 5.52 | 1.84 | 3.68 | 1.38 | 2.76 | 0.92 | 1.84 |
| 0.9 | 2.76 | 6.13 | 1.84 | 4.09 | 1.38 | 3.07 | 0.92 | 2.04 |
| 0.8 | 2.76 | 6.90 | 1.84 | 4.60 | 1.38 | 3.45 | 0.92 | 2.30 |
| 0.5 | 2.76 | 11.03 | 1.84 | 7.36 | 1.38 | 5.52 | 0.92 | 3.68 |

**Table 6.17** Operating volume flow rates of $CH_4$, $O_2$ fed to the reactor at different fuel firing rates and at stoichiometric equivalence ratio

| Fuel firing rate (kW) | $\dot{Q}_{CH_4}$ (L/min) | $\dot{Q}_{O_2}$ (L/min) |
|---|---|---|
| 0.75 | 1.38 | 2.76 |
| 1.00 | 1.84 | 3.68 |
| 1.50 | 2.76 | 5.52 |
| 2.00 | 3.68 | 7.36 |
| 2.50 | 4.60 | 9.20 |
| 3.00 | 5.52 | 11.03 |

oxygen flowrates to achieve different equivalence ratios. The volumetric flow rates of the inlet gases are calculated at room conditions, at temperature and pressure of 21 °C and 1 atm., respectively. The exhaust gas temperatures as well as steady state flame visual images were measured/recorded at each set of the experiments. The fuel lean flammability limits were studied at different fuel power and different equivalence ratios by varying percentage of $CO_2$ in the oxidizer mixture. Tables 6.16 and 6.17 show the detailed experimental plans for studying the effect of fuel firing rates on the blow-out limit at different equivalence ratios. The lean stability limits were determined by starting stable flames at a fixed equivalence ratio (fixed mass flowrates of fuel and oxygen) while increasing the diluent ($CO_2$) flowrate up to the blow out limits.

The temperature distributions in the combustor were measured using an R-type thermocouple with temperature range of up to 1600 ± 0.6 °C. Three-dimensional temperature measurements were conducted near the outlet of the combustion chamber by inserting the thermocouple probe downstream the flame, from the exhaust of the combustion chamber, as illustrated in Fig. 6.53. The temperatures were measured horizontally, vertically and longitudinally at scheduled locations shown in Fig. 6.54. The temperatures were measured for a constant fuel-firing rate of 1.5 kW.

### 6.7.2 *Flame Shape*

The flame shapes are the visible images of the oxy-methane combustion flames. Two rectangular flame-sheets were observed anchored to the upper and lower breadths of the fuel $+CO_2$ inlet section and elongated along the longitudinal axis of the combustor as shown in Fig. 6.55. The black zone represents the portion of the combustor where there was no visible flame in the combustor, while the white zone attached to the fuel $+CO_2$ inlet represents the flame regime. For the pure methane combustion case, without the $CO_2$ addition (0% $CO_2$), the flame remains highly whitish glow indicating the luminous intensity of the light at all the lean equivalence ratios under study. At the three values of equivalence ratio studied in this laminar non-premixed combustion, the flame length increases with increase in the percentage of $CO_2$ in the mixture. That is in contrary to the effect of $CO_2$ addition in a swirl stabilized turbulent flows [142–144], where the flame length decreases and become more compact due to turbulent mixing with increase in percentage $CO_2$ in the mixture. The observed increase in length in laminar combustion studies is a result of $CO_2$ being supplied together within the fuel inlet. This increases the fuel side Reynolds number causing

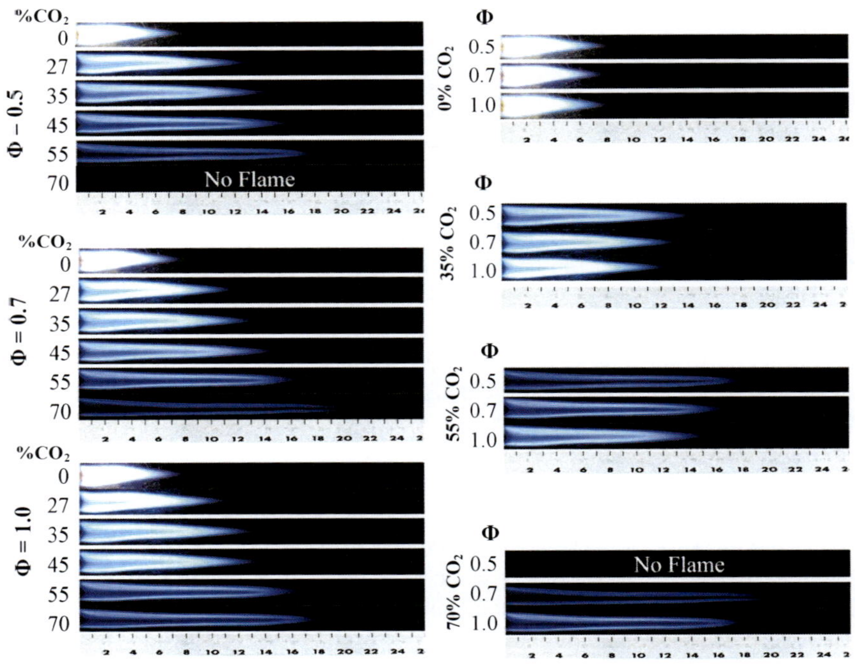


(a) Effect of $\%CO_2$ in the oxidizer mixture at three different equivalence ratios

(b) Effect of equivalence ratio at different $\%CO_2$ in the oxidizer mixture

**Fig. 6.55** Effects of operating conditions on the flame structure

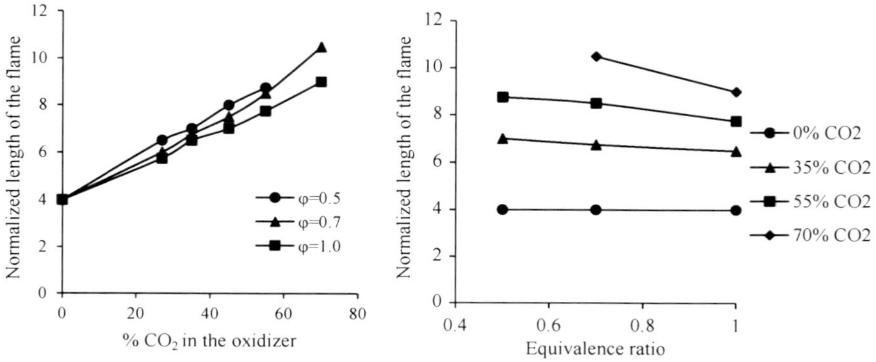


(a) % $CO_2$ in the oxidizer (b) Equivalence ratio

**Fig. 6.56** Effects of operating conditions on the normalized flame length ($l_{f,n}$ = flame length/fuel channel width)

a jet behavior with higher momentum that extend the mixing and combustion zones. This has been visualized in all cases as shown in Fig. 6.55a, with the case of highest percentage of $CO_2$ in the oxidizer mixture (70%) having the most elongated flame. As can be seen from Fig. 6.55, the flame is slightly asymmetrical. This behavior of the flame is most likely due to buoyancy effect, which is significant at lower flow rates across the combustor.

In Fig. 6.56a, the flame length has been normalized with the width of the rectangular cross-sectional area of the fuel $+CO_2$ inlet section, which represents the characteristic length of the flame that will conceptually allow the same mass and momentum flux as the actual length [145]. The effect of $CO_2$ addition elongates the normalized flame length to more than the double of its length without the $CO_2$ addition at all the equivalence ratios studied. Whereas there was significant flame length increase due to $CO_2$ addition, the flame length slightly decreases with increase in equivalence ratio when $CO_2$ was added as shown in Fig. 6.56b. This is due to better mixing of the species as well as decrease in excess oxygen that lowers the flame temperature. The flame extinguished at equivalence ratio of 0.5 with 70% $CO_2$ addition due to low fuel percentage in the mixture (4.7% by volume) to support the flame. Visually, the flame luminous intensity decreases with increase in $CO_2$ percentage in the mixture as shown in Fig. 6.55 at each equivalence ratio. The addition of $CO_2$ increases the convection and radiation heat loss from the reaction zone, which entails lower flame temperatures. Since the flame luminosity is directly correlated to the flame temperature, the cases with higher %$CO_2$ had lower luminous intensities due to lower flame temperatures. Other researchers have reported similar luminous behavior of the flame [144]. In addition, Fig. 6.56b shows that higher equivalence ratio results in higher the luminous intensity at a constant %$CO_2$. This is attributed to the increase in flame temperatures.

### 6.7.3 Combustion Temperature

The horizontal temperature distributions were measured across $z$-axis at five symmetric locations as shown in Fig. 6.54a. The results of the measurements are presented in Fig. 6.57a–c for three equivalence ratios at various percentages of $CO_2$ addition. The temperature decreases with increase in horizontal distance from center of the combustor outlet irrespective of the equivalence ratio and %$CO_2$ in the mixture, as shown in Fig. 6.57. This was anticipated due to radiation heat loss via the quartz glass plates at the two sides of the combustor to the cold ambient (at about 293 K). The variations are also symmetrical with the peak temperature at the center due to the symmetry of the inlets of the species. In all of the three cases, higher the temperature exhibit higher the percentage of inlet $CO_2$ the. Even though the pure oxy-combustion case (0% $CO_2$) was characterized by having the highest normalized flame temperature as evident in the highest luminous intensity, however the exhaust gases measured at the outlet of the combustor showed lower temperatures in comparison to the $CO_2$ added counterparts. This occurred due to flame radiation as the result of longer flames as discussed earlier. The measured temperatures were normalized by the non-dimensional flame length to cater for the effect of flame elongation on the actual temperature measurements. In all the three cases, the higher the percentage

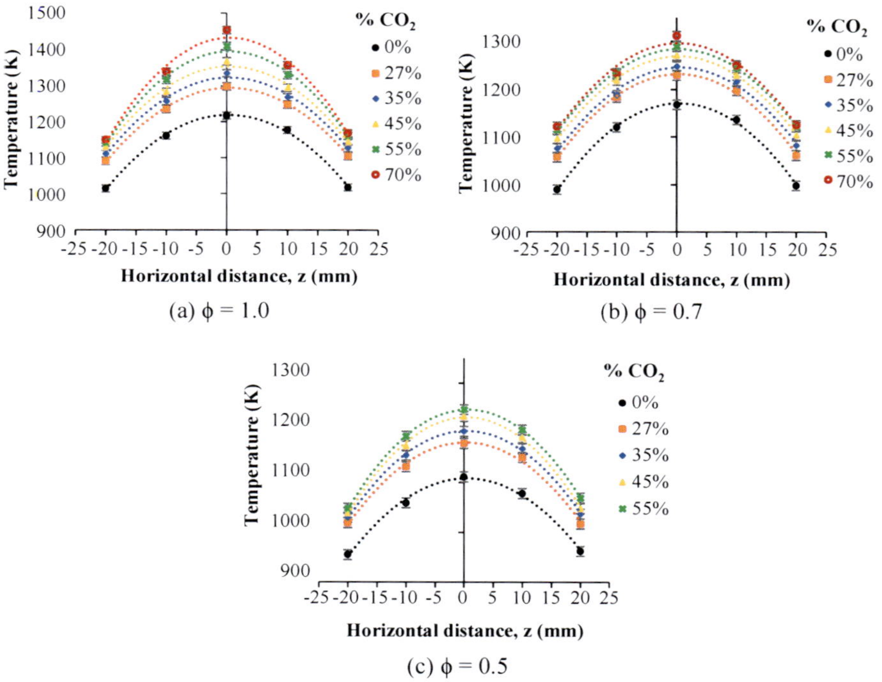


**Fig. 6.57** Normalized horizontal temperature variation at the outlet for fuel firing rate of 1.5 kW

of inlet $CO_2$ the lower the normalized flame temperature due to the thermal dilution effect characterized by $CO_2$ recirculation. The uncertainties indicated by the error bars were computed based on the standard deviation of the measured data. The higher the equivalence ratio the higher the exhaust gas temperatures were observed in all the cases despite decrease in the flame length. The higher flame temperatures at higher equivalence ratio overwhelmed the radiation effect due the slight increase in the flame lengths. This was more vividly clear for the case without $CO_2$ addition, where the flame length increase, due to decrease in equivalence ratio, was insignificant, as shown in Fig. 6.56b.

When temperature distributions were measured vertically across *y*-axis at the seven symmetric nodal locations, the results were presented in Fig. 6.58. Similar parabolic profiles were obtained as in the case of horizontal measurements for the stoichiometric oxy-methane combustion case (Fig. 6.57a). Whereas uniform temperature distributions were observed vertically at the center of the combustor outlet when $CO_2$ was added for fuel lean cases, as shown in Fig. 6.58b, c, parabolic profiles were obtained at stoichiometric mixtures. The maximum temperature occurred at the center irrespective of the percentage of $CO_2$ in the mixture. For the fuel lean conditions, the temperature profiles tend to be flattened at the center when the %$CO_2$ reached 35% or greater (Fig. 6.58b, c), and even approached an M-shape at 55%

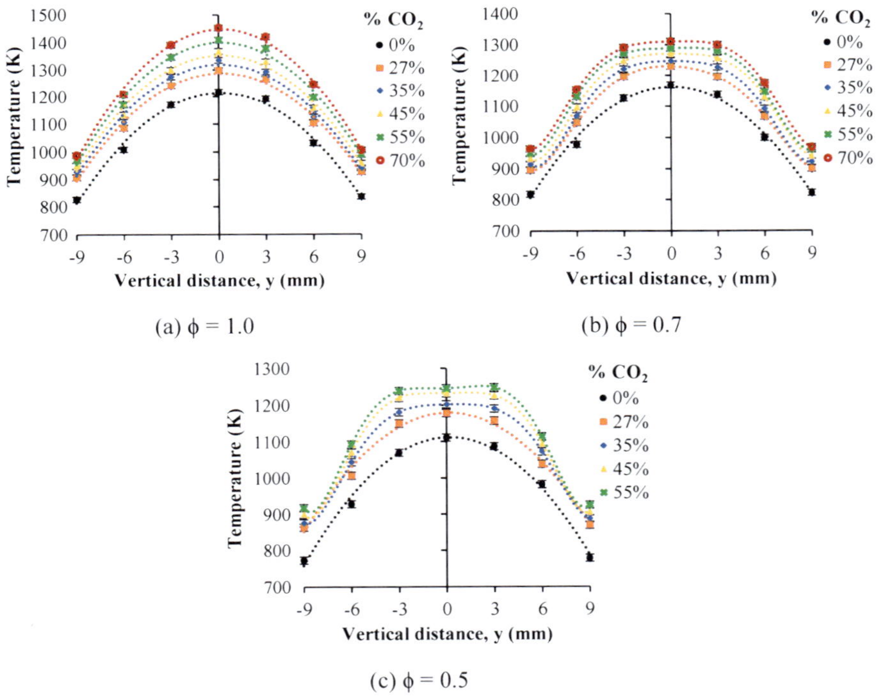


**Fig. 6.58** Normalized vertical temperature variation at the exhaust, for fuel firing rate of 1.5 kW

$CO_2$ and 0.5 equivalence ratio. This was due to upper and lower flame sheets being more distinctly separated at higher %$CO_2$ addition and longer flames as shown in Fig. 6.55. The higher the equivalence ratio the higher the temperatures as stated earlier. The vertical temperature profiles were fully parabolic during pure oxy-methane combustion as well as stoichiometric mixtures with $CO_2$ addition at all the three equivalence ratios under study. When the oxygen content was increased (fuel lean cases) while the percentage of $CO_2$ also added, the profiles becomes more uniformly distributed at the center, with more two distinct peaks at higher %$CO_2$.

The longitudinal temperature measurements were taken from the outlet of the combustor upstream along $x$-axis for the seven nodal earmarked points illustrated in Fig. 6.54b. The thermocouple's maximum temperature range limits the measuring distance upstream of the flame. The measured temperatures were normalized by the non-dimensional flame length and plotted in Fig. 6.59. All measurements elucidated linear temperature drops downstream of the flame zone along the $x$-axis (Fig. 6.57). The flame temperature behavior corroborates that of the luminous intensities of the flame images. The addition of $CO_2$ decreases the temperature due the decrease in flame temperatures as seen in the previous sections. Whereas, increasing the percentage of $CO_2$ added to the fuel stream decreases the temperature downstream of the combustor due to the thermal dilution effect of the cold $CO_2$. At the

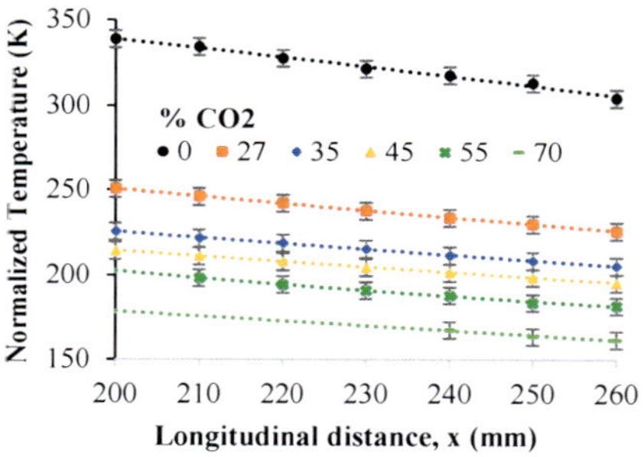


(a) ϕ = 1.0

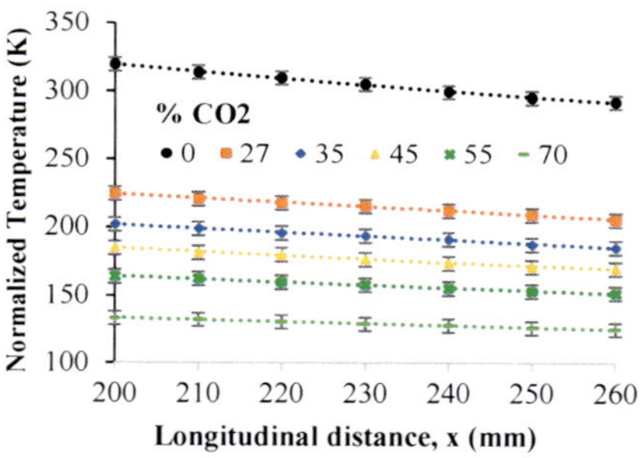


(b) ϕ = 0.7

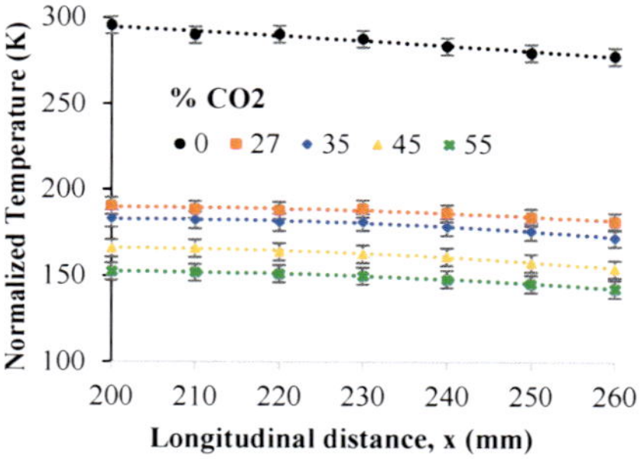


(c) ϕ = 0.5

**Fig. 6.59** Normalized longitudinal temperature variation towards the exhaust for fuel firing rate of 1.5 kW

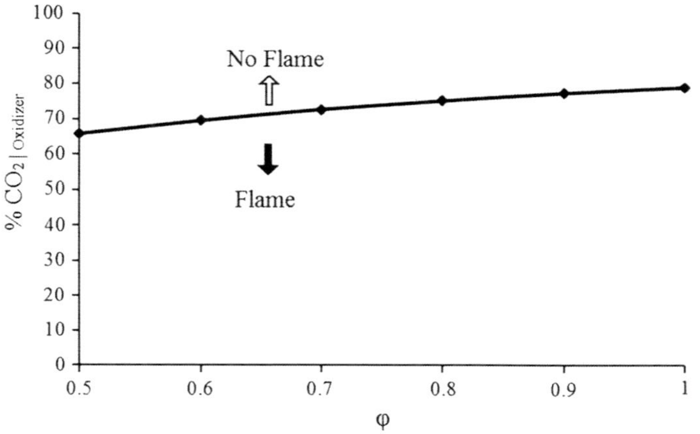


**Fig. 6.60** Effect of operating equivalence ratio on blow-out limits at fixed fuel firing rate of 1.5 kW

steady stoichiometric combustion (Fig. 6.59a) the temperature drops downstream of the combustor with an average slope of 2.4 K/mm irrespective of the % $CO_2$ in the mixture. As the amount of oxygen was increased to the too lean conditions the step temperature decrease reduces to 1.8 and 1.2 K/mm at equivalence ratios of 0.7 and 0.5 respectively, while still maintaining the fuel firing rate at 1.5 kW.

### 6.7.4 Lean Blowout Limits

The effect of equivalence ratio on the blow-out limits at fixed fuel firing rate of 1.5 kW is shown in Fig. 6.60 for a range of equivalence ratio, from 0.5 to 1.0. The results indicate increase in the LBO limit with the increase in the equivalence ratio. The flame blown out at about 65% $CO_2$ concentration in the oxidizer at the lowest operating equivalence ratio ($\varphi = 0.5$). This percentage increases to about 79% $CO_2$ at stoichiometric operating condition. A similar trend has been reported by Sayad et al. [146]. Decreasing the equivalence ratio lowers the reaction temperatures and consequently decreases the capacity to overcome the activation energy for reactions to occur, and hence decrease in flame speed. As reported by Kondo et al. [140], the flame blow-out occurs when the flow speed exceeds the flame speed. In addition, Chan et al. [147] concluded that for methane combustion, the laminar flame speed decreased with increasing $CO_2$ dilution because of decrease in the net rate of reaction due to low concentrations of reactants. Moreover, $CO_2$ addition acts as a heatsink, by lowering the combustion temperature and consequently decreasing the capacity to overcome the activation energy for reactions to occur [147]. Dirrenberger et al. [148] reported that for methane-air combustion, the laminar flame velocity is directly proportional to the equivalence ratio up to stoichiometric equivalence ratio. Hence the leaner the mixture the lower the flame speed and the higher the flow speed, which causes flame blow-out at the critical condition. The minimum obtained limit of $O_2$ percentage in the oxidizer to allow for a visible flame was found to be about 21% (79% $CO_2$) at stoichiometric equivalence ratio as shown in Fig. 6.60. These findings related to blow-out limits are not only limited to small size burner, but also can be generalized for any burner design. This can be attributed to the dependence of the blow-out limits on the radiative heat losses.

The combustor can be operated with visible flame within the flame zone. Similarly, the effect of fuel firing rates on blow-out limits has been obtained at different equivalence ratios as shown in Fig. 6.61. It was found that the percentage of $CO_2$ in the oxidizer at blow-out limit decreases with the decrease in equivalence ratio irrespective of the fuel firing rate. The fuel-firing rate was found to influence the concentration of $CO_2$ at the blow-out limit at fixed equivalence ratio. This can be seen in Fig. 6.62 where the increase in the fuel-firing rate at stoichiometric equivalence ratio decreases the percentage of $CO_2$ in the oxidizer to cause flame blow-out

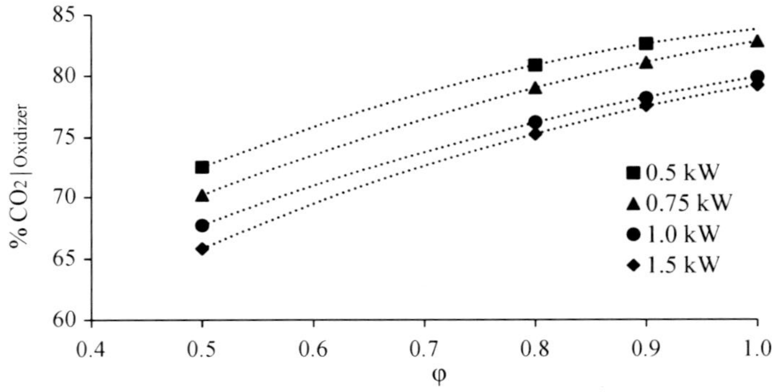


**Fig. 6.61** Effect of fuel firing rate on blow-out limits over a range of operating equivalence ratio

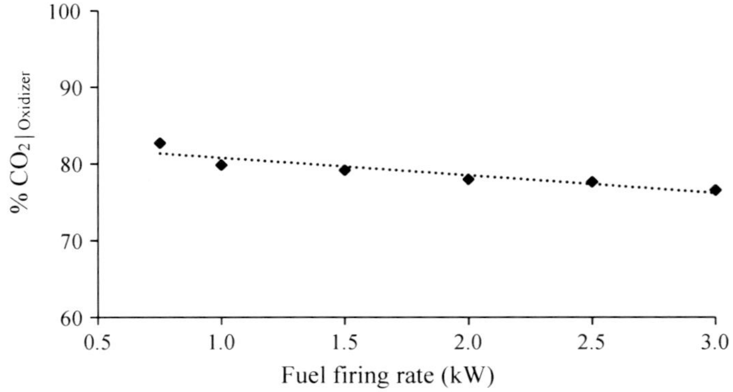


**Fig. 6.62** Effect of fuel firing rate on blow-out limits at stoichiometric operating equivalence ratio

due to attainment of the required flow speed that overwhelmed the increase in the flame speed.

## 6.8 Concluding Remarks

In this chapter, modern reactors handling oxy-combustion flames are introduced for clean energy production applications. The concepts of oxygen separation using polymeric and ceramic membranes are introduced. In addition, the procedure how the two kinds of membranes can be integrated together to improve the oxygen permeation flux for the combustion process is also introduced. This is followed by presenting detailed numerical and experimental investigations on membrane reactors that can utilize oxygen separation membranes to separate the required oxygen for the combustion process within the same unit. Design and performance of such reactors are introduced. In the first study discussed in this chapter, a numerical investigation is presented on the characteristics of air separation using polymeric membranes to generate oxygen-enriched air for clean combustion applications. An asymmetric hollow fiber polymeric membrane unit has been utilized. The results showed significant improvements in $O_2$ and $N_2$ permeation fluxes while increasing the sweep flow rate. Increasing the feed gas flow rate at zero sweep flow rate resulted in increase in $O_2$ permeation flux and reduction in $N_2$ permeation flux due to the changes in their partial pressure distributions in feed side associated with flow field modifications due to the increase in feed flow rate. Approximately linear relations between feed pressure and $O_2$ and $N_2$ permeation fluxes have been reported. Different polymer types including TR-PBO-co-PPL 28 and TR-PI-g-CD 450 have been considered to show the effect of membrane material on $O_2$ and $N_2$ permeation fluxes. Due to its high selectivity and lower permeability, the highest $O_2$ mass fraction at permeate outlet and the lowest permeation flow rate are obtained by the TR-PBO-co-PPL 28 membrane. Multi-stage gas separation has been investigated targeting higher $O_2$ concentrations at permeate outlet. Higher stage cut ratio is gained in the third stage as compared to preceding stages because of the higher permeation flow rate relative to the feed flow. The obtained $O_2$ mass concentrations at permeate outlets of first, second and third stages are 0.517, 0.8 and 0.914. Higher $O_2$ flux at higher outlet concentrations can be obtained by recycling the retentate flows of the third and second stages to the feed side in the first stage as those retentate flows have higher $O_2$ mass fractions than normal air introduced in the first stage.

In the second study discussed in this chapter, a numerical investigation is presented on the characteristics of air separation using polymeric membranes to be the first stage of developing a hybrid polymeric/ceramic membrane unit. Ceramic oxygen transport reactor (OTR) has been integrated numerically with the polymeric membrane to form the hybrid polymeric/ceramic membrane unit for oxy-fuel combustion. A tubular oxygen transport membrane (OTM) has been utilized to perform the numerical simulations under oxy-fuel combustion conditions. The calculated outputs of the polymeric membrane have been utilized as inputs to the OTR unit. The results indicated a significant rise in $O_2$ permeation flux and OTM temperature as a result of increasing the sweep flow rate. Increasing the feed flow rate resulted in decrease in $O_2$ permeation flux and membrane temperature. Further increase in the feed flow rate resulted in incomplete combustion and sharp reduction in the total permeation. The effect of the fed $O_2$ mass fraction on the OTR power output was investigated. Higher $O_2$ fraction in the $O_2$-enriched air stream resulted in higher power of the OTR. The main advantage of the using the hybrid system over the normal ambient air fed OTR is realized in terms of higher power output due to the associated reduction in the preheating load. A proposed hybrid system design is introduced; 4 shells of the polymeric membranes have been connected to one OTM shell to produce 50.46 W power output at stoichiometric conditions. One polymer membrane shell consists of 325 fibers while the single OTM shell contains 29 OTM's.

In the third study discussed in this chapter, an investigation is presented on using a low-power porous-plate reactor to study experimentally and numerically the flow field and oxy-fuel combustion characteristics over a range of operating equivalence ratio. The reactor is used to mimic the processes of oxygen permeation and oxy-combustion inside high-temperature membrane reactors (HTMRs). Numerical simulations supported by experiments have been conducted to investigate the flow behavior while oxygen permeation through the porous-plates under cold/non-reactive flow conditions. Enhanced mixing zone was created near the trailing edge of the porous plate affected by oxygen permeation. It was found that the higher the set (global) equivalence ratio, the shorter the axial position where it is achieved, and the lower the mixture fraction. Based on non-reactive flow analyses, the set equivalence ratio was achieved locally downstream of the porous plates, which predicts the flame to also be located downstream of the porous plates in the reactive flow. The combustion simulations provided more clear insight of the oxy-combustion characteristics and the location of the flame inside the porous-plate reactor. The cold flow predictions for locations of enhanced mixing zones and flame downstream of the porous plates were confirmed in the reactive flow from the distributions of species and temperature within the reactor. The creation of enhanced flow mixing zone beside the porous plate enhanced flame anchoring and stabilization. Location of the flame downstream of the porous plates helps protecting the membrane against high flame temperature in case of HTMR operation. The predicted temperature beside the porous plate lies within the operating limits of ceramic membranes. This proves the applicability of HTMRs. Also, a hot zone is created through the entire length of the porous plate, which helps keeping high oxygen flux of the membrane when the results are projected on HTMR operation. These results support the operation of the reactor under stoichiometric condition from performance points of view, especially when the case is projected on the HTMR operation.

In the last presented study in this chapter, the combustion characteristics and lean blow-out limits of non-premixed oxy-methane flames in a two face-to-face porous plate combustor have been investigated experimentally. The reactor is used to mimic the processes of oxygen permeation and oxy-combustion inside high-temperature membrane reactor (HTMR). Visual flame studies have been conducted to determine the flame shapes, length and luminous intensities at the slated operating conditions.

It was found that addition of $CO_2$ elongates the normalized flame length. The flame luminous intensity is decreasing with increase in $CO_2$ percentage in the mixture at all the equivalence ratios due to lower flame temperatures as the result of increase in the convection and radiation heat loss from the reaction zone. The temperature distributions measured at the combustor outlet are radially parabolic and longitudinally flattened at all stated conditions. The concentration of $CO_2$ in the oxidizer at lean blow-out limit is proportional to equivalence ratio with maximum percentage of 79% up to stoichiometric condition. Increasing the fuel-firing rate at stoichiometric equivalence ratio resulted in decreasing the percentage of $CO_2$ in the oxidizer to instigate flame blow-out.

**Acknowledgements** The authors wish to acknowledge the support received from King Fahd University of Petroleum & Minerals under Grant number BW191002 for the preparation of this book chapter.

## References

1. Habib MA, Nemitallah MA, Ben-Mansour R (2013) Recent development in oxy-combustion technology and its applications to gas turbine combustors and ITM reactors. Energy Fuels 27:2–19
2. Ben-Mansour R, Habib MA, Badr HM, Azharuddin IF, Nemitallah MA (2012) Characteristics of oxy-fuel combustion in an oxygen transport reactor. Energy Fuels 26:4599–4606
3. Shah K, Moghtaderi B, Zanganeh J, Wall T (2013) Integration options for novel chemical looping air separation (ICLAS) process for oxygen production in oxy-fuel coal fired power plants. Fuel 107:356–370
4. Moghtaderi B, Zanganeh J, Shah K, Wu H (2012) Application of concrete and demolition waste as $CO_2$ sorbent in chemical looping gasification of biomass. Energy Fuels 26:2046–2057
5. Shah K, Moghtaderi B, Wall T (2013) Effect of flue gas impurities on the performance of a chemical looping based air separation process for oxy-fuel combustion. Fuel 103:932–942
6. Buhre BJP, Elliott IK, Sheng CD, Gupta RP, Wall TF (2005) Oxy-fuel combustion technology for coal-fired power generation. Progress Energy Combust Sci 31:283–307
7. Lin YS (2001) Microporous and dense inorganic membranes: current status and prospective. Sep Purif Technol 25:39–55
8. Wang Z, Yang N, Meng B, Tan X, Li K (2009) Preparation and oxygen permeation properties in hollow fiber membranes. Ind Eng Chem Res 48:510–516
9. Tan X, Wang Z, Liu H, Liu S (2009) Enhancement of oxygen permeation through $La_{0.6}Sr_{0.4}Co_{0.2}Fe_{0.8}O_{3-\delta}$ hollow fiber membranes by surface modifications. J Membr Sci 324:128
10. Kim J, Lin YS (2000) Synthesis and oxygen permeation properties of thin YSZ/Pd composite membranes. Aiche 46:1521
11. Chen CC, Prasad R, Gottzmann CF (1999) Solid electrolyte membrane with porous catalytically-enhancing constituents, assigned to praxair technology. US Patent 5:938822
12. Schwartz M, White JH, Sammels A (2000) Solid state oxygen anion and electron mediating membrane and catalytic membrane reactors containing them, assigned to elytron research inc. US Patent 6:033632
13. Zebao R, Yongdan L, Lin YS (2009) Analysis of oxygen permeation through dense ceramic membranes with chemical reactions of finite rate. Chem Eng Sci 64:172–179

14. Nemitallah MA, Habib MA, Ben-Mansour R (2012) Investigations of oxy-fuel combustion and oxygen permeation in an IM reactor using a two-step oxy-combustion reaction kinetics model. J Membr Sci 432:1–12
15. Nemitallah MA, Habib MA, Ben-Mansour R, Ghoniem AF (2014) Design of an ion transport membrane reactor for gas turbine combustion application. J Membr Sci 450:60–71
16. Tan X, Li K (2007) Oxygen production using dense ceramic hollow fiber membrane modules with different operating modes. Aiche 53:838
17. Kaiser A, Foghmoes S, Chatzichristodoulou C, Sogaard M, Glasscock JA, Frandsen HL, Hendriksen PV (2011) Evaluation of thin film ceria membranes for syngas membrane reactors-preparation, characterization and testing. J Membr Sci 378:51–60
18. Ben-Mansour R, Nemitallah MA, Habib MA (2013) Numerical investigations of oxygen permeation and methane-oxy-combustion in a stagnation flow ion transport membrane reactor. Energy 1:1–11
19. Akin F, Lin YS (2004) Oxygen permeation through oxygen ionic or mixed-conducting ceramic membranes with chemical reactions. J Membr Sci 231:133–146
20. Mancini ND, Mitsos A (2011) Ion transport membrane reactors for oxy-combustion-part II: analysis and comparison of alternatives. Energy 36:4721–4739
21. Ahmed P, Habib MA, Ben-Mansour R, Kirchen P, Ghoniem AF (2014) CFD (computational fluid dynamics) analysis of a novel reactor design using ion transport membranes for oxy-fuel combustion. Energy 77:932–944
22. Gunasekaran S, Mancini ND, Mitsos A (2014) Optimal design and operation of membrane-based oxy-combustion power plants. Energy 70:338–354
23. Kotowicz J, Michalski S (2014) Efficiency analysis of a hard-coal-fired supercritical power plant with a four-end high-temperature membrane for air separation. Energy 64:109–119
24. Habib MA, Nemitallah MA (2015) Design of an ion transport membrane reactor for application in fire tube boilers. Energy. https://doi.org/10.1016/j.energy.2015.01.029
25. Abdulhameed MA, Othman MHD, Ismail AF, Matsuura T, Harun Z, Rahman MA, Puteh MH, Jaafar J, Rezaei M, Hubadillah SK (2017) Carbon dioxide capture using a superhydrophobic ceramic hollow fiber membrane for gas-liquid contacting process. J Clean Prod 140:1731–1738
26. Khalilpour R, Mumford K, Zhai H, Abbas A, Stevens G, Rubin ES (2015) Membrane-based carbon capture from flue gas: a review. J Clean Prod 103:286–300
27. Zhang Z (2016) Comparisons of various absorbent effects on carbon dioxide capture in membrane gas absorption (MGA) process. J Nat Gas Sci Eng 31:589–595
28. Barba FC, Sánchez GM, Seguí BS, Darabkhani HG, Anthony EJ (2016) A technical evaluation, performance analysis and risk assessment of multiple novel oxy-turbine power cycles with complete $CO_2$ capture. J Clean Prod 133:971–985
29. Anheden M, Yan J, De Smedt G (2005) Denitrogenation or oxyfuel concepts. Oil Gas Sci Technol 60(3):485–495
30. Smith AR, Klosek J (2011) A review of air separation technologies and their integration with energy conversion processes. Fuel Process Technol 70(2):115–134
31. Augelletti R, Conti M, Annesini MC (2017) Pressure swing adsorption for biogas upgrading. A new process configuration for the separation of biomethane and carbon dioxide. J Clean Prod 140:1390–1398
32. Matteucci S, Yampolskii Y, Freeman BD, Pinnau I (2006) Transport of gases and vapors in glassy and rubbery polymers. In: Freeman B, Yampolskii Y, Pinnau I (eds) Materials science of membranes for gas and vapor separation, 1st edn. Wiley, pp 1–42
33. Park JY, Paul DR (1997) Correlation and prediction of gas permeability in glassy polymer membrane materials via a modified free volume based group contribution method. J Membr Sci 125:23–39
34. Maeda Y, Paul DR (1987) Effect of antiplasticization on selectivity and productivity of gas separation membranes. J Membr Sci 30:1–9
35. Robeson LM (2008) The upper bound revisited. J Membr Sci 320:390–400

36. Park HB, Jung CH, Lee YM, Hill AJ, Pas SJ, Mudie ST, Van Wagner E, Freeman BD, Cookson DJ (2007) Polymers with cavities tuned for fast selective transport of small molecules and ions 318(5848):254–258
37. Sanders DF, Smith ZP, Guo R, Robeson LM, McGrath JE, Paul DR, Freeman BD (2013) Energy-efficient polymeric gas separation membranes for a sustainable future: a review. Polymer 54(18):4729–4761
38. Woo KT, Dong G, Lee J, Kim JS, Do YS, Lee WH, Lee HS, Lee YM (2016) Ternary mixed-gas separation for flue gas $CO_2$ capture using high performance thermally rearranged (TR) hollow fiber membranes. J Membr Sci 510:472–480
39. Liu Q, Paul DR, Freeman BD (2016) Gas permeation and mechanical properties of thermally rearranged (TR) copolyimides. Polymer 82:378–391
40. Jo HJ, Soo CY, Dong G, Do YS, Wang HH, Lee MJ, Quay JR, Murphy MK, Lee YM (2015) Thermally rearranged poly (benzoxazole-co-imide) membranes with superior mechanical strength for gas separation obtained by tuning chain rigidity. Macromolecules 48(7):2194–2202
41. Calle M, Lozano AE, Lee YM (2012) Formation of thermally rearranged (TR) polybenzoxazoles: effect of synthesis routes and polymer form. Eur Polymer J 48(7):1313–1322
42. Jiang X, Kumar A (2005) Performance of silicone-coated polymeric membrane in separation of hydrocarbons and nitrogen mixtures. J Membr Sci 254:179–188
43. Coombe HS, Nieh S (2007) Polymer membrane air separation performance for portable oxygen enriched combustion applications. Energy Convers Manag 48(5):1499–1505
44. Alsari A, Kruczek B, Matsuura T (2007) Effect of pressure and membrane thickness on the permeability of gases in dense polyphenylene oxide (PPO) membranes: thermodynamic interpretation. Sep Sci Technol 42(10):2143–2155
45. Feng X, Ivory J, Rajan VS (1999) Air separation by integrally asymmetric hollow-fiber membrane. AIChE J 45(10):2142–2152
46. Dixon-Garrett SV, Nagai K, Freeman BD (2000) Sorption, Diffusion, and Permeation of Ethylbenzene in Poly(1-trimethylsilyl-1-propyne). J Polym Sci 38(8):1078–1089
47. Ahmad F, Lau KK, Shariff AM, Murshid G (2012) Process simulation and optimal design of membrane separation system for $CO_2$ capture from natural gas. Comput Chem Eng 36:119–128
48. Zhao L, Riensche E, Menzer R, Blum L, Stolten D (2008) A parametric study of $CO_2/N_2$ gas separation membrane processes for post-combustion capture. J Membr Sci 325:284–294
49. Kawachale N, Kumar A, Kirpalani D (2008) Numerical investigation of hydrocarbon enrichment of process gas mixtures by permeation through polymeric membranes. Chem Eng Technol 31:58–65
50. Binns M, Lee S, Yeo YK, Lee JH, Moon JH, Yeo JG, Kim JK (2016) Strategies for the simulation of multi-component hollow fiber multi-stage membrane gas separation systems. J Membr Sci 497:458–471
51. Dong G, Woo KT, Kim J, Kim JS, Lee YM (2015) Simulation and feasibility study of using thermally rearranged polymeric hollow fiber membranes for various industrial gas separation applications. J Membr Sci 496:229–241
52. Zhang Z, Yan Y, Wood DA, Zhang W, Li L, Zhang L, Van der Bruggen B (2015) Influence of the membrane module geometry on $SO_2$ removal: a numerical study. Ind Eng Chem Res 54:11619–11627
53. Choi W, Ingole PG, Li H, Kim J-H, Lee H-K, Baek I-H (2016) Preparation of facilitated transport hollow fiber membrane for gas separation using cobalt tetraphenylporphyrin complex as a coating material. J Clean Prod 133:1008–1016
54. ANSYS® Academic Research, Release 12.1
55. Nemitallah MA, Habib MA, Ben-Mansour R (2013) Investigations of oxy-fuel combustion and oxygen permeation in an ITM reactor using a two-step oxy-combustion reaction kinetics model. J Membr Sci 432:1–12
56. Habib MA, Nemitallah MA (2015) Design of an ion transport membrane reactor for application in fire tube boilers. Energy 81:787–801

57. Patankar SV (1980) Numerical heat transfer and fluid flow. Hemisphere Publishing Corporation, Washington D.C
58. Li W, Liu JJ, Chen CS (2009) Hollow fiber membrane of yttrium-stabilized zirconia and strontium-doped lanthanum manganite dual-phase composite for oxygen separation. J Membr Sci 340:266–271
59. Choi JI, Jung CH, Han SH, Park HB, Lee YM (2010) Thermally rearranged (TR) poly (benzoxazole-co-pyrrolone) membranes tuned for high gas permeability and selectivity. J Membr Sci 349:358–368
60. Xiao Y, Chung TS (2011) Grafting thermally labile molecules on cross-linkable polyimide to design membrane materials for natural gas purification and $CO_2$ capture. Energy Environ Sci 4:201–208
61. Katoh T, Tokumura M, Yoshikawa H, Kawase Y (2011) Simulation of multicomponent gas separation by hollow-fiber membrane module: nonideal mixing flows in permeate and residue sides using the tanks-in-series model. Sep Purif Technol 76(3):362–372
62. Wall TF (2007) Combustion processes carbon capture. Proc Combust Inst 31:31–47
63. Buhre BJP, Elliott LK, Sheng CD, Gupta RP, Wall TF (2005) Oxy-fuel combustion technology for coal-fired power generation. Prog Energy Combust Sci 31:283–307
64. Haslbeck J, Capicotto P, Juehn N, Lewis E, Rutkowski M, Woods M et al (2007) In: bituminous coal to electricity. Washington D.C., 1:DOE/NETL-1291
65. Balachandran U, Kleefisch MS, Kobylinski TP, Morissette SL, Pei S (1997) Oxygen ion-conducting dense ceramic membranes. US Patent, 5, 639-437
66. Schwartz M, White JH, Sammels AF (2000) Solid state oxygen anion and electron mediating membrane and catalytic membrane reactors containing them. US Patent, 6, 033-632
67. Chen CC, Prasad R, Gottzmann CF (1999) Solid electrolyte membrane with porous catalytically-enhancing constituents (assigned to praxair technology). US Patent, 5, 938-822
68. Kim J, Lin YS (2000) Synthesis and oxygen permeation properties of thin YSZ/Pd composite membranes. AIChE J 46:1521
69. Xu SJ, Thomson WJ (1999) Oxygen permeation rates through ion-conducting perovskite Membranes. Chem Eng Sci 54:3839–3850
70. Habib MA, Ben-Mansour R, Nemitallah MA (2013) Modeling of oxygen permeation through a LSCF ion transport membrane. Comput Fluids 76:1–10
71. Hong J, Kirchen P, Ghoniem AF (2012) Numerical simulation of ion transport membrane reactors: oxygen permeation and transport and fuel conversion. J Membr Sci 71–85
72. Ben-Mansour R, Nemitallah MA, Habib MA (2013) Numerical investigation of oxygen permeation and methane oxy-combustion in a stagnation flow ion transport membrane reactor. Energy 54:322–332
73. Hong J, Kirchen P, Ghoniem AF (2013) Interactions between oxygen permeation and homogeneous-phase fuel conversion on the sweep side of an ion transport membrane. J Membr Sci 428:309–322
74. Hunt A, Dimitrakopoulos G, Kirchen P, Ghoniem AF (2014) Measuring the oxygen profile and permeation flux across an ion transport ($La_{0.9}Ca_{0.1}FeO_{3-\delta}$) membrane and the development and validation of a multi-step surface exchange model. J Membr Sci 468:62–72
75. Kirchen P, Apo DJ, Hunt A, Ghoniem AF (2013) A novel ion transport membrane reactor for fundamental investigations of oxygen permeation and oxy-combustion under reactive flow conditions. Proc Combust Inst 34:3463–3470
76. Park HB, Han SH, Jung CH, Lee YM, Hill AJ (2010) Thermally rearranged (TR) polymer membranes for $CO_2$ separation. J Membr Sci 359:11–24
77. Calle M, Lozano AE, Lee YM (2012) Formation of thermally rearranged (TR) polybenzoxazoles: effect of synthesis routes and polymer form. Eur Polym J 48(7):1313–1322
78. Gambini M, Vellini M (2012) Oxygen transport membranes for ultra-supercritical (USC) power plants with very low $CO_2$ emissions. J Eng Gas Turbines Power 134(8):081801
79. Xu SJ, Thomson WJ (1999) Oxygen permeation rates through ion-conducting perovskite membranes. Chem Eng Sci 54(17):3839–3850

80. Habib MA, Ahmed P, Ben-Mansour R, Badr HM, Kirchen P, Ghoniem AF (2013) Modeling of a combined ion transport and porous membrane reactor for oxy-combustion. J Membr Sci 446:230–243
81. Ghoniem AF, Mitsos A, Shao-Horn Y, Habib MA, Mezghani K, Ben-Mansour R (2015) Integrated polymeric-ceramic membrane based oxy-fuel combustor. U.S. Patent 9,004,909
82. Yantovski E, Gorski J, Smyth B, Ten Elshof J (2004) Zero-emission fuel-fired power plants with ion transport membrane. Energy 29(12):2077–2088
83. Hong J, Kirchen P, Ghoniem AF (2013) Laminar oxy-fuel diffusion flame supported by an oxygen-permeable-ion-transport membrane. Combust Flame 160(3):704–717
84. Simpson AP, Simon AJ (2007) Second law comparison of oxy-fuel combustion and post-combustion carbon dioxide separation. Energy Convers Manag 48(11):3034–3045
85. Burdyny T, Struchtrup H (2010) Hybrid membrane/cryogenic separation of oxygen from air for use in the oxy-fuel process. Energy 35(5):1884–1897
86. Habib MA, Nemitallah MA, Afaneh D, Mezghani K (2017) Characteristic of air separation in hollow-fiber polymeric membrane for oxygen-enriched air clean combustion applications. J Clean Prod 143:960–972
87. Matteucci S, Yampolskii Y, Freeman BD, Pinnau I (2006) Transport of gases and vapors in glassy and rubbery polymers. Materials science of membranes for gas and vapor separation, John Wiley and Sons Ltd, the Atrium, Southern gate, Chichester, West Sussex PO19 8SQ, England
88. Tan X, Li K (2002) Modeling of air separation in a LSCF hollow-fiber membrane module. AIChE J 48(7):1469–1477
89. ANSYS Academic Research, Release 16.1. ANSYS Inc. April 2015. Southpointe. 2600 ANSYS Drive, Canonsburg, PA 15317
90. Xiao Y, Chung TS (2011) Grafting thermally labile molecules on cross-linkable polyimide to design membrane materials for natural gas purification and $CO_2$ capture. Energy Environ Sci 201–208
91. Figueroa JD, Fout T, Plasynski S, McIlvried H, Srivastava RD (2008) Advances in $CO_2$ capture technology—the US Department of Energy's Carbon Sequestration Program. Int J Greenhouse Gas Control 2(1):9–20
92. Marland G, Boden TA, Andres RJ, Brenkert AL, Johnston CA (2007) Global, regional, and national fossil fuel $CO_2$ emissions. Trends Compendium Data Glob Change 37831–6335
93. Agency IE (2016) World energy statistics. IEA/OECD
94. Agency IE (2016) World energy outlook. IEA/OECD
95. Cebrucean D, Cebrucean V, Ionel I (2014) $CO_2$ capture and storage from fossil fuel power plants. Energy Procedia 63:18–26
96. Cuéllar-Franca RM, Azapagic A (2015) Carbon capture, storage and utilization technologies: a critical analysis and comparison of their life cycle environmental impacts. J CO2 Utilization 9:82–102
97. Sreenivasulu B, Gayatri DV, Sreedhar I, Raghavan KV (2015) A journey into the process and engineering aspects of carbon capture technologies. Renew Sustain Energy Rev 41:1324–1350
98. Habib MA, Badr HM, Ahmed SF, Ben-Mansour R, Mezghani K, Imashuku S, la O' GJ, Shao-Horn Y, Mancini ND, Mitsos A, Kirchen P (2011) A review of recent developments in carbon capture utilizing oxy-fuel combustion in conventional and ion transport membrane systems. Int J Energy Res 35(9):741–764
99. Kutne P, Kapadia BK, Meier W, Aigner M (2011) Experimental analysis of the combustion behavior of oxyfuel flames in a gas turbine model combustor. Proc Combust Inst 33(2):3383–3390
100. Runyon J, Marsh R, Valera-Medina A, Giles A, Morris S, Pugh D, Sevcenco Y, Bowen P (2015) Methane-oxygen flame stability in a generic premixed gas turbine swirl combustor at varying thermal power and pressure. In: ASME Turbo Expo 2015: turbine technical conference and exposition 2015 June 15. American Society of Mechanical Engineers, pp V04BT04A043–V04BT04A043

101. Xie Y, Wang J, Zhang M, Gong J, Jin W, Huang Z (2013) Experimental and numerical study on laminar flame characteristics of methane oxy-fuel mixtures highly diluted with $CO_2$. Energy Fuels 27(10):6231–6237
102. Andersson K, Johnsson F (2006) Process evaluation of an 865 MWe lignite fired $O_2/CO_2$ power plant. Energy Convers Manag 47:3487–3498
103. Habib MA, Salaudeen SA, Nemitallah MA, Ben-Mansour R, Mokheimer EMA (2016) Numerical investigation of syngas oxy-combustion inside a LSCF-6428 oxygen transport membrane reactor. Energy 2016(96):654–665
104. Meng X, Bi X, Meng B, Yang N, Tan X, Liu L, Liu S (2016) $H_2/CH_4/CO_2$-tolerant properties of SrCo0.8Fe0.1Ga0.1O3-δ hollow fiber membrane reactors for methane partial oxidation to syngas. Fuel Process Technol 161:265–272
105. Kirchen P, Apo DJ, Hunt A, Ghoniem AF (2013) A novel ion transport membrane reactor for fundamental investigations of oxygen permeation and oxy-combustion under reactive flow conditions. Proc Combust Inst 34(2):3463–3470
106. Habib MA, Tahir F, Nemitallah MA, Ahmed WH, Badr HM (2015) Experimental and numerical analysis of oxy-fuel combustion in a porous plate reactor. Int J Energy Res 39(9):1229–1240
107. Ko GH, Hamins A, Bundy M, Johnsson EL, Kim SC, Lenhert DB (2009) Mixture fraction analysis of combustion products in the upper layer of reduced-scale compartment fires. Combust Flame 156(2):467–476
108. McDermott R, McGrattan K, Hostikka S (2008) Fire dynamics simulator (version 5) technical reference guide. NIST Spec Publ 1018:5
109. Dharavath M, Manna P, Chakraborty D (2015) Numerical exploration of mixing and combustion in ethylene fueled scramjet combustor. Acta Astronaut 117:305–318
110. Han S, Kim J, Bae C (2014) Effect of air–fuel mixing quality on characteristics of conventional and low temperature diesel combustion. Appl Energy 119:454–466
111. Raffel M, Willert CE, Scarano F, Kähler CJ, Wereley ST, Kompenhans J (2018, April 3) Particle image velocimetry: a practical guide. Springer
112. Research V. Phantom (PCC) Camera control help—AMETEK material analysis division. Software Revision: 2.14.727.0 2012
113. Dynamics D (2012) Dynamic studio manual—user's guide. Dantec Dynamics A/S, 121, Tonsbakken 18, DK-2740
114. Lu L, Sick V (2013) High-speed particle image velocimetry near surfaces. J Visualized Exp JoVE 76
115. Sick V (2013) High speed imaging in fundamental and applied combustion research. Proc Combust Inst 34(2):3509–3530
116. Fluent (2013) ANSYS 15.0 fluent theory guide
117. Pitsch H (2015) Combustion theory and applications in CFD. Lecture note, Tsinghua-Princeton Summer School on Combustion.Tsinghua University, Beijing, China
118. Peters N (2010) Combustion theory. CEFRC Summer School, Lecture note, Princeton University, USA
119. Sivathanu Y, Faeth G (1990) Generalized state relationships for scalar properties in nonpremixed hydrocarbon/air flames. Combust Flame 82(2):211–230
120. Ghenai C (2010) Combustion of syngas fuel in gas turbine can combustor. Adv Mech Eng 2:342–357
121. Andersen J, Rasmussen CL, Giselsson T, Glarborg P (2009) Global combustion mechanisms for use in CFD modeling under oxy-fuel conditions. Energy Fuels 23:1379–1389
122. Nemitallah MA, Habib MA (2013) Experimental and numerical investigations of an atmospheric diffusion oxy-combustion flame in a gas turbine model combustor. Appl Energy 111:401–415
123. Chui E, Raithby G (1993) Computation of radiant heat transfer on a nonorthogonal mesh using the finite-volume method. Numer Heat Transf 23(3):269–288
124. Habib MA, Ben-Mansour R, Badr HM, Ahmed SF, Ghoniem AF (2012) Computational fluid dynamic simulation of oxyfuel combustion in gas-fired water tube boilers. Comput Fluids 56:152–165

125. Farooqui AE, Habib MA, Badr HM, Ben-Mansour R (2013) Modeling of ion transport reactor for oxy-fuel combustion. Int J Energy Res 37(11):1265–1279
126. Chen WY (2012) Handbook of climate change mitigation. Springer
127. Burdyny T, Struchtrup H (2010) Hybrid membrane/cryogenic separation of oxygen from air for use in the oxy-fuel process. Energy 35:1884–1897
128. Banaszkiewicz T, Chorowski M, Gizicki W (2014) Comparative analysis of oxygen production for oxy-combustion application. Energy Procedia 51:127–134
129. Simpson AP, Simon A (2007) Second law comparison of oxy-fuel combustion and post-combustion carbon dioxide separation. Energy Convers Manag 48:3034–3045
130. Kakaras E, Koumanakos A, Doukelis A, Giannakopoulos D, Vorrias I (2007) Oxyfuel boiler design in a lignite-fired power plant. Fuel 86:2144–2150
131. Athayde DD, Souza DF, Silva AM, Vasconcelos D, Nunes EH, da Costa JCD, Vasconcelos WL (2016) Review of perovskite ceramic synthesis and membrane preparation methods. Ceram Int 42:6555–6571
132. Murali RS, Sankarshana T, Sridhar S (2013) Air separation by polymer-based membrane technology. Sep Purif Rev 42:130–186
133. Habib MA, Tahir F, Nemitallah MA, Ahmed WH, Badr HM (2015) Experimental and numerical analysis of oxy-fuel combustion in a porous plate reactor. Int J Energy Res 39:1229–1240
134. Moore NJ, Kribs J, Lyons KM (2011) Investigation of Jet-Flame Blowout with Lean-Limit Considerations. Flow, Turbulence and Combustion 87:525–536
135. Moore N, McCraw J, Lyons K (2008) Observations on jet-flame blow-out. J Combust 2008
136. Karbasi M, Wierzba I (1998) Prediction and validation of blowout limits of co-flowing jet diffusion flames—effect of dilution. Trans Am Soc Mech Eng J Energy Resour Technol 120:167–171
137. Broadwell JE, Dahm WJ, Mungal MG (1985) Blowout of turbulent diffusion flames. In Symposium (international) on combustion, pp 303–310
138. Dahm W, Mayman A (1990) Blowout limits of turbulent jet diffusion flames for arbitrary source conditions. AIAA J 28:1157–1162
139. Dowling DR, Dimotakis PE (1990) Similarity of the concentration field of gas-phase turbulent jets. J Fluid Mech 218:109–141
140. Kondo S, Takizawa K, Takahashi A, Tokuhashi K (2006) Extended Le Chatelier's formula and nitrogen dilution effect on the flammability limits. Fire Saf J 41:406–417
141. Noble DR, Zhang Q, Shareef A, Tootle J, Meyers A, Lieuwen T (2006) ASME Paper No. GT2006-90470
142. Habib MA, Rashwan SS, Nemitallah MA, Abdelhafez A (2017) Stability maps of non-premixed methane flames in different oxidizing environments of a gas turbine model combustor. Appl Energy 189:177–186
143. Aliyu M, Nemitallah MA, Said SA, Habib MA (2016) Characteristics of H2-enriched $CH_4O_2$ diffusion flames in a swirl-stabilized gas turbine combustor: Experimental and numerical study. Int J Hydrogen Energy 41(44):20418–20432
144. Sanusi YS, Mokheimer EM, Shakeel MR, Abubakar Z, Habib MA (2017) Oxy-combustion of hydrogen-enriched methane: experimental measurements and analysis. Energy Fuels 31:2007–2016
145. Oh J, Noh D (2013) Lifted flame behavior of a non-premixed oxy-methane jet in a lab-scale slot burner. Fuel 103:862–868
146. Sayad P, Schönborn A, Clerini D, Klingmann J (2012) ASME 2012 gas turbine India conference, pp 815–826
147. Chan Y, Zhu M, Zhang Z, Liu P, Zhang D (2015) The effect of $CO_2$ dilution on the laminar burning velocity of premixed methane/air flames. Energy Procedia 75:3048–3053
148. Dirrenberger P, Le Gall H, Bounaceur R, Herbinet O, Glaude P-A, Konnov A, Battin-Leclerc F (2011) Measurements of laminar flame velocity for components of natural gas. Energy Fuels 25:3875–3884

MIX
Papier aus verantwortungsvollen Quellen
Paper from responsible sources
FSC® C105338

Printed by Books on Demand, Germany